FRONTIERS OF ENTREPRENEURSHIP RESEARCH 2025

PROCEEDINGS OF THE FORTY-FIFTH ANNUAL
BABSON COLLEGE ENTREPRENEURSHIP RESEARCH CONFERENCE

Sponsor
Babson College, USA

Doctoral Consortium Sponsors
The Ewing Marion Kauffman Foundation, USA
Society for Advancement of Management Studies (SAMS), UK

Edited by
Andrew Zacharakis
Babson College

Tom Allison
Texas Christian University

Petra Andries
Ghent University

Anne Domurath
Wilfrid Laurier University

David Gras
University of Tennessee, Knoxville

Denis Grégoire
HEC, Montreal

Annelore Huyghe
Esade University

Dan Lerner
IE University

Alex McKelvie
Syracuse University

Mona Mensmann
University of Cologne

Mariarosa Scarlata
University of Bergamo

Monika Schuhmacher
Justus-Liebig-University Giessen

Ben Spigel
Babson College

Siddharth Vedula
Technical University of Munich

Printed from e-media with permission by:

Curran Associates, Inc.
57 Morehouse Lane
Red Hook, NY 12571

Some format issues inherent in the e-media version may also appear in this print version.

Babson College
231 Forest Street
Babson Park, MA
02457

Phone: 781 239 4992
Fax: 781 239 3829

www.babson.edu

Additional copies of this publication are available from:

Curran Associates, Inc.
57 Morehouse Lane
Red Hook, NY 12571 USA
Phone: 845-758-0400
Fax: 845-758-2633
Email: curran@proceedings.com
Web: www.proceedings.com

We extend our gratitude to the authors for their assistance and collaboration in the production of FRONTIERS OF ENTREPRENEURSHIP RESEARCH BCERC PROCEEDINGS 2025

ACKNOWLEDGEMENTS

Babson College and the editors of the Frontiers of Entrepreneurship Research, BCERC Proceedings (FER) wish to thank our 2025 Babson College Entrepreneurship Research Conference (BCERC) sponsors, Babson College.

Thank you to Andrew Zacharakis, director of the 2025 BCERC, as well as the staff of Babson College, for their warm hospitality and for sharing their beautiful campus. The organizing team contributed to creating a wonderful environment to be able to network and socialize with colleagues.

We value the contribution of our Board of Reviewers, an enthusiastic team of devoted scholars, for their support and inspiration in the leadership and the future of BCERC and the *FER*. Thank you to Tom Allison, Petra Andries, Anne Domurath, David Gras, Denis Grégoire, Annelore Huyghe, Dan Lerner, Alex McKelvie, Mona Mensmann, Mariarosa Scarlata, Monika Schuhmacher, Ben Spigel, and Siddharth Vedula.

We would like to extend our thanks and appreciation to the sponsors of the Doctoral Consortium, the Ewing Marion Kauffman Foundation, and the Society for Advancement of Management Studies (SAMS), UK.

We are deeply grateful to Eliana Crosina and William Gartner for their outstanding guidance and enthusiasm as Co-directors of the 2025 Doctoral Consortium. In addition, special thanks to Candida Brush, Babson College, and Andrew Zacharakis, Babson College, for their doctoral consortium research translation expertise and instruction.

Many thanks to the Babson College Management Team of Beth Peterson, Tamara Lamenzo, Tanvi Kasliwal and Monowar Sani Shuvo. Your dedication, enthusiasm, and meticulous attention to detail were crucial in making this conference a resounding success. Thank you to the amazing Babson volunteers: Fyllis Berg-Elton, Maurice Blaustein, McKenzie Cloutier, Clare Currie, Alexandra Dunk, Tanya Dupont, Mackenzie Hayes, Safae Jdid, Ilona Klausener, Cindy Klein-Marmer, Lisa Kokidko, Jennifer MacDonald, Sarah Miesse, Sue Nealon, Tracey O'Rourke, Vannessa Rodriguez, Caryn Sowa and Kim Stanhouse.

The management of BCERC and the publication of the 2025 Frontiers of Entrepreneurship Research, BCERC Proceedings, was carried out at Babson College under the supervision of Beth Peterson. A special thank you to Elizabeth Mackin and Cadmium Event Management Software Platform team for their help and support with BCERC.

We would also like to extend our gratitude to the following 2025 Best Paper Award Sponsors for their generosity and continued support of BCERC. The best papers are published in this edition of *Frontiers of Entrepreneurship Research, BCERC Proceedings* and listed at the beginning of the Detailed Table of Contents.

2025 BEST PAPER AWARD SPONSORS

National Federation of Independent Business (NFIB) Award (1997–2025)
Babson College Bertarelli Family Award (2014–2025)
Babson College Tariq Farid Franchise Institute Award (2025)
Illinois State Means Center for Entrepreneurial Studies Award (2025)
Babson College Institute for Social Innovation Award (2025)

Lastly, we would like to thank the conference participants and our colleagues and friends for their quest for entrepreneurship research, for you have made this issue of *Frontiers of Entrepreneurship Research, BCERC Proceedings* achievable. These papers exemplified the most recent developments in entrepreneurship research composed by the world's topmost scholars.

2025 BCERC DIRECTOR
Andrew Zacharakis
Babson College

1981 – 2025 BCERC DIRECTORS
William D. Bygrave '94, '95, '01, '02, '03
Neil C. Churchill* '90, '91, '92, '93
John A. Hornaday* '83
Bruce A. Kirchhoff* '87, '88, '89
Paul D. Reynolds '96, '97, '98, '99, '00
Robert Ronstadt '86
Jeffry A. Timmons* '84, '85
Karl H. Vesper '81, '82
Andrew Zacharakis '06 – '25
Shaker A. Zahra '04, '05
* Dece sed

BCERC LIFETIME MEMBERS
Per Davidsson, 2023
Howard E. Aldrich, 2013
Robert A. Baron, 2017
Candida G. Brush, 2016
William D. Bygrave, 2008
Neil C. Churchill,* 1998
Arnold C. Cooper,* 2003
Per Davidsson, 2023
William J. Dennis, Jr., 2014
William Gartner, 2020
Murray Gillin, 2004
Lewis N. Goslin, 2000
Gerald E. Hills, 2011
John A. Hornaday,* 1997
Donald F. Kuratko, 2024
Tom Lumpkin, 2021
Irene M. McCarthy,* 2001
G. Dale M er, 2005
Daryl G. Mitton,* 1995
Georgia Papavasiliou, 2018
Robert G. Schwartz, 2006
Donald L. Sexton,* 1994
Karl H. Vesper, 1997
William E. Wetzel, Jr.,* 1996
* Dece sed

A special thank you to the Doctoral Consortium Directors and Instructors for enhancing both the curriculum and the overall conference experience for the 2025 Doctoral Consortium

2025 BCERC DOCTORAL CONSORTIUM INSTRUCTORS

Eliana Crosina, *Babson College*

William Gartner, *Babson College*

2025 BCERC DOCTORAL CONSORTIUM INSTRUCTORS

Oana Branzei
Ivey Business School

Candida Brush
Babson College

Jose Cerecedo Lopez
Elon University

Andrew Corbett
Babson College

Eliana Crosina
Babson College

William Gartner
Babson College

Cédric Gutierrez
Bocconi University

Diana Hechavarria
Babson College

Keith Hmieleski
Texas Christian University

Suntae Kim
Johns Hopkins University

Mirjam Knockaert
Ghent University

Carina Lomberg
Technical University of Denmark

Alexander McKelvie
Syracuse University

Emily Neubert
Texas Christian University

Beth Peterson
Babson College

Angela Randolph
Babson College

Maija Renko
Aalto University

Stella Seyb
Texas Christian University

Raja Singaram
University of Galway

Ben Spigel
Babson College

David Townsend
Virginia Tech

Siddharth Vedula
Technical University of Munich

Ben Warnick
Washington State University

Johan Wiklund
Syracuse University

Andrew Zacharakis
Babson College

We would like to express our heartfelt gratitude to the following individuals for their invaluable contributions, dedicated time, and enthusiastic service as external Ad hoc Reviewers.

2025 BCERC AD HOC REVIEWERS

Funke Adedeji
Obafemi Awolowo University

Petrit Ademi
University of St. Gallen

Dr. Sucheta Agarwal
GLA University

John Aiyede
University of Louisville

Naveed Akhter
Jönköping University

Aziza Al Ghafri
Malardalens University

Yannick Amend
Justus-Liebig University Gießen

Amir Amini Sedeh
St. Edward's University

Maria-Irene Angelidis
Baruch College

Fernando Angulo-Ruiz
MacEwan University

Samuel Wayne Appleton
University of Bergamo

Jonathan Arthurs
Oregon State University

Rebecca Arwine
University of Cincinnati

Olga Belousova
University of Groningen

Mohammed Bendaanane
Technical University of Munich

Gentrit Berisha
University of Prishtina

Sneha Bhatia
Curtin University Dubai

Mark Bolinger
Appalachian State University

Zahra Booyavi
Rutgers University

James Bort
DePaul University

Sirine Boukhabza
Cadi Ayyad University Marrakech

Malin Brännback
Abo Akademi University School of Business and Economics

Seth Butler
University of North Carolina Wilmington

Joseph Cabral
Louisiana State University

Jose Cerecedo
Elon University

Aardra Chandra Mouli
UCL School of Management

Sanjay Chaudhary
Jindal Global Business School

Jessie Cheung
Nanyang Technology University

Daniel Clark
Ivey Business School

Karla Corres Luna
University of Massachusetts Boston

Eduardo Cuervo
University of Tennessee, Knoxville

Vinod Das
Bayer US LLC

Fabian Diaz
University of Louisville

Fardeen Dodo
University of Bologna

Sean Dwyer
University of Oklahoma

Begonya Ebot
University of Wolverhampton

Matthew Farrell
Austin Peay State University

Christian Felzensztein
Clarkson University

Arjan Frederiks
University of *Groningen*

Casey Frid
University of St. Thomas

Cintya Gajardo Vejar
University of Massachusetts Lowell

Tera Galloway
University of St. Thomas

Aishwarya Gayen
IE Business School Madrid

Mohamed Genedy
Jönköping International Business School

Shalini Gopalkrishnan
Golden Gate University

Kim Green
University of *West Georgia*

Ramona Grimm
University of Bayreuth

Vishal Gupta
The University of Alabama

Dr. Kamal Gupta
Maharaja Agrasen Institute of Management Studies

Daria Hakola
Jyväskylä University School of Business and Economics

Mohammad Hawily
University of Bologna

Nathaniel Hill
University of South Carolina

Laura Hill Cabrera
Ghent University

Rick Hunt
Virginia Tech

Rida Ijaz
Jönköping International Business School

Pepin Ilonga Nkupo
University of Mons

Alina Mariuca Ionescu
Alexandru Ioan Cuza University of Iasi

Andrew Jay Isaak
Vienna University of Economics and Business

Stela Ivanova
ESCP Business School

Zahra Jamshidi
University of Calgary

Joao Maria Jorge
Lund University School of Economics and Management

Huiqing Ju
Iowa State University

Imge Kaya-Sabanci
IE Business School

Charles Kirschbaum
Insper

Johannes Kleinhempel
Copenhagen Business School

Franziska Koefer
University of Twente

Sonia Koller
King's Business School

Ann-Sophie Kowalewski
IESE Business School

Irfan Saleem
Sohar University

Jung H. Kwon
University of Denver

Teresia Kyalo
Karatina university

Andrea Lane
IMC Krems

Sanggeun Lee
Syracuse University

Qinglin Liu
University of International Business and Economics

Anna Long
California State University San Bernardino

Srikant Manchiraju
Florida State University

Laure Marty
ESCP Business School

Blake Mathias
Indiana University

Eduardo Meléndez
University of Oklahoma

Jimena Mendez Merino
Technical University of Munich

Sönke Mestwerdt
ESCP Business School

Michele Migliuolo
Nova Southeastern University

Ahlam Nasser
University of Cape Town

David Nawrath
Technical University of Munich

Sonal Nayak
Indian Institute of Management Bangalore

Thi (Alice) Ngo
University of Sheffield

Roderick Shane Snipes
Borough of Manhattan Community College

Martin Nordell
Åbo Akademi University School of Business and Economics

Juliet Nwakaego Nwachukwu
University of the Western Cape

Anna-Greta Nyström
Åbo Akademi University School of Business and Economics

Cristian Eugenio Orellana Pino
Universidad de Ohiggins

Ine Paeleman
University of Antwerp

Christodoulos Pavlou
American University of Beirut - Mediterraneo

Alana Pierce
HEC Montréal

Maud Pindard-Lejarraga
IE Business School, IE University

Audra Quinn
University of Southern California

Anne Radunski
University of Potsdam

Judy Rady
Virginia Tech

Charu Rastogi
Indian Institute of Management Bangalore

Laurence Rijssegem
Ghent University

Ashley Roccapriore
Auburn University

Rossella Rocchino
Technical University of Munich

João Fernando Rossi Mazzoni
Baylor University

Kristin Sabel
Jönköping International Business School

Vijay Venkataraman
Indian Institute of Management Bangalore

Aakash Sapru
Iona University

Anna-Katharina Schaper
Technical University of Munich

Verena Scheffczyk
IE University

Leon Schjoedt
The University of Texas, Rio Grande Valley

Monique Schlömmer
WU Vienna

Lukas Schnell
Johannes Kepler University Linz

Benedikt Seigner
IE University

Chelsea Sherlock
Mississippi State University

Sadek Showkat
Loyola University Chicago

Hallur Thor Sigurdarson
Reykjavik University

Raja Singaram
University of Galway

Rai Siddhant Sinha
NEOMA Business School

Vitaliy Skorodziyevskiy
University of Louisville

Simran Sodhi
Institute of Management Technology-Ghaziabad

Devin Stein
University of Alabama

Pekka Stenholm
University of Turku

Meike Stephan
University of Siegen

Marjana Subotic
KU Leuven

Imran Syed
Ball State *University*

Marselia Tan
ESSEC Business School

Urwa Tariq
UAE University

Kai Cher Tay
University of *Louisville*

Nele Terveen
University of *Auckland*

Jewel Thompson
Ashesi University

Varkey Titus
University of Nebraska, Lincoln

Viktoria Luise Unger
LMU Munich

Evy Van Lancker
Ghent University

Jake Waddingham
Texas State University

Mingchu Wang
University of Greenwich

Fangjie Wang
The University of Warwick

Miranda Welbourne Eleazar
University of Iowa

Jim Whitbeck
Oregon State University

Marcus Wolfe
University of North Texas

David Wunder
Aalto University

Nataliia Yakushko
University of Tennessee, Knoxville

Justin Yan
Baylor University

Qiongrui (Missy) Yao
York College of Pennsylvania

Mehmet Ali Yetim
University of Scranton

Chanyong Yoo
University of South Carolina

Yi Zhang
Oklahoma State University

Haibo Zhou
University of Nottingham Ningbo China

TABLE OF CONTENTS

INTERACTIVE PAPERS

RESEARCH METHODS

SUMMARIES

INTERACTIVE PAPERS

SOCIAL ENTREPRENEURSHIP

THE MEANS CENTER FOR ENTREPRENEURIAL STUDIES AWARD (ILLINOIS STATE) FOR BEST PAPER ON SOCIAL ENTREPRENEURSHIP

SUMMARIES

INTERACTIVE PAPERS

STRATEGY

TEAM & GOVERNANCE

THE G.DALE MEYER AWARD FOR THE BEST PAPER BY A DOCTORAL STUDENT WHO IS THE LEAD AUTHOR

SPATIAL ANALYSIS OF ENTREPRENEURS' CRIME EXPERIENCE IN AN INFORMAL TOWNSHIP ECONOMY

Mohammed Bendaanane, Technical University of Munich, Germany
Siddharth Vedula, Technical University of Munich, Germany
Robert Nason, McGill University, Canada

ABSTRACT

Crime is a salient yet under-theorized phenomenon in informal economies. While prior research has largely essentialized the environment of informal economy entrepreneurs as homogenously dangerous and pervasive with crime, we draw on environmental criminology and micro-geography studies to explore important variation in entrepreneurs' crime experience in an informal economy. We leverage a unique 'small area census' dataset of entrepreneurs in a South African township to examine how place (micro-geographic location) and space (proximity to physical environment) influences entrepreneurs' experiences with crime. First, we pinpoint geographically bound crime hotspots that are centered around high ambient populations, and interestingly, formal economic activity. Second, we find variance in crime experience within versus outside of crime hotspots. Some entrepreneurs within hotspots feel relatively safe; nevertheless, due to the ways in which their businesses' proximal space is constructed, a phenomenon that we attribute to spatial fortification. Third, we also find that some entrepreneurs outside of crime hotspots still feel threatened by crime due to space considerations as well, a phenomenon that is attributed to spatial vulnerability. As a whole, our paper builds generative links across the literatures of entrepreneurship, space and place, and environmental criminology, which enriches understanding of how crime relates to informal economy entrepreneurship.

INTRODUCTION

Crime constitutes a ubiquitous and often overwhelming challenge for entrepreneurs in informal economies—environments characterized by weak state presence, volatile institutional arrangements, and limited recourse to legal protection (Sutter et al., 2013; Bothello, Nason, & Schnyder, 2019). In these settings, entrepreneurs face a confluence of pressures, among which crime features prominently as both a direct threat to livelihood and an enduring illegitimate institutional pressure (Sutter et al., 2013).

While crime is undoubtedly a salient feature in the informal economy, there is reason to question its essentialization in extant literature. Criminologists emphasize that crime is not universally experienced (Lymperopoulou & Bannister, 2022; Malleson & Andresen, 2015) – even by those in high-crime neighborhoods (Graif & Sampson, 2009). Indeed, there is a strong micro-geographic component to crime as it disproportionately occurs in small areas (De Nadia et al., 2020), including places such as particular segments of streets (Weisburd et al., 2004), public locations (Spelman, 1995), and parks (Taylor et al., 2019). The built environment also plays an important role as spaces that attract risky behavior (Murray & Roncek, 2008) or have visible signs of neglect, such as damaged buildings or poor public lighting (Kelling & Coles, 1997) are associated with increased crime.

The distinct role of place and space in criminology aligns with a growing body of literature in organizational studies examining how space and place shapes institutional environments (Lawrence & Dover,

2015). In this line of research, entrepreneurs are increasingly seen as interconnected with their spatial surroundings (Dahl & Sorenson, 2012; Plummer & Pe'er, 2010; Sorenson, 2018). For instance, recent research has shown how unique elements in the towns of Arco Italy enable sport-climbing entrepreneurs (Aversa et al., 2022), physical bridges increase the founding of new ventures (Dutta et al., 2022), and the physical milieu of a co-working space can facilitate a sense of community and identification as an entrepreneur (Crosina, 2024). This literature has shown that space, in particular, not only dictates an environment but can be proactively used to counter broader institutional pressures.

In this paper, we incorporate spatial perspectives in criminology and organization theory to advance an understanding of how entrepreneurs in the informal economy experience crime. To examine the role of place and space, we mobilize a unique "small-area census" dataset covering the entire population of informal economy entrepreneurs in the South African township of Delft in Cape Town in 2020 (Charman et al., 2015; Nason et al., 2024). Like many townships, criminal activity is high in Delft – the township was even recently given the notorious moniker of South Africa's murder capital (Nobesuthu, 2023). The primary data consists of a comprehensive survey of entrepreneurs (N=1,273) with individual and firm-level variables, including firms' geographic location data and information about the entrepreneur's physical environment. We complement this primary data with secondary sources of spatial data, including high street network, and. taxi routes. We undertake an exploratory analysis of crime experience by employing an innovative analytical approach that integrates logistic regressions with techniques from geographic information analysis (GIS). In particular, we utilize spatial buffers to delineate entrepreneurs' proximity to key physical features in the township environment and employ a Getis-Ord Gi statistic to identify spatial hot spots of crime experience (Ord & Getis, 1995).

Through our abductive exploration, we make several observations with important theoretical implications. First, we demonstrate that crime is not universally distributed across the township economy but rather concentrated in crime hotspots. These hotspots are centered in places with high ambient population and, interestingly, around formal economic activity (e.g., a large grocery store). Second, we probe variation further by examining asymmetry in crime experience vis-à-vis the hotspots. We find that some entrepreneurs feel relatively safe within a crime hotspot, a phenomenon we attribute to spatial fortification since the relative safety appears to be spatially influenced. For instance, those very close to taxi ranks felt safer, as did those operating in more clandestine building structures (i.e., their residences). We also find that some entrepreneurs feel relatively threatened by crime even in areas where those around them feel safe, a phenomenon that we attribute to spatial vulnerability.

Our spatial examination of how informal economy entrepreneurs experience crime makes several contributions to existing literature. First, we move beyond essentialized characterizations of the informal economy as crime-laden and unpack important variations in the experience of crime within an informal economy. The idea of spatial fortification and spatial vulnerability provides evidence and theoretically grounded constructs that are likely to be useful for future research examining the ways in which crime and entrepreneurship influence each other (Mallon & Fainshmidt, 2020; Sanchez-Ruiz et al., 2023; Sutter et al., 2013). Second, by infusing literature on environmental criminology, we provide a generative bridge to organizational perspectives on place and space (Crosina, 2024; Lawrence & Dover, 2015) and the constitutive nature of micro-geography (Dahl & Sorenson, 2012; Plummer & Pe'er, 2010; Sorenson, 2018). The integration of these literatures provides new insights into how geographic and spatial characteristics both come to constitute an institutional environment and can be used agentically to counter institutional pressures. Finally, we contribute methodologically by employing geospatial methods that have rarely been used in organizational and entrepreneurship research. The resulting hotspots and spatial findings from this analysis also have important practical and policy implications for understanding the everyday reality of informal economy entrepreneurs.

LITERATURE BACKGROUND

Research on informal economies often essentializes these environments as inherently rife with criminality. Scholars commonly depict the informal sector as a space where the lack of formal regulation naturally breeds criminal practices. This view implies that the very characteristics that enable informal economic activity—such as flexibility and reliance on personal networks—create conditions where illicit behavior becomes the norm. In this narrative, crime and informal economic activity are not seen as separate but mutually reinforcing. Researchers argue that the strategies employed by informal entrepreneurs—operating outside formal legal frameworks—are closely linked to practices that blur the boundary between legal and illegal actions. As a result, crime becomes not just an occasional byproduct but a core element of everyday business in these spaces. Indeed, research on entrepreneurship in informal economies often focuses on specific locations that are seen as inherently dangerous, such as urban slums (Gras & Nason, 2015), favelas (Pongeluppe, 2022), or townships (Nason et al., 2024).

Some scholars describe the informal sector as a space where the lack of formal oversight allows pervasive illicit practices to flourish. In this framing, the flexible operations and reliance on personal networks that define informality are seen as factors that naturally normalize criminal behavior. This perspective views the link between informal economic activity and crime as mutually reinforcing. Informal entrepreneurs, by operating outside formal regulatory systems, adopt adaptive strategies that blur the boundaries between legal and illegal conduct. Rather than being an external contaminant, crime is presented as a core element of everyday economic interactions within these settings.

Such a pervasive characterization, however, oversimplifies the complex realities of informal economic life. While the literature frequently emphasizes criminality as a defining feature, it risks neglecting the diverse experiences and strategic responses of those operating in the informal sector. A growing stream of entrepreneurship research situates itself in understanding new economic activity in the midst of turbulent and adverse conditions such as violent protests (Dimitriadis, 2024), pervasive organized crime (Mallon & Fainshmidt, 2020), and widespread unwanted criminal attention (Kistruck et al., 2015; Nason et al., 2024; Sanchez-Ruiz et al., 2023). While this stream of literature also tends to essentialize the environment in order to justify the uniqueness of the context as crime-laden, it has also begun to reveal that crime is not uniformly experienced by entrepreneurs in an informal economy. Factors such as firm registration (Kistruck et al., 2015), visibility (Nason et al., 2024; Sanchez-Ruiz et al., 2023), financial success (Sanchez-Ruiz et al., 2023), and demographic characteristics (Bendaanane et al., in press; Nason et al., 2024; Slade Shantz et al., 2018) can play a role in determining who is targeted by criminal activity. In order to try to understand this variance in the experience with crime amongst informal economy entrepreneurs further, we look to a field that studies crime explicitly– criminology.

Criminologists point to the heterogeneous nature of crime and have come to widely recognize the role of geographic places and physical spaces in patterns of criminal activity (Brantingham & Brantingham, 1991; Jones & Pridemore, 2019; Wortley & Townsley, 2016). The specific domain of environmental criminology challenges the assumption that crime occurrence can be explained solely by the socio-economic characteristics of the offenders and the victims and argues that spatial environment plays an equally important role (Brantingham & Brantingham, 1991; Walker, 2011; Wortley & Townsley, 2016).

As the field of environmental criminology has grown, studies have examined both the role of broader geographic places as well as more proximal physical spaces. The criminology of place perspective (Brantingham & Brantingham, 1995; Weisburd et al., 2012) stipulates that areas near large shopping malls, major public transport routes, and areas known for illegal activities like drug dealing and prostitution are more likely to attract crime (Brantingham & Brantingham, 1995). Crime is also more likely to happen at edges marking clear distinction of land use such as the limit between commercial and residential areas (Brantingham & Brantingham, 2011), and in areas prone to risky behavior such as bars (Murray & Roncek, 2008). In addition, routine activities theory posits that crime is tied to the convergence in time and the physical location of an offender, a victim, and the absence of guardianship. That is, areas without guardianship tend to become crime hotspots (Cohen & Felson, 1979; Jones & Pridemore, 2019). This research has drawn attention to the way in which criminal activity can be highly localized. For instance, Spelman (1995, p. 116) recounts, "Calls-for-service data in Boston and Minneapolis show that the addresses producing the most repeat calls account for over 60% of all calls" (Pierce et al., 1988; Sherman et al., 1989).

The criminology of place perspective advances a micro-geographic examination of crime patterns taking into consideration various level of analysis, conceptualizing place as a "behavior setting" where human interactions occur (Weisburd et al., 2004, p. 40). These levels of analysis vary from street segments (Weisburd et al., 2004) to specific location of addresses (Eck, 2002; Goldsmith et al., 2000). In this vein, the conceptualization of place in the criminology of place perspective is closely aligned with definition of places as "meaningful locations" that are "continuously constructed as people do things in and around them" (Lawrence & Dover, 2015, p. 273).

Other perspectives in environmental criminology, notably that of situational crime draw attention to the more proximal role of space, such as physical structures, in crime patterns. For instance, the "broken window" view (Lanfear et al., 2020) argues that visible signs of neglect and public social disorder (broken building windows) can be a source of increased crime because the neglect is perceived as a lack of effective guardianship (Kelling & Coles, 1997; Wilson & Kelling, 1982). This perspective asserts that the presence of even light physical neglect would induce higher criminal activity (Savolainen, 2007). By extension, this school of thought points out that behavioral public incivilities in the form of appropriation of physical space by drug users, rowdy teenagers, and bothersome drunks are associated with increased criminal activity (Tarkhanyan, 2014; Taylor 1999). Furthermore, this perspective underscores that proximity to physical built environment such as bus stops (Loukaitou-sideris, 1999), banks or shopping centers can be an attractor to criminal acts (Clarke, 1997).

All in all, these insights suggest that the spatial features in which entrepreneurs in an informal economy operate are likely to influence their crime experiences. We link the environmental criminology literature on the role of place to space to growing literature on the topic in organization theory and entrepreneurship (Dutta et al., 2022; Pongeluppe, 2022; Wright et al., 2023). Specifically, we ask the research question: "How do place and space influence entrepreneurs' experience with crime in an informal economy?"

METHOD

Our empirical setting is the township of Delft in Cape Town, South Africa—a site of vibrant informal entrepreneurial activity but also significant crime (Charman et al., 2020; Nobesuthu, 2023). We utilize a unique "small-area census" dataset of 1,273 informal firms collected in early 2020 by the Sustainable Livelihoods Foundation (Charman et al., 2015b; Nason et al., 2024).

The dataset contains rich firm-level and individual-level information, including coordinates for geolocation. We complemented this with spatial data from the City of Cape Town and OpenStreetMap, enabling us to identify features such as high streets, pedestrian crossings, taxi routes, public lighting, and speed bumps. Crime experience was measured through a four-point scale derived from open-ended survey responses, then converted into a binary variable (low vs. high crime experience) after intercoder reliability procedures (O'Connor & Joffe, 2020).

We used logistic regression models to examine predictors of high crime experience, as well as spatial buffer techniques and Getis-Ord Gi statistics (Ord & Getis, 1995) to identify and analyze crime hotspots. Subsample models were also developed to probe within-hotspot and outside-hotspot experiences, and to examine mismatches between subjective safety and objective crime intensity.

RESULTS

Our empirical analysis revealed distinct and statistically robust spatial patterns in entrepreneurs' experiences of crime within the informal economy of Delft township. Through the application of Getis-Ord Gi* spatial statistics (Ord & Getis, 1995), we identified clear crime hotspots within the study area—clusters predominantly located along high streets and densely populated alleys. These clusters were not random but spatially bounded, confirming that certain micro-geographic contexts are more prone to concentrated crime exposure.

Logistic regression analyses showed that entrepreneurs located within these crime hotspots had significantly higher odds of reporting crime experience. Yet, the spatial context alone did not fully explain variation. Individual and firm-level attributes also played critical roles and, importantly, interacted with spatial factors in meaningful ways. For instance, foreign-born entrepreneurs were statistically more likely to experience crime—an effect that held both within and outside hotspots. This aligns with the "liability of foreignness" thesis (Zaheer, 1995) and recent findings that migrants in informal economies often lack embeddedness and community support (Yenkey, 2018; Bendaanane et al., in press).

Gender also emerged as a significant moderator of crime exposure, but with an important spatial nuance: female entrepreneurs reported greater vulnerability only when operating inside crime hotspots. Outside these areas, the gender effect diminished, suggesting an intersection between demographic risk factors and environmental context (Shahriar & Shepherd, 2019). Firm-level features—such as operating from sidewalk locations, using electricity, and transacting exclusively in cash—were also correlated with elevated crime experience, echoing prior work on visibility and resource attractiveness (Sanchez-Ruiz et al., 2023; Sutter et al., 2013).

A particularly novel contribution of our study was the identification of "safety mismatches" and "crime mismatches." The safety mismatch refers to entrepreneurs who report low crime experience despite being located in statistically high-crime zones. These entrepreneurs were more likely to operate residential firms, suggesting that proximity to home may offer informal guardianship and community protection (Sutter et al., 2013). Conversely, crime mismatches involved entrepreneurs who reported high crime experience even though they were outside spatially defined hotspots. In these cases, business characteristics such as older firm age or operation from sidewalk spaces likely served as attractors, regardless of location-level crime density.

Finally, across all subsamples, entrepreneurs' perception of police effectiveness significantly reduced the likelihood of reporting high crime experience. This finding suggests that institutional trust or belief in enforcement presence functions as a psychological buffer, even in environments where actual protection may be minimal. It also reinforces the importance of perceived legitimacy and public trust in shaping institutional outcomes in informal settings.

DISCUSSION AND IMPLICATIONS

Our findings yield several conceptual, methodological, and practical implications for the study of crime in informal economies. First and foremost, we challenge the dominant geography-neutral approach in entrepreneurship research by demonstrating that the spatial context—defined in terms of micro-geographic characteristics—exerts a significant and independent influence on crime experience. This supports and extends the core premises of environmental criminology (Brantingham & Brantingham, 1991; Wortley & Townsley, 2016), which emphasize the patterned distribution of crime in urban landscapes and its relationship to physical infrastructure and human activity.

Conceptually, our work advances a multi-level understanding of institutional pressure by illustrating how space, identity, and firm structure intersect to shape entrepreneurial vulnerability. We show that physical features—such as pedestrian crossings, high streets, or sidewalk operations—interact with social identities (e.g., gender, migrant status) and firm choices (e.g., registration, cash use) to either exacerbate or mitigate exposure to crime. This triangulated view enriches existing accounts of institutional voids by situating them within specific, mappable geographies (Godfrey, 2011; De Castro, Khavul, & Bruton, 2014).

Methodologically, we contribute to the emerging spatial turn in organizational studies (Dutta, Armanios, & Desai, 2022; Wright, Irving, Zafar, & Reay, 2023) by deploying geospatial tools alongside logistic regression to uncover place-based patterns. The use of crime hotspots and spatial buffers enabled us to capture not just whether an entrepreneur experienced crime, but where and under what spatial circumstances this occurred. This approach enables future research to operationalize space not merely as background context, but as an analytically active variable in shaping institutional dynamics.

Practically, our results offer policy-relevant insights for urban governance, economic development, and crime prevention in informal settlements. Specifically, they suggest that localized interventions—such as enhanced lighting, improved pedestrian infrastructure, or strengthened community policing—could be more effective than broad-based strategies. Policymakers should also consider how informal governance bodies (e.g., taxi associations) may serve as de facto protectors or regulators in spaces where state capacity is limited (Eaglin, 2023).

Additionally, the identification of mismatches in crime perception and spatial crime data underscores the need for trust-building between enforcement agencies and informal entrepreneurs. Promoting subjective perceptions of safety—through community-based initiatives or participatory urban design—may have tangible effects on entrepreneurs' willingness to operate, expand, or formalize.

In sum, our study demonstrates that crime in informal economies is not merely a matter of firm type or demographic risk but is also profoundly shaped by the physical environment in which entrepreneurs operate. Recognizing and responding to this spatialized nature of institutional pressure is crucial for both academic theory and effective intervention.

CONTACT: Mohammed Bendaanane; m.bendaanane@tum.de; Technical University of Munich – School of Management; Arcisstraße 21, 80333 Munich, Germany

PASSION'S DARK SIDE: DEPLETION AND THE PATH FROM PASSION TO UNETHICAL BEHAVIOR

Alexander B. Hamrick, University of Richmond, United States
Tim Michaelis, Northern Illinois University, United States
Mike Lerman, Iowa State University, United States
Charles Y. Murnieks, University of Missouri-Kansas City, United States
Marina Astakhova, University of Richmond, United States

ABSTRACT

Entrepreneurs are often portrayed as active agents who, driven by their passion, maintain control over their actions and use their passion to influence desired outcomes. This perspective overlooks the possibility that passion may deplete entrepreneurs such that subsequent unethical behaviors stem from their impaired judgment rather than willful malice. Drawing from theory on bounded ethicality, we examine how obsessive entrepreneurial passion (OEP) enhances depletion, increasing entrepreneurs' unethical behaviors, whereas harmonious entrepreneurial passion (HEP) reduces depletion, decreasing entrepreneurs' unethical behaviors. Additionally, we explore how entrepreneurs' family-to-business support buffer against these depletion effects which, in turn, result in lower levels of unethical behaviors. We test these relationships across two studies with entrepreneurs.

INTRODUCTION

Scholars readily acknowledge entrepreneurial passion—a strong inclination toward self-defining entrepreneurial activities—as a critical driver of entrepreneurial decisions. (Cardon et al., 2009; Murnieks et al., 2020). Indeed, entrepreneurial passion mobilizes energy necessary for starting a new business, it engenders persistence (Cardon & Kirk, 2015), "stimulates entrepreneurs to overcome obstacles and remain engaged" (Cardon et al., 2009; p. 512), and drives individual, corporate, and social entrepreneurial intentions (Biraglia & Kadile, 2017). As passion research has grown, scholars recognize that not all passions are identical. In fact, the Dualistic Model of Passion (DMP) differentiates passion into two types: harmonious and obsessive (Vallerand et al., 2003). Harmonious entrepreneurial passion (HEP) emanates from autonomous internalization of entrepreneurial activities into one's identity, enabling self-determined engagement; conversely, obsessive entrepreneurial passion (OEP) spawns from controlled internalization of entrepreneurial activities into one's identity, which compels rigid engagement (Fu et al., 2022). While passion often drives many beneficial behaviors, emerging research highlights the potential negative outcomes that can arise from passion, and OEP, in particular (Newman et al., 2021).

Entrepreneurs frequently face situations that may challenge their ethical and moral judgements (Hagg et al., 2024; Morris et al., 2002), which, combined with the entrepreneur's strong motivation for financial gain (Baron et al., 2015), are conducive of unethical behaviors that can have harmful effects (e.g., Baron et al., 2015; Tacke et al., 2023). However, the literature on ties between entrepreneurial passion and unethical behavior remains extremely sparse, highlighting the need for further investigations (Newman et al., 2021). A deeper understanding of the link between entrepreneurial passion and unethical behavior is essential for several reasons. Entrepreneurial passion, by its nature, is intense and can fuel a relentless commitment to goals (Palmer et al., 2021), leading entrepreneurs to overlook ethical considerations in favor of achieving their objectives. Additionally, as entrepreneurial passion is celebrated and often idealized in culture (Cardon & Kirk, 2015), there is a risk that society and entrepreneurs themselves may overlook the potential for this passion to inadvertently encourage moral compromises, which may be detrimental given the potentially "devastating effects [of entrepreneurs' unethical decision-making] on their companies, stakeholders, and themselves" (Baron et al., 2015; p. 107). This context underscores the imperative for research to identify the predictors of unethical behaviors among entrepreneurs as well as underlying mechanisms and boundary conditions that may shape those relationships.

To extend our understanding of the mechanisms linking entrepreneurial passion and unethical behavior, we draw from theory on bounded ethicality (e.g., Chugh et al., 2005; Kern, & Chugh, 2009). In contrast to the conventional view that entrepreneurs' passion is intentional and calculated (Barsky, 2011), we propose a different path to unethical outcomes which is rooted in the entrepreneur's impaired judgment rather than willful malice. Specifically, we argue that entrepreneurial passion, and obsessive entrepreneurial passion, in particular, will lead to a state of depletion, which refers to a feeling of mental exhaustion or fatigue and a perceived lack of personal willpower (Baumeister et al., 1998). This, in turn, will render entrepreneurs susceptible to bounded ethicality, wherein they deviate "in systematic and predictable ways from their values while being unaware that they are doing so" (Chugh et al., 2005, p. 28). We further bound our theorizing by considering when and for whom the relationship between entrepreneurial passion and unethical consequences is amplified or muted. Specifically, we argue that an entrepreneur's family-to-business support—defined as the social support received from family members for one's entrepreneurial activities (Powell & Eddleston, 2013)— can provide instrumental and affective assistance (Edelman et al., 2016) that mitigate feelings of depletion and, by extension, lower the entrepreneur's proneness to ethical fading.

HYPOTHESES DEVELOPMENT

According to the model of bounded ethicality, people cannot be fully ethical all the time, proposing instead that we are all prone to ethical failures (Chugh, Banaji, & Bazerman, 2005), and systematic and ordinary psychological processes are responsible for such failures (Chugh et al., 2005; Tenbrunsel, 2005). Similar to bounded rationality, ethical boundedness can be the outcome of the limited information and/or time available to the individual, or the limited cognitive capacity of the individual's mind (Chugh et al., 2005). As a result, individuals become blind to their own ethical transgressions, which can lead to a range of ethical misjudgments and behaviors. This blindness can be exacerbated by contextual influences, social pressures, and ingrained biases, all of which can cloud judgment and distort ethical perceptions (Chugh & Kern, 2016; Palazzo et al., 2012; Tenbrunsel & Messick, 2004). Ultimately, the bounded ethicality model underscores the complexity of human behavior, illustrating that ethical failures are not merely a result of malicious intent, but rather an inherent physiological limitation in our ability to consistently engage in moral reasoning and decision-making. Within the bounded ethicality framework, researchers identified depletion as a key factor leading to unfiltered unethical actions (Chugh & Bazerman, 2007). Under certain conditions, fatigue can impair cognitive functioning, leading individuals' rational decision-making processes to become impaired and allowing unethical choices to occur unintentionally, outside of the individual's awareness (Kern & Chugh, 2009). Individuals become so fatigued that they struggle to differentiate between right and wrong, inadvertently allowing unethical behaviors to arise without conscious thought (Chugh & Bazerman, 2007). The depletion-bound argument may be particularly applicable to explain why passionate entrepreneurs would engage in unethical behaviors, and whether they may be doing so without recognizing their actions as unethical.

We first propose that the relationship between entrepreneurial passion and depletion could be either negative or positive, depending on whether the nature of the entrepreneur's passion is harmonious or obsessive. Entrepreneurs are driven by their entrepreneurial passion to devote a substantial amount of time, attention, energy, and other cognitive resources toward venturing and being an entrepreneur (Cardon et al., 2009; Murnieks et al., 2011). However, the motivations behind entrepreneurs' passion and their resource allocation decisions are different depending on the type of passion they possess (Kakarika et al., 2022; Shepherd et al., 2022). HEP arises from autonomous internalization, where individuals engage in entrepreneurial activities out of intrinsic interest and personal enjoyment (Vallerand, 2015). This alignment between the activity and individuals' values reduces the cognitive dissonance and mental strain, which are depleting for individuals. HEP is characterized by positive emotions and affect (Murnieks et al., 2020), which are known to act as a buffer against cognitive depletion by reducing the perceived effort involved in tasks and replenishing psychological energy (Baumeister et al., 2000). Lastly, entrepreneurs with HEP tend to pursue their goals flexibly, adapting to setbacks without undue stress or rigidity (Vallerand et al., 2023), which make them less prone to feeling depleted.

In contrast, OEP stems from controlled internalization, driven by external pressures or self-esteem contingencies (Vallerand, 2015). The continuous need to address these pressures results in an increased cognitive load, compelling the individual to remain perpetually alert, which contributes to feelings of exhaustion. Indeed, obsessively passionate entrepreneurs place excessive weight on their passion, investing all their effort into their ventures (Lex et al., 2022), which puts entrepreneurship in conflict with the rest of their life and predisposes them to continuous resource loss (Amarnani et al., 2020). In general, the compulsive nature of obsessive passion can lead entrepreneurs to rigidly persist even when diminishing returns or exhaustion set in (Stroe et al., 2018). This overexertion and rigid persistence drain physical and psychological reserves, resulting in feeling of depletion (Vallerand, 2015). We therefore hypothesize the following:

> Hypothesis 1a: Harmonious entrepreneurial passion is negatively related to entrepreneurs' depletion.
> Hypothesis 1b: Obsessive entrepreneurial passion is positively related to entrepreneurs' depletion.

Entrepreneurs dedicate substantial effort to their work to demonstrate legitimacy and achieve success, given the liabilities of newness and smallness that entrepreneurs must encounter and overcome (Lohrke et al., 2022; Singh et al., 1986; Stinchcombe, 1965). These liabilities represent a challenge to entrepreneurs' quest for legitimacy, ultimately hindering entrepreneurs' ability to acquire the resources needed for new venture success (e.g., capital, support, employees, suppliers, etc.; Williamson et al., 2002; Zimmerman & Zeitz, 2002). As such, entrepreneurs may attempt to demonstrate legitimacy and achieve success by engaging in unethical behaviors, such as legitimacy lies and opportunistic behaviors. Both behaviors, despite having the potential for serious long-term consequences, including damaged reputations and loss of trust among stakeholders, can be beneficial in the short term and result in short-term gains (Das, 2006), such as enhancing the new venture's resources, legitimacy, and financial standing, which are especially important for new ventures (Jiang et al., 2018; Rutherford et al., 2009).

Building on the bounded ethicality model (Chugh et al., 2005) as well as our theorizing behind the differential effects of HEP and OEP on entrepreneurs' depletion, we argue that depletion is a critical mechanism explaining why obsessively passionate entrepreneurs engage in unethical behaviors while harmoniously passionate entrepreneurs do not. As noted earlier, theory on bounded ethicality suggests that one's state of depletion may create an impediment to the recognition and awareness of one's behaviors (e.g., legitimacy lies and opportunism) as unethical, and thus, unethical action may occur outside one's conscious deliberation (Chugh et al., 2005). Once entrepreneurs feel drained or depleted, their ethicality becomes more bounded and they may struggle to self-guard their actions and weigh potential negative consequences of wrongful behaviors, thus potentially resorting to shortcuts, opportunism, or other unethical actions in their efforts to achieve their goals, without recognizing these actions as unethical. Indeed, when individuals are depleted, they have less energy and willpower to devote towards ethical decision-making such that their decision-making becomes more biased towards goal attainment. Thus, as individuals become more depleted, they become more susceptible to failures in recognition of what is (and is not) ethical behavior, and, therefore, unethical action may occur outside one's conscious deliberation (Chugh et al., 2005). As HEP and OEP have different impacts on the entrepreneur's depletion, we also propose different outcomes regarding ethical behaviors for each of the two passions.

More specifically, HEP, characterized by an autonomous and intrinsic motivation to engage in entrepreneurship, is expected to conserve cognitive and energic resources. Entrepreneurs with HEP are likely to experience their work as fulfilling and balanced, which enables them to manage their cognitive resources effectively and avoid depletion (Ryan & Deci, 2008; Shah, Friedman, & Kruglanski, 2002). Indeed, the more balanced and voluntary engagement in entrepreneurial activities associated with HEP can help protect against the depletion. This efficient resource management associated with HEP means that harmoniously passionate entrepreneurs are better equipped to consciously consider and resist temptations and adhere to ethical standards. Moreover, the intrinsic satisfaction derived from their entrepreneurial activities may help them prioritize long-term success over short-term gains, further mitigating the likelihood of engaging in legitimacy lies and opportunistic behaviors. Thus, we hypothesize:

> Hypothesis 2a: Harmonious entrepreneurial passion will have a negative indirect effect on entrepreneurs' legitimacy lies via entrepreneurs' depletion.

Hypothesis 2b: Harmonious entrepreneurial passion will have a negative indirect effect on entrepreneurs' opportunistic behaviors via entrepreneurs' depletion.

In contrast, if OEP is depleting, as we predict above, then the depletion that entrepreneurs experience may lessen their ability to see short term-focused behaviors, like legitimacy lies and opportunistic behaviors, as unethical. Indeed, depletion can seriously hinder one's ethical awareness (Gino et al., 2011). OEP is characterized by an uncontrollable urge to engage in entrepreneurial activities, often at the expense of other aspects of life. This intense focus and overcommitment can lead to excessive resource expenditure, resulting in depletion (Amarnani et al., 2020; Milyavskaya et al., 2015). As entrepreneurs become depleted, their ethical decision-making becomes bounded such that their ability to think about their actions and recognize "right" from "wrong" diminishes, making them more prone to behave unethically.

Entrepreneurs with high levels of OEP may find themselves increasingly depleted as they constantly strive to meet their own and others' expectations. This is compounded by the fact that OEP often leads to a single-minded focus on success, increasing the appeal of any means that promise quick results (Vallerand et al., 2003). As entrepreneurs become more fatigued, the likelihood of engaging in unethical behaviors increases. These behaviors may be perceived as necessary shortcuts to demonstrate legitimacy and achieve immediate success. As such, the pressures to gain legitimacy quickly may drive obsessively passionate entrepreneurs to engage in unethical behaviors. Indeed, in line with theory on bounded ethicality, after long hours of hard work pushing for their ventures to succeed, entrepreneurs may become more biased towards goal attainment (i.e., growing their venture), which can impede ethical decision making as individuals in search of quick results and success become worse at recognizing whether the behaviors they engage in to try and reach those goals are ethical or not. Thus, the depletion caused by obsessive passion makes it more difficult for entrepreneurs to recognize these short-term temptations as unethical, increasing the likelihood of engaging in these unethical behaviors without any intentionality to behave unethically. Therefore, we hypothesize:

Hypothesis 3a: Obsessive entrepreneurial passion will have a positive indirect effect on entrepreneurs' legitimacy lies via entrepreneurs' depletion.
Hypothesis 3b: Obsessive entrepreneurial passion will have a positive indirect effect on entrepreneurs' opportunistic behaviors via entrepreneurs' depletion.

The degree to which entrepreneurs' passion causes depletion and ultimately impacts their engagement in unethical behaviors is not solely a function of the type of passion (OEP vs HEP) but might also be a function of the family-to-business support available to entrepreneurs. Family-to-business support for entrepreneurs can come in a multitude of forms. For example, family members can provide instrumental support through financing, cheap labor, network connections, or household assistance which relieves founders from duties and responsibilities outside their venture (e.g., Cogan et al., 2022). Family members can also provide affective support through emotional nurturing, empathy, reassurance, and affirmation (e.g., Powell & Eddleston, 2013; Greenhaus & Parasuraman, 1994). Stated broadly, instrumental and affective support promote effective coping processes while entrepreneurs are deeply engaged in their ventures (Shirokova et al., 2024). To that extent, we expect that high levels of instrumental and affective family-to-business support can buffer entrepreneurs against the depletion that occurs from exercising their entrepreneurial passion (e.g., Edelman et al., 2016) which, in turn, will reduce their likelihood of engaging in unethical behavior. Thus, when entrepreneurs receive high family-to-business support, they will be better equipped to manage the demands that accompany their passion. In contrast, when family-to-business support is low, entrepreneurs may lack access to resources that could shield them from depleting effects of engaging in passion-related venturing activities (Cardella et al., 2020; Edelman et al., 2016). In such contexts, entrepreneurs have less instrumental and affective support, which places greater burdens on them to manage venture demands, likely intensifying their depletion.

In the case of HEP, family-to-business support will fortify the ability of this type of passion to resist depletion, further diminishing the likelihood of entrepreneurs engaging in unethical behaviors. Family emotional support may reinforce positive affect associated with HEP, fostering resilience and motivation amidst the challenges of entrepreneurial engagement. When harmoniously passionate entrepreneurs feel understood and supported by their families, their healthy and well-functioning passion (Astakhova et al., 2022) can flourish even more freely, enabling entrepreneurs to confront setbacks with a sense of renewed optimism and determination. As a result, family-to-business support will further amplify the abundant resources that HEP provides, bullet-proofing entrepreneurs from engaging in unethical behaviors. In the case of OEP, family-to-business support may work to assuage the tendency of this consuming type of passion to lead to depletion, and as such, help entrepreneurs become less susceptible to falling prey to the enticement of unethical behaviors. The rigid nature of OEP requires constant engagement (Amarnani et al., 2020). Family-to-business support may help to ameliorate this depleting engagement in various ways. Instrumental support can assist the venture directly with resources and capital which help overcome roadblocks; this makes the daily work of the entrepreneur less effortful and tiresome (Cogan et al., 2022; Edelman et al., 2016). Affective support is nurturing and reaffirms the entrepreneur's sense of self (Cogan et al., 2022; Powell & Eddleston, 2017). Research indicates that support received by family members leads individuals to feel accepted and more secure (Carson et al., 1988). This reassurance bolsters an entrepreneur's self-concept, and likely reduces the need to rely on engaging their obsessive passion to satisfy self-esteem needs (e.g., Mageau et al., 2011). Entrepreneurs may still be extremely passionate about their venture, but will ruminate less and feel less anxiety when engaging the activity at the heart of their passion, which subsequently curtails their emotional energy consumption. Thus, through both instrumental and affective pathways, higher levels of family-to-business should reduce the tendency of OEP to lead to feelings of depletion. By extension, as family-to-business support buffers the depletion effects from OEP, it will also lessen the tendency of entrepreneurs to engage in unethical behaviors.

Hypothesis 4: The negative relationship between harmonious entrepreneurial passion and entrepreneurs' depletion is moderated by family support such that as family support increases, the negative relationship becomes stronger.
Hypothesis 5: The positive relationship between obsessive entrepreneurial passion and entrepreneurs' depletion is moderated by family support such that as family support increases, the positive relationship becomes weaker.
Hypothesis 6: The negative indirect effect of harmonious entrepreneurial passion on (a) legitimacy lies and (b) opportunistic behaviors, via entrepreneurs' depletion, is moderated by family support such that as family support increases, the negative indirect effect becomes stronger.
Hypothesis 7: The positive indirect effect of obsessive entrepreneurial passion on (a) legitimacy lies and (b) opportunistic behaviors, via entrepreneurs' depletion, is moderated by family support concerns such that as family support increases, the positive indirect effect becomes weaker.

METHOD AND RESULTS

Sample: Study 1 (n = 364 entrepreneurs, Prolific); Study 2 (n = 405 entrepreneurs, CloudResearch)
DVs = legitimacy lies (Stevenson et al., 2024) and opportunism (Provan & Skinner, 1989) scales
IVs = OEP and HEP scales (Murnieks et al., 2020)
Controls = gender, age, entrepreneurial experience, firm industry, and social desirability (all excluded from Table 1, though included in analyses) and moral disengagement
Statistical Analysis = OLS regression and causal mediation analyses (Imai et al., 2010)

The hypotheses were tested using OLS regression and causal mediation analyses (Imai et al., 2010). To summarize, across both studies, HEP reduced depletion (supporting H1a), and OEP increased depletion (H1b). Depletion mediated the effects of HEP and OEP on opportunistic behavior (supporting H2b, H3b), but not legitimacy lies (H2a, H3a not supported). In Study 2, family support moderated the OEP–depletion link (H5 supported), and weakened OEP's indirect effect on opportunistic behavior (H7b supported). No moderation effects emerged for HEP (H4, H6 not supported; see Table 1 for results).

DISCUSSION & IMPLICATIONS

Our findings advance theory in several ways. First, passion-driven unethical behavior may stem less from deliberate choice than from depletion that limits ethical awareness, consistent with bounded ethicality. This challenges portrayals of entrepreneurs as intentionally manipulative and highlights unconscious pathways to misconduct. Second, by linking passion to unethical behavior, we show how OEP exemplifies the destructive sides of entrepreneurship (Shepherd, 2019). Lastly, we show that HEP and OEP differentially influence depletion and unethical behavior, and that family-to-business support buffers OEP's harmful pathway. These findings highlight boundary conditions in passion's ethical consequences. Overall, our work shows that entrepreneurial passion can foster unethical behavior not only through intent but also via depletion and bounded ethicality, offering a more nuanced view of passion's light and dark sides.

CONTACT: Alexander B. Hamrick; alex.hamrick@richmond.edu; University of Richmond, United States

Table 1

Study 2 Mediation and Moderated-Mediation Analysis for Opportunistic Behaviors

Dependent variable	Depletion			Opportunistic Behaviors			Depletion			Opportunistic Behaviors		
	b	*t*	*p*	*b*	*t*	*p*	*b*	*t*	*p*	*b*	*t*	*p*
(Intercept)	3.14***	4.69	.000	1.01*	2.23	.026	3.76***	4.08	.000	2.12**	3.04	.003
Moral disengagement	.23*	2.44	.015	.76***	12.41	.000	.24*	2.58	.010	.51***	7.34	.000
Harmonious passion HP	-.26***	-4.25	.000	-.01	-.27	.789	-.36*	-2.31	.021	-.05	-.42	.677
Obsessive passion OP	.12*	2.12	.034	.04	1.23	.221	.40**	2.85	.005	.11	1.01	.314
Depletion				.13***	3.89	.000				.08*	2.10	.037
Family support							.05	.21	.835	-.03	-.18	.859
HP*Family support							.03	.73	.466	.00	.01	.990
OP*Family support							-.10*	-2.22	.027	.01	.18	.854

Moderated-Mediation Analysis	*c'*	*95% CI*	*p-value*
HP → Depletion → Opportunistic Behaviors	-.03***	(-.06, .-01)	.000
OP → Depletion → Opportunistic Behaviors	.02*	(.01, .03)	.036
OP*Low (-1 SD) Family Support → Depletion → Opportunistic Behaviors	.02*	(.001, .04)	.046
OP*High (+1 SD) Family Support - Depletion → Opportunistic Behaviors	.00	(-.01, .01)	.920

N = 405

Note. *$p < .05$, **$p < .01$, ***$p < .001$; c' = Indirect effect; CI = Confidence Interval, Indirect effects tested with n = 1,000 bootstrapped draws

UNDERSTANDING THE ROLE OF JOINERS: EXAMINING THE IMPACT OF EX-ENTREPRENEURS ON NEW VENTURE PERFORMANCE

Luise Kaufmann, Technical University of Munich, Germany
Miriam Bird, Technical University of Munich, Germany

ABSTRACT

While research on founders and new venture team composition has flourished over the past few decades, the characteristics of joiners remain largely unexplored. Thus, we analyze how ex-entrepreneurs influence new venture performance in their role as joiners. We employed a longitudinal database, combining several registers provided by Statistics Sweden, and estimated fixed effects regression models. Our findings reveal a negative, statistically significant relationship between the ex-entrepreneur joiners and new venture performance. We explain this result using resource mobilization theory, which suggests that although ex-entrepreneurs may possess valuable resources, those may not be fully leveraged. Our findings contribute to the new venture hiring and joiner literature.

INTRODUCTION

Research on team composition in new ventures is growing and gaining relevance (Lazar, Miron-Spektor, Agarwal, Erez, Goldfarb, & Chen, 2020; Shepherd, Souitaris, & Gruber, 2021). However, joiners—defined as "the first employees who assist founders in (Roach & Sauermann, 2015) launching a new company" (Roach & Sauermann, 2015, p. 2162)—remain largely unexplored so far. Existing studies suggest that joiners play a crucial role in shaping the early trajectory of new ventures (Giraudo, Rocha, & Grilli, 2019) by providing innovative ideas and overseeing essential work activities (Roach & Sauermann, 2015). The success of new ventures is often closely tied to the quality and fit of their early hires, who play a critical role in navigating uncertainties (Eesley & Roberts, 2012). Employing arguments from resource mobilization theory (McCarthy & Zald, 1977), employees with prior entrepreneurial experience can provide employers with a competitive advantage through creative and entrepreneurial skills (Zhan, Zhang, Li, & Wu, 2024). They contribute valuable resources, such as knowledge, networks, and strategic insights, that can strengthen a venture's resource base (Clough, Fang, Vissa, & Wu, 2019). However, ex-entrepreneurs are more complex to retain as employees than ones without entrepreneurial experience (Zhan et al., 2024) due to lower organizational commitment (Botelho & Chang, 2023), with turnover costs outweighing their benefits (Zhan et al., 2024). Additionally, ex-entrepreneurs are often perceived as behaving insubordinately and as organizational misfits by recruiters, and thus being less favored as employees compared to employees without prior entrepreneurial experience (Waddingham, Zachary, & Walker, 2024).

Ex-entrepreneur joiners often experience greater organizational commitment and job satisfaction when they can contribute actively to the development of a venture. However, if their previous entrepreneurial experience conflicts with the responsibilities or expectations of their new role, it may lead to frustration and reduced satisfaction (Breugst, Domurath, Patzelt, & Klaukien, 2012). This theorizing raises an important research question: *How do ex-entrepreneurs joining a new venture influence new venture performance?*

We examine the relationship between hiring ex-entrepreneurs and venture performance by proposing two competing assumptions: First, the existence of ex-entrepreneurs has a negative impact on new venture performance due to their potentially lower job satisfaction. Second, their entrepreneurial skills and networks are a unique resource that could enhance new venture performance. These competing hypotheses allow us to unpack the complex dynamics of integrating ex-entrepreneurs into new venture teams. We test these hypotheses using a longitudinal database from 1990 until 2021 provided by Statistics Sweden. Our findings indicate a significant negative relation between the number of ex-entrepreneurs joining a new venture and new venture performance. With those findings, we contribute to the literature on new venture hiring (Giraudo et al., 2019; Hietaniemi, Santamaria, Kacperczyk, & Peltonen, 2024) and joiners (Engel, Lewis, Cardon, & Hentschel, 2023; Roach & Sauermann, 2015).

HYPOTHESES DEVELOPMENT

Resource mobilization and ex-entrepreneurs

Resource mobilization—the process through which entrepreneurs acquire, access, and deploy tangible and intangible resources—is foundational to the survival and growth of new ventures (Aldrich & Fiol, 1994; Barney, 1991). In the context of new ventures, the capacity to mobilize critical resources significantly drives performance (Aldrich, Zimmer, & Jones, 1986; Qin, Wright, & Gao, 2017). Ex-entrepreneurs transitioning from leading their own new ventures to taking employee roles bring resources that can benefit new ventures (Amaral, Baptista, & Lima, 2011; Waddingham et al., 2024). For example, individuals with entrepreneurial experience know how to identify and exploit resources through their knowledge and networks (Hitt, Ireland, Camp, & Sexton, 2001). However, how new ventures mobilize and leverage resources may vary, leading to divergent influences on joiner commitment and new venture performance (Zott & Huy, 2007).

Resource mobilization and job satisfaction

One key determinant of an employee's effectiveness as a resource is their level of job satisfaction (Bakker & Demerouti, 2007), a highly relevant resource for new ventures (Judge, Bono, Erez, & Locke, 2005). From a resource mobilization perspective, human capital is a resource whose effectiveness depends on psychological and emotional factors, including job satisfaction. In particular, satisfied employees are more engaged and perform better (Anitha, 2014). Employees in new ventures often face greater job demands and uncertainty. Thus, and through the lack of resources, new ventures must also compensate their employees through psychological rewards, learning opportunities, and recognition, making satisfaction a particularly important determinant of their employees' long-term engagement and contribution (Cardon & Stevens, 2004). Specifically, as joiners are quicker to be included in meetings and social events in new ventures compared to more established ventures, job satisfaction and productivity of joiners are often greater in new ventures (Cardon & Stevens, 2004). Thus, understanding the role of job satisfaction in resource mobilization is particularly relevant for ex-entrepreneurs who transition into employee roles in new ventures. If these individuals exhibit lower job satisfaction, their effectiveness as an organizational resource may be compromised, ultimately affecting venture performance. Applying resource mobilization theory, we argue that ex-entrepreneurs' count in new ventures may have a dual impact depending on how well their resources align with the venture's needs.

The negative contribution of ex-entrepreneurs

Ex-entrepreneurs contribute significant resources through their understanding and experience of how new venture processes work and how to mobilize relevant resources (Eesley & Roberts, 2012). However, their ability to mobilize these resources may be constrained. In particular, ex-entrepreneurs may have difficulties adjusting to subordinate roles, especially in the early stages of new ventures where power dynamics are still being solidified, and there might be an organizational misfit (Waddingham et al., 2024). Hoang and Gimeno (2010) found that individuals must imprint their founder identity to become a founder, making it difficult to transition to an employee role. In their previous roles as founders, ex-entrepreneurs often exercised high levels of autonomy and decision-making authority. Transitioning into an employee role may, therefore, lead to a role conflict (Hopp & Stephan, 2012). This role conflict can diminish their job satisfaction, as ex-entrepreneurs may feel their prior entrepreneurial experience is either undervalued or insufficiently utilized in their new position (Jansen, 2005). Therefore, we theorize that ex-entrepreneur joiners may feel alienated if their contributions are not aligned with the strategic direction of the new venture. If the ex-entrepreneur's resources are underutilized, it leads to job dissatisfaction, negatively influencing new venture performance.

H1a: The number of ex-entrepreneurs in new ventures is negatively associated with new venture performance.

The positive contribution of ex-entrepreneurs

Individuals with entrepreneurial experience positively impact new ventures when their resources are effectively utilized (Shane, 2003). Having previously navigated the challenges of starting and growing a venture, ex-entrepreneurs bring knowledge about, for example, strategic decision-making, resource acquisition, and network building (Ucbasaran, Lockett, Wright, & Westhead, 2003). They have the skills and abilities to mobilize resources effectively within the new venture, making themselves a valuable resource for new ventures (Hoang & Gimeno, 2010). In particular, entrepreneurial networks and knowledge of resource acquisition strategies can be invaluable to new ventures facing uncertainty and limited resources (Shane, 2003). Besides, paid employees are more satisfied with their jobs than self-employed employees (Millán, Hessels, Thurik, & Aguado, 2013). Such job satisfaction is important for new ventures because employee job satisfaction positively influences organizational effectiveness (Koys, 2001). We theorize that if ex-entrepreneurs' resources are effectively mobilized and valued in their new role as employees, their job satisfaction will likely increase, positively impacting new venture performance.

H1b: The number of ex-entrepreneurs in new ventures is positively associated with new venture performance.

METHOD

Data

The data for this study comes from a panel database combining several registers provided by Statistics Sweden (Bird & Wennberg, 2016), covering the period from 1990 to 2021. Statistics Sweden collects data on the Swedish population aged 16 and above and Swedish ventures, providing all demographic information, job histories, and venture metrics. For this study, we focused on entrepreneurs, their new ventures, and employees from all industries, and geographic regions to ensure that our sample was representative of the impact of ex-entrepreneurs joiners on new venture performance. The final sample consisted of 8,023 new venture observations.

Measures

Dependent variable. We measured new venture performance using the *profitability ratio,* calculated through EBIT divided by fixed assets. Profitability ratios provide insight into a venture's ability to generate returns from its assets, which is essential for assessing the success of new ventures. While similar in purpose to return on assets, the profitability ratio focuses explicitly on the venture's operating results before interest and tax expenses, isolating the effects of financial leverage and tax structures.

Independent variable. We defined the *count of ex-entrepreneurs* by the annual number of employees who at least once were entrepreneurs by founding and managing their own venture from 1990 until the year before they were registered as employees but now are employed by a different new venture. During the time of employment, ex-entrepreneurs did not manage their own venture and solely acted in their employee role.

Control variables. We controlled for venture characteristics (i.e., *venture age, venture size*), financial characteristics (e.g., *fixed assets, earnings*), and corporate governance characteristics (i.e., *owner size, average owner age, percentage of female owners, board size, average board member age, percentage of female board members, percentage of female employees, industry experience of employees*). These controls are essential to isolate the actual effect of ex-entrepreneur joiners on new venture performance by accounting for key factors that independently influence profitability.

Data analysis

To assess the impact of ex-entrepreneur joiners on new venture performance, we estimated fixed effects regression models (Brymer & Rocha, 2024). Fixed effects models allow us to control for time-invariant venture-specific characteristics, mitigating concerns about unobserved heterogeneity that could bias our results (Wooldridge, 2010). Given our panel dataset, this approach allowed us to test our hypotheses while accounting for both within-venture and between-venture variability.

RESULTS

Testing hypotheses H1 and H2

We examined the relationship between the count of ex-entrepreneurs in new ventures and new venture performance, specifically focusing on the profitability ratio, with fixed effects included to account for unobserved heterogeneity at the venture level. Our findings indicate a significant negative relationship between ex-entrepreneur joiner's count and new venture performance ($\beta = -0.09$, $p = 0.005$). This result supports Hypothesis H1a, suggesting that the count of ex-entrepreneurs in new ventures is negatively associated with venture performance.

Robustness checks

To test the robustness of our findings, we reestimate our model. The ex-entrepreneurs' impact remains negative and significant ($\beta = -0.02$, $p = 0.05$), confirming that new ventures with more ex-entrepreneur joiners have lower new venture performance. Among controls, venture age, and debt ratio positively affect new venture performance, while investment intensity remains strongly negative. These results reinforce our main findings that ex-entrepreneurs negatively impact new venture performance, supporting H1.

DISCUSSION AND IMPLICATIONS

Our findings contribute to the literature on new venture hiring (Giraudo et al., 2019; Hietaniemi et al., 2024), emphasizing the pivotal role of early hiring decisions in shaping new venture performance. Prior research examining how the role of founder characteristics (Honoré, 2022; Gruber et al., 2024), employee skills (Chen, 2005; Dencker et al., 2009), and team composition (Agarwal et al., 2016; Jung et al., 2017) influences new venture performance. However, much of the new venture hiring literature has concentrated on entrepreneurs in their role as founders rather than examining them as employees. As a result, the diverse backgrounds and characteristics of early hires have often been overlooked. In particular, our study extends resource mobilization theory (Aldrich & Fiol, 1994; Barney, 1991; McCarthy & Zald, 1977) by demonstrating that resources—such as stemming from entrepreneurial experience—are not inherently beneficial. While ex-entrepreneur joiners might be a valuable resource in the early hiring processes, we show that they tend not to transform their resources into supporting new ventures. We explain this by the lack of job satisfaction of ex-entrepreneurs in their employee roles.

Additionally, we contribute to the literature on joiners (Engel et al., 2023; Roach & Sauermann, 2015) and how their entrepreneurial experience manifests in new venture performance. Our findings reveal the negative impact of the number of ex-entrepreneur joiners on the profitability ratio of new ventures. Research still suggests the benefits individuals with entrepreneurial experience can provide to a new venture, such as network and knowledge of resource mobilization (Hoang & Gimeno, 2010; Shane, 2003; Ucbasaran, Lockett, Wright, & Westhead, 2003). Our findings support the stream of literature that highlights that job satisfaction is important to the willingness to provide their resources (Cardon & Stevens, 2004) and that ex-entrepreneurs struggle to adapt to the employee role (Hoang & Gimeno, 2010; Hopp & Stephan, 2012). Individuals transitioning from entrepreneurial to employee roles often experience a decline in job satisfaction due to less autonomy (Hyytinen & Ruuskanen, 2007). We find that this is particularly evident among early hires, who play a critical role in shaping the path of new venture formation (Eesley & Roberts, 2012; Giraudo et al., 2019; Roach & Sauermann, 2015). Rather than leveraging their prior experience, ex-entrepreneurs may become disengaged, less invested, or even resistant to organizational structures, leading to lower overall performance. Our results highlight the important organizational paradox that while ex-entrepreneurs bring valuable skills, their integration into structured ventures may create tensions.

Beyond theoretical contributions, our research has practical relevance for founders and investors alike, offering insights into how the composition of joiners shapes new ventures and enlightening them on the best practices in early-stage hiring. In doing so, we build on and extend the findings of Rocha and Brymer (2021), who highlighted the importance of affiliation-based hiring for new venture performance. While their work focused on the value of hiring from prior affiliations, we show that ex-entrepreneurs potentially hinder resource mobilization as new team members. This signals founders to carefully integrate ex-entrepreneur joiners in their new ventures to avoid adverse effects on them.

While our study identifies the venture-level impact of hiring joiners with entrepreneurial experience, future research should investigate whether the founder-joiner fit influences the relationship, providing critical insights into team composition (Ruef, Aldrich, & Carter, 2003) and its importance on new ventures.

CONTACT: Luise Kaufmann; luise.kaufmann@tum.de; Technical University of Munich, Bildungscampus 9, 74076 Heilbronn, Germany

SUMMARY

WHEN HUSTLE HURTS: EXAGGERATED ENTREPRENEURIAL HUSTLE AS AN IMPRESSION MANAGEMENT TECHNIQUE AND ITS EFFECTS ON EXPECTED INVESTMENT

Gianfranco Ciurlia, Technical University of Dortmund, Germany
Bastian Kindermann, Technical University Braunschweig, Germany
Julie Saesen, Technical University Dortmund, Germany

Principal Topic

Impression management (IM) techniques enable founders to positively shape stakeholder interactions (Nagy et al., 2012) and to secure funding (e.g., Croce et al., 2013). Entrepreneurs can use Entrepreneurial Hustle (EH) self-presentations as an IM technique (Kindermann et al., 2023). Prior research highlights the positive effects of EH on perceived founder competence and organizational attractiveness (Fisher et al., 2020; Kindermann et al., 2023). However, popular media criticize excessive forms of hustle (e.g., Forbes, 2023). Focusing mainly on moderate forms of hustle, research has not yet explored exaggerated EH and its potential backlashes. We address this gap by introducing the notion of exaggerated EH and exploring its association with expected investment. We propose that "investor trust" mediates the negative relationship between exaggerated EH and expected investment. Additionally, we examine the two moderators "founder prior experience" and "investor analytical thinking".

Method

To introduce the concept of exaggerated EH, we conducted a qualitative pre-study and interviewed 10 venture capital investors, defining exaggerated EH traits. We manipulated exaggerated EH by adapting a pitch based on the three traits identified in our qualitative pre-study. We conducted a pre-registered, randomized between-subject experiment using experimental vignette methods (Auguinis & Bradley, 2014). A final sample size of 583 participants, recruited via Prolific, evaluated a fictional entrepreneur's pitch. The treatment included a manipulated audio file and a CV. The participants first answered logical questions to assess analytical thinking, then they made an investment decision and provided feedback on their perceptions of the founder and themselves. We tested our hypotheses by conducting Ordinary Least Squares regressions and bootstrapping following Preacher and Hayes (2008).

Results and Implications

Bootstrapping reveals a significant indirect effect, indicating a negative mediation between exaggerated EH, trust, and expected investment. The study reveals a positive moderation effect of prior founding experience on the relationship between exaggerated EH and investor trust. Analytical thinking does not show a statistically significant moderating effect. This study makes three key theoretical contributions to EH, general entrepreneurship, and IM literature. First, it introduces and conceptualizes exaggerated EH as an assertive IM technique. Second, it demonstrates the indirect negative relationship between exaggerated EH and expected investment through trust, enhancing the understanding of EH. Third, it advances IM theory by highlighting the role of actor-based characteristics, showing that prior founding experience moderates the relationship between exaggerated EH and trust. We provide practical guidance on effective self-presentation, recommending that entrepreneurs refrain from excessive self-promotion.

CONTACT: Gianfranco Ciurlia; gianfranco.ciurlia@tu-dortmund.de; Technical University of Dortmund; Färbergraben 4, Munich, Bayern 80331, Germany

SUMMARY

UNLOCKING ENTREPRENEURIAL POTENTIAL: HOW PERSONAL RESOURCES AND COMMUNITY SUPPORT DRIVE ENGAGEMENT AND FUEL VENTURE GROWTH

Willy Das, Lehigh University, United States
Daniela Gimenez Jimenez, Tu Dortmund, Germany

Principal Topic

Entrepreneurs face significant stressors—high uncertainty, intense workloads, resource constraints, and commitment pressures—which can deplete psychological resources, leading to stress, burnout, and ill-being (Williamson et al., 2021). Despite advancements in the literature on entrepreneurial well-being (Wiklund et al., 2019; Stephan et al., 2023), there remains limited understanding of the individual resources needed to mitigate this challenging context and their influence on founders' growth aspirations. This study aims to address this gap by examining how entrepreneurs' access to financial capital, social support, and well-being resources influences their engagement and business growth expectations, applying Conservation of Resources (COR) theory.

Method

This study employs a mixed-methods approach, integrating quantitative survey data and qualitative interviews. Conducted in collaboration with the Nasdaq Entrepreneurial Center, the survey launched in August 2024 has gathered 340 responses, supplemented by Prolific. It assesses well-being, perceived capital acquisition, and community support, incorporating the RICH (Resource Induced Coping Heuristic) inventory to analyze how entrepreneurs acquire, protect, and develop resources. Qualitative data was collected through focus groups at the Nasdaq Entrepreneurial Center and individual interviews at Plug and Play, a leading innovation platform. To date, 12 interviews have been conducted, with additional sessions planned. This mixed-methods approach provides a comprehensive understanding of how individual and community resources influence entrepreneurial engagement and business growth.

Results and Implications

Preliminary findings reveal that perceived financial capital and well-being conservation significantly predict higher growth expectations, but social support does not exhibit a direct impact. Entrepreneurial engagement is a critical mediator, amplifying the effects of financial and well-being resources on business growth expectations. Entrepreneurs who actively engage in their ventures while maintaining well-being practices report greater resilience and sustained business efforts. Qualitative insights highlight the role of peer mentorship, mental health discussions, and community support in mitigating entrepreneurial stress. Entrepreneurs emphasized that structured support networks alleviate isolation and improve well-being, enabling them to sustain innovation and business growth. A central challenge identified was balancing productivity with self-care, as many founders prioritize business growth at the expense of mental health, leading to burnout. However, those who proactively seek mental health resources and social support experience improved adaptability to uncertainty. This study extends prior research (Williamson et al., 2021; Stephan et al., 2019) by demonstrating how financial, social, and well-being resources interact to influence entrepreneurial engagement and venture success. The findings provide actionable insights for policymakers, entrepreneurship support organizations, and founders on fostering resilient entrepreneurial ecosystems.

CONTACT: Willy Das; wid221@lehigh.edu; Lehigh University; 510 E 3rd St, Bethlehem, Pennsylvania 18015, United States

SUMMARY

UNLOCKING THE IMPACT OF ENTREPRENEURIAL COACHING: THE INTERPLAY OF SELF-EFFICACY, COACHABILITY, AND VENTURE PHASE

Cintya L. Gajardo-Vejar, University of Massachusetts Lowell, United States

Principal Topic

This study examines how entrepreneurial self-efficacy and coachability interact with the phase of the entrepreneurial lifecycle to influence coaching outcomes. While entrepreneurial coaching has been associated with increased business success (Gray et al., 2011; Keith et al., 2016), prior research suggests that its effectiveness varies across entrepreneurs (Schermuly et al., 2021). Entrepreneurial self-efficacy, defined as an individual's belief in their ability to perform entrepreneurial tasks (Chen et al., 1998), is linked to persistence and resilience (Cassar and Friedman, 2009). However, research has primarily focused on nascent entrepreneurs (Hechavarria et al., 2012; Renko et al., 2012), leaving gaps in understanding its role in later stages. Similarly, coachability, or the extent to which an entrepreneur seeks and integrates feedback, is critical for coaching success (Ciuchta et al., 2018). This study explores how these characteristics interact across different entrepreneurial phases to impact coaching outcomes.

Method

This study employs a multi-wave survey approach, collecting data from an entrepreneurial coaching program in Chile that involved 50 entrepreneurs and 50 coaches over nine months. Using hierarchical linear regression analyses, the study examines the effects of entrepreneurial self-efficacy and coachability on coaching outcomes across different entrepreneurial stages (nascent, new business, and established business). Surveys were administered at multiple points to both coaches and entrepreneurs, measuring self-efficacy (McGee et al., 2009), coachability (Ciuchta et al., 2018), and coaching outcomes (Brinkley & Le Roux, 2018).

Results and Implications

Findings indicate that entrepreneurial self-efficacy is a strong predictor of positive coaching outcomes, particularly for entrepreneurs in the nascent phase. However, entrepreneurs with high self-efficacy achieved significantly better outcomes when their ventures were established businesses. Coachability moderated this relationship with highly coachable entrepreneurs experiencing greater benefits, particularly in later-stage ventures. A three-way interaction between self-efficacy, coachability, and entrepreneurial phase suggests that coaching is most beneficial for entrepreneurs at more advanced stages. These findings challenge previous research (e.g., Audet & Couteret, 2012), which emphasized coaching benefits for early-stage entrepreneurs. The study provides insights into how coaching interventions should be tailored based on an entrepreneur's phase and individual characteristics, contributing to the broader discussion on entrepreneurial coaching effectiveness.

CONTACT: Cintya L. Gajardo-Vejar; cintya_gajardovejar@student.uml.edu; University of Massachusetts Lowell; 10 Gershom Ave, Lowell, Massachusetts 01854, United States

SUMMARY

THE DUAL IMPACT OF SUPPORT: EXPLORING THE DYNAMICS OF COMMUNITY AND GOVERNMENT SUPPORT ON INNOVATIVENESS IN NASCENT VENTURES

Kurian George, Tilburg University, Netherlands
Hooman Abootorabi, Towson University, United States

Principal Topic

Understanding the influence of support systems on innovativeness of nascent ventures is critical for advancing entrepreneurial success and promoting economic development. Community support serves as a crucial mechanism in providing resources, networks, and legitimacy; however, its impact on innovativeness is complex, potentially acting both as an enabler and a constraint. Similarly, government support may either reinforce or counteract the dynamics created by community engagement, introducing another layer of complexity. Drawing on Self-Categorization Theory (SCT), this study investigates the nuanced interplay between community and government support in shaping the innovativeness of nascent ventures. We theorize that community support exhibit a U-shaped relationship with innovativeness, wherein strong community support enhances in-group identification, encouraging intergroup comparisons that drive innovation as a means of differentiation from strong actors. Conversely, low community support reduces in-group identification, encouraging entrepreneurs to emphasize their individuality, leading to distinct and often unconventional innovations. Further, we propose that perceived government support weakens this curvilinear relationship by buffering against the extremes of community-driven innovation.

Method

We utilize individual-level longitudinal data from the Panel Study of Entrepreneurial Dynamics (PSED) II dataset, spanning the years 2005-2011 to test our hypotheses. The PSED II database contains valid and reliable data on the process of business formation, capturing key data on the characteristics of nascent entrepreneurs, the activities undertaken during the start-up process, and the attributes of start-up efforts that evolve to new firms. Our final sample consists of 1,251 nascent entrepreneurs who are identified prior to launch of their firms, and were tracked through gestation, the launch phase, and the subsequent growth or death of their firms. To analyze our data, we employ fixed-effects multinomial logistic regression and fixed-effects OLS regression.

Results and Implications

Our findings reveal a curvilinear relationship between perceived community support and innovativeness, indicating that both high and low levels of support foster greater innovation, while moderate levels are associated with diminished innovativeness. Our results also suggest that perceived government support mitigates the extremes of community-driven innovation. These findings contest the assumption that higher support uniformly enhances innovation, instead demonstrating complex and sometimes offsetting dynamics at play. Our study contributes to the literature on community dynamics and entrepreneurship and offers insights into the socio-cognitive and institutional dynamics of entrepreneurial ecosystems. It also advances theoretical understanding and provides practical guidance for creating support structures that balance inclusivity and individuality to effectively stimulate innovation.

CONTACT: Kurian George; k.george@tilburguniversity.edu; Tilburg University; Warandelaan 2, Tilburg, Noord-Brabant 5037AB, Netherlands

SUMMARY

THE CONTINGENT RELATIONSHIPS BETWEEN FOUNDING CEO DARK TRIAD PERSONALITY CHARACTERISTICS AND NEW VENTURE GROWTH

Keith Hmieleski, Texas Christian University, United States

Principal Topic

Dark triad personality traits—narcissism, Machiavellianism, and psychopathy—are often perceived negatively but may provide advantages in entrepreneurial settings. These traits are linked to assertiveness, risk-taking, and strategic manipulation, which can enhance new venture performance in competitive environments. However, their effects on venture growth likely depend on individual and contextual factors. Drawing from social cognitive theory and social role theory, this study examines how entrepreneurial self-efficacy (ESE) and gender moderate the relationship between dark triad traits and new venture growth. Founding CEOs with high ESE may experience amplified benefits from these traits. Additionally, societal perceptions of leadership may shape how these traits influence firm performance differently for men and women.

Method

The study began with a sample of 3,000 U.S. new ventures sourced from Mergent Intellect, with survey responses received from ~300 founding CEOs. Personality traits were assessed using established dark triad measures, while ESE was measured using six items evaluating confidence in entrepreneurial tasks. New venture growth was measured by revenue increases over two years. A moderated regression analysis was conducted using SPSS 29.0, with interactions examined at high (+1 SD) and low (-1 SD) levels of ESE and gender. Control variables included firm age, size, and founder characteristics such as experience and education.

Results and Implications

Findings indicate that dark triad traits positively predict new venture growth, but only when founders possess high entrepreneurial self-efficacy. Founding CEOs with both high dark triad traits and high ESE exhibited the greatest firm growth, suggesting that confidence on one's entrepreneurial abilities enables more effective use of assertiveness, risk-taking, and strategic manipulation for business success. Gender significantly moderates these effects. Narcissism and psychopathy were more beneficial for male founders, consistent with gender role congruity theory, which suggests that dominant, self-serving behaviors align more closely with societal expectations for male leaders. In contrast, female founders high in these traits experienced weaker growth effects, potentially due to resistance against such leadership styles. Machiavellianism, however, was beneficial for both genders, indicating that strategic manipulation serves as a universally effective entrepreneurial tool. These findings demonstrate that dark triad traits can enhance venture success, but their effectiveness depends on an entrepreneur's self-efficacy and societal gender norms. Practical implications suggest that entrepreneurial training programs should emphasize ESE development while acknowledging gender-specific challenges in leadership perception.

CONTACT: Keith Hmieleski; k.hmieleski@tcu.edu; Texas Christian University; TCU Box 298530, Fort Worth, Texas 76129, United States

SUMMARY

MENTAL HEALTH AND ENTREPRENEURSHIP: EXPLORING THE IMPACT OF HYPOMANIA ON BEHAVIORAL DYNAMICS DRIVING VENTURE SUCCESS

Yasmin Khalfallah, RWTH Aachen University, Germany
Sebastian Kruse, RWTH Aachen, Germany

Principal Topic

Entrepreneurs' mental health significantly influences venture success by shaping cognitive processes, emotional responses, and behavior (e.g., Stephan, 2018). Hypomania, a core symptom of bipolar disorder, is highly prevalent among entrepreneurs and linked to traits such as goal orientation, self-esteem, and positive affect (Wolfe et al., 2020). While research highlights the overrepresentation of mental health conditions among entrepreneurs (Freeman et al., 2018), it has primarily focused on ADHD while largely neglecting hypomania. Additionally, studies have overlooked how these conditions shape actual entrepreneurial behavior and the moderating role of entrepreneurial experience (Wiklund et al., 2018). This study examines how hypomania influences key entrepreneurial behaviors—opportunity recognition, entrepreneurial persistence, and acquisition of resources—and how entrepreneurial experience moderates these relationships. We hypothesize that hypomania enhances opportunity recognition and persistence, with entrepreneurial experience weakening these effects, while it hinders resource acquisition due to hubris and risk-taking, with entrepreneurial experience strengthening this effect.

Method

Our sample contains 1182 German entrepreneurs surveyed online between November 2022 and January 2023 using the AMADEUS database. Hypomania was measured using a shortened Hypomanic Personality Scale (Eckblad & Chapman, 1986; Pushkarskay et al., 2020). Opportunity recognition (Shane & Nicolaou, 2014), entrepreneurial persistence (Baum & Locke, 2004), acquisition of resources (Spreitzer, 1996), and entrepreneurial experience (Stuart & Abetti, 1990) were assessed with validated constructs. Structural equation modeling (SEM) was used and robustness ensured through checks including hierarchical regression analysis.

Results and Implications

Our results highlight the impact of hypomania on entrepreneurial behavior. We confirm its positive association with opportunity recognition and entrepreneurial persistence. Unexpectedly, we find a positive association with acquisition of resources, possibly due to the associated positive affect, creativity and optimism outweighing hubris and risk-taking concerns in human interactions. Additionally, entrepreneurial experience weakens hypomania's influence on opportunity recognition and persistence, but not on resource acquisition. Our study has essential contributions: We advance the research stream into mental health conditions beyond ADHD and show that individuals with hypomania symptoms and traits have specific abilities corresponding to unique demands of entrepreneurship. We extend scarce literature on hypomania beyond traits towards entrepreneurial behaviors responsible for venture success. Lastly, we provide a lens into the continuity of the effects of mental health conditions on entrepreneurial behaviors through the introduction of the moderating effect of entrepreneurial experience. Practically, our results suggest that mental health conditions can benefit behavior critical to entrepreneurial success and guide ventures and investors in founder selection, role definition and task allocation.

CONTACT: Yasmin Khalfallah; khalfallah@time.rwth-aachen.de; RWTH Aachen University; Klenzestr. 87, Munich, Bayern 80469, Germany

BEYOND THE INDIVIDUAL ENTREPRENEUR: HOW EMOTIONAL STABILITY OF THE NEW VENTURE TEAM PREDICTS EARLY-STAGE VENTURE PERFORMANCE

Felix Kuhn, Technical University Munich, Germany
Jannik Nolden, Technical University of Munich, Germany

Principal Topic

The individual entrepreneurs' personality significantly affects the performance of early-stage ventures (Rashid, Alzafari, & Kratzer, 2020). However, venture capitalists rarely consider individual founders' proposals, but focus more on the skill and experience of an entire founding team (Kamm, Shuman, Seeger, & Nurick, 1990). Thus, the personality composition of founding teams is decisive for venture performance (Jin, Madison, Kraiczy, Kellermanns, Crook, & Xi, 2017; Kollmann, Stöckmann, Linstaedt, & Kensbock, 2021; Zhou, Hu, & Zey, 2015). This study examines the dynamic impact of personality composition of new venture dyads, particularly of emotional stability, on early-stage venture performance beyond other personality traits. Building on research on trait activation theory, we develop and test our hypothesis that emotional stability in new venture dyads has a dynamic effect on early-stage venture performance in a longitudinal study design (Freiberg & Matz, 2023).

Method

Our sample consists of German-speaking dyadic new venture teams who registered their venture between 1 and 12 months before the study (N = 230). We invited them to an on-site lab session where they had to complete a questionnaire about their personality (Soto and John, 2017). We tracked their venture performance over two years. It is measured as the year-over-year growth of website performance data provided by a publicly available data provider. Early-stage ventures without website traffic are filtered out, accounting for 42%.

Results and Implications

Our results show a significant dynamic effect of the team's mean score of emotional stability on venture performance over two years. The standardized effect sizes per month vary between $\beta_{min} = 0.26$ and $\beta_{max} = 0.37$, all significant ($p < 0.05$). Effect sizes increase as the venture matures. Furthermore, we find a significant dynamic effect of team diversity in extraversion on the venture performance within the first months of the venture, with β varying from $\beta_{min} = 0.24$ to $\beta_{max} = 0.29$ ($p < 0.05$), which diminishes afterward. As one of the first studies in entrepreneurship, we contribute to the investigation of this temporal dimension in early-stage venture performance by applying a longitudinal design to answer our research questions. Our findings suggest that the dynamic interplay of personality and venture performance provides a compelling reason why previous studies on early-stage ventures seem to tell different sides of the same story – depending on how 'early-stage' got interpreted.

CONTACT: Felix Kuhn; felix.kuhn@tum.de; Technical University Munich; Arcisstrasse 21, Munich, Bayern 80333, Germany

SUMMARY

DOES WORKAHOLISM PREDICT UNETHICAL PRO-ORGANIZATIONAL BEHAVIOR IN ENTREPRENEURS? THE MODERATING ROLE OF GROWTH MINDSET

Lennard Lang, RWTH Aachen University, Germany

Principal Topic

In entrepreneurship, where dedication and tireless effort are deemed essential for success, the phenomenon of workaholism is highly prevalent. Yet, existing research primarily focuses on employees in established organizations (Simmons et al., 2023). Research indicates that workaholics resort to morally reprehensible behavior, such as cheating or misrepresentation, to satisfy their psychological need for competence and superior performance (Farasat et al., 2021; Van Beek et al., 2012). For entrepreneurs, personal achievements are reflected in the success of their ventures (Cardon et al., 2009). Given the fatal consequences of ethical misconduct in entrepreneurship, growing interest exists in why some entrepreneurs behave unethically to promote company interests, while others do not (Scheaf & Wood, 2022; Tacke et al., 2023). Drawing on Self-determination theory (Ryan & Deci, 2000), we suggest that workaholism positively influences entrepreneurs' intention to engage in unethical pro-organizational behavior (UPB; actions intended to benefit the organization and violate standards of proper conduct), with growth mindset moderating this relationship.

Method

This research study utilizes cross-sectional survey data from 314 entrepreneurs from Germany, Switzerland and Austria. Our independent variable was measured with the Dutch Workaholism Scale (Schaufeli et al., 2009). UPB was measured according to Umphress et al.,(2010) and growth mindset following Dweck et al.,(1995). Consistent with existing UPB research, we controlled for individual- and firm-level factors to ensure robustness of results. Hierarchical regression analyses tested the relationships and moderation effects.

Results and Implications

Our results demonstrate that entrepreneurs' workaholism positively relates to UPB and that growth mindset moderates this relationship. In anticipation of performance advantages, workaholic entrepreneurs show a higher intention of UPB to validate their performance-based self-esteem. Possessing a growth mindset, the belief that abilities can be developed through learning and perseverance, provides workaholic entrepreneurs with the confidence to achieve goals through their own efforts without resorting to unethical shortcuts that put their venture at risk. We contribute to workaholism research in three ways. First, we identify workaholism as a key driver of UPB. Second, we challenge the dominant view that workaholics' behavior is solely oriented towards sustaining compulsive working habits and suggest that their behavior is primarily driven by the associated instrumental value (i.e., achievements). Third, we propose that cultivating a growth mindset mitigates the workaholism-UPB relationship, thereby reducing its negative consequences. From a practical perspective, we identify UPB as a risky behavior adopted by workaholic entrepreneurs that threatens long-term venture success and identify a trainable skill to mitigate this relationship.

CONTACT: Lennard Lang; lennard.lang@time.rwth-aachen.de; RWTH Aachen University; Buttermelcherstr 8, Munich, Bayern 80469, Germany

SUMMARY

ENTREPRENEURIAL ACTION AND STAKEHOLDER ENROLLMENT BY ENTREPRENEURS WITH DISABILITIES

Wyatt Lee, Auburn University, United States
Franz Lohrke, Auburn University, United States
Ashley Roccapriore, Auburn University, United States

Principal Topic

In the U.S., nearly 25% of working-age adults have disabilities, yet many struggle to gain or maintain traditional employment due to visible or audible impairments. This leads many to pursue entrepreneurship despite the significant challenges entrepreneurship poses to persons with disabilities (PWDs). Extant research largely ignores entrepreneurs with disabilities, focusing on traditional employment. AV-disabled entrepreneurs face discrimination, lower wages, and fewer career opportunities, pushing them toward self-employment. However, little is known about context-specific ways in which they operate their businesses, their actions, and how their relationships with stakeholders impact their outcomes. As such, we qualitatively explore entrepreneurial action and stakeholder enrollment by entrepreneurs with disabilities (EWDs) to understand entrepreneurship in non-traditional contexts better.

Method

We use an interview-based, grounded theory approach to data collection and analysis to explore these concepts. 18 entrepreneurs have been interviewed about their experiences owning and operating a business, their interactions with stakeholders, the actions they've taken to start and manage their businesses, and how their disabilities specifically impact them as entrepreneurs. By using grounded theory, we can gain insights into how and why a traditionally under-represented population of entrepreneurs does what they do. Data was then coded to identify emergent concepts and themes.

Results and Implications

This project aims to contribute to entrepreneurship literature by examining how entrepreneurs with audible or visible (AV) disabilities operate their businesses. This research addresses a notable gap in the intersection of physical disability and entrepreneurship, an area often overshadowed by studies focusing on mental health disabilities or other DEI topics. By spotlighting this underexplored population, we expand the theoretical and practical understanding of entrepreneurship. Moreover, analyzing these qualitative data through a management theory lens contributes to broader theoretical development. Finally, the findings offer insights for entrepreneurs and stakeholders, shedding light on mutual perceptions and encouraging more supportive, productive relationships in the entrepreneurial ecosystem, fostering a more inclusive future for entrepreneurs and their businesses.

CONTACT: Wyatt Lee; wel0008@auburn.edu; Auburn University; 1965 Downs Way, Auburn, Alabama 36832, United States

SUMMARY

PASSION COMES WITH A PRICE: RELATING DUALISTIC ENTREPRENEURIAL PASSION TO ENTREPRENEURS AND THEIR EMPLOYEES' UNETHICAL BEHAVIORS

Qinglin Liu, University of International Business and Economics, China (People's Republic)

Principal Topic

The purpose of this paper is to explore the dark and destructive side of entrepreneurial passion. Specifically, we investigate how dualistic (i.e., obsessive vs. harmonious) entrepreneurial passion might impact entrepreneurs' and their employees' unethical pro-organizational behavior (UPB), a particular type of unethical behavior aimed at benefiting an organization or its members. From a goal-shielding theoretical perspective (Bélanger et al., 2013, 2019), we expect that obsessive entrepreneurial passion will lure individuals (both entrepreneurs and employees) to engage in UPB, whereas harmonious entrepreneurial passion will reduce the likelihood of such activities. Furthermore, we propose that the entrepreneur's obsessive and harmonious passion may inspire employees to experience similar passionate experiences through emotional contagion, thus affecting employees' engagement in UPB. We also explore the boundary condition during this process.

Method

To test our model, we conducted a three-wave, multi-source survey from Chinese employee responses nested within Chinese entrepreneurs. We used scales with high reliabilities to measure our key constructs. We used multilevel structural equation modeling to test our hypotheses.

Results and Implications

Our results show that the two facets of dualistic entrepreneurial passion are all contagious. Furthermore, the contagion of such dualistic passion from entrepreneurs to their subordinates could influence the employee's UPB engagement. The results suggest that passion plays a critical role in how entrepreneurs influence their employees' unethical behaviors through emotional contagion. Interestingly, contrary to our expectations, neither the entrepreneur's obsessive passion nor harmonious passion directly impacts the entrepreneur's UPB. This unique difference could raise further discussion about whether passion's effect might differ among different objects. We are now undergoing further surveys and experiments to explore the boundary conditions and check the robustness of our findings. This study makes several contributions. First, focusing on how entrepreneurial passion could drive unethical behaviors, our research echoes the current "researching the dark and destructive side of entrepreneurship" guidepost (Shepherd, 2019) and gives us a more comprehensive understanding of entrepreneurial passion. Second, we advance entrepreneurial passion research by exploring how entrepreneurial passion influences others and increasing emotional contagion theory's range of application (Newman et al., 2021). Finally, our study also contributes to the entrepreneurial ethics research (Vallaster et al., 2019; Uriarte et al., 2024).

CONTACT: Qinglin Liu; mplql1991@gmail.com; University of International Business and Economics; No.10, Huixin Dongjie, Chaoyang District, Beijing, China, 100029, China (People's Republic)

SUMMARY

UNDERSTANDING INTERGENERATIONAL PERSISTENCE OF ENTREPRENEURSHIP

Orietta Marsili, University of Bristol Business School, United Kingdom
Matt Dickson, Institute for Policy Research, University of Bath, United Kingdom
Georgios Tsiachtsiras, University of Bristol Business School, United Kingdom

Principal Topic

This paper explores the persistence of entrepreneurship across generations from a life-course perspective, examining the extent to which children's entrepreneurial decisions are impacted by the time-points in their upbringing that their parent(s) were entrepreneurs. We further investigate the transmission pathways by comparing the impact of having a biological parent who is an entrepreneur, with the impact of a step-parent entrepreneur, and whether this differs depending on the gender of the child, the time in childhood when the parent is an entrepreneur, and on whether the parent and child live in the same household.

Method

We use a cohort dataset of children born in England in 1991/92 (N=3,931) and estimate a series of models in which the dependent variable is an indicator of whether the 'child' is an entrepreneur at age 29. We create several parental entrepreneurship variables at each of three key development points in the child's life: birth, childhood (age 8), and adolescence (age 18). We control for a range of parental characteristics including age, education, socioeconomic status, financial situation, and health.

Results and Implications

Among children without entrepreneurial parents, 7.5% are entrepreneurs at age 29. Having an entrepreneurial parent at any point from birth to age 18 increases this by 6.9 percentage points. After controlling for a wide range of familial characteristics the effect slightly reduces to +6.6pp. The parental entrepreneurial influence is more pronounced among boys (+9.7pp) than girls (+4.9pp), and exploring further we find that ever having an entrepreneurial mother increases the likelihood of boys becoming entrepreneurs (+8.6pp) but not girls, while the father's influence is important for both boys (+7.3pp) and girls (+5.0pp). The most influential factor is the presence of an entrepreneurial biological father in the household. This effect is consistent across genders and at all three development stages. The mother's influence is only observed for boys but is consistent at all points, while the effect of having an entrepreneurial step-father in the household is only ever significant for girls. In sum, maternal and paternal transmission effects are both important over the child's life course but present different pathways. Maternal transmission is dependent on the child's gender, while biological paternal transmission is conditional on the father's living arrangement. Non-biological paternal transmission is weaker and depends on child's age and gender. Our results suggest an interdependence between intergenerational transmission of entrepreneurship, family structure, and the child's lived experience of parental relationships.

CONTACT: Orietta Marsili; orietta.marsili@bristol.ac.uk; University of Bristol Business School; Howard House, Bristol, England BS8 1SD, United Kingdom

SUMMARY

PART-TIME FOUNDERS, FULLY FUNDED? ANALYZING INVESTORS' PERCEPTIONS OF HYBRID ENTREPRENEURS.

Torben A. Nübel, TU Dortmund University, Germany
Simon Hensellek, TU Dortmund University, Germany

Principal Topic

Hybrid entrepreneurship—starting a venture while employed—is common, with 62% of German founders following this path (Gänser-Stickler et al., 2022). Research suggests hybrid entrepreneurs gain a survival advantage by aligning skills with business ideas and building competencies (Raffiee & Feng, 2014). Also, low entry barriers enable experimentation with high-risk technologies (Andries et al., 2021), while risk tolerance and resources enhance credibility (Baumol & Strom, 2007). However, concerns remain about long-term growth and investor perception. Using signaling theory, we examine whether investors view hybrid entrepreneurs as less committed, leading to lower investment and potentially limiting their venture's success.

Method

Using generalized estimating equations, we examined how founders' hybrid entrepreneurship status affects funding success (i.e., the amount of capital raised) based on 27,669 firm-month observations from 1,361 founders across 864 German startups. We combined data from the Startupdetector database with LinkedIn profiles, capturing founders' professional and educational backgrounds. We measured hybrid entrepreneurship in four ways at the founding team level: (1) binary hybrid team indicator, (2) proportion of hybrid founders, (3) duration of hybrid employment, and (4) number of additional positions. We also tested whether prior founding experience moderates these relationships. Drawing on signaling theory, we time-lagged our independent variables and used 12-month averages to capture founders' pre-investment signals.

Results and Implications

Our findings show that hybrid entrepreneurship negatively impacts startup funding, implying that investors penalize founders for hybrid engagement. While the binary hybrid team indicator is insignificant, more nuanced measures—hybrid duration rate and hybrid wage employment prevalence—exhibit significantly negative effects. Hybrid duration rate has the strongest effect size, and hybrid wage employment prevalence is most statistically significant, indicating that prolonged, intensive hybrid engagement sends a stronger adverse signal. We also find that prior founding experience moderates these effects positively, suggesting that practical skills gained from previous ventures can alleviate investor concerns regarding hybrid status. By introducing hybrid ratio, hybrid duration rate, and hybrid wage employment prevalence, our study moves beyond a binary view of hybrid entrepreneurship. Each new measure conveys a distinct negative signal: (2) the proportion reflects the share of founders employed elsewhere, (3) the duration indicates how long founders remain in parallel employment, and (4) the number shows how widespread parallel employment is within the team. These factors highlight how divided attention and commitment raise doubts about the venture's prospects. Ultimately, these findings underscore the importance of founders' signals and call for deeper examination of how employment commitments shape startup investment trajectories.

CONTACT: Torben A. Nübel; torben.nuebel@tu-dortmund.de; TU Dortmund University; Kreuzstraße 36, Münster, Nordrhein-Westfalen 48143, Germany

SUMMARY

BRIDGING THE GAP: THE IMPACT OF INTERIM EMPLOYMENT ON ENTREPRENEURIAL SUCCESS

Catherine J. Paek, Boise State University, United States

Principal Topic

Entrepreneurial careers often involve multiple transitions between venture creation and traditional employment, forming complex career patterns rather than unidirectional paths. While research has focused on initial entry into entrepreneurship and the transition back to paid employment, limited attention has been given to entrepreneurs who return to entrepreneurship after a period of traditional employment. This study examines how interim employment experiences between ventures shape subsequent entrepreneurial performance. Drawing on entrepreneurial career and human capital literature, I investigate how organization type (startup vs. established firm) and industry similarity (same vs. different) during the interim period affect future venture performance, contingent on prior venture outcomes (success vs. failure).

Method

To test my hypotheses, I conducted logistic regression analyses using Crunchbase data from entrepreneurs who founded at least two ventures in the U.S. between 1980 and 2020. The dependent variables were binary, measuring whether the venture secured Series A funding and whether it achieved a successful exit (acquisition or IPO). To address potential selection bias, I employed Coarsened Exact Matching to balance covariate distributions between entrepreneurs with interim work experience and those without.

Results and Implications

Results show that the relationship between interim employment choices and subsequent venture performance depends significantly on prior venture outcomes. Entrepreneurs with successful first ventures perform better in subsequent ventures when they join established firms during their interim period, benefiting from exposure to professional management practices and operational systems that complement their existing skills. Conversely, entrepreneurs whose first ventures failed perform better when they remain within the same industry, allowing them to deepen industry-specific knowledge, learn from past mistakes, and build relevant networks. Failed entrepreneurs joining startups may not significantly benefit their future venture performance as this provides limited exposure to new organizational processes. Similarly, successful entrepreneurs remaining in the same industry might not see substantial improvement due to limited exposure to new ideas and innovation opportunities. Notably, for successful entrepreneurs, organization type during the interim period is more critical, while industry continuity proves more important for entrepreneurs who previously experienced failure. This research contributes to career development theories by highlighting how interim employment experiences influence subsequent entrepreneurial performance. For entrepreneurs, it provides practical guidance on optimizing career moves between ventures based on prior outcomes. For policymakers, it suggests ways to design more effective support structures that facilitate knowledge transfer between established organizations and entrepreneurial ventures, ultimately contributing to economic growth and innovation.

CONTACT: Catherine J. Paek; catherinepaek@boisestate.edu; Boise State University; 2130 S Longmont Ave, Boise, Idaho 83706, United States

SUMMARY

THE ENTREPRENEURIAL MASQUERADE: HOW NEUROTICISM AND CHARISMA MOVE IN STEP TO SHAPE FUNDING SUCCESS

Oleg Petrenko, University of Oklahoma, United States
Nathan Hayes, University of North Carolina at Charlotte, United States
Diana Hechavarria, Babson College, United States
Sean Lux, Texas Tech University, United States
Johnny Smith, University of Arkansas, United States

Principal Topic

This paper offers a novel conceptualization on neuroticism and charisma in which we test the effects on equity funding success among entrepreneurs. Traditionally, charismatic, emotionally stable leaders are seen as ideal candidates for attracting investment. However, our research adds nuance to this view by suggesting that neuroticism—a trait linked with anxiety and emotional instability—can, in certain high-pressure pitch environments, be perceived as charisma by investors. Our findings support a curvilinear, U-shaped relationship, where entrepreneurs with either very low or very high levels of neuroticism tend to secure more investment than those exhibiting moderate neuroticism.

Method

We analyzed a sample of 300 lead founders who secured their first round of equity funding between 2017 and 2018, tracking the total funding raised until May 2024. Personality traits, specifically charisma and neuroticism, were assessed using videometric analysis. Complementary firm-level data—such as funding status, industry, firm age, number of investors, and founder demographics—were sourced from Crunchbase, while team-level characteristics were gathered from LinkedIn and Truthfinder. An ordinary least squares regression was used to evaluate how these personality traits interact and influence funding outcomes, controlling for a variety of firm- and team-level factors.

Results and Implications

The regression analysis revealed a significant interaction between charisma and neuroticism ($b = 22.6$ million, $p = .023$) in predicting total equity funding. Specifically, founders with high charisma and either very low or very high neuroticism raised substantially more funding than those with moderate levels of neuroticism, supporting our hypothesized U-shaped relationship. Conversely, entrepreneurs with lower charisma consistently attracted less investment, regardless of their neuroticism level. These results suggest that in the high-stakes environment of investor pitches, traits traditionally viewed as liabilities can be reframed as signals of intense commitment and drive. The study challenges conventional, linear models of leadership by demonstrating that non-traditional personality characteristics—when combined effectively—can enhance entrepreneurial appeal. For investors, these insights encourage a broader, more nuanced evaluation of founder profiles, potentially leading to more diverse and informed investment decisions. For entrepreneurs, the findings imply that embracing a balance between presenting emotional stability and passionate intensity might offer an edge in securing funding.

CONTACT: Oleg Petrenko; Oleg.petrenko@ou.edu; University of Oklahoma; 307 West Brooks, Norman, Oklahoma 73019, United States

SUMMARY

THE ROLE OF FOUNDER HUMILITY IN REDUCING STRESS AMONG FOUNDING TEAMS: THE MEDIATING ROLE OF WORKLOAD

Julia Teresa Prefi, RWTH Aachen University, Germany
Sebastian Kruse, RWTH Aachen University, Germany

Principal Topic

In the last decade, the view of an ideal leader has switched from a narcissistic, arrogant leader to a more collaborative and humbler leader (Zhang, Ou, Tsui & Wang 2017). Humble leaders are willing to view themselves accurately, admit mistakes, appreciate others' strengths and contributions, and demonstrate techability (Owens & Hekman 2012). While the topic has been mostly examined by prior research in corporate settings how it affects interactions between top-level leaders and followers, we want to discover the effectiveness and dynamics of leader humility at the founding team level. Specifically, the study aims to investigate whether founder humility decreases stress levels of the co-founder.

Method

The online survey was conducted in summer 2024 and resulted in 354 responses after data cleaning. To test our hypotheses, we employ hierarchical regression analysis. Our research relies on well-established measurement scales previously validated in other studies. To ensure respondents correctly understood all items, our questionnaire was pre-tested with experienced scholars and entrepreneurs. To measure leader humility, we use the scale by Owens, Johnson & Mitchell (2013), as it shows best validity and reliability. To mitigate biases, the founder's humility was evaluated by the co-founder. We draw upon the perceived stress scale by Cohen et al.,(1983) to measure perceived stress. Further, we include the scale by Moore (2000) for our mediator workload. To ensure robustness of results, we deploy various processual remedies, control variables, and ex-post analyses, e.g., to avoid common method bias and address social desirability and single informant bias.

Results and Implications

The findings of our study indicate that higher levels of founder humility correlate with significantly lower levels of stress among co-founders, with workload serving as a mediator. The study contributes to previous research in two vital ways. Firstly, it extends the application of humility research into the entrepreneurial context, and area that has received little attention. Secondly, we introduce co-founder and team dynamics within the founding team into humility research. Our research makes three important practical contributions. First, it aids founders in recognizing the importance of expressed humility as a strategic asset to reduce turnover and enhance the survivability of ventures. Secondly, it equips co-founders with insights regarding how the leadership style of their founder can decrease stress levels and promote their overall mental well-being. Lastly, it provides valuable guidance for investors looking to assemble balanced founding teams, thereby fostering a sustainable and healthy work environment within ventures.

CONTACT: Julia Teresa Prefi; prefi@time.rwth-aachen.de; RWTH Aachen University; Severinstrasse 126, Aachen, Nordrhein-Westfalen 52080, Germany

SUMMARY

UNVEILING THE SILENT STRUGGLE: DEFINING AND UNDERSTANDING ENTREPRENEURIAL LONELINESS

Sergio A. Rifo, Universidad de Concepción, Chile

Principal Topic

Entrepreneurship is often portrayed as a rewarding yet challenging endeavor marked by independence, risk-taking, and innovation (Shane & Venkataraman, 2000). However, entrepreneurs also facc a significant cmotional toll that people discuss less frequently: entrepreneurial loneliness. Unlike general or workplace loneliness, entrepreneurial loneliness arises from unique demands such as long working hours, high stress, and the absence of traditional support systems (Cardon & Arwine, 2024). It negatively impacts mental health, motivation, and business outcomes (Torrès et al., 2022; Zhu et al., 2023). Despite its significance, research on entrepreneurial loneliness remains fragmented, lacking a clear definition and theoretical framework. This study addresses these gaps by defining entrepreneurial loneliness, identifying its core dimensions, and mapping its evolution across the entrepreneurial process.

Method

First, we reviewed the research on loneliness based on psychology, entrepreneurship, and identity literature and interviewed 25 entrepreneurs to identify the dimensions (or facets) of entrepreneurial loneliness. We then applied facet theory (Guttman, 1968; Levy, 1994) and similar structure analysis to a sample of 120 entrepreneurs to test the core dimensions of entrepreneurial loneliness and ensure construct clarity. Entrepreneurial loneliness was also mapped throughout the entrepreneurial process, from opportunity recognition to venture scale-up, to examine its evolution. This structured approach ensures that the construct is conceptually precise and empirically based, contributing to a clearer understanding of entrepreneurial loneliness at different stages of entrepreneurship.

Results and Implications

Our findings reveal that entrepreneurial loneliness comprises three core facets: the affective facet (emotional exhaustion from unmet relational needs), the identity facet (loneliness tied to the entrepreneur's identity), and the contextual facet (shaped by societal expectations and cultural norms). These facets define entrepreneurial loneliness as: "A negative affective state experienced by entrepreneurs, stemming from perceived relational deficiencies, exacerbated by the high-stakes, identity-laden nature of entrepreneurial tasks, and influenced by external environmental and cultural factors." Entrepreneurial loneliness evolves across the entrepreneurial process, manifesting as a lack of social validation during opportunity recognition, stress during venture development, and heightened pressure during scale-up. Key antecedents include high personal responsibility and lack of peer support, while consequences encompass reduced well-being, impaired decision-making, and negative business outcomes. Our findings have important implications. For theory, our framework enables the development of validated measurement tools and testing of entrepreneurial loneliness's impact on cognition, behavior, and business outcomes. For practice, our findings underscore the need for peer support networks and mental health resources to mitigate entrepreneurial loneliness, enhancing emotional well-being, decision-making, and resilience in navigating entrepreneurial challenges.

CONTACT: Sergio A. Rifo; sergiorifo@udec.cl; Universidad de Concepción; Juan Antonio Coloma 0201, Los Ángeles, Bio-Bio 4440000, Chile

SUMMARY

FROM GREED TO EXPLORATION: HOW SITUATIONAL STRENGTH SHAPES ENTREPRENEURIAL BEHAVIOR

Friedrich Felix Tacke, Technical University Munich, Germany

Principal Topic

Greed is an insatiability focusing on material and immaterial desires (Seuntjens et al., 2015), and greedy entrepreneurs often pursue them through their ventures (Tacke et al., 2023). While exploration is key to entrepreneurial success (Junni et al., 2013; Ries, 2011) and, thus, essential for greedy entrepreneurs striving to fulfill greedy desires, greed is situational (Lambie & Haugen, 2019). Therefore, we ask: When do greedy entrepreneurs engage in explorative behavior? Drawing on situational strength theory (Mischel, 1977), we posit that industry experience (H1), cognitive trust (H2), and venture size (H3) moderate the relation between entrepreneurial greed and explorative behavior.

Method

We tested these hypotheses using online survey data from 233 entrepreneurs in 111 team-led ventures across various industries. Data was collected at two points, with a three-month gap. The first survey measured greed using the scale by Seuntjens et al.,(2015) and the moderators. The second survey measured explorative behavior using the scale by Mom et al.,(2007). Hierarchical linear modeling accounted for the nested data structure.

Results and Implications

Our analysis supports H1 and H2 but not H3. We contribute to the literature on explorative behavior in organizations (e.g., Fini et al., 2022). While extant research indicates that new ventures should engage in explorative behavior contingent on their size (Parida et al., 2016), we find that greedy entrepreneurs engage in explorative behavior irrespective of their ventures' size, suggesting they may engage in explorative behavior when not needed. Moreover, we provide insight into why greedy entrepreneurs avoid exploration. When they perceive their current setup as promising – due to high industry experience and cognitive trust – they may focus on executing the existing plan. Second, we contribute to the greed literature. We challenge prior assumptions that greed is purely short-term oriented (Seuntjens, 2016) by showing that greedy entrepreneurs engage in long-term oriented explorative behavior (Mom et al., 2007) to maximize their outcome. Research also indicates differences in greed-driven behavior between new ventures and established organizations (Tacke et al., 2023). We show that, unlike greedy managers exploiting existing organizations (Haynes et al., 2017), greedy entrepreneurs explore new opportunities. Third, we add to situational strength literature in new ventures (e.g., Schlaegel et al., 2021) by showing that individual experiences influence entrepreneurs' situational strength. Contrary to existing theories (Meyer et al., 2018), we find no evidence that venture size affects situational strength. Venture-level factors may be too distant to influence how greed translates into behavior.

CONTACT: Friedrich Felix Tacke; f.tacke@tum.de; Technical University Munich; Donnersbergerstraße 57, Munich, Bayern 80634, Germany

SUMMARY

THE ENTREPRENEURIAL HEART: A STUDY OF DIFFERENT ENTREPRENEURIAL PASSION AND EUDAIMONIC WELL-BEING DIMENSIONS

Yi Zhang, Oklahoma State University, United States
Mi (Kelly) Tran, Syracuse University, United States

Principal Topic

Eudaimonic well-being, focusing on self-realization, comprises six dimensions: autonomy, environmental mastery, personal growth, positive relations, purpose in life, and self-acceptance. Despite these distinct, noninterchangeable aspects of self-realization, existing studies often treat eudaimonic well-being as a holistic construct, overlooking the nuanced distinctions across its individual dimensions. Recent studies highlight the importance of examining these dimensions separately because certain aspects may resonate more profoundly with entrepreneurs and thereby influence their activities in varied ways. To address this theoretically significant and practically meaningful question, we draw on the three forms of entrepreneurial passion—inventing, founding, and developing—to explore how the heterogeneity of entrepreneurial passion shapes the heterogeneity of eudaimonic well-being, uncovering the psychological trajectories of entrepreneurs with diverse pursuits.

Method

Using a two-stage survey involving 437 entrepreneurs in the U.S. and Australia, we investigate the impacts of varied entrepreneurial passion on overall eudaimonic well-being and its six dimensions.

Results and Implications

The results support our hypotheses, showing that entrepreneurial passions for inventing, founding, and developing have positive effects on overall eudaimonic well-being and its six dimensions: autonomy, environmental mastery, personal growth, positive relations, purpose in life, and self-acceptance. Our findings make several contributions. First, this study advances the literature on eudaimonic well-being by responding to recent calls in the literature to conceptualize entrepreneurship as a dynamic eudaimonic process. By separating the six dimensions of well-being, our research sheds light on the diverse psychological outcomes of varied entrepreneurial passions regarding self-fulfillment, thereby complementing the traditional holistic approach to studying eudaimonic well-being in entrepreneurship. Second, our research offers a fresh perspective on the psychology of entrepreneurship. We extend existing work by probing the psychological progress of entrepreneurs, suggesting how their entrepreneurial passion translates into well-being. This deeper understanding of entrepreneurs' psychological trajectories helps to reveal the dynamic nature of well-being as they go through their entrepreneurial journeys. Moreover, we advance the entrepreneurial passion literature by moving beyond traditional outcomes such as intention, commitment, and stakeholder engagement. Our study suggests that entrepreneurial passion catalyzes an upward spiral of psychological growth. Entrepreneurs derive eudaimonic fulfillment through varied passions, which in turn reinforces their entrepreneurial activities. From a practical perspective, for established entrepreneurs, understanding how their specific passions contribute to different dimensions of psychological well-being can help them match their entrepreneurial activities with personal fulfillment. Additionally, by viewing how role models translate their entrepreneurial passions into self-realization and life fulfillment, aspiring entrepreneurs may gain the confidence to take the leap into entrepreneurship.

CONTACT: Yi Zhang; yizhangm@okstate.edu; Oklahoma State University; 4622 W Country Club Dr, Stillwater, Oklahoma 74074, United States

INTERACTIVE PAPER

VENTURE DURATION AND FINANCIAL PERFORMANCE: A COMPARATIVE STUDY OF SERIAL AND NOVICE ENTREPRENEURS.

Omar Alrubaiaan, Rensselaer Polytechnic Institute, United States
Murali Chari, Rensselaer Polytechnic Institute, United States

Principal Topic

We contribute to the entrepreneurship literature by examining the performance differences between serial and novice entrepreneurs both in terms of the duration of their ventures and financial performance. Specifically, we investigate (1) whether the duration of the first ventures of serial entrepreneurs is shorter or longer than novice entrepreneurs' ventures, (2) whether serial entrepreneurs experience progressively longer venture durations with each subsequent venture (3) whether serial entrepreneurs experience greater financial performance than novice entrepreneurs. Prior research has explored the impact of entrepreneurial experience on venture performance. Parker (2013) found that prior financial success positively influences subsequent ventures but diminishes over time. Lafontaine and Shaw (2016) showed that prior experience increases venture survival. Research suggests serial entrepreneurs actively seek better opportunities by closing prior ventures and launching higher-quality (Plehn-Dujowich, 2010), implying that their first ventures may be shorter-lived than those of novice. We hypothesize that serial entrepreneurs' first ventures are shorter in duration. Serial entrepreneurs are believed to learn from experience (Lafontaine & Shaw, 2016), improving their ability to identify better opportunities overtime. We therefore hypothesize that the duration of serial entrepreneurs' ventures increases with subsequent ventures. Regarding financial performance, serial entrepreneurs may possess superior skills or learning advantages, enabling them to achieve higher financial performance in their first ventures compared to novice entrepreneurs, and their subsequent ventures are likely to outperform their earlier ones.

Method

To test our hypotheses, we conduct a longitudinal study using data from PSID, the longest-running longitudinal household survey. The data has been widely employed in entrepreneurship research (Thomas & Chen, 2014; Greidanus & Liao, 2021; Parker, 2013). For our empirical strategy, we consider both Weibull and Cox proportional hazards duration models to test our survival hypothesis. The literature suggests a negative duration dependence for firm survival where older firms have lower exit rates, and the Weibull model can capture negative duration dependence in the data (Lafontaine & Shaw, 2016). Our data indeed shows negative duration dependence, and hence, the Weibull model is the appropriate model for our data. We use OLS and panel regressions to test our financial hypotheses.

Results and Implications

Consistent with our arguments, we find that the first ventures of serial entrepreneurs have greater financial performance and shorter survival than the ventures of novice entrepreneurs. We also find that the subsequent ventures of serial entrepreneurs survive longer than their previous ventures, however, the financial performance of their subsequent ventures are not better than their earlier ventures.

CONTACT: Omar Alrubaiaan; oalrubiaan@gmail.com; Rensselaer Polytechnic Institute; 27 Esplande St, Selkirk, New York 12158, United States

INTERACTIVE PAPER

ENTREPRENEURIAL TEAMS UNDER PRESSURE: A LONGITUDINAL STUDY ON PHYSIOLOGICAL PREDICTORS OF SATISFACTION AND EXIT INTENTIONS

Jennifer Ettner, Technical University Munich, Germany
Johannes Lützenrath, Technical University Munich, Germany
Isabell Welpe, Technical University Munich, Germany

Principal Topic

Entrepreneurial team conflict is a key predictor for early venture exits (Mol et al., 2015) and therefore ultimately early-stage venture failure (Preller et al., 2023). While entrepreneurial team conflict has been a focus of prior research (Brattström, 2024; Ensley et al., 2002), the role of physiological reactions as predictors of founders' exits remains unexplored. Drawing on insights from marriage research, which demonstrate that physiological arousal and linkage during conflict deteriorate relationships (Gottman & Levenson, 1992; Levenson & Gottman, 1985), we hypothesize that these physiological factors negatively impact team satisfaction and, over time, contribute to founders' exit from entrepreneurial ventures.

Method

To test the hypothesis, a longitudinal observational lab study was conducted with 136 dyadic founding teams. Participants first completed an online demographic questionnaire before attending an on-site lab session. The session included a baseline measurement of physiological state at rest, followed by a 15-minute conflict discussion. Throughout both phases, galvanic skin response (GSR) and interbeat intervals (IBIs) were continuously recorded using a Shimmer3 GSR+ device. Six months later, team satisfaction (Hendrick, 1988) and exit intentions were assessed via an online questionnaire. Physiological linkage was analyzed using OLS regressions, while Linear Mixed-Effects Models (LMEMs) were used for analyzing longitudinal venture outcomes.

Results and Implications

Team satisfaction remained high ($M = 5.24$), while exit intentions were relatively low ($M = 2.41$), with 37% never considered leaving and 17% having already exited. Physiological measures indicated increased stress during conflict, as IBIs decreased by an average of 54.07 ms. Founders' physiological responses were synchronized, with IBI and GSR linkage explaining 20% and 43% of variance in co-founders' physiological responses, respectively. LMEMs revealed that physiological arousal negatively affected team satisfaction ($\beta = 0.247$, $p = 0.044$), while physiological linkage showed no significant effect. Physiological arousal, measured via IBI, also led to decreased exit intentions ($\beta = -0.436$, $p = 0.013$). Contrary to expectations, IBI linkage tended to reduce exit intentions ($\beta = -0.163$, $p = 0.057$). These findings contribute to entrepreneurial team literature (Brattström, 2024; Patzelt et al., 2021) by identifying stress responses as early indicators of team stability. While aligning with marriage studies, the results reveal a contrasting effect of physiological linkage. Additionally, this study offers practical implications for entrepreneurs, incubators, and venture capitalists. By recognizing heightened physiological arousal during conflict as an early predictor of venture exit, we enable proactive interventions, ultimately fostering team longevity and venture success (Bolzani et al., 2019; Patzelt et al., 2021).

CONTACT: Jennifer Ettner; jennifer.ettner@tum.de; Technical University Munich; Oettingenstr. 33A, Munich, Bayern 80538, Germany

INTERACTIVE PAPER

SELF-COMPASSION IN ENTREPRENEURS: HOW TO ASK FOR HELPFUL HELP

Lucia Johanna Kuhn, Technical University of Dortmund, Germany
Tessa Flatten, Technical University Dortmund, Germany
Corinna Vera Hedwig Schmidt, Technical University Dortmund, Germany

Principal Topic

While entrepreneurs receiving help experience significant benefits, like improved venture growth and reduced failure rates, typical entrepreneurial characteristics like self-reliance counter help-seeking behavior. Especially, autonomous help-seeking, referring to seeking help to solve problems independently, consistently links to improved performance. However, actionable insights are missing, which mechanisms outbalance anti-help-seeking behaviors. Drawing on the extended threat-to-self-esteem model (TTSEM), we propose that self-compassion – defined as treating oneself kindly during failure – increases autonomous help-seeking in entrepreneurs. Following recent mediation typology, we investigate if two mediators – help-seeking cost and self-improvement motivation – complementary mediate self-compassion and autonomous help-seeking intent. We choose intent as the closest predictor to behavior.

Method

With 10 entrepreneurs we developed several problem scenarios, which were evaluated by 30 entrepreneurship scholars based on credibility, complexity and severity. The overall highest scoring scenario was selected: an employee attrition problem. In this pre-registered RCT, we recruited an entrepreneurial sample (n = 316) by using strict pre-screening criteria and additional controls via prolific.com. We asked entrepreneurs to visualize their own venture, imagine the employee attrition problem and describe their emotions (min. 200 characters). We randomized entrepreneurs into manipulations to either increase self-compassion or positive affect as control group with three writing prompts each. We used a well-established autonomous help-seeking scale and integrated each item into a standardized E-mail template. We introduced Taylor Jones, a topic expert they recently met. Entrepreneurs rated the likelihood of sending each E-mail to Taylor. Validated scales for manipulation checks, mediators and controls ensued.

Results and Implications

Bootstrapped mediations with 5,000 repetitions reveal no total direct effect of self-compassion on autonomous help-seeking intent. We find support for both mediations showing indirect-only effects from self-compassion to autonomous help-seeking via, one, help-seeking cost and, two, self-improvement motivation. To explore if the mediators cover overlapping aspects, we conduct supplemental, serial mediation based on the extended TTSEM pathway from self-compassion to help-seeking cost to self-improvement motivation to autonomous help-seeking. Results reveal a significant, serial indirect-only mediation via both mediators. We contribute to the entrepreneurship and help-seeking literature and expand the TTSEM: While Nadler and Fisher (1986) posit that improvement motivation can only stem from self-threatening messages, our serial mediation suggests that in the case of self-compassion, improvement motivation can also stem from self-supportive messages. Practically, our findings highlight self-compassion as a valuable resource that can be easily induced within ventures and increases entrepreneurs' intent to seek performance-enhancing help.

CONTACT: Lucia Johanna. Kuhn; lucia.kuhn@tu-dortmund.de; Technical University of Dortmund; Arndtstrasse 34, Berlin, Berlin 10965, Germany

INTERACTIVE PAPER

THE DOUBLE-EDGED SWORD EFFECT OF ENTREPRENEUR WORKAHOLISM ON INTRAGROUP CONFLICT. THE MODERATING ROLE OF A TOLERANCE FOR FAILURE CULTURE

Lennard Lang, RWTH Aachen University, Germany

Principal Topic

Workaholism, characterized by excessive and obsessive work investment, represents a commonly observed disposition among entrepreneurs and leaders (She et al., 2024; Simmons et al., 2023). Despite scholars highlighting the crucial role entrepreneurs play as leaders in shaping organization culture and employee dynamics (Schein, 1983), we know little about the impact of workaholism among entrepreneurs on their organizations (Clark et al., 2016). Existing studies associated workaholism with perfectionism, non-delegation and aggressive workplace behavior, suggesting multiple sources of conflict when working for and with workaholics (Balducci et al., 2012). Research agrees that whether intragroup conflict is beneficial depends on the type of conflict. While task conflict (disagreements regarding the accomplishment of tasks) positively relates to higher performance, relationship conflict (interpersonal tension) is commonly viewed as detrimental to the organization (Chun & Choi, 2014). Drawing on social exchange theory, we propose that entrepreneurs' workaholism positively affects task and relationship conflict and suggest that tolerance for failure positively moderates these relationships.

Method

Our study is based on cross-sectional survey data of 215 entrepreneurs of small ventures from Germany, Switzerland and Austria. We measured workaholism with the Dutch Workaholism Scale of Schaufeli et al.,(2009). Task and relationship conflict were assessed with the three-item scales of Jehn and Mannix (2001) and tolerance for failure following Danneels (2008). We tested our hypothesis using Structural Equation Modeling in AMOS. To ensure robustness of results, we included established controls from previous research.

Results and Implications

Our findings show that entrepreneurs' workaholism has a double-edged sword effect by positively influencing task and relationship conflict. Workaholic entrepreneurs stimulate the productive discourse on efficient task accomplishment at the cost of creating interpersonal tensions. Furthermore, a tolerance for failure culture strengthens the positive impact of entrepreneurs' workaholism on task conflict, implying that embracing mistakes as development opportunities reinforces leader-induced exchanges in task resolution. Our hypothesis that tolerance for failure moderates the impact on relationship conflict was not supported. We contribute to the literature in three ways. First, we underscore the two-sided effect of workaholic leaders in small organizations. Second, we expand the existing perspective on leader workaholism, emphasizing that in their central leadership role, workaholic entrepreneurs influence the entire organization beyond direct subordinates. Third, we suggest that cultivating a tolerance for failure culture enhances the positive effect of leader workaholism in entrepreneurial settings. Practically, our findings reveal the dual impact of entrepreneurial workaholism on employee dynamics and highlight the value of a tolerance for failure culture.

CONTACT: Lennard Lang; lennard.lang@time.rwth-aachen.de; RWTH Aachen University; Buttermelcherstr 8, Munich, Bayern 80469, Germany

INTERACTIVE PAPER

CROSS-CULTURAL EXPERIENCE AND ENTREPRENEURIAL HUSTLE: A GLOBAL NOMAD-BEHAVIORAL LENS

Robert J. Pidduck, Old Dominion University, United States
Kyoungmin Lee, University of Oklahoma, United States
Margaret Shaffer, University of Oklahoma, United States
Yejun Zhang, Texas State University, United States

Principal Topic

Entrepreneurship research has traditionally assumed that cognition precedes behavior, meaning individuals become entrepreneurs through deliberate intention and planned opportunity pursuit. However, this view overlooks how many successful entrepreneurs emerge without explicit intent, suggesting that entrepreneurial careers can be unintentional and experimental. This paper challenges the cognition-first paradigm by applying Cognitive Behavioral Theory to study global digital nomads—young individuals who blend work and travel for extended periods. Through cross-cultural tensions and detachment from their home environments, these individuals develop entrepreneurial behaviors, even without identifying as entrepreneurs. Using a time-lagged design, the study demonstrates that these behaviors, particularly entrepreneurial hustle, enhance performance in tasks requiring entrepreneurial cognition. The findings suggest that fostering entrepreneurial behaviors in non-explicit contexts, such as international travel, can cultivate key entrepreneurial aptitudes, providing preliminary evidence that "doing" entrepreneurship before "thinking" like an entrepreneur may be a viable path to venture creation and success.

Method

In study 1, we conduct a structural equations model analysis on a primary sample of global digital nomads (n = 520). We assess all known dimensions of international/cross-cultural experience including the breadth, depth, cultural distance, and meaningfulness of experience traveling or living in foreign countries (cf. Maddux et al., 2020). Entrepreneurial behaviors are measured using Burnell et al's (2024) hustle construct—urgency and unorthodoxy behaviors. In study 2, with a 3-month interval we collect the same data from this sample to first replicate our hypotheses (see section below) but link these behaviors to scores on the new venture idea assessment (cf. Davidsson et al., 2021). Control variables include factors known to influence international travel self-selection (e.g., openness to experience, risk-taking personality, among others).

Results and Implications

Our preliminary analyses support our theoretical model, showing that the breadth of international experience among global digital nomads significantly enhances entrepreneurial hustle dimensions—urgency and unorthodoxy—while depth of experience has weaker effects. Study 2 broadly replicates these findings, with the expected link between hustle behaviors and new venture idea assessment pending analysis. We also examine cultural distance as a moderating factor. Collectively, our studies challenge the assumption that cognition precedes behavior in entrepreneurship, highlighting how experiential learning fosters entrepreneurial skillsets before individuals explicitly choose to pursue ventures, advancing research on cross-cultural experiences as drivers of entrepreneurial ideation and behavior.

CONTACT: Robert J. Pidduck; rob.pidduck@gmail.com; Old Dominion University; 9525 6th Bay St, Norfolk, Virginia 23518, United States

INTERACTIVE PAPER

NAVIGATING UNCERTAINTY: HOW EXPRESSED FOUNDER HUMILITY INFLUENCES DECISION-MAKING IN ENTREPRENEURIAL VENTURES

Julia Teresa Prefi, RWTH Aachen University, Germany

Principal Topic

The ideal leader archetype has evolved from one characterized by narcissism and arrogance to a more collaborative and humble approach (Zhang et al.,2017). Humble leaders recognize their limitations, acknowledge mistakes, appreciate others' strengths, and exhibit techability (Owens & Hekman 2012). While prior research links humble leadership to enhanced start-up performance (Ren et al.,2020), the underlying mechanisms driving this effect remain unclear, particularly in entrepreneurial contexts. Specifically, little is known about how leader humility shapes entrepreneurial decision-making, particularly in terms of causal versus effectual approaches. Grounded in effectuation theory (Sarasvathy 2001), we propose that founder humility positively influences both decision-making styles. Furthermore, we investigate the moderating role of charismatic leadership to offer a more nuanced understanding of these relationships.

Method

We conducted an online survey in summer 2024, yielding 354 responses after data cleaning. To test our hypotheses, we applied hierarchical regression analysis using well-established, validated measurement scales. To ensure clarity, the questionnaire was pre-tested with experienced scholars and entrepreneurs. Leader humility was measured using the scale by Owens, Johnson and Mitchell (2013), chosen for its strong validity and reliability. To mitigate bias, founders' humility was assessed by their co-founders. Effectual and causal decision-making were measured using Chandler et al.'s (2011) scale, while charismatic leadership, our moderator, was assessed with Waldman et al.'s (2001) scale. To enhance robustness, we implemented processual remedies, control variables, and ex-post analyses to address potential biases, including common method bias, social desirability, and single informant bias.

Results and Implications

Our findings indicate that founder humility has a positive association with both causational and effectual decision-making. Furthermore, we observe that charismatic leadership moderates these relationships in distinct ways. Specifically, charismatic leadership weakens the positive relationship between founder humility and effectual decision-making, while strengthening the relationship between founder humility and causal decision-making. Theoretically, this study advances research on leader humility in entrepreneurship, an area that remains underexplored. It also contributes to the understanding of entrepreneurial decision-making by demonstrating how humility shapes strategic choices in uncertain environments, offering new insights into the mechanisms linking leadership traits to decision-making processes.Practically, our findings suggest that investors may view humility as a valuable founder trait, as it helps balance structured planning with adaptability, key for navigating uncertainty. Additionally, incubators and accelerators can incorporate humility training into leadership programs, equipping entrepreneurs with the soft skills needed for effective decision-making and long-term success.

CONTACT: Julia Teresa Prefi; prefi@time.rwth-aachen.de; RWTH Aachen University; Severinstrasse 126, Aachen, Nordrhein-Westfalen 52080, Germany

DO WE ALWAYS NEED A STORY? THE ROLE OF CATEGORY STRUCTURES IN ENTREPRENEURIAL COMMUNICATION AND SOCIAL EVALUATIONS

Donato Cutolo, IE University, Spain
Antonio Paco Giuliani, University of Bologna, Bologna, Italy
Simone Ferriani, University of Bologna/ City University, Italy/UK

ABSTRACT

This study explores the contingent role of storytelling in entrepreneurial communication, emphasizing how its efficacy varies across market contexts. Storytelling, a foundational mechanism for human cognition and meaning making, enables entrepreneurs to frame opportunities, engage stakeholders emotionally, and signal intent. However, its universal effectiveness is questioned, particularly in highly analytical or empirically focused environments. The paper introduces two critical dimensions—category fuzziness and symbolic versus material orientation—to assess the impact of storytelling in entrepreneurial settings. Through a dataset of 74,654 Etsy seller profiles, the research reveals that storytelling is more effective in fuzzy and symbolic categories, such as wedding products, where narratives enhance emotional resonance and legitimacy. Conversely, clear and material categories, like office supplies, favor expository communication for precision and trust. This work challenges binary legitimacy assessments, offering a nuanced framework that aligns communication strategies with market structures, thereby optimizing stakeholder engagement and entrepreneurial success.

INTRODUCTION

Storytelling is intrinsically tied to the way individuals process information, construct meaning, and navigate social realities. It transcends mere communication to serve as a cornerstone of human cognition and cultural evolution (Bruner, 1987). Narratives shape our understanding of the world, providing frameworks through which actions are interpreted, and decisions are made. In the realm of entrepreneurship, storytelling functions as a strategic tool that goes beyond rhetoric to influence perception, mobilize resources, and align stakeholder interests (Ibarra & Barbulescu, 2010). Indeed, entrepreneurs often utilize stories to frame their ventures in compelling and memorable ways. These narratives are not merely tools of persuasion but mechanisms for signaling intent, competence, and the unique value proposition of a venture (Lounsbury & Glynn, 2001; Garud & Giuliani, 2013). Through evocative storytelling, entrepreneurs can engender trust, enhance credibility, and foster emotional connections that galvanize support from investors, customers, and other stakeholders. As McLean et al., (2007) and Cornelissen & Werner (2014) observe, the affective dimension of storytelling often proves instrumental in mobilizing critical resources for entrepreneurial success. However, storytelling is not universally effective. While narratives can evoke powerful emotional responses, they may inadvertently obscure factual details or set unrealistic expectations, potentially alienating analytically oriented stakeholders (Krause & Rucker, 2020). Storytelling, when perceived as exaggerated or disingenuous, can erode trust, undermining its intended purpose (Chapple et al., 2022; Murray & Fisher, 2022). This introduces a strategic tension: when should entrepreneurs harness storytelling's emotive allure, and when should they prioritize clarity and precision through expository communication?

Further complicating this dynamic are the social evaluations that entrepreneurs must navigate. Stakeholders do not assess ventures in isolation but contextualize them within prevailing norms, expectations, and societal standards. Social evaluations often operate within dichotomous frameworks, labeling ventures as either good or bad, legitimate or illegitimate (Navis & Glynn, 2011; Pollock et al., 2019). This binary tendency constrains nuanced understanding, forcing entrepreneurs to balance emotional resonance with evidentiary rigor to avoid skepticism or distrust. These challenges underscore the need for a nuanced understanding of communication strategies that blend emotional appeal with empirical support, tailored to different audiences and contexts.

This paper advances the argument that storytelling's efficacy in entrepreneurial communication is contingent upon the structural characteristics of the market categories. Two dimensions—category fuzziness and symbolic versus material orientation—are posited as critical mediators. By linking these dimensions to social evaluations, this study proposes a framework that transcends binary assessments, fostering a multidimensional approach to evaluating entrepreneurial ventures.

HYPOTHESES DEVELOPMENT

Market categories exhibit substantial variation in their degree of fuzziness, defined as the ambiguity or absence of clear boundaries (Negro et al., 2011). Highly fuzzy categories, such as emerging industries like plant-based proteins or decentralized finance, lack consensus regarding standards or benchmarks (Durand & Paolella, 2011). Entrepreneurs in such categories often employ storytelling to infuse meaning into their offerings, guiding audience interpretation amid uncertainty (Garud et al., 2013). For example, during the organic food movement's early stages, companies like Whole Foods leveraged narratives emphasizing health, sustainability, and environmental consciousness, effectively differentiating their products and cultivating a loyal customer base. These narratives created a sense of purpose and alignment with consumer values, positioning Whole Foods as a pioneer in its category. Conversely, low-fuzziness categories—such as consumer electronics—are characterized by established norms and clear comparability. In these contexts, expository communication emphasizing detailed factual information often proves more effective. Apple, for instance, highlights technical specifications and functional innovation in its product launches to appeal to an audience well-acquainted with the category's standards. By prioritizing data and precision, Apple ensures that its messaging resonates with a technically savvy audience, reinforcing its reputation for reliability and excellence.

The symbolic versus material orientation of a market category further shapes the effectiveness of communication strategies. Symbolic categories—such as luxury goods, haute couture, or artisanal foods—derive value from cultural resonance, identity alignment, and emotional engagement. Gucci, for example, constructs narratives around heritage, exclusivity, and artisanal craftsmanship, fostering a deeply emotional connection with its audience. These narratives are not merely promotional tools but essential components of brand identity, reinforcing the symbolic value of Gucci products. Material categories, such as industrial commodities or office supplies, prioritize functional attributes. Even a company as innovative as Tesla frequently emphasizes technical metrics like battery efficiency and sustainability to underscore credibility and deliver value through factual precision. By doing so, Tesla balances its visionary aspirations with tangible, measurable achievements, ensuring alignment with stakeholder expectations. Social evaluations intersect with these dimensions, influencing how audiences perceive legitimacy and coherence. For symbolic categories, audiences scrutinize the alignment between narratives and the cultural codes underpinning the category (Glynn & Navis, 2013). In material categories, stakeholders expect factual correspondence between claims and measurable outcomes, leaving little room for embellishment. By identifying these intersections, this study provides a pathway to undo binary categorizations in social evaluations, encouraging more nuanced assessments. It highlights the potential for entrepreneurs to navigate these complexities by tailoring their communication strategies to the unique demands of their market categories.

METHOD

Our dataset consists of 78,684 "About Pages" from entrepreneurs on Etsy, the largest digital marketplace for handmade and craft items.

To systematically quantify the degree to which the content of a given About Us page reflects a narrative or a commercial, fact-based communication, we applied LLM-based classification using OpenAI's GPT-4o via its API. Although still nascent as a methodological innovation, preliminary evidence indicates that large language models enable the categorization of textual data with markedly greater reliability than alternative text analysis approaches (e.g., Le Mens, Kovács, Hannan, & Pros, 2023; Rathje et al., 2024). In line with established guidelines for optimizing the effectiveness of LLM prompts (e.g., Rathje et al., 2024), we explicitly assigned a role to the LLM, furnished contextual information pertaining to the specific task, delineated our definitions of both narrative and factual communication, and requested that the LLM provide an explanatory rationale prior to returning the score. To ensure the reliability of the measure, one of the authors manually assessed a random sample of 100 "About Us" pages and cross-compared these evaluations with the ratings produced by the LLM, with findings indicating that the LLM-based approach yielded highly consistent classification results. The dataset was then segmented using two primary dimensions: category fuzziness and symbolic versus material orientation. Fuzziness was quantified by assessing audience consensus and the definitional clarity of category boundaries by identifying the number of atypical sellers in each category. Symbolic versus material orientation instead was determined through linguistic analysis of subcategories. Our dependent variable is the number of products sold by each entrepreneur.

RESULTS

Our preliminary findings reveal that storytelling's efficacy is contingent on market category characteristics. In fuzzy and symbolic categories—such as "Books, Movies & Music" or "Wedding-related" products—narratives proved pivotal in contextualizing offerings and imbuing them with meaning. Etsy sellers in the "Wedding" category who emphasized love, craftsmanship, and personal aspects resonated deeply with audiences, fostering trust and engagement. These sellers successfully positioned their products as not merely items but symbols of meaningful life events, enhancing their emotional appeal. In contrast, clear and material categories, such as "Office Supplies," exhibited a preference for expository communication. Sellers who emphasized product durability, cost-efficiency, and functionality consistently outperformed those who relied on narrative-driven approaches. Social evaluations amplified these patterns, as stakeholders in material categories penalized deviations from empirical benchmarks, underscoring the importance of factual precision. These results demonstrate the critical role of aligning communication strategies with audience expectations to optimize legitimacy and success.

DISCUSSION AND IMPLICATIONS

This study advances the understanding of entrepreneurial communication by elucidating the nuanced relationship between narrative strategies and market category structures. Our findings highlight the contingent nature of communication strategies, emphasizing the importance of context in shaping their effectiveness. By linking narrative effectiveness to category structure, we contribute to a more nuanced understanding of how entrepreneurial communication strategies are shaped by both structural and cultural contexts (Aldrich & Fiol 1994; Navis & Glynn 2011; Garud et al., 2014). Our empirical results highlight that narratives and market categories function as cultural mechanisms that either reinforce or counteract each other (Giorgi et al., 2015), thereby challenging the prevailing assumption that storytelling is universally beneficial in entrepreneurial communication. Entrepreneurs can adopt tailored communication strategies that encourage stakeholders to evaluate ventures along a continuum of legitimacy rather than through binary classifications. This approach fosters a richer understanding of entrepreneurial efforts, emphasizing the interplay of context, strategy, and audience perception. Moreover, it underscores the potential for dynamic and adaptive communication practices that evolve in response to shifting market and stakeholder demands.

We might also highlight some practical implications. Entrepreneurs in fuzzy categories should invest in emotionally resonant narratives to clarify and elevate their offerings. Those in symbolic categories can leverage storytelling to reinforce brand identity and cultural resonance. In contrast, entrepreneurs in material categories should prioritize data-driven approaches that emphasize functionality and reliability to meet audience expectations. By strategically aligning their communication methods with market characteristics, entrepreneurs can navigate complex evaluative landscapes more effectively.

CONTACT: Donato Cutolo; donato.cutolo@ie.edu; IE University; C/ María de Molina, 13, 28006 Madrid, Spain

WHAT ENTREPRENEURS CAN LEARN FROM STAND-UP COMEDIANS: THE IMPACT OF INVESTOR LAUGHTER DURING ENTREPRENEURIAL STORYTELLING ON FUNDING DECISIONS

Siwei Zhu, Minnesota State University Moorhead, Minnesota, United States
Baris Morkan, Guven Partners LLC, New Jersey, United States

ABSTRACT

This study investigates how investor laughter influences entrepreneurial storytelling for funding decisions, integrating qualitative insights from stand-up comedians with quantitative data from Shark Tank. We find that the frequency of investor laughter during entrepreneurial storytelling positively impacts funding decisions. The duration of investor laughter moderates the relationship between laughter frequency and funding decisions, while the reflection of investor laughter does not show a moderating effect. Our findings reveal that investor laughter as an emotional and cognitive mechanism provides entrepreneurs with real-time feedback to refine their pitches and enhance their chances of securing funding.

INTRODUCTION

Entrepreneurial storytelling is a powerful tool for conveying innovative ideas and incorporating salient cues to facilitate knowledge transfer from pitching entrepreneurs to evaluating investors (Chen et al., 2009; Garud et al., 2014). Researchers have drawn on insights from various disciplines to examine how specific storytelling techniques help entrepreneurs overcome investor resistance and enhance their likelihood of securing funding (Parhankangas & Ehrlich, 2014; Zellweger & Zenger, 2021). A core technique, emotional display, involves using storytelling tactics to conceptualize knowledge as part of idea pitching (Clarke et al., 2019). Entrepreneurs thus engage investors more effectively and accelerate their sensemaking processes.

While the logic behind the emotional display is sound, existing research predominantly focused on the emotions of pitching entrepreneurs (e.g., Jiang et al., 2019; Oo et al., 2023), neglecting the impact of investor reactions during storytelling. In practice, investors assess entrepreneurial storytelling from the opposite side of the table with their primary role being evaluation. However, investor reactions are not isolated from the storytelling; they carry complementary meanings that can inform the pitches crafted by entrepreneurs (Falchetti et al., 2022). Yet, little is known about how pitching entrepreneurs leverage emotions from evaluating investors to improve their chances of securing funding.

Stand-up comedians serve as valuable role models in this dynamic as they extensively interact with their audience's laughter to refine their performance (Carter, 2001). Both stand-up comedians and entrepreneurs are highly attuned to audience reactions, making this parallel particularly compelling. This connection led us to explore how principles from stand-up comedy can be applied to entrepreneurial storytelling. It's crucial for entrepreneurs to transform investor laughter into actionable knowledge for receiving better funding (Srivastava et al., 2023). This study dives into two key questions: What is the role of investor laughter in entrepreneurial storytelling and how do laughter patterns influence their funding decisions?

Given the absence of previous research involving investors' reactions in entrepreneurial storytelling, we adopt a mixed-methods approach. This study began with a video ethnography involving 10 stand-up comedians. We recorded and observed their training sessions and performances to uncover distinct laughter patterns - frequency, duration, and reflection - in how stand-up comedians interact with audience laughter. We visited the literature on organizational aesthetics (Baldessarelli et al., 2022) to support our findings within or extend the existing body of knowledge. To validate each laughter pattern and its potential impact on entrepreneurial storytelling, we conducted interviews with four entrepreneurs. These insights led to the development of a conceptual model with testable hypotheses. We then analyzed 264 pitches from the reality show, Shark Tank, to test our model. Our analysis revealed strong empirical evidence that investor laughter during storytelling significantly influences their funding decisions.

This study makes a contribution to entrepreneurial storytelling by broadening the understanding of how evaluating investors engage in entrepreneurial storytelling. This study demonstrates that entrepreneurial storytelling is not merely a one-way form of communication (Clarke et al., 2019); investor reactions, particularly laughter, serve as an emotional and cognitive mechanism in shaping entrepreneurs' pitches. This study also contributes to the theory of organizational aesthetics by bridging stand-up comedy principles with entrepreneurship. Responding to the call by Baldessarelli et al., (2022) for a deeper understanding of how aesthetics is evoked in organizational life, this study identifies three distinct types of laughter and demonstrates their relevance between stand-up comedy and entrepreneurial storytelling.

ENTREPRENEURIAL STORYTELLING

Existing research highlights two main tactics of storytelling used by pitching entrepreneurs. One approach demonstrates that entrepreneurs pitch like scientists, adopting a positivist view of linguistics (Cho et al., 2014; Oo et al., 2023). In contrast, another approach shows that entrepreneurs pitch like artists, drawing from the agile view of cognitive psychology (Cohn et al., 2012). Storytelling, therefore, requires a blend of science and art from entrepreneurs.

Research on creativity assessment offers multiple perspectives of how investors formulate their judgments on entrepreneurs and their stories. Berg (2016) finds that investors with an "innovator" orientation value stories that exhibit divergent thinking and reflect on the pitch's creative aspects. Conversely, Mueller et al., (2018) argue that investors taking on a "manager" role may prioritize economic perspectives, focusing on the pitch's practical and financial implications. Huang and Pearce (2015) suggest that investor assessments are dynamic, involving shifts between emotional and cognitive judgments described as "gut feel".

The roles of entrepreneurs and investors in storytelling are established, yet entrepreneurial storytelling is more than just a series of isolated pitching and evaluating events. There is a sensemaking process between the two parties. Recent studies argue that a core technique, emotional display by entrepreneurs, matters for funding decisions. Chen et al., (2009) underscore how entrepreneurs' passion and preparedness positively influence venture capitalists in business plan presentations. Elsbach and Bechky (2018) explore the impact of employee crying in workplaces. Hmieleski and Cole (2022) explore the role of humor as a key indicator for CEOs engaging stakeholders in new venture success. However, existing research primarily provides a one-sided assessment focused on pitching entrepreneurs, overlooking the role of investor reactions during entrepreneurial storytelling. Investors' felt emotions may prompt adaptive storytelling tactics and influence funding outcomes (Srivastava et al., 2023). Understanding investor reactions is, therefore, crucial for entrepreneurs.

QUALITATIVE STUDY: METHOD, RESULTS, AND HYPOTHESES

Ethnographic Study on Stand-up Comedian

Two research assistants observed 10 stand-up comedians in a three-week training camp hosted by a comedy club in 2019 at Beijing, China. They captured over 100 hours of video footage on nine training sessions and three live performances. The data collection began by observing the first performance, an open-mic night, where novice comedians faced the challenge of jokes failing to land with the audience. Over the next three days, they received instruction on joke structure and storytelling techniques. In the second open mic for the new audience, the comedians displayed notable improvement. They performed the same jokes but with better-structured storytelling. The following training sessions focused on helping comedians categorize jokes by audience reaction and refine their material. By the third performance, comedians effectively used audience laughter as pauses and interactions, demonstrating substantial progress even with the same jokes for new audiences. A final three-day training session followed, further enhancing their storytelling skills.

Literature Review of Organizational Aesthetics

The ethnographic study provides valuable insights into how audience laughter influences stand-up comedians to refine their performances. Research in organizational aesthetics has embraced this process, treating aesthetics as triggers for emotional and cognitive mechanisms that facilitate individual understanding and behavior (Baldessarelli et al., 2022).

One perspective frames aesthetics as a knowledge tool, where reactions arise from immersion in evocative environments and can range from positive to negative, such as like/dislike (e.g., Creusen & Schoormans, 2005) or attraction/repulsion (e.g., Mack, 2007). These reactions are often immediate and spontaneous, differing from more deliberate forms of response (Dane & Pratt, 2007). Laughter serves as a rapid, momentary judgment that stands out as one of the most salient experiences for audiences. Miron-Spektor et al., (2023) illustrate a related dynamic in their analysis of humor in TED talks. Female speakers can nurture audience reactions to challenge competence-related stereotypes and deliver impactful talks.

A second perspective views aesthetics as a direct stimulus, where reactions have enduring effects and encompass the full spectrum of positive and negative effects, which in turn can provoke behaviors (Massa et al., 2017) and commitments (Siebert et al., 2017). Reactions can shape observers' attitudes, thus influencing their judgments in evaluation processes (Eisenman, 2013), their perceptions of product novelty (Goode et al., 2013), and their assessments of market value (Sgourev, 2021). Jiang et al., (2019) examine the impact of emotional temporal dynamics on funding decisions that peak emotional moments in interpersonal contexts play a particularly salient role in affecting investment decisions.

A third perspective treats aesthetics as an open-ended outcome. Research in this stream highlights how individuals engage intellectually with aesthetics, often experiencing multiple emotions across different stages (Taylor & Hansen, 2005). This perspective emphasizes how aesthetic encounters reflect cognitive interpretations (Endrissat & Noppeney, 2013). Brown et al., (2020) argue that shifting distinctions between meaningful sounds are fundamental in an organizational acoustic environment: a sound can be interpreted as dangerous in one context but as a source of happiness in another.

In summary, research in organizational aesthetics has shed light on the sensory nature of laughter (Marsh & Śliwa, 2022) and implicitly demonstrates that both entrepreneurs and stand-up comedians are attentive to aesthetic experiences, engaging dynamically with their audiences.

Interviews with Entrepreneurs

In our effort to identify similar uses of laughter in stand-up comedy and entrepreneurial storytelling, we conducted interviews with two stand-up comedy club owners who navigate dual identities as both comedians and entrepreneurs (referred to as I#1 and I#2 throughout the paper), as well as two entrepreneurs who are not comedians—one from the pet industry (referred to as I#3) and another from the tech industry (referred to as I#4). One interviewee (I#1) drew the parallels between entrepreneurs and stand-up comedians, emphasizing their shared attributes:

> *"They are creators and performers – in charge of their own destiny. No one can rescue them and always find their own way to succeed. Isn't entrepreneurship like stand-up comedy? Indeed, it is."*

Respondents were selected based on pre-determined inclusion and exclusion criteria to enable the selection of interviewees with the necessary background to answer our questions. Using semi-structured questions, such as "Can you describe the connection between stand-up comedians and entrepreneurs?" and "Do investor reactions affect your perceptions? Please elaborate." Interviews were professionally translated into English for analysis.

Coding and Analysis

Our coding and analysis followed an inductive approach (Corbin & Strauss, 2015). We began by extracting quotes from video recordings related to the descriptions of laughter. Open coding revealed that comedians often describe laughter from the audience as a mechanism that informs their understanding of jokes. Using axial coding, we categorized these quotes into first-order codes, clustering statements with similar keywords and themes. After iterative refinement, we grouped the data into three overarching codes representing distinct laughter aspects: frequency, duration, and reflection of laughter. Finally, we conducted selective coding and integrated findings into a roadmap. We then used a process similar to the constant comparative method that involved taking iterative steps among the roadmap, existing literature, and interview data (Glaser & Strauss, 2017) to validate our learnings from video recordings. Through several iterations, we defined the three laughter aspects: frequency of laughter represents the total count of laughter events; duration of laughter refers to the length of time that laughter persists; reflection of laughter captures the time taken for the audience to process and respond with laughter, reflecting the cognitive digestion of humor. We linked the aggregate codes of laughter to three aesthetic perspectives: the frequency of laughter as a "Knowledge Tool", the duration of laughter as a "Directed Stimulus", and the reflection of laughter as an "Open-ended Outcome".

Findings and Hypotheses Development

Investor laughter serves as a sensory cue that entrepreneurs can interpret as a measure of relatability (Bloch, 1995). A high frequency of laughter indicates aesthetic knowledge. In this sense, In this sense, laughter frequency acts not only as a sensory cue but also as a knowledge tool for real-time adjustments to refine tone and content to secure favorable evaluations (Brush et al., 2001). Similarly, entrepreneurs can leverage investor laughter as an implicit cue to tailor their storytelling strategies. This adaptability is particularly critical in high-stakes environments such as investment pitches, where sustained attention and resonance can influence evaluative judgments. I#2 stated:

> *"Funny is not the new sexy; it has always been this way! Sometimes, people may laugh at your joke once just to be polite, but if they smile continuously, it means you've already achieved a milestone."*

We posit that entrepreneurs employ investor laughter frequency to enhance funding decisions:

> *Hypothesis 1: During entrepreneurial storytelling, frequency of investor laughter is positively associated with their funding decisions.*

Assessing an event's impact requires considering both its intensity and duration (Morgeson et al., 2015). The duration of laughter amplifies the persuasive power of storytelling (Lu et al., 2019).
Comedians can manage laughter and deepen its impact by establishing a shared identity with audiences. Similarly, in discussing the role of laughter in investor engagement, entrepreneurs can also gauge investors' interest levels (Cornelissen et al., 2014). I#4 articulated the role of laughter in strengthening the entrepreneur-investor relationship as follows:

> *"If your target audience laughs longer, I believe every extra nanosecond strengthens the invisible bond between you and them. And with stronger ties, your chances of securing the funding you are after only get better."*

We thus propose that prolonged laughter from investors serves as a directed stimulus that reinforces the emotional connection between entrepreneurs and investors, thereby strengthening the influence of laughter frequency on funding decisions:

> *Hypothesis 2: The relationship between frequency of investor laughter and their funding decision is positively moderated by duration of investor laughter.*

Laughter can also trigger investors' reflective thinking. It's interesting to note that stand-up comedians often evoke personal experiences, many of which are negative, but delivered in a humorous way (Carter, 2001). These unique interventions allow the audience to reinterpret ideas for open-ended outcomes, creating reflective laughter. During storytelling, entrepreneurs also try to convince investors that they are unique enough to distinguish them from other competitors (Mueller et al., 2018). I#3 draws an analogy between humor and persuasion, suggesting that the delayed response may indicate that the idea has resonated on a deeper level, making it more likely to be digested with the audience:

> *"Nowadays' fast-food culture wants quick, easy laughs, but the best comedy—and of course the best pitches—take time to settle. When an investor keeps thinking about your joke after the moment has passed, that's like the slow-cooked meal of persuasion. It means your idea isn't just funny—it's memorable.*

The reflection of laughter serves as a valuable digestion for investors' reflective thinking. We argue that entrepreneurs should maintain a heightened awareness of how laughter influences investor reflections:

> *Hypothesis 3: The relationship between frequency of investor laughter and their funding decision is positively moderated by reflection of investor laughter.*

QUANTITATIVE STUDY: METHOD, MODEL ASSESSMENT, AND RESULTS

To assess the generalizability of our findings in the qualitative study, we collected quantitative data from the well-known entrepreneurial reality show, "Shark Tank." Data was gathered from season 1 to season 4 of "Shark Tank". The method of coding from media sources has been widely adopted by professional students and has demonstrated its effectiveness in previous research (Lavanchy et al., 2022; Pollack et al., 2012). A total of 264 pitches were randomly assigned to the two research assistants. They were trained to understand the concepts of different types of laughter and tested to ensure the quality of the generated data. Each research assistant independently watched 132 pitches and coded the data. The results were subsequently discussed to ensure a shared understanding of the coding scheme (Martin et al., 2003).

The DV of this study is whether sharks decided to invest in the project. We recorded the number of offers made by sharks as the dependent variable on a scale from 5, as all the sharks offered, to 0, as none of the sharks offered. The IVs and moderating variables are the three types of laughter exhibited by sharks during the pitch. We reviewed the Humor Styles Questionnaire by Martin et al., (2003) and measured frequency of investor laughter, duration of investor laughter, and reflection of investor laughter in three steps. This study included control variables. The industry type was recorded because sharks' funding decisions are likely to change as a consequence of projects working in a certain industry. We created a dummy variable to control whether the sharks negotiated a special offer. We controlled their gender as male, female, or mixed team and also set the season number as a control.

Our analyses use negative binomial regressions to correct for overdispersion. The correlation between laughter frequency and laughter duration is at 0.68 and the correlation between laughter frequency and laughter reflection is at 0.53. It seems logical that duration and reflection of laughter are likely to see similar improvements with frequency of laughter. The variance inflation factor (VIF) of each variable is lower than 1.45 (Greene, 2003). Multicollinearity thus does not seem to be a problem. Control variables of gender and special offers show significance.

Results indicated a significant and positive effect of frequency of investor laughter on funding decisions ($b = 0.12$, $SE = 0.04$, $p < 0.01$), supporting H1. Duration of laughter is highly and significantly related to funding decisions ($b = 0.31$, $SE = 0.11$, $p < 0.05$) and reflection of laughter also shows significance on funding decisions ($b = 0.56$, $SE = 0.27$, $p < 0.05$). Furthermore, results showed a significant moderation effect of duration of laughter ($b = 0.08$, $SE = 0.03$, $p < 0.05$) on laughter frequency for funding decisions, supporting H2. However, the interaction term between laughter frequency and laughter reflection was not significant ($b = 0.09$, $SE = 0.08$, ns). H3 was not supported. To ensure the robustness of our results, we analyzed the data using an alternative dependent variable: a binary outcome that indicates whether a pitch received at least one offer (1 = at least one offer, 0 = no offer). The findings by logistic regression are confirmed. We also included a sensitivity analysis using different measures of moderators. Instead of using the average of all laughter events, we identified the top 5 longest laughter events and the top 5 longest reflections to calculate duration and reflection. Our findings are confirmed as well.

DISCUSSION AND IMPLICATIONS

This study contributes to the literature on entrepreneurial storytelling by emphasizing the active role of evaluating investors in the storytelling. Previous research has predominantly examined entrepreneurial storytelling as one-way communication, focusing on how pitching entrepreneurs use emotional displays to manage investor resistance and convey their ideas (Clarke et al., 2019). These studies often position investors as passive evaluators who listen to pitches within their own frameworks, relying on increasingly ambiguous judgment criteria (Mueller et al., 2018). In contrast, our study shifts the lens to explore the reciprocal nature of this interaction, highlighting how investors actively shape entrepreneurial storytelling through their reactions. Specifically, we demonstrate how investor laughter serves as an emotional and cognitive mechanism that informs and transforms entrepreneurial storytelling in real time. The way investors laugh during pitches influences the feedback entrepreneurs attend to, the improvements they make, and ultimately, the success of their funding decisions. We reveal three key insights. First, the frequency of investor laughter acts as a knowledge tool for entrepreneurs and allows them to adjust the tone and content of their stories (Brush et al., 2001). Second, the duration of investor laughter functions as a directed stimulus and offers entrepreneurs a means to establish stronger connections with investors (Cornelissen et al., 2014). Third, while our results show no significant moderating effect of the laughter reflection on funding decisions, it still opens up important theoretical considerations. The reflection of laughter may operate on a longer timescale than a pitch (Marsh & Śliwa, 2022). Together, these insights position investor laughter as a new form of emotional display for entrepreneurial storytelling.

This study advances the concepts of organizational aesthetics within the field of entrepreneurship. While prior research has focused on formal feedback mechanisms between entrepreneurs and investors, such as divergent thinking (Berg, 2016) or practical suggestions (Mueller et al., 2018), the role of organizational aesthetics in facilitating sensemaking has been largely overlooked. Through an ethnographic lens with semi-structured interviews, we integrate insights from stand-up comedy into entrepreneurial storytelling. This exploration offers valuable theoretical lessons for entrepreneurship, positioning laughter not only as a measure of engagement but also as a mechanism for creating real-time feedback. The quantitative component of our study strengthens this theoretical contribution. We demonstrate that laughter is not merely a byproduct of storytelling but a critical aesthetic phenomenon that influences funding decisions. Specifically, we show that feedback in entrepreneurial contexts can be remarkably short, such as the real-time response of investor laughter, yet it provides entrepreneurs with actionable cues to refine their narratives. This finding challenges conventional notions of feedback in entrepreneurship, which often emphasize delayed, formal, or structured responses (Huang & Pearce, 2015).

For practical implications, entrepreneurial storytelling is a highly skill-dependent activity, yet its quality remains challenging to quantify. While ideas are the core of entrepreneurial storytelling, many good ideas fail to gain traction during pitches because of ineffective delivery. Similar to the Lean Startup Method—a popular practical theme in entrepreneurship education that emphasizes a build–measure–learn feedback loop (Reis, 2011), our study highlights how skills from stand-up comedy can be applied to enhance entrepreneurial storytelling and underscores the importance of iterative refinement in storytelling. Humor not only makes presentations more memorable but also strengthens emotional connections with investors. Entrepreneurs can attend training programs offered by comedy schools or workshops to develop skills for crafting engaging and interactive pitches. These programs can teach techniques such as timing, delivery, and reading audience reactions—skills that are directly transferable to entrepreneurial storytelling.

CONTACT: Siwei Zhu; siwei.zhu@mnstate.edu; 1104 7th Avenue South, Center for Business, Q207, Moorhead, MN, 56563

SUMMARY

SHARING DREAMS OR SPREADING BULLSHIT? UNPACKING ENTREPRENEURIAL NARRATIVES

Eduardo Cuervo, University of Tennessee, Knoxville, United States
Daniel Aldave, Auburn University, United States

Principal Topic

Entrepreneurial discourse is often characterized by persuasive but unverifiable claims, aligning with Frankfurt's (2005) theory of bullshit, which distinguishes between falsehoods and statements made without regard for truth. Unlike fields where institutional oversight enforces factual accuracy—business narratives often rely on evasive and argumentative bullshit, designed to sound persuasive without being falsifiable. This study examines whether entrepreneurship tweets flagged by Community Notes differ from those in fields with overseeing institutions such as science and health, predicting that business tweets will be less likely to contain bald-faced bullshit (apparent lies) and more likely to rely on bullshit lies (mix of falsehoods with elements that appear credible) and evasive bullshit (vague, irrelevant, or overly general information), which may be harder to fact-check.

Method

This study analyzes flagged tweets from the Community Notes database, focusing on tweets categorized under entrepreneurship and other fields. We used variables from the Community Notes database and categorized tweets according to their misleading characteristics. We categorized tweets as bald-faced bullshit those flagged with factually erroneous statements, as evasive bullshit those flagged with statements rated as challenging to validate and lacking necessary context, and bullshit lies for those flagged tweets rated as containing misleading unverified claims and lacking a credible source. Logistic regression models test whether business tweets are more likely to be flagged for evasive and argumentative bullshit than bald-faced bullshit and whether bullshit type predicts lower helpfulness ratings and higher engagement. Interaction effects test whether these patterns are more substantial in business content than in fields with overseeing institutions.

Results and Implications

Preliminary results suggest that business/entrepreneurship tweets are flagged for evasive bullshit and bullshit lies rather than bald-faced bullshit, supporting the idea that business misinformation is more about persuasion than deception. Moreover, evasive and argumentative bullshit receives lower helpfulness ratings, reinforcing that bullshit is harder to fact-check than outright lies due to its unverifiable nature. Argumentative and evasive bullshit also attract higher engagement than bald-faced bullshit, highlighting the persuasive appeal of misleading business content. These findings have critical implications for misinformation detection and content moderation. First, fact-checking systems struggle more with business misinformation because evasive bullshit lacks falsifiability. Second, platforms may inadvertently amplify persuasive but misleading business content, given its high engagement levels. Finally, entrepreneurial discourse may uniquely exploit algorithmic incentives, raising concerns about the spread of unverifiable entrepreneurial claims. Addressing business-related misinformation may require different intervention strategies than those applied to fields with stronger institutional oversight.

CONTACT: Eduardo Cuervo; fcuervog@vols.utk.edu; University of Tennessee, Knoxville; 3026 Legacy Pointe Way, Knoxville, Tennessee 37921, United States

SUMMARY

PITCH PERFECT: CULTURAL SYMBOLISM AND MULTIMEDIA USAGE IN ENTREPRENEURIAL PITCHING

Rehab I. Elkhiyat, University of Tennessee, Knoxville, United States

Principal Topic

I study how entrepreneurs leverage symbolic resources in their pitches as they make sense of investor expectations in an emerging entrepreneurial ecosystem. Given the state-led push for entrepreneurship in Saudi, entrepreneurs need to navigate local cultural expectations and global investor norms. The process of symbolic enactment—where entrepreneurs utilize cultural cues, dress, and multimedia elements—becomes a critical part of their pitching strategy. The short and high-stake nature of pitching demand that entrepreneurs maximize the use of all available resources to resonate with investors (Kalvapalle et al., 2024). Notably, sensemaking exist at the heart of entrepreneurial pitching as a social activity (Weick, 1995). Investors operate in a context of high ambiguity especially in early-stage investments (Daellenbach et al., 2016), requiring entrepreneurs to frame their business ideas to align with or shape investor expectations. However, the prominent interpretive view in the pitching often fail to capture for the intersubjective state, where this interaction is based on the involved parties' sensemaking of the social and cultural context, (Soubliere & Lockwood, 2022). Moreover, much of the existing research overlooks how cultural contexts shape and inform the pitch process.

Method

My research context is a multi-year accelerator program in Saudi, it started in 2021 and is still ongoing with its eighth cohort. The program attracted 97 entrepreneurs from Arab and non-Arab countries, providing a rich setting to explore the research question. I take a multi-sensory qualitative approach in analyzing entrepreneurs' use of symbolic cues and artifacts in their pitches through video recording of the pitches, interviews with entrepreneurs and participants, and archival data. I am adopting a case study methodology coupling within and across pitches analysis (Eisenhardt, 1989). I am following a processual view of the data, considering the history of the event taking place across multiple years (Langley, 1999).

Results and Implications

The analysis reveals a progressive process for how entrepreneurs' symbolic choices reflect broader institutional changes in Saudi. Entrepreneurs adjust their pitching strategies in response to evolving investor expectations and accelerator interventions. Over time the pitches evolved from a standardized format to more interactive, multimodal and culturally hybrid performance. In this study I highlights pitching as a dynamic process, where entrepreneurs continuously refine their sensemaking and sensegiving strategies. Additionally, how entrepreneurs enact resourcefulness (Fisher, 2024) through cultural symbols and multimedia in their pitches. By framing pitching as an iterative process of institutional translation, I contribute to research on entrepreneurial legitimacy, sensemaking ,and multimodal communication in emerging ecosystems.

CONTACT: Rehab I. Elkhiyat; relkhaya@vols.utk.edu; University of Tennessee, Knoxville; 1213 Laurel Ave, Knoxville, Tennessee 37916, United States

INTERACTIVE PAPER

BLURRED LINES: LEGITIMATION NARRATIVES OF CRIMINAL ENTREPRENEURSHIP

Eduardo Cuervo, University of Tennessee, Knoxville, United States
Ava Haddox, University of Tennessee, Knoxville, United States
Ondina Sevilla, Clemson University, United States

Principal Topic

Entrepreneurs are often celebrated as visionaries who push boundaries, disrupt industries, and create economic growth. Many, however, engage in illegal activities such as fraud, regulatory evasion, and deceptive business practices to gain a competitive advantage. Despite violating laws, these entrepreneurs are frequently forgiven or even admired by the public. This study examines why people justify and tolerate entrepreneurial misconduct, focusing on the psychological and social mechanisms that shape societal responses. Using moral decoupling theory, this research explores how economic impact, innovation, and media narratives influence public perceptions, leading to the selective reordering of moral priorities, where entrepreneurial success is ranked above legal and ethical concerns.

Method

This study employs a multiple case study design. Using theoretical sampling, we analyze ten entrepreneurs who engaged in illegal activities but maintained public support. Data was collected from social media platforms, news articles, investor and stakeholder reactions, and online forums to assess how society responded to each case. We used these sources to create analytic artifacts for each case in order to conduct a deep within-case analysis. Following the within-case analysis, we engaged in a cross-case search for patterns. Iterating between the data and the moral decoupling literature, we extract the psychological and social mechanisms that shape public responses to criminal entrepreneurship.

Results and Implications

Preliminary findings suggest that society forgives entrepreneurial misconduct when business success is framed as a greater social good, leading people to view legal and ethical violations as justifiable means to a productive end. Two key mechanisms explain this pattern. First, moral decoupling allows people to separate an entrepreneur's competence from their ethical violations, leading them to overlook rule-breaking if the entrepreneur is seen as a highly skilled innovator. Second, moral rationalization occurs when illegal business practices are justified by their perceived benefits, such as job creation, industry disruption, or economic progress. Society increasingly views challenging authorities, pushing the limits, or rule-breaking as a necessary cost of progress. This reflects a broader cultural shift where entrepreneurial success is framed as inherently good, even if it comes at the expense of fairness or integrity. Public forgiveness of entrepreneurial misconduct reflects a cultural shift. As society adopts more entrepreneurial values, legal and ethical norms become more flexible in the face of perceived innovation and progress. This has implications for corporate regulation, business ethics, and accountability, as it suggests that society's moral boundaries align and accommodate the ideology of entrepreneurship.

CONTACT: Eduardo Cuervo; fcuervog@vols.utk.edu; University of Tennessee, Knoxville; 3026 Legacy Pointe Way, Knoxville, Tennessee 37921, United States

SURVIVAL-DRIVEN CROWDFUNDING IN TIMES OF CRISIS: THE FREE LUNCH OF KEEP-IT-ALL CROWDFUNDING CAMPAIGNS

Syrine Adala, Technical University Darmstadt, Germany
Carolin Bock, Technical University Darmstadt, Germany
Konstantin Kurz, Technical University Darmstadt, Germany

ABSTRACT

Crowdfunding has become an important funding channel for early-stage ventures. However, the COVID-19 pandemic shifted this landscape, as financially distressed businesses also turned to crowdfunding for survival. This dual-purpose ecosystem created a competitive environment between value-driven (aspirational) and survival-driven (crisis-response) campaigns. Drawing on impression management (IM) theory, we investigate how campaign type and strategic self-presentation influence crowdfunding success. Analyzing 3,477 campaigns from Startnext, a German crowdfunding platform, we find that survival-driven campaigns outperform value-driven ones in terms of backer engagement and funding raised. However, the effectiveness of self-presentation strategies is context-dependent: while self-promotion benefits value-driven campaigns, it can backfire for survival-driven ones. Conversely, supplication – expressing vulnerability – strengthens survival-driven campaigns but reduces the appeal of value-driven ones. These findings offer theoretical insights into the nuanced role of impression management and practical guidance for entrepreneurs navigating crowdfunding in crisis contexts.

INTRODUCTION

Access to capital is a persistent challenge for entrepreneurs, particularly during crises that heighten risk aversion and disrupt traditional financial channels (Berger & Udell, 1998; Hall & Lerner, 2010). The COVID-19 pandemic exemplified this disruption, leading many entrepreneurs to turn to crowdfunding – a financing method that allows direct appeals to the public without financial intermediaries – as an alternative source of funding (Mollick, 2014). While crowdfunding has traditionally been associated with value-driven campaigns aimed at developing innovative products and creative ventures, the pandemic catalyzed a surge in survival-driven campaigns, launched by existing businesses seeking immediate financial relief (Chandler et al., 2021). This shift created a complex funding environment where two fundamentally different campaign types – value- and survival-driven – competed for backers' support.

Although crowdfunding's role during crises is increasingly recognized, prior research has typically examined either value- or survival-driven campaigns in isolation. Few studies investigate how these types operate in parallel or how their presentation strategies affect relative success within the same resource-constrained environment. This oversight is especially salient in crisis contexts, where heightened uncertainty, financial anxiety, and shifting social priorities influence how backers assess and respond to funding appeals. In such high-stakes situations, entrepreneurs must not only communicate their needs but strategically manage how they are perceived by an audience navigating its own crisis-related constraints. The way campaigns are framed, what impression they convey and how well that impression aligns with backer expectations can play a pivotal role in determining funding outcomes.

Impression management (IM) theory, rooted in Goffman's (1959) dramaturgical framework, provides a valuable lens for understanding these dynamics. IM posits that individuals and organizations actively shape the way they are perceived by others. In crowdfunding, campaign creators act as "performers" on a "digital stage," carefully crafting their self-presentation to influence backer perceptions. From this perspective, campaign success hinges on the degree to which a presentation aligns with what backers consider credible, compelling, and worthy of support (Goffman, 1959; Gardner & Martinko, 1988). Because crowdfunding backers often lack objective data and instead rely heavily on narrative cues (Du et al., 2021), self-presentation tactics such as self-promotion (projecting competence and confidence) or supplication (emphasizing vulnerability and need) become especially relevant. However, audience expectations are not static – they evolve in response to context. During crises, heightened uncertainty and financial constraints may shift what backers value, requiring campaign creators to adapt their presentation strategies accordingly.

This study investigates how value- and survival-driven campaigns perform within the same crowdfunding environment during the COVID-19 pandemic and how different IM strategies – specifically self-promotion and supplication – moderate their success. By analyzing 3,477 campaigns on Startnext, the largest German-speaking reward-based crowdfunding platform, we explore how the alignment between campaign type, presentation style, and backer expectations shapes funding outcomes. This research contributes to the literature on entrepreneurial finance, crowdfunding, and crisis management by offering a framework that accounts for both the dual nature of campaigns and the role of strategic self-presentation in uncertain times. Beyond theoretical contributions, our findings provide practical insights into how crowdfunding can serve as an adaptive financial tool in times of crisis, offering entrepreneurs and platforms strategic guidance on how to present campaigns to maximize success.

HYPOTHESIS DEVELOPMENT

According to IM theory, individuals and organizations strategically construct their self-presentation to influence audience perceptions (Goffman, 1959; Bolino et al., 2008). In crowdfunding, where financial disclosures and due diligence are limited, impression management becomes central to attracting support. As Gardner and Martinko (1988) note, audience expectations are shaped by the broader socio-economic environment and evolve in response to external conditions. Crises, in particular, fundamentally alter the cognitive landscape in which decisions are made (Behr & Storr, 2022).

Prior research in psychology and economics identifies two primary behavioral shifts during crises: heightened prosocial behavior and increased risk aversion (Rodríguez et al., 2006; Solnit, 2010; Cassar et al., 2017; Bellavitis et al., 2022). We posit that these shifts influence how crowdfunding campaigns are evaluated. Survival-driven campaigns are more likely to align with these expectations. Their urgent goals – such as covering rent, wages, or operational costs – offer a direct and transparent link between contribution and outcome. Furthermore, they are often perceived as lower risk, as they aim to preserve existing operations rather than initiate new ones. Their established business history and operational footprint reduce the "liability of newness" (Stinchcombe, 1965) and signal a greater likelihood of successful execution. In contrast, value-driven campaigns often promote innovation, creativity, or long-term value creation. While these goals may be inspirational, they are typically aspirational, contingent on future outcomes, and may lack immediate social relevance. This perceived uncertainty and distance from crisis-urgency may dilute backers' emotional engagement and increase perceived risk (Mollick, 2014; Kuppuswamy & Bayus, 2017; Kahneman & Tversky, 1979). Accordingly:

Hypothesis 1: In times of crisis, value-driven crowdfunding campaigns will achieve lower success compared to survival-driven campaigns.

While campaign type affects alignment with backer expectations, impression management techniques can serve as strategic tools to shape perceptions. Self-promotion and supplication are two key IM tactics available to campaign creators (Jones & Pittman, 1982; Bolino et al., 2008). We theorize that these tactics can moderate the relationship between campaign type and success, depending on how well they align with crisis-specific expectations for impact and risk reduction.

A prominent form of self-promotion is the use of the All-Or-Nothing (AON) funding model. Unlike Keep-It-All (KIA), AON signals confidence and commitment by requiring campaigns to meet a predefined funding goal before receiving any funds. Prior research identifies AON as a credibility-enhancing mechanism (Cumming et al., 2020). We argue that the choice of the AON model is particularly valuable for value-driven campaigns that often face higher skepticism due to their uncertain outcomes. In contrast, survival-driven campaigns may not benefit as much from AON, as the conditional nature of the model could contradict the urgency of their needs and reduce perceived immediacy of impact. Thus:

Hypothesis 2: The choice of the All-or-Nothing campaign model during crises will positively moderate the relationship between campaign type and crowdfunding success, with a stronger effect for value-driven campaigns compared to survival-driven campaigns.

Another self-promotion strategy involves the use of positive emotional language in campaign narratives. Such language – emphasizing optimism, confidence, and resilience – has been shown to enhance backers' perceptions of competence and viability (Parhankangas & Ehrlich, 2014; Anglin et al., 2018). In value-driven campaigns, where uncertainty is high and project realization is long-term, these cues can reduce perceived risk and bolster appeal. However, for survival-driven campaigns, positive emotional language may weaken perceptions of urgency or financial distress, undermining backers' desire to support those perceived as "deserving victims" (Jenni & Loewenstein, 1997; Allison et al., 2024). Accordingly:

Hypothesis 3a: The use of positive emotional language during crises will positively moderate the relationship between value-driven campaigns and crowdfunding success.

Hypothesis 3b: The use of positive emotional language during crises will negatively moderate the relationship between survival-driven campaigns and crowdfunding success.

Conversely, supplication tactics highlight vulnerability and dependence, aiming to elicit empathy and prosocial responses (Jones & Pittman, 1982; Mohamed et al., 1999). In crowdfunding, this often manifests as negative emotional language, signalling urgency, anxiety, or distress. For survival-driven campaigns, such language aligns with crisis-induced backer expectations and reinforces perceptions of need and legitimacy (Kenworthy et al., 2023). However, for value-driven campaigns, negative emotional cues may contradict expectations of competence and control, thereby increasing perceived risk and reducing campaign appeal (Anglin et al., 2018; Majumdar & Bose, 2018). Therefore:

Hypothesis 4a: The use of negative emotional language during crises will negatively moderate the relationship between value-driven campaigns and crowdfunding success.

Hypothesis 4b: The use of negative emotional language during crises will positively moderate the relationship between survival-driven campaigns and crowdfunding success.

METHOD

To test our hypotheses, we use a dataset from Startnext, the largest German-speaking reward-based crowdfunding platform (Crosetto & Regner, 2018; Bürger & Kleinert, 2021). Startnext serves as an ideal empirical context due to its proactive response to the COVID-19 crisis. In March 2020, the platform launched several initiatives, such as "Startnext Corona Hilfsaktion" and "Support your local bar", aimed at supporting small businesses and entrepreneurs affected by lockdowns and economic disruptions. These programs led to a substantial influx of campaigns seeking immediate financial relief, thus increasing the prevalence of survival-driven crowdfunding. At the same time, Startnext continued to host traditional value-driven campaigns focused on the creation of new products, services, and creative endeavors. In response to the evolving financial needs of users, the platform also introduced a second funding model, the KIA model, alongside the existing AON model, providing a natural experiment setting for analyzing campaign performance under different strategic choices. Our final dataset includes 3,477 campaigns launched between March 11, 2020 – the date the World Health Organization declared COVID-19 a global pandemic – and February 28, 2022, when Startnext concluded its major crisis-response efforts (WHO, 2020; Startnext, 2022).

Crowdfunding success is measured through two outcomes: the number of backers and total funds raised, both log-transformed to address skewness. The key independent variable, campaign type, was manually coded from narrative content into a binary indicator: 0 for survival-driven and 1 for value-driven campaigns. This coding followed established content analysis procedures, with over 96% inter-rater agreement. To test the moderating role of IM strategies, we include three moderators. The first, funding model, is a dummy variable where AON campaigns are coded as 1 and KIA as 0. The second and third moderators capture linguistic signals: positive and negative emotional language, respectively, both measured via LIWC (Linguistic Inquiry and Word Count) software. Positive emotional language represents self-promotion (e.g., confidence, optimism), while negative emotional language reflects supplication (e.g., urgency, distress). We control for multiple campaign and creator characteristics. At the campaign level, we include duration, narrative length, platform promotion, number of offered rewards, and campaign category dummies. At the creator level, we control for gender, crowdfunding experience, and engagement indicators such as blog updates and pinboard activity during the campaign period.

RESULTS

We estimate a series of ordinary least squares (OLS) regressions with robust standard errors to examine the relationship between campaign type, impression management strategies, and crowdfunding success. Both dependent variables – log of number of backers and log of total funds raised – are modeled separately. Variance inflation factor (VIF) checks confirmed no multicollinearity issues among predictors. The results provide strong support for the first hypothesis: Survival-driven campaigns significantly outperform value-driven campaigns in both backer engagement and total funds raised, affirming that campaigns centered on immediate financial need are better aligned with backer expectations during crises (Hypothesis 1 supported).

Regarding IM strategies, the use of the All-Or-Nothing (AON) model positively moderates the relationship between campaign type and success, particularly enhancing the performance of value-driven campaigns, which benefit from the signaling effect of conditional funding (Hypothesis 2 supported, see Figure 1). With respect to linguistic cues, we observe differential effects depending on campaign type. Positive emotional language significantly improves outcomes for value-driven campaigns, mitigating perceived risk and increasing both backer support and funding volume (Hypothesis 3a supported, see Figure 2). Conversely, its use in survival-driven campaigns is negatively associated with success, suggesting that overly optimistic tones may undermine urgency and reduce perceived need (Hypothesis 3b supported). In contrast, negative emotional language strengthens the performance of survival-driven campaigns, enhancing perceptions of legitimacy and urgency (Hypothesis 4b supported). However, value-driven campaigns using similar language experience significantly worse outcomes, likely due to a mismatch between supplication tactics and the expectations for competence and control in aspirational projects (Hypothesis 4a supported, see Figure 3).

Interaction plots (see Figures 1-3) and robustness checks across alternative model specifications confirm the consistency of these effects. Together, the findings underscore the importance of the strategic alignment between campaign type, campaign presentation, and backer expectations for the success of crowdfunding campaign in times of crisis.

DISCUSSION AND IMPLICATIONS

This study advances our understanding of crowdfunding in crisis contexts by examining how campaign type and strategic self-presentation interact to influence funding outcomes. Drawing on IM theory, we show that the effectiveness of self-presentation techniques – such as the use of emotional language or the choice of funding model – is contingent on the campaign's underlying objective and the broader socio-economic context.

Our findings highlight a central tension within the crowdfunding ecosystem during crises: while survival-driven campaigns align more naturally with backer expectations for immediate impact and lower risk, value-driven campaigns must work harder to establish credibility. However, when equipped with the right impression management strategies – such as the AON model or positive emotional framing – value-driven campaigns can partially overcome this relative disadvantage. This reveals that backers are not uniformly biased against aspirational campaigns during crises, but rather respond to cues that reduce uncertainty and affirm campaign legitimacy. Conversely, survival-driven campaigns benefit from supplication techniques, particularly the use of negative emotional language, which enhances perceptions of urgency and need. However, these same cues prove detrimental to value-driven campaigns, likely due to a perceived mismatch between emotional tone and long-term aspirational goals.

Our study offers three main contributions. First, we introduce a framework that captures the coexistence of value- and survival-driven campaigns competing for resources during crises, advancing our understanding of how crowdfunding evolves under conditions of uncertainty. Second, we highlight the importance of aligning campaign presentation with backer expectations. Using IM theory, we show that campaign success depends on how well self-presentation strategies resonate with audience expectations in a crisis context. Third, we highlight an important boundary condition for the effectiveness of self-promotion strategies. While positive emotional language enhances the success of value-driven campaigns, it appears to backfire for survival-driven campaigns, potentially diminishing perceptions of financial distress and urgency. These findings challenge the assumption that positive emotional appeals are universally effective and underscore the need for context-sensitive communication strategies. Practically, our findings offer guidance for entrepreneurs, crowdfunding platforms, and policymakers on how to design, support, and communicate campaigns based on their purpose and the broader crisis environment.

CONTACT: Carolin Bock; bock@ent.tu-darmstadt.de; Technical University Darmstadt; Hochschulstraße 1, 64289 Darmstadt, Germany

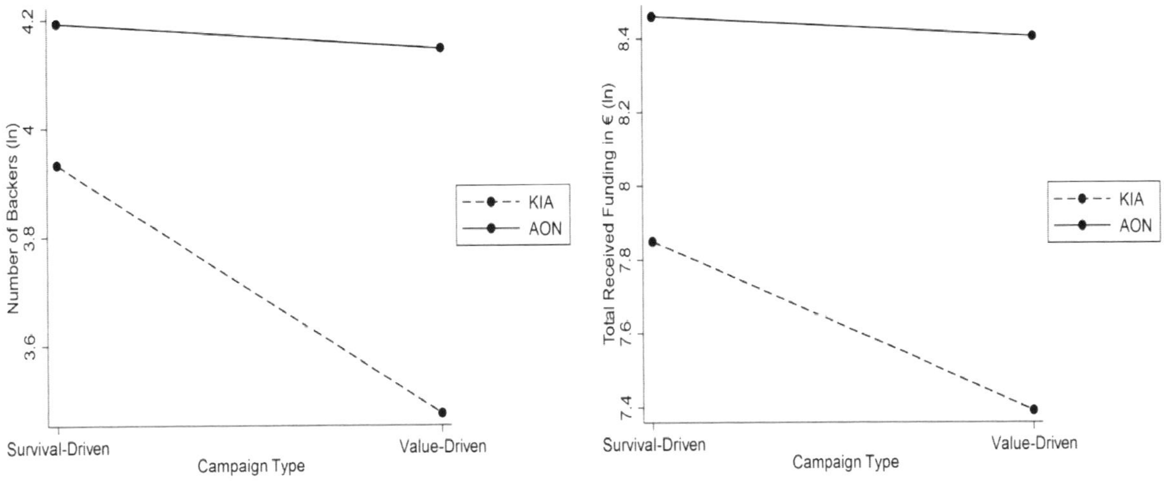

Figure 1. Interaction effects of the AON funding model as a self-promotion IM technique.

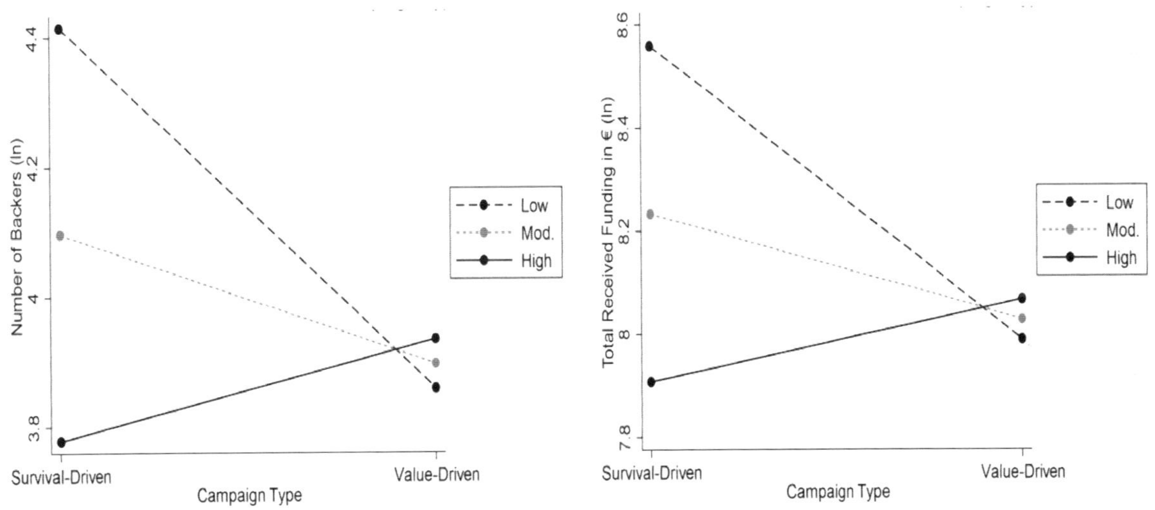

Figure 2. Interaction effects of positive emotional language (+/- 2SD) as a self-promotion IM technique.

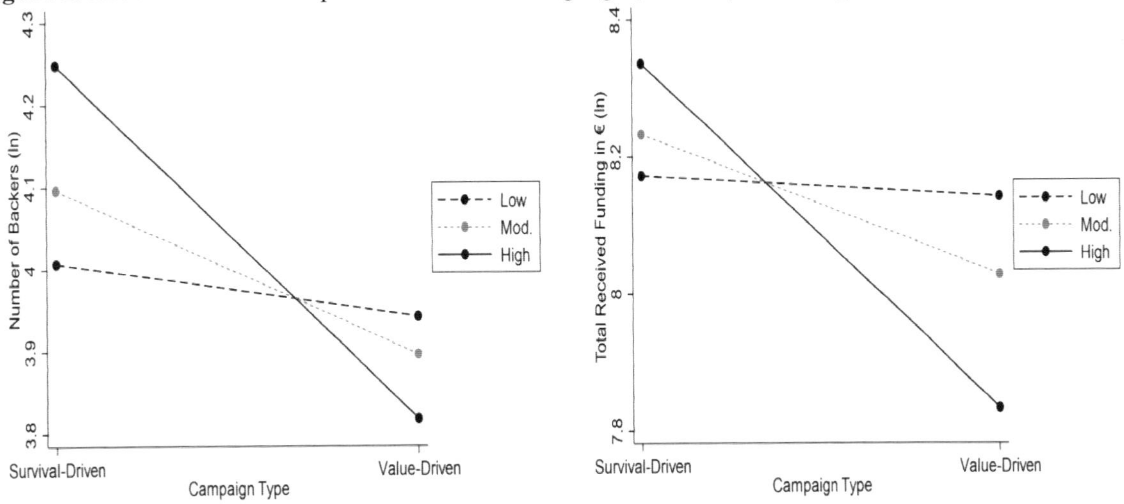

Figure 3. Interaction effects of negative emotional language (+/- 2SD) as a supplication IM technique.

SUMMARY

INVESTORS' REJECTION CRITERIA: EXPLORING NEGATIVE SIGNALS IN EQUITY CROWDFUNDING CAMPAIGNS

Daniel Berliner, University of Agder, Norway
Eli Gimmon, Tel-Hai College, Israel

Principal Topic

Equity crowdfunding (ECF) has emerged as a significant alternative source of equity finance for early-stage ventures, allowing companies to raise capital by selling small equity shares to numerous investors via online platforms. While extensive research has examined the factors contributing to ECF campaign success, limited attention has been given to the reasons investors reject investment opportunities. This study addresses this gap by investigating the rejection criteria utilized by ECF investors, offering insights into the decision-making process in online investment contexts. Unlike business angels (BAs) and venture capitalists (VCs), who conduct thorough due diligence and negotiate investment terms, ECF investors make decisions based on limited publicly available platform information. Consequently, their rejection criteria may differ from those of traditional equity investors.

Method

This study explores the rejection criteria used by ECF investors through a qualitative-exploratory verbal protocol analysis, capturing real-time decision-making rather than self-reported. This method has been widely used in BA and VC research. We developed an online demo platform simulating real-world ECF investments to examine investors' rejection criteria. Using judgment-purposive sampling, we selected nine experienced ECF investors to evaluate campaigns while thinking aloud, providing authentic insights. In total, 29 campaigns were evaluated by the interviewees. We applied the Gioia Methodology to systematically organize the findings. Interview data were reviewed, common themes identified, and insights categorized into first-order, second-order, and aggregate concepts, ensuring rigor analysis.

Results and Implications

Our analyses yielded nine rejection criteria utilized by ECF investors: lack of investors' personal interest or understanding, too-high share price, too-high minimum ticket, small market size and growth, highly competitive landscape and no competitive advantage, management team lacking experience, knowledge and experience, clarity and honesty of the offer, a company's unclear growth plan and use of proceeds, and lastly a campaign's low funding goal. The study revealed that ECF investors reject opportunities that present too high agency and market uncertainties. Additionally, our findings show that ECF investors employ rejection criteria which are similar not only to BAs and VCs rejection criteria but also to IPO and individual stock market investors' rejection criteria. This study contributes to entrepreneurial finance by highlighting the role of negative signals in investment decision-making, complementing existing signaling theory research. The findings offer practical insights for entrepreneurs seeking ECF funding and platform managers aiming to improve campaign success by addressing key investor concerns.

CONTACT: Daniel Berliner; daniel.berliner@uia.no; University of Agder; Universitetsveien 25, Kristiansand, Vest-Agder 4630, Norway

SUMMARY

NEW VENTURE LEGITIMACY: TEMPORAL AND CONTEXTUAL CONTINGENCIES OF DISTINCTIVENESS IN CROWDFUNDING

Vincenzo Butticè, Politecnico di Milano, Italy
Xiaojuan Hu, Tonji University, China (People's Republic)

Principal Topic

New ventures face a fundamental trade-off in their market positioning when seeking financial resources. Increasing distinctiveness—defined as the degree to which a venture differentiates itself from competitors—can enhance visibility and attract resource providers. However, distinctiveness can also introduce challenges by making ventures less comprehensible, thus increasing legitimacy pressures and hindering investment. Prior research on optimal distinctiveness suggests that ventures benefit from a balance between standing out and conforming to industry norms, but this assumption relies on the idea that investors evaluate a venture's value proposition as a primary decision-making factor. While this may hold for established firms, early-stage ventures often have flexible, untested, and evolving value propositions, making distinctiveness a less reliable investment criterion. This study develops a theory that contextualizes the role of value proposition stability in moderating the effect of distinctiveness on resource attraction. We argue that distinctiveness influences funding decisions only when the venture's value proposition is stable and serves as a credible signal. When the value proposition is fluid, investors may shift their attention to alternative signals, such as third-party endorsements or transparency in communication, rather than relying on distinctiveness as an evaluative criterion.

Method

We test our hypotheses using a dataset of 1,233 crowdfunding campaigns from Kickstarter, a platform where ventures vary in distinctiveness and can signal credibility through third-party endorsements and transparency. Our longitudinal approach follows campaigns over time, assessing how these variables shape the relationship between distinctiveness and funding success at different stages of the campaign. We employ computational text analysis, machine learning-based distinctiveness measures, and regression models to test for variations in funding performance based on these factors.

Results and Implications

Our results reveal an inverted U-shaped relationship between distinctiveness and resource acquisition in the early stage of crowdfunding campaigns: moderate levels of distinctiveness yield optimal funding outcomes, while extreme distinctiveness (either too high or too low) reduces success. However, this effect is contingent on value proposition stability. When ventures provide third-party endorsements or transparent disclosures distinctiveness remains a significant predictor of funding. Conversely, in the absence of such stabilizing signals, distinctiveness exerts little to no influence on investor decisions. These findings contribute to the optimal distinctiveness literature by introducing temporal and contextual contingencies into the theory, showing that distinctiveness is not universally beneficial but depends on available investor information. Additionally, the study advances crowdfunding research by illustrating how distinctiveness functions differently throughout a campaign and how investors adjust their decision-making criteria over time.

CONTACT: Vincenzo Butticè; vincenzo.buttice@polimi.it; Politecnico di Milano; Via Lambruschini 4, Milano, Lombardia 20156, Italy

SUMMARY

GENERATIVE AI AND THE COMPLEXITY OF ENTREPRENEURIAL PITCHES: SIGNAL SET FORMATION IN CROWDFUNDING

Vincenzo Butticè, Politecnico di Milano, Italy
Francesca Di Pietro, BICOCCA, Italy
Francesca Tenca, Università del Piemonte Orientale, Italy

Principal Topic

Entrepreneurial pitches are critical in crowdfunding success, as they shape investor perceptions and decision-making. Advances in generative artificial intelligence (AI) have enabled entrepreneurs to craft pitches that mimic successful elements from existing campaigns, raising questions about how AI-generated content influences investor attention and information processing. Drawing on the Elaboration Likelihood Model (ELM), we argue that AI-generated pitches reduce reliance on central route processing—where investors critically engage with message content—and instead activate the peripheral route, where decisions are based on heuristics like tone, structure, and presentation style. AI-generated pitches tend to lack the iterative fine-tuning found in human-created content, leading to messages that are either overly simplistic or excessively complex. This structural imbalance alters signal set formation, increasing the number of signals investors consider while reducing the depth of their engagement.

Method

We adopt a multi-method empirical approach combining archival data analysis and experimental research. Our first study analyzes a dataset of European crowdfunding campaigns on Kickstarter, comparing pre- and post-ChatGPT pitch complexity using measures of lexical diversity, sentence complexity, and syntactic cohesion. The second study employs an experiment with eye-tracking and EEG data, randomly assigning participants to evaluate either (1) a human-generated campaign, (2) an AI-generated version of the same pitch, or (3) an AI-generated pitch modified to match human linguistic complexity. Eye-tracking measures focus and engagement duration, while EEG assesses cognitive processing via alpha and theta wave activity. A final online experiment replicates these tests in a real-world setting to strengthen ecological validity.

Results and Implications

Preliminary findings from a pilot study (N=124) suggest that while investment likelihood does not significantly differ between AI- and human-generated pitches, the reasoning behind investor decisions changes. AI-generated content appears to shift attention toward different heuristic cues, increasing the number of signals considered while reducing scrutiny per signal. EEG data indicates that AI-generated pitches are less likely to activate deep cognitive engagement, reinforcing reliance on peripheral processing. Our findings have important implications for entrepreneurs, investors, and crowdfunding platforms. Entrepreneurs using AI to craft pitches may need manual fine-tuning to maintain optimal message complexity. Investors should be aware of the cognitive biases introduced by AI-generated content, which may affect decision accuracy. Platforms might consider disclosure requirements for AI-generated content to enhance transparency. More broadly, this study contributes to the literature on AI-mediated communication, entrepreneurial finance, and investor decision-making.

CONTACT: Vincenzo Butticè; vincenzo.buttice@polimi.it; Politecnico di Milano; Via Lambruschini 4, Milano, Lombardia 20156, Italy

SUMMARY

PROVIDING GOOD OR PREVENTING HARM: THE ROLE OF REGULATORY FOCUS IN THE EFFECTIVENESS OF INNOVATION CLAIMS ON CROWDFUNDING PERFORMANCE

Ming Chen, University of Huddersfield, United Kingdom
Qihai Huang, University of Huddersfield, United Kingdom
Jialin (Snow) Wu, University of Huddersfield, United Kingdom

Principal Topic

The effect of innovation claims on reward-based crowdfunding performance was found to be primarily negative, contradicting the legitimacy of reward-based crowdfunding to encourage innovation. Drawing from regulatory focus theory and signaling theory, we propose an interaction effect between innovation claims and the regulatory focus of the campaign narratives on crowdfunding performance. Claiming innovation with a promotion-focused narrative can have a less negative effect on crowdfunding performance, while innovation claims with a prevention-focused narrative result in a worse crowdfunding performance. Furthermore, innovation claims with a promotion-focused narrative can lead to even better crowdfunding performance when the campaign narrative shows a future focus.

Method

We collected data of 959 campaign narratives exclusively in the Technology category from Kickstarter, the biggest reward-based crowdfunding website in the world. This study adopts the computer-aided Text Analysis (CATA) approach, which measures theoretical constructs by applying two software packages, DICTION and LIWC, to capture the variables in texts to capture the frequency of word occurrence with validated dictionaries (McKenny et al., 2018). Specifically, crowdfunding performance, the dependent variable, is measured as crowdfunding success. This study adopts the dictionary developed by Seigner et al., (2022), Gamache et al., (2015), and the pre-installed dictionary in the LIWC to measure innovation claim, regulatory focus, and future focus.

Results and Implications

The result showed that a prevention-focused narrative negatively moderates the effectiveness of innovation claims on crowdfunding performance (β=-.048, P=.014). However, a promotion-focused narrative does not significantly moderate the relationship between innovation claims and crowdfunding performance (β=.002, P=.931). Nevertheless, when the campaign narrative shows a high level of future focus (making up 2.96% of the campaign words), the promotion-focused narrative can positively improve the effectiveness of innovation claims (β=.051, P=.041). This study extends the prior research on innovation claims in reward-based crowdfunding. While previous studies have an inconclusive view of whether reward-based crowdfunding supported innovative ideas, we suggest that entrepreneurial narratives can largely shape the effectiveness of claiming innovation on crowdfunding performance. This study also provides practical implications for entrepreneurs strategically manipulating their wording when presenting innovative ideas.

CONTACT: Ming Chen; Ming.Chen@hud.ac.uk; University of Huddersfield; University of Huddersfield, Huddersfield, England HD1 3DH, United Kingdom

SUMMARY

LEVERAGING AMBIGUITY: HOW ENTREPRENEURS BALANCE COMMUNITY ENGAGEMENT AND FINANCIAL MOBILIZATION IN DIGITAL PLATFORMS USING AUDIENCE FEEDBACK

Daniela Defazio, University of Bath, United Kingdom

Principal Topic

This study examines how creative and social entrepreneurs adjust their communication strategies when receiving negative audience feedback during role transitions. These entrepreneurs when mobilizing resources from their audiences must shift between their original identity (e.g., artists, mission-driven leaders) to a new fundraising role. This transition often leads to identity tensions, as audiences may disengage when entrepreneurs emphasize financial appeals over their original social or creative endeavors. Drawing on role communication accommodation theory, this study investigates whether entrepreneurs strategically use linguistic ambiguity to manage audience expectations and engagement. By analyzing 4,286 social media posts from 142 crowdfunding campaigns, the study explores how different types of negative feedback—transition-specific (targeting fundraising appeals) and persistent negative feedback (ongoing disengagement)—shape entrepreneurs' communication strategies.

Method

The research employs a quantitative analysis of social media interactions during crowdfunding campaigns on the Italian platform Produzioni dal Basso (PdB). The study tracks Facebook engagement metrics (likes) as indicators of audience feedback. The construction of the measure of ambiguity follows a three-step methodology inspired by Gong et al.,(2018). First, we estimated the probability of each word in a post being associated with different topics using Latent Dirichlet Allocation (LDA). LDA assigns probability distributions over words to define latent topics and provides word-topic probabilities, allowing us to identify words spanning multiple meanings. Next, we measured topic entropy to quantify word ambiguity. Topic entropy captures the uncertainty in a word's topic distribution—higher entropy indicates greater ambiguity, while lower entropy reflects a more distinct topic association. Entropy is computed using the probability of a word belonging to each topic, summing over all topics. We used fixed-effects regression models to test our model.

Results and Implications

Findings show that negative feedback increases linguistic ambiguity, with stronger effects when feedback is transition-specific or persistent. Entrepreneurs receiving immediate, transition-specific feedback adjust their language instantly by making their messages more ambiguous to maintain audience engagement. However, if negative feedback is persistent, entrepreneurs gradually increase ambiguity, even for non-financial content, as they recognize long-term disengagement. This study advances entrepreneurial role transition research by demonstrating how language is actively used to negotiate identity conflicts. Unlike prior research focusing on static storytelling, this study shows that entrepreneurial communication is dynamic and iterative, shaped by real-time audience reactions. By integrating communication, role transition, and social media engagement, this study offers new insights into how entrepreneurs navigate audience expectations and legitimacy challenges in digital contexts.

CONTACT: Daniela Defazio; dd486@bath.ac.uk; University of Bath; 10 EAST 2.33, Bath, Bath, England BA1, United Kingdom

SUMMARY

THE USER'S RESOLVE: A ROLE IDENTITY PERSPECTIVE ON THE LINK BETWEEN USER ENTREPRENEURSHIP AND PERSISTENCE AFTER CROWDFUNDING FAILURE

Stephanie B. Escudero, University of Texas at Arlington, United States
Pyayt Oo, University of Texas at Arlington, United States

Principal Topic

This study seeks to answer the questions: why do some entrepreneurs persist after crowdfunding failure while others do not? and under what conditions does this persistence vary? Specifically, we focus on the distinction between user entrepreneurs—those who create ventures to solve their own problems (Shah and Tripsas, 2007)—and non-user entrepreneurs. Using role identity theory (Stryker and Burke, 2000) as a framework, we propose that user entrepreneurs possess dual identities as both users and entrepreneurs, which helps buffer them from the negative effects of crowdfunding failure. While failure threatens their entrepreneurial role identity, their user role identity remains intact, reinforcing their motivation to persist. In contrast, non-user entrepreneurs, whose connection to the venture is based solely on their entrepreneurial identity, face a more significant challenge, as failure undermines the only identity tying them to the venture, making them less likely to persist. Further following the principles of role identity theory, which posits that both contextual and psychological factors shape role identities and role-related behaviors (Anglin et al., 2022), we also explore two moderators: market validation, representing a potent element of the crowdfunding context (Stevenson et al., 2022), and the Big Two dimensions (agency and communion), representing key psychological traits (Abele & Wojciszke, 2007). Market validation is expected to reinforce the entrepreneurial role identity, mitigating the effects of failure, particularly for non-user entrepreneurs. Meanwhile, agency enhances persistence for user entrepreneurs by reinforcing their self-oriented user identity while communion diminishes this persistence by shifting focus away from personal problem-solving.

Method

The data comes from 300 failed Kickstarter campaigns in the apps category along with the corresponding app store data for each crowdfunding campaign in our sample. Our independent variable (user entrepreneurship) and moderators (agency and communion) were derived from the entrepreneurial narratives and creator biographies of the Kickstarter campaigns using a combination of manual qualitative coding, trained large language models, and computerized content analysis. Market validation was measured as the percentage of fundraising goal raised (following Stevenson et al., 2022). Persistence after failure was measured as the number of days between the campaign's unsuccessful completion and the last update made by the creator to the app store (in the five-year period following their failure). We tested our hypotheses using cox proportional hazards regression.

Results and Implications

Results largely support our hypotheses. We find a positive relationship between user entrepreneurship and persistence after failure, with this effect strengthened by agency and weakened by both communion and market validation.

CONTACT: Stephanie B. Escudero; stephanie.escudero@uta.edu; University of Texas at Arlington; 2759 Avon, Dallas, Texas 75211, United States

SUMMARY

BREAKING BAD NEWS: IMPRESSION OFFSETTING AND CROWDFUNDER REACTIONS IN THE AFTERMATH OF NEGATIVE INFORMATION DISCLOSURES

Jasmine N. Kelley, Auburn University, United States
Jacob Waddingham, Texas State University, United States
Miles Zachary, Auburn University, United States
Michelle Zorn, Auburn University, United States

Principal Topic

Scholars have drawn attention to negative information disclosures and impression management in the management literature. Extant crowdfunding literature has examined the impact of financial and social drivers on investment decisions (e.g., Colombo et al., 2015). However, there is heavy focus on reward-based over equity crowdfunding (Vulkan et al., 2016). There is an opportunity for scholars to better understand the underlying mechanisms of the relationship between negative information disclosures and crowdfunding outcomes via equity crowdfunders. We wish to contribute the following: 1) Extend knowledge of how negative information disclosures impact crowdfunding investor behavior, and 2) how forms of impression offsetting influence crowdfunding outcomes. We aim to help firms understand investor decision-making and how to deploy effective impression management and fundraising strategies.

Method

We conducted three studies. The first study is a novel policy-capturing experiment (e.g., Connelly et al., 2016). We designed a neutral company description like those on the popular equity crowdfunding platform, Wefunder. We made hypothetical scenarios of "negative news" (i.e., negative information disclosures) and positive news (anticipatory & reactive impression offsetting) based on popular press releases. We consulted an expert pretest panel of 14 PhD-holding business professors from three southern U.S. public universities to guarantee scenario validity and reliability. For Study 2, we collected Prolific interviews from 39 crowdfunders. We used NVivo to conduct qualitative content analysis and draw emergent themes. For Study 3, we plan to conduct a follow-up policy-capturing experiment with perceived severity of the company news as a moderator.

Results and Implications

We found in Study 1 of 497 crowdfunding investors that negative information disclosures increase divestment and decrease confidence. We see that anticipatory and reactive impression offsetting can reduce these negative effects. Further, our results indicate that reactive impression offsetting after negative information disclosures was more effective than anticipatory offsetting. For Study 2, we collected 39 equity crowdfunder interviews and drew emergent themes surrounding crowdfunder considerations: 1) They care about the source of the company and executive news, 2) they care about the tone of the company and executive news (i.e., positive or negative), and 3) they care about receiving company and executive news updates. Our findings provide unique insight into the investment decisions of equity crowdfunding investors. Our results suggest that peak-end theory may have important predictive power in equity crowdfunding.

CONTACT: Jasmine N. Kelley; jnk0024@auburn.edu; Auburn University; 700 West Magnolia Avenue, Auburn, Alabama 36832-4530, United States

SUMMARY

A FAMILIAR TARGET MARKET? EXPLORING A GENDER-BASED HEURISTIC IN EQUITY CROWDFUNDING

Christina Koutouroushi, Durham University Business School, United Kingdom
James Hayton, Warwick Business School, United Kingdom
Mona Mensmann, University of Cologne, Germany
Ahmed Maged Nofal, emlyon Business School, France

Principal Topic

Investors' gendered life experiences responsible for the formation of the familiarity heuristic, can subsequently affect investment behavior. We argue that individuals' familiarity with firms catering to their gender -developed during their role as consumers- transfers to their role as investors, whereby they view firms which target same-gender consumers favorably. Based on the Stereotype Content Model, we theorize that a sequential mechanism of warmth and competence perceptions explains the effect of gender-based familiarity on investment.

Method

We examined our hypotheses with a multi-study design. In Studies 1 and 2, we established ecological validity by using archival data from two platforms (FundedByMe: 13,944 obs.; Seedrs: 4,463 obs.). In Study 3, an online experiment tested the proposed sequential mechanism of warmth and competence (2,846 participants from the US, UK and other European countries). In Study 4, to further isolate the effect of familiarity from that of homophily, minimize the plausibility of omitted variable bias and to rule out alternative explanations (i.e., other than warmth and competence perceptions), we ran a two-wave experiment (908 US participants; 1,444 observations).

Results and Implications

We found support for Hypothesis 1, showing a significant positive relationship between a female target market and the attraction of a female investor. There was some support for Hypothesis 2, suggesting that firms with a female target market and female founders are more likely to attract female investors. In Study 3, we found support for Hypothesis 3, which states that female investors exhibit higher willingness to fund firms which target female consumers through perceiving these familiar firms as warm, and then competent. In Study 4 we found that this serial mediation effect holds, even if we account for alternative explanations (i.e., product knowledge, connection to same-gender consumers). We found no significant results for male investors in any of our models. Our research contributes to the entrepreneurial finance literature by demonstrating that a gender-based familiarity heuristic can motivate investment. Gender-based familiarity emerges as an automated response when observing or assuming a firm's same-gender target market. Our studies demonstrate the complementary effect of two gender-based heuristics on investment decisions (i.e., familiarity and homophily). Finally, we find support for our theoretical model. Our findings show that there is an alternative route to appearing competent from demonstrating essential competence characteristics, which is to first appear warm. We extend familiarity theory by proposing that familiarity is transferrable from one market role (i.e., consumer) to another (i.e., lay investor) (Zajonc, 1968).

CONTACT: Christina Koutouroushi; christina.koutouroushi@durham.ac.uk; Durham University Business School; 8 Cameronian Square, Worsdell Drive, Gateshead, England NE8 2DB, United Kingdom

SUMMARY

TAKING A CONFIGURATIONAL APPROACH TO STUDY THE EMERGENCE OF OUTLIER TECHNOLOGY ENTREPRENEURSHIP IN CROWDFUNDING

Xuanye Li, New Mexico Tech University, United States
Christopher Crawford, The University of Tampa, United States

Principal Topic

Technology entrepreneurship (TE) plays a key role in developing innovative products and services that push the boundaries of science, engineering, and design. Outlier TE disproportionately influences the larger environment (Crawford, Aguinis, Lichtenstein, Davidsson, & McKelvey, 2015), has a higher probability of cascading effects to other domains (Boisot & McKelvey, 2011), and greater likelihood of external funding (Thunstrom, 2021). Statistically, these "interesting outliers"(Aguinis et al., 2013) are data points located distantly from the majority of observations and may contain valuable knowledge by following heavy-tail distributions. Their extreme success remains mysterious, yet exciting to study. Addressing this gap requires research approaches that embrace the complexity, interdependency, and nonlinearity inherent in outlier outcomes. Our study focuses on TE in the context of reward-based crowdfunding, where campaign founders receive financial support from individuals in return for product rewards (Mollick, 2014). Crowdfunding data offers an ideal context to explore outlier TE outcomes because it could capture nonlinear relationships between input factors and success levels (Leppänen et al., 2023). Thus, our research investigates two primary questions: (1) which endowment and engagement variables are associated with outlier outcomes in the context of reward-based crowdfunding? and (2) How can firms configure these variables to increase the likelihood of achieving outlier outcomes? Using the configurational approach (Meyer et al., 1993) and leveraging insights from studies about heavy-tail distributions and their generative mechanisms (Andriani & McKelvey, 2009; Crawford & McKelvey, 2018), this study identifies critical configurations—combinations of inputs that trigger nonlinear, outlier-level outcomes (Clark et al., 2023).

Method

A robust dataset of 39,908 observations from 2014 to 2022, scraped from the population of Kickstarter.com campaigns, serves as the empirical foundation for this research. To uncover the generative mechanisms of outlier outcomes, we utilize three analytical methods, including fuzzy-set qualitative comparative analysis (fsQCA), Machine Learning—Word Embedding Models, and the analysis of heavy-tailed distributions.

Results and Implications

Results confirm that outliers are positioned at the tail of log-normal type heavy-tailed distributions, driven by multiple critical configurations of inputs sufficient to achieve extreme funding success. Shared sufficient conditions across configurations include platform endorsement and engagement through FAQs, while factors such as past campaign experience, business experience, and demonstrations of innovativeness, autonomy, and competitive aggressiveness appear in specific configurations. The absence of some endorsement or engagement can prevent outlier outcomes, even when campaigns exhibit innovativeness. Entrepreneurs can benefit from our research by formulating resource allocation strategies that emphasize platform endorsement and backer engagement for outlier-level funding success.

CONTACT: Xuanye Li; xuanye.li@nmt.edu; New Mexico Tech University; 207 Bernard Street, Apartment D, Socorro, New Mexico 87801, United States

SUMMARY

INCENTIVIZING WOMEN'S ENTRY INTO ENTREPRENEURIAL FINANCE: A RANDOMIZED FIELD EXPERIMENT

Anna Lukkarinen, Ivey Business School, Canada
Markku Kaustia, Aalto University, Finland

Principal Topic

Digital entrepreneurial finance offers to democratize entrepreneurs' access to capital. However, due to investor homophily, realizing this democratization potential requires increasing the participation of women as investors. We consider an innovative approach to encouraging women's entry onto digital funding platforms: incentivizing existing platform users to make gift investments on behalf of women. Receiving their first investment as a gift may serve as a gateway for women's subsequent self-investing. In this gateway approach, the critical question becomes how existing users can be incentivized to make the initial gifts. Proposing that gifting could be motivated through modest economic incentives, we ask: How does an incentivized cue to invest for women affect investors' behavior?

Method

We conducted a randomized field experiment with the email newsletter of a crowdinvesting platform. The newsletter includes links through which users can access and invest in campaigns open on the platform. Our focal newsletter, sent on International Women's Day, encouraged investors to make a gift investment. The control version–which the platform would have sent to all recipients outside our experiment–did not specify gift recipients' gender in its subject line: "Happy Women's Day – Make a gift investment and receive a movie ticket." The treatment version was otherwise identical, but its subject line additionally specified that the giftee should be a woman: "Happy Women's Day – Make a gift investment to a woman and receive a movie ticket." At the time of the newsletter, six campaigns by six ventures were open on the platform. We track recipients' investments in these campaigns after the newsletter.

Results and Implications

We analyze 9,672 local-language investors to whom the newsletter was successfully delivered. Among all recipients, the probability of investing was 0.76%, and the average investment was 1,180 euros. Unexpectedly, the gender cue did not affect gift investments. Instead, it increased self-investments. The average self-investment rate increased from 0.56% without the gender cue to 0.91% with the gender cue. The effect is statistically significant ($p = 0.047$) and economically meaningful, reflecting an increase of over 60% in self-investments. While directionally present among both genders, the effect was statistically significant ($p = 0.057$) only among men. The treatment increased investments among investors who live in conservative-leaning areas ($p = 0.026$) but not among investors in liberal-leaning areas. The experiment points to an unintended consequence of diversity initiatives by investment platforms. We will evaluate possible mechanisms underlying the observed reactions, including zero-sum beliefs, identity threat, and increased risk-taking behavior.

CONTACT: Anna Lukkarinen; alukkarinen@ivey.ca; Ivey Business School; 89 Bloomfield Drive, London, Ontario N6G1P4, Canada

SUMMARY

WHO IS GUIDING WHO? CROWDSOURCING NEW PRODUCT DEVELOPMENT IN ENTREPRENEURIAL VENTURES

Shijian Wei, University of Alabama, United States
Thomas English, University of Alabama, United States
Devin Stein, University of Alabama, United States

Principal Topic

Crowdsourcing is becoming an increasingly important strategy for entrepreneurial firms to leverage the collective intelligence of external communities for problem-solving and product development. Unlike crowdfunding, which primarily provides financial support, crowdsourcing harnesses external ideas and expertise to improve product development and other forms of problem-solving. However, little is known about how entrepreneurial firms can best govern crowds to optimize product-market fit. Drawing on theories of complex adaptive systems and the problem-finding/problem-solving framework, we investigate how a balanced governance approach--combining both firm-level interventions and user-driven self-organizing--shape the efficacy of crowdsourcing in new product development.

Method

To test our theory, we constructed a longitudinal dataset from the Steam platform, the largest digital distribution platform for PC video games. We focus on "Early Access" games, where Steam allows developers to release incomplete games to users for feedback. This Early Access feature provides a unique context for examining crowdsourced product development and subsequent product-market fit, as proxied by the percentage of positive player reviews for each game. We tracked 15,610 games from 2013 to 2020, including 1,929 that transitioned from early access to full commercial release. Our analysis focuses on user engagement, developer communications, hybrid governance mechanisms, and managerial attention over each game's lifecycle. We use a difference-in-differences analysis and propensity score matching to examine the effectiveness of different governance approaches on product performance and explore how product lifecycle moderates these effects.

Results and Implications

Results indicate that early access games outperform traditional development approaches in terms of player satisfaction. Constructive player input, developer communications, and manager attention all improve product performance. Critically, we find that a balanced governance approach—blending firm-led interventions with user-driven self-organizing—yields the strongest product-market fit, especially in early stages of development. Firms that communicate frequently, respond to user feedback, and encourage community engagement benefit from richer knowledge flows and reduced information asymmetries. Our findings contribute to entrepreneurship and open innovation research by demonstrating how firms can govern collective problem-finding and problem-solving to harness the "wisdom of crowds" without stifling more emergent forms of crowdsourcing. We thus reconcile tensions between firm-level interventions and community-level self-organizing by demonstrating the importance of balanced governance. Unpacking how balanced governance approaches enable collective problem-finding and problem-solving has important implications for entrepreneurial firms which must navigate high uncertainty in establishing product-market fit during new product development.

CONTACT: Shijian Wei; fred.wei@hotmail.com; University of Alabama; 361 Stadium Drive, Tuscaloosa, Alabama 35487, United States

INTERACTIVE PAPER

FITTING IN WHILE STANDING OUT: THE CURVILINEAR INFLUENCE OF SLANG ON FUNDING PITCHES

Blakley C. Davis, Virginia Commonwealth University, United States
Thomas Allison, Texas Christian University, United States
Aaron Anglin, Texas Christian University, United States

Principal Topic

Language binds humans through communication; while language is often distinguished in terms of broader dialect, such as English, Spanish, or Mandarin, it may be further split into dialectical subgroups (White et al., 2018). One such prominent subgroup is often referred to as slang, or "an informal nonstandard vocabulary composed typically of coinages, arbitrarily changed words, and extravagant, forced, or facetious figures of speech" that are typically restricted to a particular setting or group (Meriam Webster, 2024). While the use of slang can make a person seem more relatable to a target audience, this may come at the cost of seeming unprofessional. Drawing on expectancy violation theory (Burgoon, 2015), we argue that while some use of slang will enable entrepreneurs to achieve greater funding performance, increasing the entrepreneurs' relatability and enabling them to fit in specific in-groups. At the same time, we predict that high-levels of slang use with violate expectations of professionalism, leading to negative perceptions and thus reduced funding performance.

Method

We test our hypotheses using a validated slang dictionary within a dataset of crowdfunding campaigns. We control for standard control variables including campaign goal, category, country, featured project status, length of campaign, and length of narrative.

Results and Implications

When modeling our key curvilinear effect, the linear term is positive and the squared term is significant and in the expected negative direction. Implications include the importance of including non-standard (slang) phrases in content analytic approaches to understanding entrepreneurial communication and the value of studying naturalistic language use of entrepreneurs. Moreover, our study contributes by highlighting a new tool, in the form of slang, for entrepreneurs to use when attempting to obtain in-group membership and access to needed resources.

CONTACT: Blakley C. Davis; bcdavis@vcu.edu; Virginia Commonwealth University; 510 Gardiner Rd, Richmond, Virginia 23229, United States

WE ARE FAMILY... AREN'T WE? EXPLORING THE ANTECEDENTS OF ENTREPRENEURS' AND EMPLOYEES' BOUNDARY MANAGEMENT IN START-UPS

Nadine Albrecht, University of Bayreuth, Germany
Sonja Kristin Franzke, University of Bayreuth, Germany
Viktoria Georgina Kilian, Innovation Manager at Value (Liotec GmbH), Germany

ABSTRACT

Although boundary management strategies and micro role transitions at the workplace are essential in building, maintaining, and differentiating private and professional relationships, we have limited insights into the antecedents of boundary management formation. Drawing on Rothbard et al.,'s (2022) relational boundary blurring framework, we explore the dyadic exchanges between entrepreneurs and their employees in a new venture context. Using semi-structured interviews and an open, inductive approach to data analysis, we qualitatively explore the mechanisms informing entrepreneurs' and their respective employees' boundary management practices. Emerging from the data, we develop a model bridging employees' and entrepreneurs' boundary management strategies, revealing an intricate balancing act between personal closeness and professional distance and different motives for a family-like working culture: while employees aim for close relationships with entrepreneurs and colleagues, entrepreneurs use close relationships with employees for their venture's advantage.

INTRODUCTION

In new ventures, boundaries between entrepreneurs' and employees' relationships are often blurred and intertwined, with colleagues considering each other as close friends (Bartels, van Vuuren, & Ouwerkerk, 2019). Boundary management practices and micro-role transitions are critical for establishing, maintaining, and differentiating private and professional relationships (Ashforth, Kreiner, & Fugate, 2000). However, we have limited insights into the antecedents of boundary management, specifically the individual-level motivational factors that influence entrepreneurs' and employees' boundary-setting practices (Capitano & Greenhaus, 2018).

Understanding these dyadic exchanges is essential because previous research paints an inconclusive picture (Schinoff, Pillemer, Rogers, & Petriglieri, 2025). While entrepreneurs may benefit from clear boundaries to navigate personal closeness and professional distance, they must also calibrate their communication between sharing too much or too little (Flynn & Lide, 2023). Conversely, employees may achieve well-being by establishing firm boundaries (Spieler, Scheibe, Stamov-Roßnagel, & Kappas, 2017). However, these may counteract a desired familial start-up culture, flat hierarchies, and flexible structures.

To shed new light on this puzzle, we qualitatively explore the antecedents and mechanisms informing entrepreneurs' and their respective employees' boundary management practices in a new venture context. We take a closer look at the dyadic dynamics between entrepreneurs (i.e., the venture founder) and employees (i.e., employees in the venture founders' company) and ask the research question, *"How do entrepreneurs and employees navigate boundaries between personal closeness and professional distance in new ventures?"*.

THEORETICAL BACKGROUND

Boundary management is based on boundary theory, referring to how people decide whether and how to separate (or blend) different areas of their lives and how individuals manage boundaries to structure their environment (Ashforth et al., 2000). Boundaries can be behavioral, temporal, physical, and communicative (Kreiner, Hollensbe, & Sheep, 2009) and vary in flexibility and permeability (Ashforth et al., 2000; Capitano & Greenhaus, 2018).

Boundary work contains "the practices that concretize and give meaning to mental frameworks by placing, maintaining, and challenging cultural categories" (Nippert-Eng, 1996: 563). Nippert-Eng (1996) distinguishes between two types of boundary work: those that are highly *integrating* and those that are highly *segmenting*. While segmenting individuals prefer to separate areas of life, integrators prefer to dissolve boundaries (Nippert-Eng, 1996). Ashforth et al., (2000) theoretically expand this concept by arguing for a continuum between the segregator and integrator roles. Additionally, Rothbard et al., (2022) highlight the relevance of *relational boundary blurring*, assuming that one's choice to blur the line between personal and professional life is significantly influenced by whom they interact.

Still, the optimal amount of information disclosure at the workplace remains debated (Flynn & Lide, 2023), propelling important questions on how entrepreneurs and employees navigate personal closeness and professional distance in dyadic encounters. Understanding such dyadic exchanges is especially important in an organizational context, where "individuals often derive significant meaning from the relationships they have with other individuals (e.g., supervisor, coworkers)" (Eberly, Holley, Johnson, & Mitchell, 2011: 747).

METHOD

We conduct a qualitative study to understand how entrepreneurs and start-up employees navigate the complexities of blending personal and professional boundaries (Glaser & Strauss, 1999). Through purposive sampling (Patton, 2002; Suri, 2011), we collected data from 24 interviews, consisting of dyadic matches between 12 new venture founders and 12 new venture employees in Germany (each pair originating from a different venture). Please refer to Table I in Appendix for an overview of sample characteristics.

We approached potential interviewees via personal contacts and the business platform LinkedIn. Following previous studies (Engel, Lewis, Cardon, & Hentschel, 2022; Hubner, Baum, & Frese, 2020; Laforet, 2013), we define new ventures' age up until 10 years. For new ventures, founders may be reluctant to hire full-time staff in the early stages, so we decided to sample new venture employees in a broad spectrum (e.g., interns, working students, freelancers, part-time workers).

Sample: Dyadic matches between 12 entrepreneurs and 12 employees, 24 semi-structured interviews (a total of 20:45:46 hours; an average interview lasted 52 minutes), data collection in two separate waves between May 2023-August 2023 and January 2025-April 2025 (a total of 8 months of data collection). Each pair originated from a different venture.
Qualitative analysis: Inductive coding, theory-building methodology (Gioia, Corley, & Hamilton, 2013).

After conducting first-wave interviews (6 dyads), we extensively analyzed data for emerging themes and aggregated our first findings to overall dimensions. After the first wave of interviews, it strikingly emerged that while entrepreneurs and employees aim for a friendly working atmosphere, the motives for doing so varied. To dive deeper into the antecedents of boundary management strategies in these dyadic encounters and to explore the emerging balancing act between personal closeness and professional distance in even more depth, we decided to collect additional data (additional 6 dyads), adjusted our interview guide, and analyzed individuals' motives more deeply.

RESULTS

We develop a model detailing how entrepreneurs and employees negotiate their boundaries in dyadic encounters (please refer to Figure I in Appendix). We use Goffman's (1956) frontstage/backstage concept to theorize visible and invisible boundary considerations. Generally, Goffman's (1959) theory on self-presentation involves two key areas: *frontstage,* where individuals present an idealized self to an audience, and *backstage,* where they prepare these presentations (Ringel, 2019). Backstage behavior protects secrets and involves careful preparation that often deviates from public ideals, creating a continuous effort to maintain boundaries between frontstage and backstage (Ringel, 2019). Our findings suggest that boundary management between entrepreneurs and employees happens on two levels.

Frontstage. In relation to others, entrepreneurs must balance personal closeness and professional distance. As such, they demonstrate a rather selective and strategic use of integrating versus segmenting when managing their boundaries. On the contrary, employees prefer to blur boundaries more openly, aiming for a family-like working culture. Thus, we find an *inter-relational* process of "calibrating" between entrepreneurs and employees, where they negotiate and establish a working mode for their dyadic boundary management.

Backstage. In addition, we find an invisible boundary management stage, where entrepreneurs and employees develop motives for setting or blurring boundaries that shape their willingness to adapt boundaries when circumstances change. We refer to this as the *intra-individual* dimension of boundary management. Entrepreneurs described wanting to establish and maintain their authority and competence as internal motives for boundary management. This increased conflict and separation between their frontstage closeness and backstage preference for more distance. Moreover, entrepreneurs describe how they aim to strategically use close relationships with employees for their venture's advantage. Employees, however, aimed for self-protection and closeness in their boundary management, which was closely aligned with their frontstage boundary management.

In summary, our data shows that boundary management is an ongoing, dynamic process that occurs both visibly in relation to others *(frontstage)* and invisibly in internal considerations *(backstage).* We find that some entrepreneurs adapt flexibly to changing situations, bridging both their backstage needs with frontstage demands, exploring how and why they emerge within our model.

DISCUSSION AND IMPLICATIONS

Our theoretical contributions are two-fold. *First,* we extend the existing literature on boundary management by showcasing that strategies are not dichotomous (i.e., integrating or segregating) but are enacted flexibly and adaptive in dyadic encounters (Ashforth et al., 2000; Kreiner et al., 2009; Nippert-Eng, 1996). Since boundary management between entrepreneurs and employees happens through interactions where both people can initiate changes (Eberly et al., 2011), studying these dynamic encounters is crucial. We, therefore, contribute to the organizational learning and coordination literature (Jones & Schou, 2023) by showcasing that entrepreneurs and employees adapt or switch their roles in dyadic encounters based on prior interactions and experiences.

Second, we explore boundary management at the micro-level, allowing us to portray how and why different strategies emerge. While entrepreneurs exploit a familiar company culture to attain advantages for their venture, employees blur boundaries more openly as they aim for close relationships with the entrepreneur.

CONTACT: Nadine Albrecht; nadine.albrecht@uni-bayreuth.de; University of Bayreuth, Institute for Entrepreneurship and Innovation, Nürnberger Straße 38, 95448 Bayreuth, Germany

APPENDIX - TABLES AND FIGURES

Figure I: Entrepreneurs' and employees' relational boundary management in start-ups

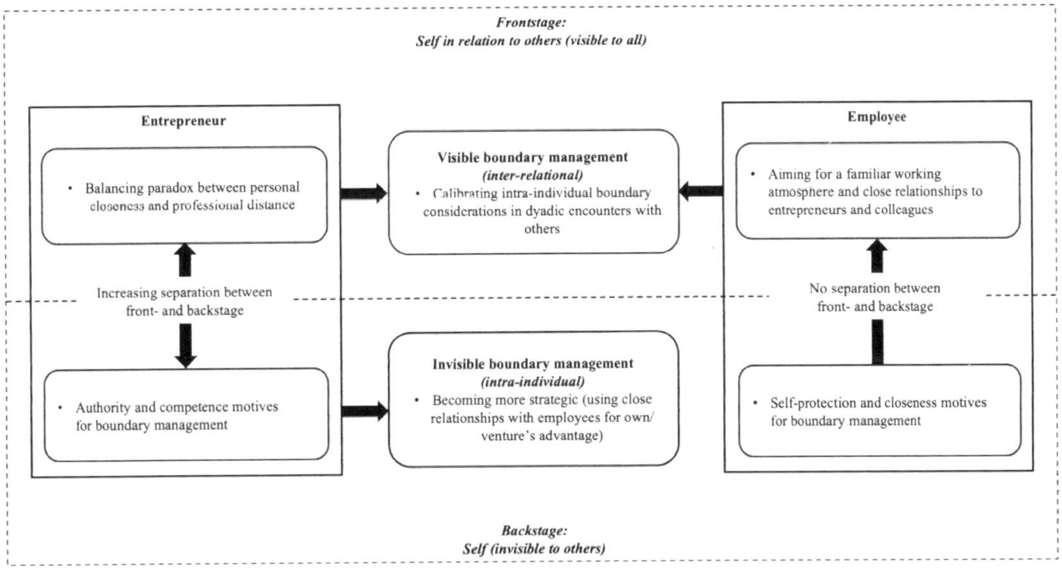

Table I: Sample characteristics

#	ID	Dyad	Pseudonym	Age*	Venture age*	Gender	Industry
Entrepreneurs							
1	ENT1	D1	Andy	29	1	Male	Software development
2	ENT2	D2	Bea	31	3	Female	Human resources (digital platform)
3	ENT3	D3	Carl	26	2	Male	Pharma tech
4	ENT4	D4	Dan	29	3	Male	Medical equipment manufacturing
5	ENT5	D5	Eric	39	7	Male	Delivery service
6	ENT6	D6	Frank	32	5	Male	Software development
7	ENT7	D7	George	39	7	Male	Retail apparel and fashion
8	ENT8	D8	Hannes	30	1	Male	Software development
9	ENT9	D9	Ingo	34	6	Male	Appliances, electrical, and electronics manufacturing
10	ENT10	D10	Jan	27	4	Male	Packaging and containers manufacturing
11	ENT11	D11	Keith	36	3	Male	Radar antennas, 3D printing
12	ENT12	D12	Liam	30	1	Male	Measuring and control instrument manufacturing
Employees					**Years in venture***		
13	EMP1	D1	Adam	20	1	Male	Software development
14	EMP2	D2	Ben	26	2	Male	Human resources (digital platform)
15	EMP3	D3	Clemens	27	1	Male	Pharma tech
16	EMP4	D4	David	31	1	Male	Medical equipment manufacturing
17	EMP5	D5	Elena	36	6	Female	Delivery service
18	EMP6	D6	Fabiola	27	1	Female	Software development
19	EMP7	D7	Gustav	29	6	Male	Retail apparel and fashion
20	EMP8	D8	Henry	26	1	Male	Software development
21	EMP9	D9	Ian	27	1	Male	Appliances, electrical, and electronics manufacturing
22	EMP10	D10	Janine	24	3	Female	Packaging and containers manufacturing
23	EMP11	D11	Kai	24	1	Male	Radar antennas, 3D printing
24	EMP12	D12	Lina	26	3	Female	Measuring and control instrument manufacturing

*Please note: ENT=Entrepreneur, EMP=Employee, D=Dyad. The average entrepreneur is 32 years old, and the average employee is 27 years old. *Person's age, venture age, and years in venture calculated according to the respective interview date.*

GREENHOUSE: ENTREPRENEURIAL DECISION-MAKING UNDER TIME COMPRESSION

Ash Ganzoury, University of Colorado, United States
Jeffrey G. York, University of Colorado, United States

ABSTRACT

Despite growing interest in venture accelerator programs and the established entrepreneurial cognition literature, we lack adequate understanding of the decision to pivot during early-stage entrepreneurship while founders are most susceptible to time, funding and community pressures. Through a qualitative study examining why and how entrepreneurs pivot in accelerator programs, we found that variations in founder identity led to pivoting behaviors different from those predicted by prior adaptation literature, which we suggest is due to community effects. Our findings suggest a social-cognitive explanation for the pivot decision-making process, which we draw on to develop a process-model. We introduce situated learning as a mechanism for entrepreneurial pivots in accelerator programs.

INTRODUCTION

A growing stream of research is interested in how entrepreneurial support organizations (ESO), particularly accelerator programs, may improve success odds for new ventures (Cohen & Hochberg, 2014). Bergman & McMullen (2022, p.690) defined ESO as "an organization whose primary purpose is to support individuals and collectives, through (in)direct and (im)material assistance, as they seek to initiate and progress through the stages of the entrepreneurial process." Unlike other ESO such as incubators and science parks, accelerator programs often batch ventures into cohorts (Huang & Pearce, 2015), which creates groups and interactions between communities. Hallen et al., (2020, p.378) underscore key members of the accelerator community, defining accelerators as "entrepreneurial programs that attempt to help ventures learn, often utilizing extensive consultation with mentors, program directors, customers, guest speakers, alumni and peers".

Bergman & McMullen (2022) have called for a deeper examination of the relationship between entrepreneurs with other entrepreneurs and ESO. Particularly in relations to pivoting behavior and prosocial motivation, Shepherd & Gruber (2021) have called for a deeper examination of The Lean Startup framework to close its subsequent academic–practitioner divide. In the popular practitioner book *The lean Startup*, Ries (2011, p.149) defined pivots as "structured course correction designed to test a new fundamental hypothesis about the product, strategy, and engine of growth".

Pivots represent observable strategic re-alignments or adaptations such as changes to mission, market, and business model (Busenitz & Barney, 1997). As reviewed in Mmbaga et al., (2020), founder identity has been established as a valuable construct to examine adaptation and pivot decision-making. Powell & Baker (2014, 9.1409) defined founder identity as "the set of identities that is chronically salient to a founder in her/his day-to-day work". For example, Zuzul & Tripsas (2020) have shown how Revolutionaries (or Discoverers) founder identity could hinder (or aid) the experimentation and adaptation processes needed for startup success. From another perspective, collective sensemaking affects a founder's identity work practices, creative revision and the incorporation of mentor feedback (Grimes, 2018). Fundamentally, the motivation and drive of an entrepreneur or any human originate in the self and extends to the group, but can also start in the group and extends to the self (Tajfel & Turner, 2010). We therefore studied in real-time how founder identity and community reciprocally influence the pivot decision-making process during two sustainable accelerator programs.

METHOD

We executed a qualitative study (Corbin & Strauss, 1990) during the years 2023 and 2024 in France at an accelerator program focused on developing ventures to address climate change. We conducted 40 semi-structured interviews with 15 founders and 3 accelerator staff. These interviews took place during the 14-week program and afterward while the founders were actively addressing real-world problems. All interviews were digitally recorded and transcribed. The quotes from the interview transcripts were compiled to form first order codes, then second order themes and aggregate theoretical dimensions (Saldaña, 2013).

The accelerator organization, functioning as a pre-seed investment fund, invested in each venture that was admitted to the accelerator program in exchange for equity. One objective of the accelerator program was to make founders venture-capital acceptable. To raise seed funding at the program's conclusion, founders pitch their ventures to potential investors and venture capitalist.

RESULTS

Overall, we observed that founders experienced the accelerator program as time-crunched learning-oriented individual and collective journeys. Image 1 illustrates the extensive sessions schedule and tight program timeline founders navigated during a typical week. The sessions and meetings had between 15-30 founders in addition to facilitators, presenters and guests. The accelerator manager explained to us "*My work is actually to juggle between tailor-made approach ... and managing a large pool of people; 120 mentors, ... living in my CRM, investors like 500 ... The alumni, I have like 100 of them now, and I still need to make sure.. that they will attend our meetings.*" This pointed our attention to the importance of these various community members of the startup-world, and the idiosyncratic nature of sustainable ventures and environmental entrepreneurs. During a specific week of the program, each venture met with 10 to 12 mentors daily. All the mentors were unpaid volunteers. The founders had to decide between an insurmountable amount of mentor feedback and directions to pivot in a very short period.

Our first finding was that variations in founder identity on the ecological-to-commercial typology (York et al., 2016) led to variations in the number and type of pivots. First, we used statements from the interview transcripts to ultimately aggregate founders into ecological-dominant, blended or commercial-dominant founder identity theoretical dimensions. Lars the founder of WorldShapers, expressed his ecological-dominant and prosocial motivations: "*that's why I started a business, so we could use business for good.*" On the other side, Anna the co-founder of BioGrade, explained clearly their commercial-dominant intentions to starting the business: "*basically selling the company. And this is probably the final point... I want to see it through... yeah, selling the company, acquisition.*" Second, we used the founders' answers to our open-ended questions to aggregate the number of adaptations into fewest-to-most pivots. Sofia, from SaveIT, told us clearly: "*it's the second time we pivoted*". While Roberto, from AgainUze, explained their perseverance and improbability to pivot: "*we are bringing packaging, one of the most polluting items in the world, towards sustainability. And I'm sure that we're going towards the right... I wouldn't change our business model for an outside investor.*" Third, by matching the founder identity to the pivot theoretical dimensions, we discovered that founders with ecological-dominant identity pivoted more times than those with commercial-dominant identity.

Our second major finding revealed explanations to why founders pivoted. Both the founders of Heartty explained that their decision to pivot market from education-tech to finance-tech was ultimately due to their career preferences "*we were building a whole education part because we believe financial education is really important, and quickly, with the [accelerator] team realized that you can't be building an ed-tech on a fin-tech. It's too much work at the same time... What do you guys want to do?.... It's more what you want to wake up and be doing in 5 years ... I would rather be managing investments in the future rather than education... And so, we're not building out education anymore.*" Helena from CarePlus attributed their decision to pivot business models from business-to-business to business-to-consumer to their newly developed sense of confidence "*we now had the confidence to actually take this step... 4 months ago there would be no way... So, because of [accelerator], we now have the confidence, that oh yeah, we could totally make this work*". On the other hand, Sofia from SaveIT explained that they were targeting small to medium size clients as their ideal customer profile, but then an accelerator member asked them to target large enterprise (corporate) instead: "*we did the brainstorming on this ... with the whole team, and it didn't come up as a corporate... we're not ready... we will fail before we can deliver. We will run out of money.*" This indicated that the decision to pivot did not initiate from inside the venture. Founders were often nudged to pivot towards investor preferences.

Our third and final finding shed light on how the pivot decision-making process progressed. The interactions between founders and the accelerator community clearly shaped the pivot decision-making process. Sarah from Heartty explained that they became *aware* of the need to pivot from the accelerator community, rather than through direct signals from clients: "*We've now received a plethora of feedback from mentors and other people, .. etc. And it's more than talking to potential customers.*" Two of the cofounders of SaveIT described how the accelerator community, particularly the peers-community in their cohort, was *empowering* and reflected shared experiences: "*we have these moments when we can share our highs and lows, and not just as entrepreneurs, but also as a person.... I've found out that many founders in [accelerator], my batch, are experiencing similarities with me.*" Helena from CarePlus explained how a mentor *enabled* them to pivot by assisting them in acquiring another brand: "*Also the CTO mentioned, you know, who was a mentor in [accelerator]. He helped us immensely to put the term sheet and the kind of whole proposal together for GoFree.*" CarePlus later hired that mentor as CTO.

DISCUSSION AND IMPLICATIONS

Accelerator programs compress iterations of entrepreneurial decision-making into relatively short timeframes. This time-compression leads founders to follow heuristics, take shortcuts and adapt quickly to the influences of the surrounding startup-world community. This rate of adaptation can be measured by the number of pivots during the accelerator program.

Although the founders with an ecological-dominant identity exhibited idealism, we did not find that this identity-induced idealism hindered flexibility and adaptation as predicted by prior literature on revolutionary founder-identity (Zuzul & Tripsas, 2020). On the contrary, founders with an ecological-dominant identity pivoted the most number of times. Founders generally received abundant, often conflicting advice from mentors (i.e., mentor whiplash). Our findings suggest that the accelerator community provided founders with psychological benefits such as identity confirmation, trust in particular advice, self-confidence to adapt and perceived integrity of their decisions to pivot. This aligns with the belief-oriented approach to entrepreneurship, which proposes that founders achieve revolutionary accomplishments by having belief and confidence in their decisions and conjectures (Elkington & Hartigan, 2008).

Our findings suggest that founders with ecological-dominant identities experienced more pressure to pivot than founders with commercial-dominant identities, explaining why the latter pivoted the least number of times. While inherent gaps, tradeoffs and synergies exist between pro-environmental, pro-social and pro-market values, beliefs and logics (Vedula et al., 2022). The startup-world functions largely as a pro-market community of practice. The number of pivots can be regarded as the distance founders travel to reconcile their membership in two communities (Brannon et al., 2015; Caplan, 1979). We assert that when overlap is larger between a founder's community of identity and community of practice, founders pivot less; when overlap is smaller, they pivot more. Figures 1A and 1B illustrate varying overlap situations. This assertion suggests situated learning (Lave, 1991) as a mechanism to explain pivots within accelerators, which supplements earlier work on organizational learning (Cohen, 2013; Hallen et al., 2020) and validated learning (Ries, 2011) in accelerators and pivoting respectively. Situated learning theory offers a dynamic lens to understanding inter-communal and intra-communal effects of communities of practice, which is particularly relevant for an accelerator program as a practice-based knowledge ESO (Hamilton, 2011; Jones et al., 2021).

Figure 2 offers a social-cognitive illustration of the interplay between communities, founder identity and pivoting in ESO.

To distill these theoretical dimensions in a timebound framework, we developed a process model of awareness, empowerment and enablement (Figure 3). Awareness was typically initiated by accelerator staff, mentors and investors, which interviewees repeatedly attributed to authority and credibility garnered through status and practical expertise. Empowerment, often supplied through cohort peers and alumni, reflected the value placed on shared identity and belonging to a community of individuals experiencing similar challenges. This second step of the pivot decision-making process often emerged after repeated opportunities to gain awareness. Enablement, the third step, developed through direct support, collective knowledge and resource mobilization, which emphasizes the key roles of social capital and community.

In conclusion, the accelerator programs we examined amplified the time, funding and community pressures that entrepreneurs often encounter elsewhere. In contrast to Ries's (2011) above definition of pivots; we found that founders were not always testing fundamental hypothesis for the venture but were often making career choices, or following identity-based and community-based decision logics. Rather than experimentation through customer feedback, in many of the cases we observed, pivot decision-making was driven by community nudging and founder social-cognitive processes. It is evident that pivots are not always structured nor course corrections but could also be detrimental to the success of the venture or founders. We underscore that customers represent just one community of influence; pivoting often occurs through interactions with startup-world communities such as ESO, investors, peers, and mentors. Our empirical findings suggest a generalizable alternative view to the entrepreneurial pivot construct compared to the Lean Startup Framework. We assert that entrepreneurial pivots are *strategic adaptations of ventures decided upon and steered by founders with communities of influence to create and/or capture value*. Our assertions are consistent with the choice-based view of entrepreneurship and Bayesian entrepreneurship, which show that founders' beliefs guide their choices and sensemaking (Cohen & Koning, 2024; Gans, 2025).

Future research and contributions

We suggest the following propositions for future researchers (illustrated in Figure 4). Proposition 1: Investor preference has a strong relationship to entrepreneurial pivot. Proposition 2A: Community memberships moderate the relationship between founder identity and entrepreneurial pivot. Proposition 2B: Community memberships moderate the relationship between investor preference and entrepreneurial pivot. Future researchers can operationalize part of the community effect as sense of belonging to community of peers, identity and/or practice. Our suggestive evidence can aid entrepreneurial finance and economies of time compression researchers explicate when and how accelerators help founders buy time at a lower cost (Graves, 1989; Hawk & Pacheco-de-Almeida, 2018; Henderson & Graebner, 2023).

We contribute to the entrepreneurial support organizations and entrepreneurial cognition literatures by introducing the effects of the accelerator communities on pivot decision-making. Community is especially important in conferring trust in pivot advice for idealistic entrepreneurs, previously assumed less adaptable by prior literature. We explicate the pivot decision-making process within an accelerator program. Our findings present generalizable explanations for pivots such as founders' career choices, investor preferences and community nudging. Finaly, we supplement our understanding of the entrepreneurial pivot construct based on empirical evidence and insights from situated learning theory.

We hope our research can support scholars, policymakers, ESO managers, entrepreneurship educators and founders in developing community as an influential input to experiential learning, the process of adaptation, and entrepreneurial success. Our paper supports accelerator programs in becoming more effective social learning systems which can aid entrepreneurs succeed, particularly those working towards solving time-sensitive grand challenges such as environmental degradation.

Appendix and illustrations:

Figure 1A. Smaller overlap
(Ecological-dominant founder identity were most to pivot)

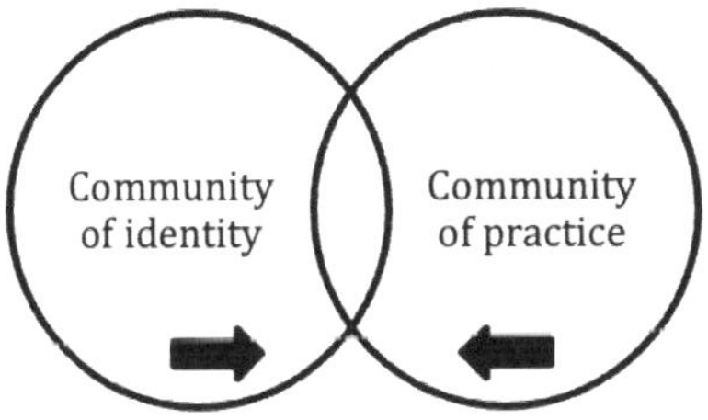

Figure 1B. Larger overlap
(Commercial-dominant founder identity were least to pivot)

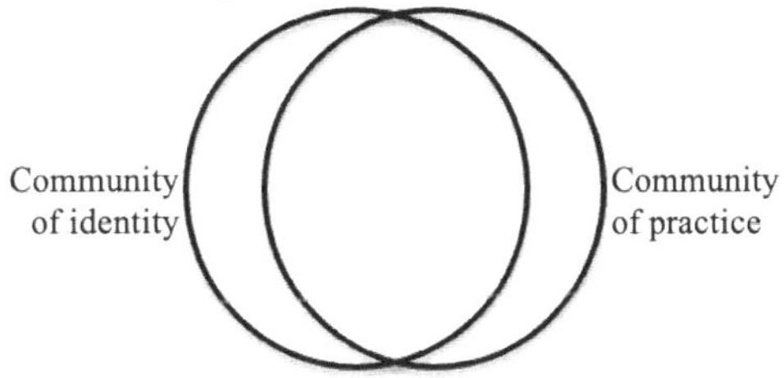

Figure 2. Founder identity, communities and pivoting

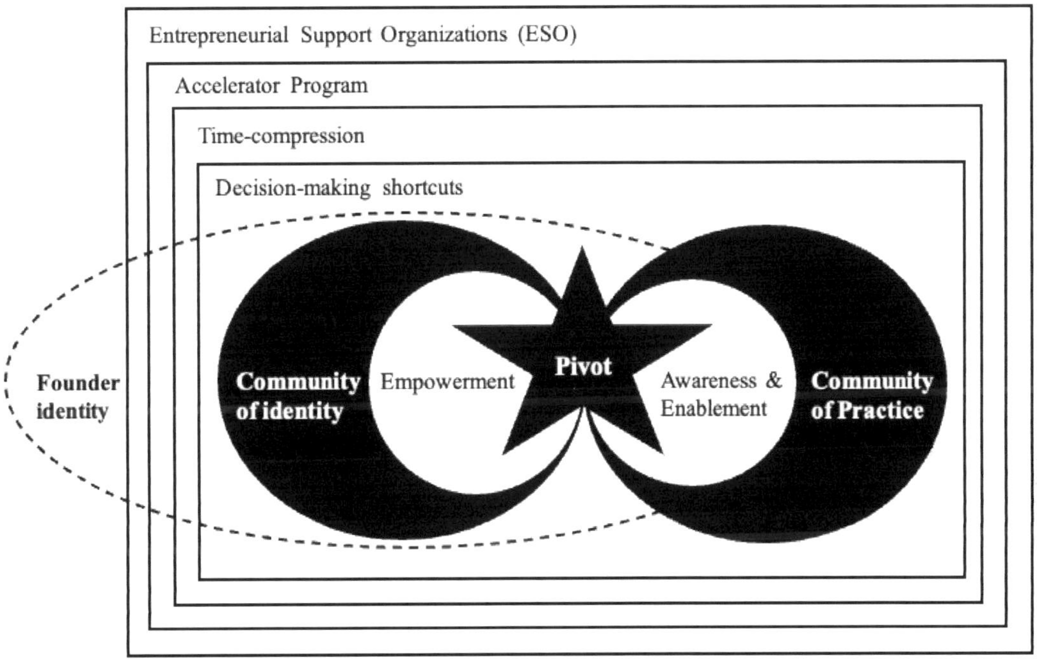

Figure 3. Pivot decision-making process within accelerator program

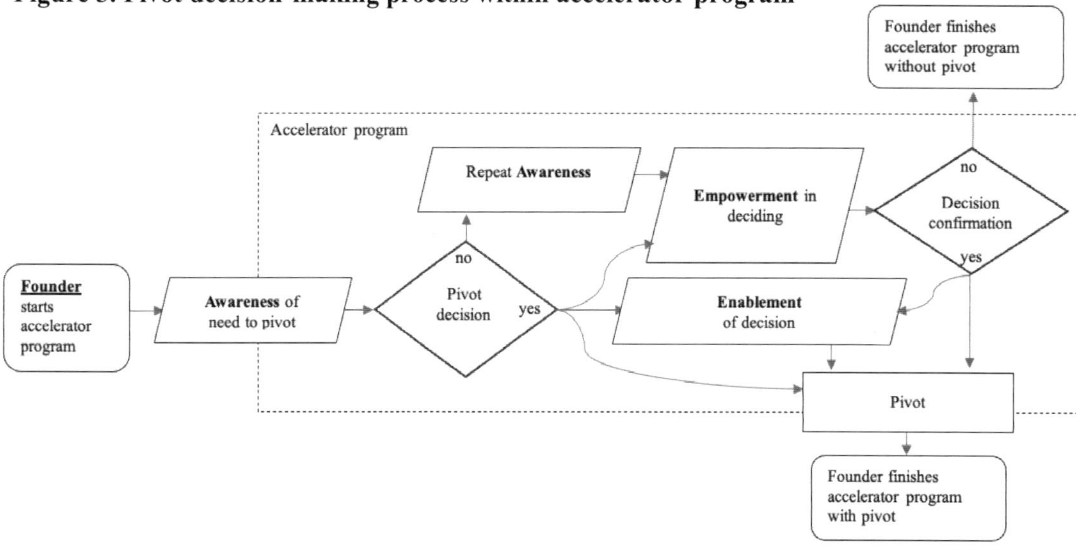

Figure 4. Propositions for future research

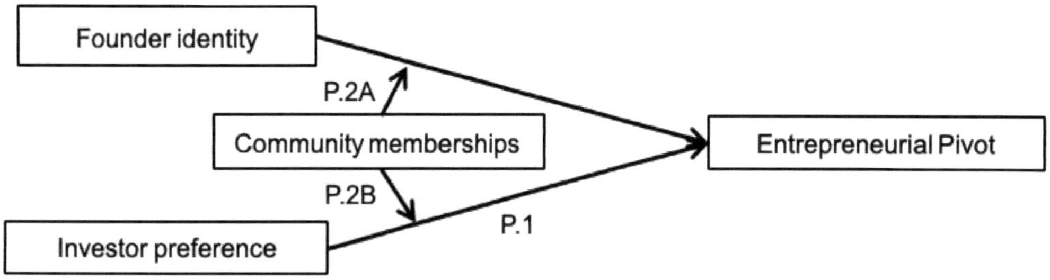

Image 1. Sample week of the founders' calendar (with accelerator program sessions in the darker shade).

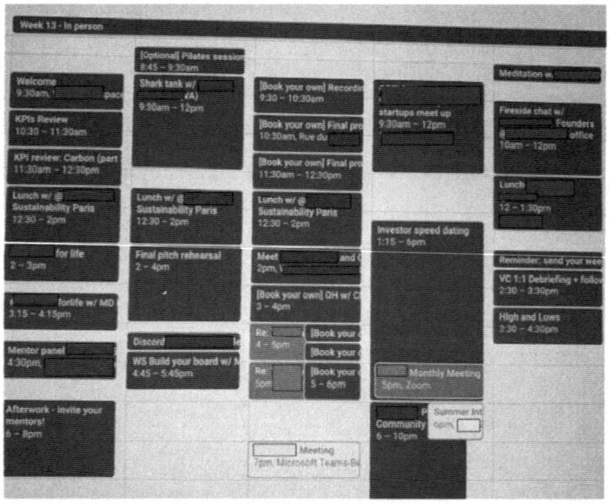

....

CONTACT: Ash Ganzoury; ash.ganzoury@colorado.edu; University of Colorado Boulder; Leeds School of Business, 995 Regent Dr, Boulder, CO 80309, USA

Acknowledgements: This research is partially funded through a fellowship from the Center for Ethics and Social Responsibility at the Leeds School of Business, University of Colorado Boulder.

MINDFUL IDEATION: HOW MINDFULNESS AND ENTREPRENEURIAL SELF-EFFICACY ENHANCE NEW VENTURE IDEATION

Alexander S. Kier, Washington State University, United States
Benjamin J. Warnick, Washington State University, United States

ABSTRACT

New venture idea generation is crucial in entrepreneurship, involving the creation of many ideas before selecting one to pursue. While cognitive factors supporting ideation are often seen as innate, mindfulness—characterized by receptive attention and awareness of the present—can enhance creativity. We theorize that trait, state, and mindfulness practice influence idea generation, with state mindfulness's positive effects amplified by mindfulness practice and entrepreneurial self-efficacy. Study 1 links trait mindfulness to idea generation, while Study 2 shows state mindfulness's causal effect on idea originality. We advance entrepreneurship and mindfulness research by demonstrating how mindfulness can be a powerful tool for entrepreneurs seeking to enhance their creative capacity in generating original new venture ideas.

INTRODUCTION

New venture ideas are a cornerstone of the entrepreneurial process, serving as the foundation upon which entrepreneurial ventures are built (Davidsson, 2015). New venture ideation—the process of generating ideas for creating a new venture—plays a critical role in this process (Kier & McMullen, 2018, 2020; Warnick et al., 2021). Idea generation is vital to new venture ideation because generating more ideas increases the likelihood of identifying a compelling creative idea (Kier & McMullen, 2018; Osborn, 1963; Perry-Smith & Mannucci, 2017). Despite its importance, the cognitive factors influencing new venture idea generation (e.g., Baron & Ensley, 2006; Gaglio, 2004) are considered innate and not easily developed, at least in the short term, raising the question of how entrepreneurs can improve their creative capacity for idea generation.

Recent research suggests that mindfulness—receptive attention to and awareness of present events and experiences (Brown et al., 2007)—may offer a promising avenue for fostering entrepreneurial creativity. Originating from ancient meditation practices, mindfulness has entered mainstream discourse as a promising way to promote functional outcomes in various facets of modern life. Research has begun to provide evidence of this promise, finding effects of mindfulness on various facets of cognition, emotion, behavior, and physiology (Good et al., 2016). From reducing stress and anxiety to enhancing self-esteem and emotional regulation (Brown & Ryan, 2003; Eberth & Sedlmeier, 2012), mindfulness has been heralded for its capacity to promote well-being and other functional outcomes across a variety of work contexts. Moreover, within entrepreneurship, mindfulness has been linked to sustainable decision-making (Engel et al., 2020), fear of failure reduction (Engel et al., 2021), and exhaustion mitigation (Murnieks et al., 2020), underscoring its promise for entrepreneurs.

Mindfulness have become increasingly prevalent such that entrepreneurs are likely using these techniques not only to improve their well-being but also to enhance their creative performance. By fostering a state of open and nonjudgmental awareness, we study how mindfulness can help entrepreneurs tap into their creative potential, leading to more productive idea generation. In addition to examining a baseline relationship between trait mindfulness and idea generation, we theorize that using mindfulness meditation to cultivate state mindfulness offers an actionable tool that encourages new venture idea generation and, in turn, idea originality. We further account for contingencies of the effect of state mindfulness on new venture idea generation and idea originality, theorizing that capabilities in mindfulness and entrepreneurship play a joint role in enabling entrepreneurs to leverage the ideational benefits of mindfulness in entrepreneurship. We theorize and test these capabilities in the form of mindfulness practice and entrepreneurial self-efficacy (ESE), respectively.

HYPOTHESIS DEVELOPMENT

Trait Mindfulness and New Venture Idea Generation

Trait mindfulness refers to an individual's dispositional propensity to be "attentive and aware of what is taking place in the present" (Brown & Ryan, 2003: 822). We argue that trait mindfulness holds considerable potential to facilitate new venture idea generation by virtue of its established cognitive effects, which creative cognition research suggests would benefit the creative process of new venture idea generation. Trait mindfulness can promote curiosity, spark creativity, open new perspectives, enhance problem solving, and facilitate a coping with uncertainty (Baas et al., 2014; Lebuda et al., 2016; Ostafin & Kassman, 2012; Vonderlin et al., 2020), all of which are important throughout the entrepreneurial journey. We focus on new venture idea generation, a critical early phase of this journey in which mindfulness holds great potential by cultivating nonjudgmental awareness that fosters divergent thinking, suggesting its potential in helping entrepreneurs generate more new venture ideas. A growing body of research supports that mindfulness is related to divergent thinking (Colzato et al., 2012; Colzato et al., 2017; Ding et al., 2015; Ding et al., 2014; Hughes et al., 2023), insight problem solving (Ostafin & Kassman, 2012), and other forms of creative achievement (Langer, 2014). Indeed, two different meta-analyses examined the mindfulness-creativity link, finding that mindfulness was significantly related to creativity with a small-to-medium effect size (Hughes et al., 2023; Lebuda et al., 2016). Given that new venture idea generation is fundamentally a creative process that requires divergent thinking (Kier & McMullen, 2018), we expect trait mindfulness will facilitate idea generation.

Furthermore, trait mindfulness facilitates idea generation because it encourages open and nonjudgmental processing (Good et al., 2016; Hülsheger et al., 2013; Murnieks et al., 2020). Therefore, when entrepreneurs engage in new venture idea generation, trait mindfulness should discourage critical evaluation that can stifle the creative process and enable greater awareness of and receptivity to new possibilities. This disengagement from evaluation is particularly important when engaging in idea generation (Perry-Smith & Mannucci, 2017). Indeed, research on brainstorming has found that evaluation during idea generation restricts creative output (De Bono, 2015; Osborn, 1963). Parnes and Meadow (1959) showed that subjects who were instructed to suspend evaluation of their ideas during brainstorming generated twice as many high-quality ideas as subjects instructed to come up with "good ideas." Supporting this, trait mindfulness encourages awareness and acceptance of observed phenomena while refraining from evaluation or self-criticism (Baer et al., 2006), allowing entrepreneurs to relax preconceived constraints and open their minds to imagined possibilities. Because mindful individuals are "more open to the environment, more open to new information," they are more likely to "find new ways to structure problems by developing a new perspective" (Corbett & McMullen, 2006: 49).

Hypothesis 1: Trait mindfulness is positively related to new venture idea generation.

State Mindfulness and New Venture Idea Generation

State mindfulness is a temporary state of heightened awareness in which individuals intentionally focus their attention on present experiences with openness and curiosity (Bishop et al., 2004). It involves observing thoughts, feelings, and bodily sensations without over-identifying with them or reacting in habitual ways (Bishop et al., 2004). Individuals engender deeper states of mindfulness during meditation (Kiken et al., 2015). Meditation typically involves sitting quietly, often with eyes closed, and directing attention to a focal point such as the breath, bodily sensations, or sounds in the environment. Mindfulness meditation is mainly characterized by observing one's experiences, not creating, or modifying them (Eberth & Sedlmeier, 2012). Mindfulness meditation cultivates a psychological state of mindfulness wherein individuals are attentive to the "here and now" (Herndon, 2008: 32) as opposed to the past or future (Dane, 2011).

We contend that state mindfulness is likely to increase entrepreneurs' idea generation because it has been linked to various forms of cognitive functioning that are associated with creativity, including (1) open, nonjudgemental awareness, (2) working memory, and (3) cognitive flexibility.

First, state mindfulness promotes open, non-judgmental awareness by allowing individuals to observe their thoughts and experiences without immediate evaluation or criticism (Good et al., 2016). This openness creates space for diverse ideas to emerge, increasing idea generation. Second, state mindfulness increases working memory (Chiesa et al., 2011; De Dreu et al., 2012; Hughes et al., 2023), which should facilitate expanded cognitive capacity to form the novel connections and means-ends relationships underpinning new venture ideas (Eckhardt & Shane, 2003). Third, state mindfulness increases cognitive flexibility—the ability to shift schemas and cognitive categories (Baas et al., 2014; De Dreu et al., 2008; Good et al., 2016; Moore & Malinowski, 2009). Cognitive flexibility is of particular importance in idea generation because it enables individuals to connect previously disconnected elements, helping them generate more ideas (Amabile, 1996; Guilford, 1967; Mednick, 1962; Perry-Smith & Mannucci, 2017).

Hypothesis 2: State mindfulness is positively related to new venture idea generation.

The Moderating Effect of Mindfulness Practice

Becoming proficient in mindfulness, like any skill, requires consistent practice and dedicated effort over time. Mindfulness practice involves setting aside time to engage in mindfulness meditation practices such as body scan meditation, sitting meditation, and mindful movement (Birtwell et al., 2019). We argue that the positive impact of state mindfulness on idea generation is amplified for those who regularly engage in mindfulness practice, proposing a moderation effect in which mindfulness practice strengthens the positive relationship between state mindfulness and new venture idea generation. The rationale is that consistent mindfulness practice enables practitioners to more effectively reap the benefits of state mindfulness and drives cognitive changes toward this end, as manifested in observable neuroplasticity changes in the brain (Hughes et al., 2023). Indeed, a large-scale meta-analysis on meditation practitioners found consistent evidence that meditation alters the anatomical structure of brain regions responsible for meta-awareness, body awareness, attention, memory, and emotional regulation (Fox et al., 2014). Additionally, Verhaeghen's (2021) meta-analysis found that long-term meditators have better attentional performance than naïve mediators. Thus, the heightened ability to regulate attention and emotion allows mindfulness practitioners to immerse themselves more fully in the present moment during mindfulness meditation, leading to a deeper and more immersive experience.

Hypothesis 3: Mindfulness practice positively moderates the relationship between state mindfulness and new venture idea generation.

The Moderating Effect of Entrepreneurial Self-Efficacy

Entrepreneurial self-efficacy (ESE) refers to the "strength of a person's belief that he or she is capable of successfully performing the various roles and tasks of entrepreneurship" (Chen et al., 1998: 295). ESE leads to various task specific effects such that entrepreneurs with high ESE hold strong beliefs in their own abilities to accomplish tasks in entrepreneurial areas, establish challenging goals, display persistence, and invest considerable effort (Miao et al., 2017). Thus, when faced with a new venture idea generation task, those with high ESE should have more confidence in the task and exert more effort, such that they are able to generate more new venture ideas. Indeed, Drnovsek et al., (2010) theorized that ESE will have a positive effect on entrepreneur's opportunity identification because individuals with high levels of ESE are more likely to perceive themselves as more competent in identifying new venture opportunities.

We argue that ESE enhances the positive relationship between state mindfulness and new venture idea generation. When individuals engage in mindfulness meditation, they cultivate cognitive resources—

such as cognitive flexibility—that are beneficial for creative tasks such as new venture idea generation. However, the extent to which these cognitive resources are effectively channeled toward idea generation depends on an individual's confidence in their entrepreneurial abilities. High ESE provides a motivational boost, encouraging individuals to engage more deeply with the cognitive insights gained from state mindfulness, ultimately increasing the likelihood of producing more innovative ideas for new ventures.

Hypothesis 4a: Entrepreneurial self-efficacy positively moderates the relationship between state mindfulness and new venture idea generation.

Finally, we expect that ESE plays a role in leveraging the benefit of state mindfulness and mindfulness practice for new venture idea generation, posing a three-way interaction. First, state mindfulness shapes cognition in a manner conducive to new venture idea generation by promoting open, non-judgmental awareness (Good et al., 2016), working memory (Chiesa et al., 2011; De Dreu et al., 2012), and cognitive flexibility (De Dreu et al., 2008; Good et al., 2016; Moore & Malinowski, 2009). Second, mindfulness practice interacts with state mindfulness in promoting new venture idea generation by allowing practitioners to immerse themselves more fully in the present moment during meditation, leading to a deeper and more immersive experience. Finally, ESE channels the creative cognition realized through state mindfulness toward productive new venture idea generation to the extent that the individual feels confident in their ability to generate new venture ideas (i.e., high ESE). In essence, ESE acts as a catalyst, channeling the creative cognitive power of mindfulness for fruitful new venture idea generation.

Hypothesis 4b: State mindfulness, mindfulness practice, and entrepreneurial self-efficacy jointly interact to predict new venture idea generation, such that the joint effect of state mindfulness and mindfulness practice results in higher new venture idea generation when entrepreneurial self-efficacy is high as opposed to low.

New Venture Idea Generation to Idea Originality

Generating many new venture ideas enhances the potential for originality. This principle is based on the idea that a higher volume of ideas increases the probability of discovering high-quality, original ideas (Simonton, 2003, 2010). Osborn's (1963) seminal work on brainstorming introduced the notion that "quantity breeds quality," emphasizing that generating numerous ideas leads to more innovative and creative solutions. Empirical studies support this view, consistently finding a positive correlation between idea quantity and quality (Diehl & Stroebe, 1987; Gallupe et al., 1992; Jung et al., 2015; Meadow et al., 1959). A larger idea pool allows individuals to explore diverse perspectives and unconventional approaches, fostering originality. As individuals continue generating ideas, they are more likely to break free from conventional thinking patterns and explore more original and innovative concepts. Therefore, increasing the quantity of new venture ideas enhances the likelihood of identifying truly original ones (Kier & McMullen, 2018).

Hypothesis 5: New venture idea generation is positively related to new venture idea originality.

METHOD

We test our hypotheses using two studies. In Study 1, we examine the relationship between trait mindfulness and new venture idea generation. To do so, we recruited a sample of 174 entrepreneurs to perform a new venture idea generation task, accounting for the relationship between trait mindfulness and idea quantity. In Study 2, we examine mindfulness behaviors, namely state mindfulness induced by meditation, and mindfulness practice, by conducting an experiment with a sample of 106 entrepreneurs randomly assigned to a state mindfulness or control condition before engaging in a new venture idea generation task. We further measured the entrepreneurs' mindfulness practice and entrepreneurial self-efficacy to test how these factors jointly channel the influence of state mindfulness on new venture idea generation, including the number of ideas generated and, in turn, the originality of their ideas.

RESULTS

Study 1 found that trait mindfulness significantly predicted new venture idea generation ($b = 0.261$, $p = .046$), supporting Hypothesis 1, which in turn predicted idea originality ($b = 0.065$, $p = .044$), supporting Hypothesis 5. Study 2 found that state mindfulness significantly predicted new venture idea generation ($b = 0.804$, $p = .048$), supporting Hypothesis 2. The two-way interaction of state mindfulness and mindfulness practice was not significant in predicting new venture idea generation, such that Hypothesis 3 is not supported. We found that the positive effect of state mindfulness on idea generation was greater to the extent that it was accompanied by ESE, supporting Hypothesis 4a (Figure 1.; $b = 1.600$, $p = .013$). State mindfulness resulted in a 43.6 percent increase in ideas generated (1.79 more ideas) for entrepreneurs with high ESE. We also found that state mindfulness, mindfulness practice, and ESE jointly predicted idea generation in the form of a three-way interaction (Figure 2; Hypothesis 4b; $b = 0.705$, $p = .040$). Entrepreneurs in the state mindfulness condition with high mindfulness practice and high ESE generated 49.1 percent more ideas (2.29 more ideas) than their counterparts in the control condition. Consistent with Study 1, new venture idea generation significantly predicted idea originality ($b = 0.157$, $p < .001$), providing additional support for Hypothesis 5. Finally, we found a positive indirect effect of state mindfulness on idea originality via new venture idea generation ($b = 0.127$, $SE = 0.075$, 95% CI [.004, .300]).

DISCUSSION AND IMPLICATIONS

The results of both studies confirm our theorizing, offering important contributions to entrepreneurship and mindfulness research. First, we extend research on the effects of mindfulness for entrepreneurs by illuminating its utility in helping entrepreneurs generate more new venture ideas. This is important because generating numerous ideas increases the likelihood of identifying a highly original idea from which entrepreneurs can further explore and develop (Kier & McMullen, 2018). Second, we illustrate the contingent effects of mindfulness on new venture idea generation, finding that state mindfulness better facilitates idea generation for entrepreneurs with greater mindfulness practice and ESE. This suggests that interventions aimed at inducing a mindful state are not equally effective across individuals. Our research demonstrates that mindfulness practice allows one to get in the creative "zone" more effectively through mindfulness meditation and thereby generate more new venture ideas, particularly for individuals with greater confidence in their entrepreneurial ability (i.e., high ESE). Finally, we contribute to the mindfulness literature more broadly by demonstrating that nuance is warranted in accounting for different mindfulness constructs in concert. Extant mindfulness research typically overlooks differences in the amount of meditation practice (Hughes et al., 2023). Our theorizing and findings illustrate why this oversight can be problematic, as we found that state mindfulness enhanced idea generation only for individuals who frequently practice mindfulness and possess high ESE. We thus demonstrate the complementary relationship between state mindfulness and mindfulness practice in fostering idea generation when ESE is present, even after accounting for individual differences in trait mindfulness. In doing so, we advance a more comprehensive understanding of mindfulness and its effects in the entrepreneurial process.

CONTACT: Alexander Kier; alex.kier@wsu.edu; Washington State University, Carson College of Business, 14204 NE Salmon Creek Ave., Vancouver, WA, USA 98686

Figure 1. Interaction of State Mindfulness and ESE on Number of Ideas

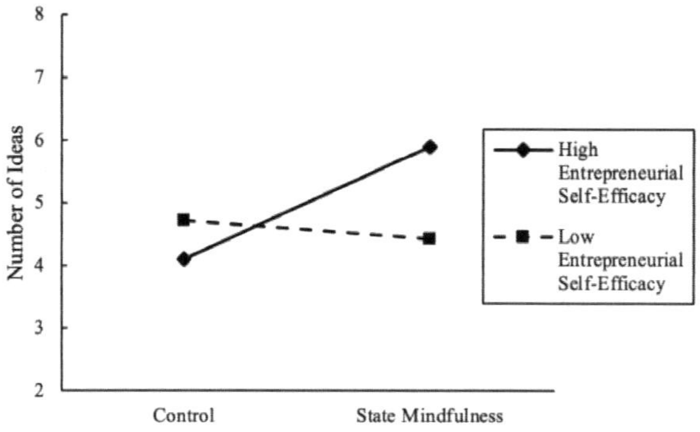

Figure 2. Interaction of State Mindfulness, Mindfulness Practice, and ESE on Number of Ideas

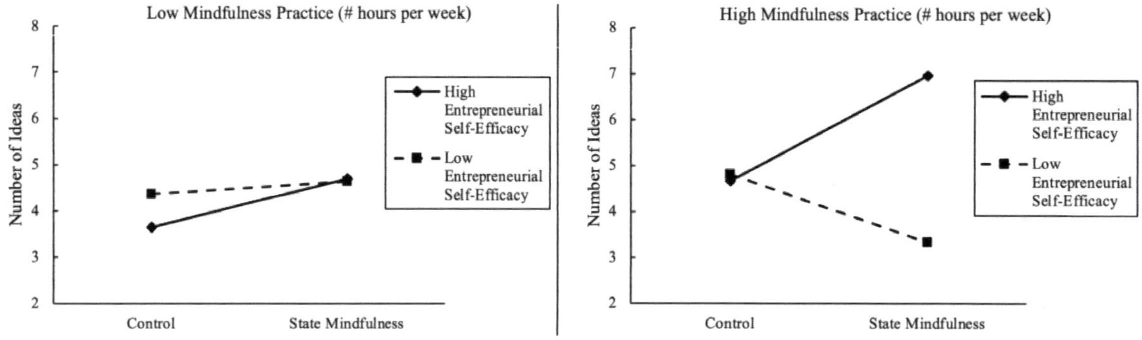

ENTREPRENEURS' EMOTIONS IN IN-PERSON AND VIRTUAL VENTURE PITCHES – HOW THEY DIFFER AND WHY IT MATTERS FOR FUNDING DECISIONS

Denis Lambert, ESCP Business School, France
Yi Dragon Jiang, ESCP Business School, France
Chang-Wa Huyen, ESCP Business School, France

ABSTRACT

How entrepreneurs communicate their ventures to potential investors is critical for funding success. While prior research links emotional expression to funding outcomes, studies often overlook how pitch format shapes emotional dynamics. We analyze 9,171 facial expressions in pitch videos (half in-person, half virtual) using computer-vision-based emotion tracking and sequence analysis. Our findings show that virtual pitches show more expressive but less variable emotional expressions, whereas in-person pitches involve more reactive emotional transitions. Entrepreneurs with emotionally stable profiles in virtual settings were more likely to receive funding. We also identify three latent emotional profiles that reflect distinct expressive strategies. This study shows that emotional strategies are strongly shaped by pitch context, contributes to the dynamic view of entrepreneurial emotions, and introduces computational tools for tracking emotions at scale.

INTRODUCTION

Emotions expressed by entrepreneurs during funding pitches play a critical role in how investors evaluate early-stage ventures (Chen et al., 2009; Cardon et al., 2017; Davis et al., 2017; Jiang et al., 2019). In these uncertain contexts where information is limited, investors often rely on emotional cues to assess the potential of a venture. However, while prior research has shown that emotions can influence funding decisions, findings are sometimes dissonant. For example, Davis et al., (2017) and Jiang et al., (2019) suggest that sustained positive emotion supports funding success, whereas Warnick et al., (2021) argues that emotional variability and moderation are more effective.

One reason for these inconsistencies may lie in a key contextual factor that has been largely overlooked: the format of the pitch. With the rise of virtual pitching, through platforms like Kickstarter and Indiegogo, entrepreneurs are increasingly communicating through mediated environments that differ significantly from in-person settings. Yet most emotion research in entrepreneurship does not account for how pitch format influences emotional expression or perception. Moreover, while studies increasingly recognize that emotions unfold dynamically, they often overlook how these emotional trajectories vary between virtual and in-person settings (Drolet & Morris, 1999; Bicchieri & Lev-On, 2007; Zant & Kray, 2014). To address this gap, we ask: (1) How do entrepreneurs express their emotions differently in in-person and virtual pitch settings? (2) Why do these differences matter for funding outcomes?

To explore these questions, we analyzed 28 pitch videos (14 in-person and 14 virtual) using computer-vision-based emotion tracking (OpenFace 2.0; Baltrušaitis et al., 2016) and a four-step analysis: rates of change (George & Jones, 2000; Lambert & Jiang, 2025), Markov chain analysis (Feller, 2008; Jiang et al., 2021), Sequential Pattern Mining (SPM; Giuntini et al., 2021; Huang et al., 2024) and Hidden Markov Models (HMM; Eddy, 2004; Quan & Ren, 2016). This method allowed us to uncover pitch-specific emotional patterns and identify latent emotional profiles.

Our findings reveal that emotional dynamics differ meaningfully across pitch formats. While in-person pitches often feature reactive emotional transitions, virtual pitches tend to prompt more expressive but less variable emotional expressions. Moreover, entrepreneurs with more stable emotional profiles in virtual settings were more likely to receive funding. These results help explain divergent findings in prior research and highlight how emotional strategies must be tailored to the pitch context.

Our study makes three contributions. First, the study advances theories of emotional dynamics by showing how emotions unfold differently across mediated and non-mediated formats. Second, the study identifies three latent emotional profiles that reflect underlying emotional strategies and their contextual effectiveness. Third, the study brings computational tools from computer science into emotion research in entrepreneurship, offering a scalable method for capturing dynamic, expressive behavior. These insights have practical value for entrepreneurs refining their pitch delivery and for investors seeking to interpret emotion more accurately across settings.

HYPOTHESIS DEVELOPMENT

Emotions are defined as subjective feelings of pleasure or displeasure that involve automatic appraisal processes where external causes are assessed, followed by a response to the assessment via a set of reactions such as subjective feelings, physiological responses, and behavioral reactions (Cardon et al., 2012; Ekman, 1992; Keltner et al., 2019; Russell, 2003; Warnick et al., 2021). Traditionally, these reactions are organized and measured using six basic emotions: anger, surprise, disgust, happiness, fear, and sadness (Ekman, 1992, 1997).

How entrepreneurs communicate their ventures to potential investors is critical for funding success. The emotions displayed during pitching influence investor support over and above the nature of the venture (Davis et al., 2017; Huang et al., 2023; Warnick et al., 2021). For example, Davis et al., (2017) showed that perceived entrepreneurial passion enhances the positive impact of product creativity by eliciting funders' positive impression, which in turn increases crowdfunding success. Jiang et al., (2019) demonstrated that brief moments of peak joy, especially at the beginning or end of a pitch, are linked to higher funding outcomes, while prolonged displays may diminish effectiveness. Building on this, Warnick et al., (2021) found that moderate expressions of happiness, anger, and fear are most effective, and that frequent emotional transitions signal authenticity and further promote funding. Huang et al., (2023) extended this perspective by showing that peak negative emotions, whether conveyed through human faces or visual elements, can boost funding by highlighting consumer pain points. Finally, Clarke et al., (2019) emphasized that emotional expression also occurs through verbal and gestural channels, finding that figurative language and expressive gestures help investors visualize the opportunity, enhancing their willingness to invest.

When studying emotions, scholars often focus on their expression through observable behaviors such as facial expression, body language, and vocal tones, providing concrete data for analysis (e.g., Ekman, 1992; Dael et al., 2012). In the field of entrepreneurship, researchers capture the expression of emotions during funding pitches by examining entrepreneurs' verbal and vocal expressions (e.g., Clarke et al., 2019; Russell, 2003), facial expressions (e.g., Ekman, 1992, 1997; Warnick et al., 2021), and body gestures (e.g., Atkinson et al., 2007; Clarke et al., 2019). These cues are interpreted as emotions such as happiness, fear, anger, and passion. Clarke et al., (2019) analyzed expressed emotions by combining forms of speech (content, rhetoric, and tone) with non-verbal communication (gesture, posture, and facial expression) to capture positive emotion through figurative language used during entrepreneurial pitches. Warnick et al., (2021) investigated emotions in crowdfunding campaigns by analyzing entrepreneurs' facial expressions, using facial recognition software to track second-by-second expressions of emotion. In contrast, Allison et al., (2022) studied emotions by analyzing entrepreneurs' vocal expressions.

Over the past two decades, entrepreneurial pitching, which was primarily conducted in physical settings, has increasingly shifted to online formats. Although the literature on entrepreneurship and funding pitches has acknowledged this shift, existing research has not fully examined the different dynamics between virtual and in-person contexts. For instance, many foundational studies have focused on in-person pitch environments, including televised competitions like Shark Tank and Dragons' Den (Pollack et al., 2012; Blaseg & Hornuf, 2024), or angel investor gatherings involving face-to-face presentations (Huang & Pearce, 2015). More recent work has shifted toward online platforms, including Kickstarter and Indiegogo (Jiang et al., 2019; Warnick et al., 2021; Huang et al., 2023), yet these studies often emphasize funding outcomes rather than the distinct emotional and interactional dynamics shaped by the medium of communication.

However, existing literature highlights that face-to-face versus digital interactions alter the quality of the communication between parties (Bicchieri & Lev-On, 2007; Zant & Kray, 2014). Notably, Drolet & Morris (1999) stress that computer-mediated interactions affect the quality of the negotiations that may follow the pitch by restricting the range of communication and ability to coordinate on mutually-beneficial outcomes. The medium of communication affects how trust is built and shapes expectations about others' behavior, as it directly influences the interaction process. Therefore, this paper examines the emotional trajectories of entrepreneurs in virtual versus in-person settings as well as how interactions with investors shape these trajectories.

Specifically, because in-person pitches involve direct interaction between entrepreneurs and investors, we hypothesize that entrepreneurs pitching in-person display a broader range of positive emotions and more frequent emotional switching in response to investor engagement. In contrast, virtual pitches may involve more constrained emotional expressions due to limited interpersonal cues. We also hypothesize that interactions with investors shape the trajectory of emotional development differently across formats. In in-person settings, the further investors' questions deviate from the pitch, the greater the expected range of emotion and frequency of emotional switching, compared to virtual settings. In virtual contexts, emotions are more difficult to convey, prompting entrepreneurs to make additional efforts to vary their emotional displays in order to communicate their feelings clearly and effectively.

Consequently, since prior research has largely examined emotional expressions as isolated events in in-person settings, little is known about how entrepreneurs' express emotions in virtual versus in-person environments, and how these emotional dynamics influence funding outcomes. To address this, we examine entrepreneurial pitches as an empirical context and ask: How do entrepreneurs express their emotions differently in in-person and virtual venture pitches, and why do these differences matter for funding decisions? Addressing this question advances our understanding of emotional interactions in high-stakes communication and highlights the contextual nature of emotional expression in entrepreneurship.

METHOD

To examine how entrepreneurs' express emotions during virtual and in-person venture pitches, we analyzed video data from startup pitch competitions conducted between 2020 and 2024 in Europe and North America.

Here are the details of our data:

Qualitative quantification
Sample: 28 participants, 9,171 emotional states
Context: video recorded startup competitions in Europe and North America
Groups: 14 in-person and 14 virtual
Variables: emotional states, funding success, gender
Coding method: OpenFace 2.0 (Baltrušaitis et al., 2016), Facial Action Coding System (FACS; Ekman et al., 2002)
Data analysis: Rates of change (George & Jones, 2000; Lambert & Jiang, 2025), Markov chain analysis (Feller, 2008; Jiang et al., 2021), Sequential Pattern Mining (Giuntini et al., 2021; Huang et al., 2024), and Hidden Markov Model (Eddy, 2004; Quan & Ren, 2016)

We used OpenFace 2.0 (Baltrušaitis et al., 2016) and the Facial Action Coding System (FACS; Ekman et al., 2002) to identify frame-by-frame emotional states from facial expressions during the pitch. Our data analysis entailed four steps and compared their dynamics across pitch conditions, outcomes and gender. First, we examined the rates of change of emotions (George & Jones, 2000;

Lambert & Jiang, 2025) to capture how frequently entrepreneurs shifted between emotional states during their pitches. Second, we applied Markov chain analysis (Feller, 2008; Jiang et al., 2021) to examine one-step transitions between emotions (e.g., happiness to contempt). Third, using a Sequential Pattern Mining (SPM) approach (Giuntini et al., 2021; Huang et al., 2024), we identified frequently recurring emotional sequences of more than two transitions (e.g., surprise to anger to surprise). Finally, we used Hidden Markov Models (HMM; Eddy, 2004; Quan & Ren, 2016) to uncover latent emotional profiles that represent underlying affective modes driving observable facial emotion sequences.

RESULTS

Rates of change in emotional expression

We found no difference in the overall rates of change of emotions between in-person and virtual pitches (12 emotion transitions per minute in both formats). However, breakdowns by investment outcome and gender revealed meaningful variation. In virtual settings, entrepreneurs who secured investment showed fewer emotion changes per minute (10) than those who did not (14), suggesting that a more stable emotional profile may be advantageous in virtual settings. In in-person settings, rates of change were similar between invested and not-invested entrepreneurs. Female entrepreneurs exhibited the highest rate of emotional change in in-person settings (15), while male entrepreneurs showed fewer transitions (10), indicating greater emotional expressiveness among women. In virtual settings, rates of change were similar across genders.

Markov chain analysis (one-step transitions)

Across both pitch formats, the most frequent transition was from happiness to contempt, suggesting a common emotional pivot from enthusiasm to skepticism during the pitch process. The transition from fear to surprise was also consistently present in both in-person and virtual formats, indicating a shared emotional pathway linked to high-stakes moments. Other transitions, such as anger to surprise, sadness to surprise, and neutral to surprise, appeared primarily in virtual settings, pointing to more varied emotional dynamics.

Sequential Pattern Mining (more than one-step transitions)

Our analysis revealed that emotional sequences during pitches rarely formed long chains of distinct emotions. Instead, these sequences tended to oscillate between two emotions, with long patterns repeating. The most frequent oscillation in in-person pitches involved transitions between contempt and disgust, reflecting a repeated return to critical or evaluative emotional states. In contrast, virtual pitches more often featured transitions between surprise and anger, indicating more reactive or fluctuating emotional displays. When examined by investment outcome, the oscillation was the same between in-person and virtual pitches. The oscillation between contempt and disgust was more common among entrepreneurs who secured investment, whereas the oscillation between contempt and happiness appeared more frequently among those who did not, suggesting a potential association between affective tone and investment success. Gender differences were more pronounced. Female entrepreneurs more often exhibited contempt with happiness in-person and anger and surprise virtually, while male entrepreneurs showed contempt and disgust in-person and contempt and happiness virtually.

Hidden Markov Modeling

The HMM identified three latent emotional profiles, each defined by distinct patterns of observed facial expressions. Profile 1, the controlled profile, was characterized by anger, disgust, and neutral expressions, with very low levels of happiness and surprise. This suggests an expressive style marked by restraint or low emotional intensity. Profile 2, the evaluative profile, was dominated by contempt and sadness, with negligible levels of other emotions, pointing to a more critical or judgmental expressive pattern. Profile 3, the expressive profile, showed the highest levels of surprise and happiness, along with moderate levels of disgust and sadness, reflecting a more emotionally dynamic and outward display. These profiles appeared in similar proportions across in-person pitches, while virtual pitches showed a greater prevalence of Profile 3 (39%) compared to Profile 1

(31%) and Profile 2 (30%). This suggests that entrepreneurs in virtual settings were more likely to engage in expressive emotional displays.

DISCUSSION AND IMPLICATIONS

Discussion

Our findings show that the emotional strategies entrepreneurs adopt during pitches are strongly shaped by the context of communication. In-person pitches, with direct investor interaction, create opportunities for immediate feedback and emotional responsiveness. In contrast, virtual pitches, while lacking such cues, appear to prompt compensatory strategies, entrepreneurs become more outwardly expressive to ensure their emotions are seen and felt. Interestingly, while emotional expression is often thought to be constrained online, we found that entrepreneurs pitching virtually often adopted more expressive emotional profiles. At the same time, those who demonstrated greater emotional stability in virtual settings were more likely to secure investment, suggesting that clarity and consistency may be especially persuasive when social feedback is limited. These contextual insights may also help explain dissonant findings in prior research. Studies based on in-person pitch settings, such as Shark Tank or Dragons' Den (Pollack et al., 2012; Blaseg & Hornuf, 2024) and live angel investor events (Huang & Pearce, 2015), highlight the importance of charisma, preparedness, and nonverbal presence, often linked to positive emotions. In contrast, studies using virtual, pre-recorded crowdfunding contexts like Kickstarter (Jiang et al., 2019; Warnick et al., 2021; Huang et al., 2023) tend to emphasize emotional clarity, modulation, and the strategic use of variation. These divergent findings suggest that the communication medium may shape which emotional strategies are most effective, pointing to the need for more research that explicitly compares affective dynamics across pitch formats.

Second, this study contributes to the dynamic perspective on entrepreneurial emotions, emphasizing how emotions evolve as sequences over time. Rather than treating emotions as discrete events (Ekman, 1992, 1997), our analysis uncovers how entrepreneurs transition between emotional states, and enact distinct expressive strategies, aligning with the theory of constructed emotions (Barrett, 2006). Sequential patterns (Giuntini et al., 2021; Huang et al., 2024), such as oscillations between surprise and anger in virtual pitches, or between contempt and disgust in in-person ones, reveal pitch-specific emotional trajectories. Our application of Hidden Markov Models (Eddy, 2004; Quan & Ren, 2016) further identifies latent profiles of emotional expression, capturing underlying emotional modes that drive observable behavior. These dynamics suggest that investors respond not only to how emotions are expressed, but also to how emotions unfold over time. Gender differences observed in emotional sequences add nuance, indicating that emotional norms may be interpreted differently across pitch contexts.

Finally, our study demonstrates the value of interdisciplinary methods, specifically by importing computational tools used for emotion analysis in computer science to advance emotion research in entrepreneurship. By applying facial recognition software (OpenFace 2.0; Baltrušaitis et al., 2016), Sequential Pattern Mining (Giuntini et al., 2021; Huang et al., 2024) and Hidden Markov modeling (Eddy, 2004; Quan & Ren, 2016), we were able to track emotional dynamics at scale and with fine granularity. These methods allow researchers to move beyond static snapshots and self-reported emotions, offering new insight into how emotions function in high-stakes interactions. As interest grows in the role of emotions in organizational behavior and management science, we advocate for deeper integration of computational techniques that can uncover the subtle, complex, and sequential nature of emotion in entrepreneurial settings.

Practical implications

This research advances theories of emotional dynamics in entrepreneurship and offers practical insights for enhancing pitch effectiveness. Entrepreneurs can use these findings to better tailor their emotional strategies to the pitch format. In virtual settings, where social cues are limited, maintaining emotional clarity and stability may be particularly persuasive. In contrast, in-person settings may reward more adaptive emotional transitions in response to investor engagement. For investors, this research provides a more nuanced framework for interpreting emotional cues across different contexts, moving beyond charisma-based judgments toward a more informed reading of emotional patterns.

Limitations and future research

While pitch competitions offer valuable insights into entrepreneurial emotion, their time-bound nature limits our understanding of longer-term emotional dynamics. This study also focuses on non-verbal facial cues, excluding vocal tone and verbal content that may enrich emotional analysis. In the following study, we will test the effects of the three emotional profiles experimentally by varying pitch format (in-person vs. virtual), using actors or controlled videos. This design would isolate how profile effectiveness varies by context, offering causal insight into emotional strategies and their impact on funding outcomes.

CONTACT: Denis Lambert, denis.lambert@edu.escp.eu, ESCP Business School, 3, rue Armand Moisant 75015 Paris, France

Figure 1
Findings

Rates of change: changes in emotions / minute

	Total	Invested	Not invested	Female	Male
In-person	12	12	12	15	10
Virtual	12	10	14	12	12

Markov tables (1 transition)

	Case 1	Case 2	Case 3	Case 4	Case 5
In-person	Happiness -> contempt	Fear -> surprise	n/a	n/a	n/a
Virtual	Happiness -> contempt	Fear -> surprise	Anger -> surprise	Sadness -> surprise	Neutral -> surprise

Sequential Pattern Mining (from 2 to n transitions, starting at any emotions)

	Total	Invested	Not invested	Female	Male
In-person	Contempt < > disgust	Contempt < > disgust	Contempt < > happiness	Contempt < > happiness	Contempt < > disgust
Virtual	Anger < > surprise	Contempt < > disgust	Contempt < > happiness	Anger < > surprise	Contempt < > happiness

Hidden Markov Model - Multinomial states

Emotion	Profile 1	Profile 2	Profile 3
Anger	0.00	0.00	0.00
Contempt	-2.85	3.17	-2.22
Disgust	-0.56	-0.47	1.81
Fear	-1.21	-8.46	-7.81
Happiness	-1.91	-8.95	2.51
Neutral	-0.53	0.09	-0.72
Sadness	-1.77	2.23	1.69
Surprise	-10.60	-0.69	3.65

Hidden Markov Model - Settings

Emotion	Profile 1	Profile 2	Profile 3
In-person	0.33	0.35	0.32
Virtual	0.31	0.31	0.39

Hidden Markov Model - Investment

Emotion	Profile 1	Profile 2	Profile 3
Invested	0.37	0.29	0.34
Not-invested	0.25	0.38	0.38

Hidden Markov Model - Gender

Emotion	Profile 1	Profile 2	Profile 3
Female	0.32	0.32	0.36
Male	0.32	0.33	0.35

HOW WORKING HOURS REGULATION INFLUENCES ENTREPRENEURS' SUBJECTIVE WELL-BEING

Sanggeun Lee, Syracuse University, United States
Johan Wiklund, Syracuse University, United States

ABSTRACT

This study examines how working-hour regulations intended for employees influence entrepreneurs' subjective well-being. Drawing on the Job Demands-Resources model, we argue that restricting employee hours imposes resource constraints, increasing entrepreneurial stress to entrepreneurs. Using longitudinal data from Korean entrepreneurs, we find that working-hour regulations negatively affect entrepreneurs' SWB immediately and exhibit significant yet inconsistent long-term effects. Our findings shift the focus from individual to institutional determinants of entrepreneur's subjective well-being, highlighting unintended consequences of policies intended to benefit employees and argue policymakers to consider entrepreneurial contexts when designing regulations.

INTRODUCTION

Recent research highlights significant growth in exploring entrepreneurs' subjective well-being (SWB), traditionally focusing on individual-level factors such as personal traits, workplace demands, and job position (Baron et al., 2016; Nikolova et al., 2023; Shir et al., 2019; Tahar et al., 2023). However, this study advances the literature by shifting the analytical lens toward an institutional-level perspective, recognizing that well-being is profoundly shaped by external conditions—such as regulatory frameworks and policy interventions—that define the operational environment for entrepreneurs and their ventures (Baumol, 1996; Stephan & Uhlaner, 2010). Entrepreneurship inherently involves navigating complex institutional challenges, including high uncertainty (Knight, 1921), and demands stemming from legal and economic structures (Baumol, 1996), which collectively influence mental and physical health, business outcomes, and overall life satisfaction (Stephan, 2018; Wiklund et al., 2019).

Moreover, recent studies have indicated that entrepreneurs are not a homogeneous group (Chatterjee et al., 2022), emphasizing the need to explore the diversity in their responses to institutional constraints (Stephan, 2018). Regulatory changes designed to benefit employees, such as working-hour restrictions, can inadvertently create unforeseen challenges for entrepreneurs by intensifying workload demands, reshaping resource allocations, and complicating work-home balance. Policies limiting employee working hours, while protective for workers, may unintentionally heighten stress and reduce SWB among entrepreneurs by increasing managerial burdens. To systematically analyze these dynamics, this study employs the Job Demands-Resources (JD-R) model—a well-established framework capturing how different working conditions affect stress, motivation, and well-being (Bakker & Demerouti, 2007; Demerouti et al., 2001). Central to the JD-R model is the dynamic interplay between demands and resources, where even minor shifts—such as reductions in available resource—can disrupt balance, leading to decreased engagement and increased stress (Bakker & Demerouti, 2007; Bakker, Demerouti, & Verbeke, 2004). Applying this model allows us to precisely examine how externally imposed institutional demands intersect with entrepreneurial resources, elucidating why institutional-level policies generate divergent well-being outcomes across different entrepreneurial contexts.

HYPOTHESES DEVELOPMENT

Drawing on the JD-R model (Bakker & Demerouti, 2007; Demerouti et al., 2001), we argue that working-hour regulations limiting employee availability effectively constrain critical resources for entrepreneurs, leaving their job demands unchanged. Employees represent essential resources for entrepreneurs (Hancock et al., 2013), and regulatory restrictions reducing employee working hours significantly constrain entrepreneurs who rely heavily on their workforce to manage operational demands.

This institutional shift directly elevates entrepreneurs' workload and managerial pressures, as they must rapidly redistribute tasks (Baker et al., 2005), reorganize workflows (Kitching et al., 2015), or personally compensate for the reduced labor input (Stroe et al., 2018) to sustain business. Consequently, entrepreneurs experience heightened psychological strain, negatively influencing their SWB (Wiklund et al., 2019). Even subtle constraints in critical resources, such as employee availability, can substantially amplify entrepreneurs' job stress and reduce their SWB (Stephan et al., 2023). Therefore, we hypothesize:

Hypothesis 1: Entrepreneurs subjected to employees' working-hour restrictions will experience an immediate decrease in subjective well-being compared to entrepreneurs not subjected to these restrictions.

Nevertheless, the JD-R model also emphasizes individuals' capacity to adapt over time, deploying personal and organizational resources to mitigate adverse effects of increased job demands (Bakker & Demerouti, 2007). Entrepreneurs, specifically known for their resilience (Yu et al., 2022) and adaptability (Stephan et al., 2021), are likely to implement coping strategies deal with business domain stressors (Bullough & Renko, 2013; Yu et al., 2022). Entrepreneurs are not only capable of managing work-related stressors, but they also adopt alternative strategies—such as engaging in resource-sharing activities (Istipliler et al., 2023), seizing opportunities (Manolova & Yan, 2002), and applying bricolage techniques (Desa, 2012)—to overcome institutional constraints (Simarasl et al., 2022; Stephan et al., 2021) and enhance their performance (Schindehutte & Morris, 2001).Thus, while entrepreneurs initially experience significant stress and reduced SWB due to resource constraints, these effects should gradually weaken as they adapt to the new regulatory environment. Therefore, we hypothesize:

Hypothesis 2: Working-hour regulation constraints exert a long-term negative impact on entrepreneurs' subjective well-being that diminishes over time.

METHOD

Setting, Data, and Sample

Prolonged working hours have been a significant issue in South Korea, consistently exceeding 2,000 hours annually per employee since 2008—significantly higher than the OECD average (1700–1800 hours). Addressing concerns about health and work-life balance, the Korean government introduced a policy in 2018 limiting weekly working hours from 68 to 52. Implementation varied by firm size: firms with over 300 employees in July 2018, 50–300 employees in 2020, and 5–50 employees in 2021. This study focuses specifically on entrepreneurs with 5–50 employees as the treated group, comparing them with entrepreneurs employing between 1–5 workers (control group).

We employed data from the Korean Labor Income Panel Study, a longitudinal dataset initiated in 1998, recognized internationally for its consistency and reliability. KLIPS tracks comprehensive economic, social, and demographic information from Korean households through both in-person and telephone interviews. This study utilizes data from 2017–2023, comprising 195 observations for entrepreneurs with 5–50 employees and 475 observations for entrepreneurs with 1–5 employees.

Measurements

Dependent Variable. *Subjective well-being* was measured using a single-item global life satisfaction question from the World Values Survey (Inglehart et al., 2014). Respondents rated their overall life satisfaction on a 5-point scale ranging from "very dissatisfied" (1) to "very satisfied" (5). The 5-point scale was chosen to minimize respondent burden while effectively capturing variability. This single-item measure is widely validated, correlates strongly with multi-item scales such as the Satisfaction with Life Scale and is internationally recognized for its reliability and validity (Andrews & Withey, 2012; Abdel-Khalek, 2006; Cheung & Lucas, 2014).

Control Variables. To test the effect of the regulation on entrepreneurs' and employees' SWB, we controlled eight different variables. Based on the life satisfaction and well-being literature we included *financial asset*

(D'Ambrosio et al., 2020) *financial income* (Clark et al., 2008), *financial debt* (Ferreira et al., 2021), *number of children* (Edwards & Klemmack, 1973), *firm size* (Fors Connolly et al., 2021), *socioeconomic status* (Navarro-Carrillo et al., 2020; Liu & Liu, 2023; Huang et al., 2017), *physical exercise* (Puce et al., 2023), and *job satisfaction* (Claes et al., 2023). The descriptive summary for each variable is shown in Table 1.

Econometric Models and Strategy

We utilized a differences-in-differences (DiD) approach, a standard empirical method to assess policy impacts by comparing treatment and control groups before and after policy implementation. DiD effectively controls for time-invariant unobserved factors, thereby addressing common endogeneity concerns in policy evaluations and strengthening causal inference. Recognizing that policy effects on SWB may take time to manifest, we incorporate multiple time periods after policy enactment. Including these periods allows us to capture immediate, delayed, and persistent effects. We employed two econometric specifications:

$$SWB_{it} = \lambda_i + \tau_t + bPost_t + \delta\,(Treat_i \times Post_t) + gX_{it} + e_{it} \tag{1}$$

Equation (1) is the standard DiD model with control variables, where it shows the general effect of the regulation on the treated group. In equation (1), i represent individuals, and t represents the year. SWB_{it} is the SWB of individual i at time t. λ_i is the individual fixed effects that control for all time-invariant characteristics of each entrepreneur. τ_t denotes the time fixed effects that capture common shocks or trends affecting all entrepreneurs in period t, such as macroeconomic conditions or widespread regulatory changes not unique to the treatment. $Treat_i$ is a binary variable equal to 1 if entrepreneur i is in the treatment group (i.e., affected by the 2021 regulation), and 0 otherwise. $Post_t$ is a binary variable equal to 1 for the post-regulation period and 0 for the pre-regulation period. The key variable $Treat_i \times Post_t$ is the interaction term capturing the differential change in SWB for the treatment group after the regulation. X_{it} is a vector of control variables that account for other factors influencing SWB. Lastly, e_{it} is the error term.

Also, known as the dynamic DiD model, equation (2) measures the long-term effect of the regulation. Here is the following equation:

$$SWB_{it} = \lambda_i + \tau_t + \textstyle\sum_{k\geq 0} \delta_k\,(Treat_i \times D_{t=k}) + gX_{it} + e_{it} \tag{2}$$

λ_i represents individual fixed effects to control for time-invariant characteristics of each entrepreneur. τ_t represents the time fixed effects to control for shocks that affect all entrepreneurs in a given period. $D_{t=k}$ is a dummy variable for each time period k relative to regulation. X_{it} denotes the control variables that has been included in the model and e_{it} is the error term. Each coefficient δ_k reflects the change in SWB for the treatment group relative to the control group in period k compared to the base period. This specification allows the model to trace the temporal dynamics of the regulation's impact, revealing whether the effect on SWB is immediate, delayed, or persistent over time.

RESULTS

Before presenting the main DiD results, it is crucial to confirm the parallel trends assumption, which ensures that the treated and control groups follow similar outcome trends before policy implementation (Roberts & Whited, 2013). Specifically, we employ an event study to assess whether entrepreneurs with over 5 and fewer than 50 employees, and those with 1 to 5 employees, exhibited comparable SWB trends before the working-hour regulation. Figure 1 displays the estimated yearly coefficients on SWB with 95% confidence intervals. Points before the treatment year are close to zero, and confidence intervals consistently include zero, indicating negligible differences between the groups and validating the parallel trends assumption. Additionally, Model 3 from Table 2 confirms these pre-treatment effects as non-significant, further supporting comparability of groups in the pre-regulation period. The main regression results are shown in Table 2. Models 1 and 2 report the average treatment effect without and with year fixed effects, respectively, derived from equation (1). Results indicate the treated group's SWB decreased by 0.73 points compared to the control group due to the regulation change (b = -0.73, p < 0.01), supporting H1, which predicted reduced SWB for affected entrepreneurs. Model 3 illustrates long-term effects from equation (2), demonstrating that from 2021 onward, the treated group consistently exhibited significantly lower SWB than controls ($b_{2021} = -1.29$, $p < 0.01$; $b_{2022} = -0.82$, $p < 0.05$; $b_{2023} = -1.09$, $p < 0.01$). However, no evidence suggests a diminishing effect over time. Thus, Hypothesis 2 is only partially supported.

DISCUSSION AND IMPLICATIONS

This study contributes to entrepreneurship and SWB research by shifting attention from individual-level determinants toward institutional-level influences, particularly working-hour regulations toward employees. While prior studies primarily emphasized personal characteristics affecting entrepreneurs' well-being (Baron et al., 2016; Nikolaev et al., 2020), our findings highlight how regulatory frameworks initially designed to enhance employee welfare can significantly and negatively influence entrepreneurs' SWB (Baumol, 1996; Stephan & Uhlaner, 2010). Specifically, working-hour restrictions immediately reduce entrepreneurs' SWB after implementation, illustrating unintended consequences across occupational contexts. Second, while previous research on the JD-R model has primarily focused on employees (Bakker & Demerouti, 2017; Zeng & Hu, 2024), our study demonstrates that regulatory changes act as institutional-level constraints impacting entrepreneurs. These regulations reduce managerial flexibility and increase workload demands by limiting the human resources available from employees. Consequently, this leads to heightened stress among entrepreneurs and potentially lower subjective well-being, illustrating how external policy interventions interplay with internal job demands and resources in entrepreneurial settings. Third, our study finds that regulatory constraints—such as limits on working hours—not only reduce entrepreneurs' subjective well-being in the short term but also have long term effect by increasing managerial demands. While previous studies have focused on immediate effects on entrepreneurs' well-being (Hahn et al., 2012; Larsson &Thulin, 2019), our study demonstrates how entrepreneurs' subjective well-being evolves over time (Wiklund et al., 2019). However, the absence of a consistent SWB trajectory underscores the need for future research into specific adaptive strategies, such as dynamic capability (Eisenhardt & Martin, 2000) employed by entrepreneurs facing institutional constraints. Practically, policymakers should consider both immediate and uncertain longer-term impacts of working-hour regulations on entrepreneurs. Given the initially negative consequences and unclear long-term outcomes, regulatory policies should be coupled with targeted support mechanisms—such as managerial training or transitional assistance—to help entrepreneurs effectively manage institutional changes and sustain SWB over time.

CONTACT: Sanggeun Lee; slee258@syr.edu; Syracuse University; 721 University Ave Suite 311, Syracuse, NY, 13244, USA

Table 1. Descriptive summary and correlation table.

Variables	Mean	S.D	1	2	3	4	5	6
(1) Financial Asset (Logged)	8.40	1.27						
(2) Financial Income (Logged)	6.25	0.65	0.29***					
(3) Financial Debt (Logged)	9.50	1.21	0.10***	0.20***				
(4) Children	0.69	0.91	-0.09***	0.14***	0.04**			
(5) Firm Size	4.21	5.13	0.14***	0.23***	0.13***	-0.01		
(6) Socioeconomic Status	3.69	0.81	-0.22***	-0.34***	-0.20***	-0.05***	-0.20***	
(7) Physical Exercise	13.12	7.09	0.05*	-0.04	-0.08**	-0.06**	-0.01	-0.11***
(8) Job Satisfaction	2.42	0.62	-0.13***	-0.36***	-0.08***	-0.08***	-0.17***	0.27***
(9) Subjective Well-being	7.09	1.34	0.19***	0.28***	0.09***	0.07***	0.15***	-0.25***

Note. *** $p<0.01$, ** $p<0.05$, * $p<0.1$

Table 2. DiD regression analysis.

	Subjective Well-being		
	Model 1	Model 2	Model 3
Overall DiD	-0.54 (0.20)***	-0.73 (0.24)***	
DiD (Base = 2017)			
2018			-0.32 (0.30)
2019			-0.25 (0.36)
2020			-0.35 (0.37)
2021			-1.29 (0.41)***
2022			-0.82 (0.40)**
2023			-1.09 (0.41)***
Financial Asset (Logged)	-0.06 (0.06)	-0.07 (0.06)	-0.07 (0.06)
Financial Income (Logged)	0.18 (0.15)	0.19 (0.15)	0.21 (0.15)
Financial Debt (Logged)	0.12 (0.09)	0.10 (0.09)	0.11 (0.09)
Children	0.11 (0.12)	0.13 (0.22)	0.12 (0.22)
Firm Size	0.04 (0.03)	0.04 (0.03)	0.01 (0.30)
Socioeconomic Status	-0.33 (0.12)***	-0.34 (0.12)***	-0.33 (0.12)***
Physical Exercise	0.005 (0.01)	-0.01 (0.01)	-0.01 (0.01)
Job Satisfaction	-0.36 (0.12)***	-0.38 (0.12)***	-0.38 (0.12)***
Constant	7.25 (1.45)***	7.54 (1.47)***	7.13 (1.46)***
Year Fixed Effect	N	Y	Y
Within R^2	0.11	0.12	0.14
F-statistics	4.33***	3.28***	2.75***

Note. *** $p<0.01$, ** $p<0.05$, * $p<0.1$
Unstandardized coefficient reported; Standard errors in parentheses

Figure 1. Parallel trends test.

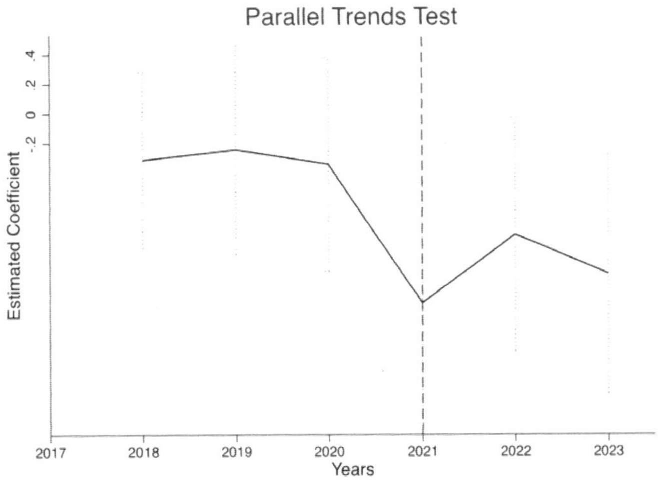

THE IMPACT OF LEISURE CRAFTING ON ENTREPRENEURS' SERENDIPITY

Brindha S. Srinivasan, Indian Institute of Management, Kozhikode
M.K. Nandakumar, Indian Institute of Management, Kozhikode
Alex Hamrick, Robins School of Business

ABSTRACT

Entrepreneurs' work and life design are given much less attention in the literature, which is quite surprising because the autonomy to design work is one of the major reasons individuals enter entrepreneurship. How entrepreneurs design their work decides the growth of the venture (Hsu et al., 2021). How they craft their time for leisure impacts their work and helps them and how they handle stress (Hamrick, 2022; Hamrick et al., 2023). However, our understanding of why and how entrepreneurs craft their work and leisure is very limited. So, in this paper, we explore crafting in an entrepreneurial context and how it impacts their personal and venture outcomes. We first conduct a qualitative study to understand when, why, and how entrepreneurs craft their work and leisure. The interviews revealed a prominent theme of how leisure crafting helps them at work. Diving into the data and literature, we develop a model of how leisure crafting helps entrepreneurs identify serendipitous instances at work. In the next study, we test this model empirically by conducting a time-lagged survey in three instances with a gap of two weeks between each survey. Our paper contributes to the literature on entrepreneurial serendipity by identifying leisure crafting as a proactive behavior and enabler. We also theorize on how leisure work spillover happens and how entrepreneurs can use it to their advantage. We contribute to the work and leisure design literature by placing entrepreneurship as a unique career context and theorizing on the various enablers and barriers they face.

INTRODUCTION

Entrepreneurs' work is characterized as uncertain and hectic in general (Stephan, 2018). However, it is quite surprising that the work design of entrepreneurs has received limited attention in the literature (Baron, 2010). Autonomy and work flexibility are important motivations for individuals to enter entrepreneurship. *But are they able to design their work and life as they wish? What are the different aspects affecting this? How does it impact personal and venture outcomes for them?*

So, in this paper, *we explore how entrepreneurs craft their time and how this impacts them.* Study 1 is an *exploratory qualitative study* to understand entrepreneurs' work and leisure crafting. We also look at how this helps them personally and at work. Through semi-structured interviews with *42 founders, we understand why and how founders craft.* They *have higher degrees of freedom than employees* and different kinds of enablers and barriers in crafting. The interview study led us to explore a nuanced outcome. During the interviews, *we repeatedly found instances of how engaging in leisure crafting benefits their work.* When founders actively craft their leisure time, it helps them identify things of value to their work quite unexpectedly. This led us to delve deeper into the literature on crafting and serendipity. *Using the findings from the interviews and the theoretical insights gained by exploring the literature, we develop a model of how leisure crafting can enable serendipity for entrepreneurs. We intend to test this model in Study 2 through a time-lagged survey by collecting data at three-time points with a gap of two weeks in between.*

We contribute to the work and life design literature by exploring the unique working context of entrepreneurs and the different enablers and barriers of crafting. We identify potential crafting outcomes for entrepreneurs in their personal and work domains. This study also contributes to the serendipity literature by identifying a unique proactive behavior of entrepreneurs that enables them to identify serendipitous instances.

This is a mixed-method study to understand the phenomena of crafting in entrepreneurs. Study 1 is a qualitative study that broadly explores both work and leisure crafting and its outcomes. In the interest of space, we do not elaborate on the qualitative study. We develop a theory around leisure crafting and serendipity from the interview data and literature. Study 2 is a quantitative time-lagged survey-based study. Partial data collection of the quantitative model is done. We are collecting data further. We analyze the data, present the results, and discuss the results.

HYPOTHESIS DEVELOPMENT

Leisure crafting and serendipity

Leisure crafting refers to deliberate and proactive engagement in leisure activities to achieve specific objectives, fostering social relationships, acquiring knowledge, and fostering personal development (Petrou & Bakker, 2016). Leisure crafting is shown to impact opportunity recognition and improve venture performance by enhancing thriving at work for entrepreneurs (Hamrick et al., 2023). It also enhances entrepreneurs' creativity and helps them manage their hindrance stressors better (Hamrick, 2022). Our interview data gave us insights into how engaging in leisure crafting can enhance serendipity. Serendipity, the concept of discovering unexpected and useful discoveries, significantly contributes to the success of both individuals and societies (Busch, 2024). Serendipity is facilitated by chance and is inherently unpredictable (Fultz & Hmieleski, 2021). Serendipity is the pleasant surprise that comes from the unexpected ways our choices and deeds pay off. Three necessary conditions of serendipity are agency, surprise, and value (Busch, 2024). The presence of agency is an essential precursor of serendipity. Leisure crafting, as an agentic behavior, has the potential to inspire serendipity. Instead of simply relaxing, leisure crafting entails consciously and intentionally designing leisure activities to acquire new skills, build new connections with valued others, and grow and develop (Petrou & Bakker, 2016).

Agency, surprise, and value are the three necessary conditions for the agency (Busch, 2024). Recreational pursuits are not deliberately oriented towards meeting work goals but are generally followed for inherent motivations. They allow individuals to psychologically detach from work and dedicate time to hobbies unrelated to their occupation (Sonnentag, 2012). It necessitates individuals to manifest initiative, agency, and proactivity to establish circumstances for realizing one's identity and cultivate a feeling of purpose. Therefore, it is more dynamic and developmental than other leisure activities, as it is motivated by individuals' inherent aspirations for self-actualization, personal progress, and fulfilling needs (Berg et al., 2010). As leisure crafting activities are not directed towards work-related aspects, finding something of value related to work will be a surprise. Surprise is a necessary condition of serendipity. And as leisure crafting does not aim to look for something of value in work, it will be a surprise if something of value comes up by engaging in these activities.

One entrepreneur discusses how she plans her time out for other activities. Though this feels like a relaxing and social activity, it helps her meet people who are somehow helpful for her work.

> *"I go to a lot of events, parties and gatherings. Not related to my work. I am a very social person. I enjoy meeting and talking to new people. But, in this kind of activity, I have met many people who have become my clients or collaborators later on. You never know whom you will meet." - Amrita*

H1: Leisure crafting is positively related to serendipity.

Leisure crafting leads to enabling qualities of serendipity

In this paper, we argue that leisure crafting leads to enabling qualities of serendipity. Busch (2024) identified four individual-level enabling qualities in a recent review on serendipity. These qualities are cognitive flexibility, perseverance, social skills, and self-efficacy. We theorize, using literature and data, on how these qualities mediate the relationship between leisure crafting and serendipity.

Role of Self-efficacy

Self-efficacy is an individual's judgment of their competence for performing in various circumstances (G. Chen et al., 2004). Individuals can experience heightened autonomy and influence over their lives through deliberate and strategic leisure time planning. The autonomy and capacity to shape experiences might enhance self-efficacy, as individuals interpret themselves as capable of influencing results in different domains of life (Garvis et al., 2011; Narayanan et al., 2023). Engaging in leisure activities is also a way to achieve their goals and an opportunity to learn and grow. This will improve one's self-efficacy (Schunk, 1990). Leisure crafting has the potential to elicit good emotional experiences, which are associated with increased self-efficacy. The experience of positive emotions can enhance psychological resources, therefore increasing persons' sense of competence in addressing obstacles encountered in their personal and business matters (Ouweneel et al., 2013; Villavicencio & Bernardo, 2016).

An entrepreneur discusses how the founders and employees spend time every day learning new things. This helps them to upskill themselves and improve their confidence and self-efficacy.

> *"We learn something new every day for two hours or so. It makes us happy and greatly improves our confidence. We eventually become better at what we do. And a lot of new things come up during this process." - Aditya*

H2a: Leisure crafting is positively related to self-efficacy.

Self-efficacy is an enabling quality of serendipity (Busch, 2024). So, we argue that self-efficacy will mediate the relationship between leisure crafting and serendipity. H2b: Self-efficacy mediates the relationship between leisure crafting and serendipity.

Role of Cognitive Flexibility

Cognitive flexibility is identified as one of the enabling qualities of serendipity. Engaging in leisure crafting will help enhance the cognitive flexibility of entrepreneurs. The degree to which people can think and respond flexibly to various points of view is known as cognitive flexibility (De Dreu et al., 2008). People aim to acquire new knowledge through leisure crafting, broaden their understanding (Berg et al., 2010), and seek challenges to evaluate their capabilities (Deci & Ryan, 2000). Consequently, they are driven to establish and strive for new, elevated objectives. This may lead them to strive to reconfigure the scope of their recreational pursuits and expand their perception (I.-S. Chen & Choi, 2024).

One founder of a FinTech company discusses how he spends time learning random stuff every day. And this has helped improve his thinking and approach a lot.

> *"I spend around one to two hours every morning, learning various things. I either read or listen to podcasts or watch documentaries. Over time, I have realized how important this is. It has helped me finetune my thinking and has helped me develop multiple perspectives. It has been immensely useful at work too." – Abhay*

We see that crafting leisure helps them be more flexible with their thinking. So we hypothesize as follows: H3a: Leisure crafting is positively related to cognitive flexibility. Cognitive flexibility is an enabling quality of serendipity (Busch, 2024; Laureiro-Martínez & Brusoni, 2018). So, we argue that cognitive flexibility will mediate the relationship between leisure crafting and serendipity. H3b: Cognitive flexibility mediates the relationship between leisure crafting and serendipity.

Role of Perseverance

Perseverance is the capacity to persist in the face of challenges, impediments, or failures. It is the resolve to remain dedicated to an objective, despite slow progress or emerging hurdles (Duckworth et al., 2007). By intentionally incorporating challenging elements into their leisure pursuits, individuals can fulfill their intrinsic need for personal development. This, in turn, cultivates a heightened sense of responsibility and perseverance, as success in leisure activities reinforces their ability to persist through difficulties in other domains of life (Tsaur et al., 2023). While these are domain-specific, similar to self-efficacy, there are signs that people will exhibit greater perseverance when they successfully manage difficult situations in one area (Bunea et al., 2023).

Entrepreneurs discuss how crafting their off-work and leisure time helps them persevere and perform well at work. An entrepreneur discusses how she needs to have a balanced life in order to do her work better. She talks about how she separates and views her life as parts and how having a balance is essential for her to be calm and persist at work.

"I view my life as work, family, personal and spiritual blocks. Of course, I spend a little more time on work in general. I need overall balance in life. Otherwise, I will not be able to work properly and perform well. If there is no balance in my life, say, I have been too involved in work, I become very cranky and unproductive every few weeks. Then, I take a space, work on other aspects, and return to work. But I mostly take care that this does not happen often. So that I can persevere and be good at work." `Aarohi

Thus, we arrive at the hypothesis, H4a: Leisure crafting is positively related to perseverance. Perseverance is an enabling quality of serendipity (Busch, 2024). So, H4b: Perseverance mediates the relationship between leisure crafting and serendipity.

Role of Social Skills

Social skills are the competencies that provide effective interaction, communication, and relationship-building with others. Pursuing human connection is one of the main aspects of leisure crafting (Petrou & Bakker, 2016). Engaging in recreational activities is a potent method for fostering social capital (Blackshaw & Long, 2005; Tsaur et al., 2023). Taking leisure activities seriously and continuously participating in them is seen to improve the social capital of leaders (Bunea et al., 2023). They might engage in different kinds of activities during leisure crafting. And when these activities are group activities or sports, which in any way involve other people, they improve their social skills. Building relationships is a common way for individuals to craft their leisure time, increasing their communication with people.

Another entrepreneur discusses how random ideas of value occur during the most unexpected discussions and how useful this is for him. He practices having multiple breaks to relax and reconnect with the team. And they also have dedicated time to sit and talk after finishing the day's work. He talks about how this is helpful for them to work better.

"We have this habit of having random chats during breaks and after work. We usually do not discuss work but just rant if we are stressed about something. During one such talk, I explained how I could not get a client to accept a model even after all the alterations and rework. This intern casually mentioned that green packets could be included in the design if he needs that structure and a little aesthetics. That was something she had just read about. So she explained it to us, and I just thought of trying it, and the client was so happy, and he immediately accepted. I was so happy that these kinds of random conversations are of help, especially when we are not looking for it." - Anmol

H5a: Leisure crafting is positively related to social skills.

Social skills enable the quality of serendipity (Busch, 2024; Busch & Barkema, 2022). So, we argue that it will mediate the relationship between leisure crafting and serendipity. H5b: Social skills mediate the relationship between leisure crafting and serendipity.

METHOD

For the qualitative study, we have conducted 43 interviews with founders from various industries. The findings show that entrepreneurs proactively design their work and leisure for various reasons. The enablers and barriers of the entrepreneurial context are very different from those of the employees' context. Taking data from interviews, we build the model for our quantitative study. In this study, we collect data from active entrepreneurs, i.e., those currently working on a venture. We collect data through a three-wave survey, with a time lag of two weeks between the waves. We take this gap between data collection to avoid participant fatigue and to reduce common method bias (Podsakoff et al., 2012). We reach out to entrepreneurs in various ways. We first use LinkedIn to identify potential contacts from the authors' network. Next, we reach out to incubators and accelerators to reach potential participants. We also mail entrepreneurs by getting their contact information from Indian Startup Databases.

Following the best practice methods for time-lagged surveys in entrepreneurship (Hamrick et al., 2023), all the measures are collected at Time T1 – including IV, DV, and mediators. At the time, T2, Mediators, moderator, and control variables are collected. At the time, T3, DV, and demographics are collected. Currently, the data collection is in progress. With the data collected so far, we analyze and present the results.

Sample = Entrepreneurs (n=92 at T1, n=72 at T2, n=56 at T3)
DV = Serendipity (T3 measurement)
IV = Leisure crafting (T1 measurement)
Mediators = Self efficacy, Perseverance, Cognitive flexibility, Social skills ((T2 measurement)
Controls = Passive leisure, Age, Gender, Education, baseline values
Analysis = Moderated Mediation analysis

RESULTS

The descriptive statistics and correlations are reported in the appendix (refer to Table 1). All alpha values of scales are above the necessary cutoff of 0.7. Proceeding with this, we tested our proposed model. A linear regression analysis revealed that LC had a statistically significant positive effect on the outcome variable, $b = 0.29$, $SE = 0.13$, $t(\text{df}) = 2.20$, $p = .033$, 95% CI [0.025, 0.562]. H1 is supported. Then, we tested the mediation models. H2a is supported. Leisure crafting positively impacts self-efficacy, b=0.18, p= 0.04. However, the mediation is not supported. The coefficients and results are reported in the table (refer to Table 2). Other hypotheses are also not supported. Although the hypothesized indirect effects in the moderated mediation model were not statistically significant, this may be attributed to the limited statistical power resulting from a small sample size (N = 55). Moderated mediation models inherently involve multiple parameters, including interaction terms and indirect paths, which require a larger sample to detect effects reliably, especially when the expected effect sizes are small to moderate. Furthermore, the use of time-separated data, while methodologically rigorous, may have introduced additional variability that further reduced power. Despite the non-significance, the direction and magnitude of the effects were generally consistent with theoretical expectations, suggesting that the proposed relationships still hold true. Data collection is in progress, and this is just the first phase of data. Further analysis will be done once data collection is complete.

DISCUSSION AND IMPLICATIONS

This study contributes to the literature on work design in entrepreneurs. We look at how entrepreneurs design their work and lives and how it impacts their personal and venture outcomes. This study specifically examines leisure crafting and how it impacts serendipity. Through interviews, we understand how crafting leisure actively enables serendipity for entrepreneurs. Leisure crafting helps them enhance self-efficacy, cognitive flexibility, perseverance, and social skills. These are all enabling qualities of serendipity. We develop a theoretical model with the interview data and literature on crafting and serendipity. We then quantitatively test this model through a time lag survey. This survey was conducted three times, with a gap of two weeks. Though the results are not as expected with smaller sample size, further data collection and analysis will be done to test the hypotheses. The results show significant potential, as they are generally aligned with the proposed direction, and the confidence interval is closer to zero and not widely scattered, which shows potential. This study sheds light on the spillover effects of leisure crafting into work, specifically how it enables certain qualities that enable serendipity.

Leisure crafting can be used strategically by entrepreneurs to improve both their business and their personal results. By deliberately designing their leisure activities, they can acquire abilities including self-efficacy, cognitive flexibility, perseverance, and social skills—important enhancers of serendipity. These help them solve problems and raise their chances of finding unanticipated prospects that might advance their businesses. Policymakers can include leisure crafts in training courses and support programs for businesses to assist them in developing resilience and adaptation, promoting innovation and long-term success.

CONTACT: Brindha Srinivasan; brindhas15phd@iimk.ac.in; Indian Institute of Management, Kozhikode; IIMK Campus P. O, Kunnamangalam, Kerala 673570, India.

APPENDIX

Table 1

Variable	Mean	SD	1	2	3	4	5	6	7	8	9	10	11
Leisure crafting	5.09	1.10	1.00										
Self Efficacy (T2)	5.79	0.73	0.21	1.00									
Social Skills (T2)	5.48	0.98	0.09	0.31	1.00								
Cognitive flexibility (T2)	5.46	0.77	0.10	0.51	0.26	1.00							
Perseverance (T2)	6.04	0.72	0.10	0.41	0.45	0.35	1.00						
Job crafting	6.34	0.58	0.09	0.40	0.41	0.41	0.67	1.00					
Gender	1.31	0.47	0.04	-0.06	-0.02	-0.16	0.07	0.03	1.00				
Age	33.72	9.75	-0.16	0.10	0.24	0.28	0.25	0.30	0.12	1.00			
Education	3.69	0.72	-0.07	-0.16	0.17	-0.14	0.14	0.08	0.07	-0.02	1.00		
Passive leisure	3.83	1.09	0.22	-0.22	0.07	0.12	0.15	0.19	0.01	-0.04	0.06	1.00	
Serendipity (T3)	4.84	1.09	0.35	0.18	0.13	0.14	0.09	0.10	0.07	-0.10	-0.09	0.25	1.00

Table 2

	Direct effect (b and CI)	Indirect effect (b and CI)
LC à Self-Efficacy à Serendipity	0.299 (.0228, .5769)	0.02 (-.0695, .1440)
LC à Social Skills à Serendipity	0.32 (.0569, .5853)	0.003 (-.0639, .0638)
LC à Cognitive flexibility à Serendipity	0.31 (.0465, .5787)	0.01 (-.0322, .1116)
LC à Perseverance à Serendipity	0.31 (.0518, .5860)	0.005 (-.0658, .0532)

TURNING CRISIS INTO INNOVATION: HOW EXPERIENCE DRIVES ENTREPRENEURS' RESPONSES TO PATENT LITIGATION AND SUBSEQUENT INNOVATION

Miranda J. Welbourne Eleazar, University of Iowa, United States
Yiying Zhu, Montclair State University, United States

ABSTRACT

Facing patent infringement lawsuits can be a crisis for entrepreneurial new ventures when they threaten the ventures' limited products and resources. We develop the concept of *entrepreneurial human capital risk calculus* to theorize how entrepreneurial teams' human capital determines how they perceive and respond to the various risks involved in patent infringement litigation and subsequent patent decisions. Drawing on the literature on human capital, we theorize and test how entrepreneurs' specific human capital affects how they balance the risks and costs of pursuing litigation vs. settlement. We then consider how patent litigation experience combined with entrepreneurial-specific human capital affects entrepreneurs' patenting decisions. We contribute to the entrepreneurship literature on crisis, such as patent litigation, human capital, and entrepreneurial team decision-making.

INTRODUCTION

Entrepreneurial ventures (ventures) have limited resources, requiring them to apply those resources effectively to develop their product (Symeonidou et al., 2022) or put their venture at risk of failure (Symeonidou et al., 2022; Symeonidou & Nicolaou, 2018). Given their limited resources, it can be particularly challenging when they face costly crises such as lawsuits (Welbourne Eleazar & Park, 2022). Patent infringement lawsuits that attack the very core of the venture can put the venture itself at risk, particularly since most ventures rely on one or few products. Even when ventures have their own patents, firms may still bring patent infringement lawsuits against the entrepreneurs, arguing that the ventures infringed on the firms' patents.

Despite the critical nature of patent infringement lawsuits to ventures, limited research has considered how entrepreneurs respond to such lawsuits, other than noting potential negative consequences to innovation of patent trolls, or non-practicing entities (NPEs), whose sole business model is to acquire patents that they can use to sue other entities for royalties, settlements, or other profits (Chen et al., 2023; Cohen et al., 2019; Huang et al., 2024; Reitzig et al., 2007). Further, regardless of its outcome, litigation can be detrimental to ventures by requiring the venture to spend necessary resources on the lawsuits that could be used elsewhere. The lawsuits can also put ventures at risk of losing proprietary knowledge through the litigation process (Awate & Makhija, 2022). Although all firms may face patent litigation, the entrepreneurial context is unique because due to their limited resources, entrepreneurs often make decisions as to how to respond to crises such as patent litigation without the input of an in-house legal department that is employed by the venture (Quinn, 2023). Entrepreneurs also make innovation and patent decisions (Chen et al., 2024), so their patent litigation experience may affect the ventures' decisions whether to patent their subsequent innovation. Thus, we consider: *how do entrepreneurs manage the crisis of patent infringement litigation and patent subsequent innovation?*

We theorize that entrepreneurs' human capital will affect how they respond to lawsuits, and that human capital combined with patent litigation experience will affect whether they patent subsequent innovation. Drawing on literature that has noted that in general, expectations and personal knowledge affect how individuals bring and respond to lawsuits (MacFarlane, 2001), and integrating that literature with the human capital entrepreneurship literature, particularly specific human capital (Dimov, 2010), we theorize that entrepreneurs' human capital, particularly their specific human capital gained from specific types of experience, will shape how entrepreneurs prioritize risks and resource costs in responding to patent infringement litigation and patenting subsequent innovation. This expands on recent research that has suggested that additional insight is needed into how human capital influences decision-making in workers more generally (Deming & Silliman, 2024).

We develop the concept of an *entrepreneurial human capital risk calculus*, through which different types of experience affect how entrepreneurs respond to patent litigation. Expanding on findings related to specific human capital affecting entrepreneurial outcomes (Bruderl et al., 1992; Preisendorfer & Voss, 1990), we theorize how experience in areas related to the venture itself, law, STEM (science, technology, engineering, and math), entrepreneurship, and investing, will affect how entrepreneurs view the risks and costs involved in continuing to engage with patent litigation as opposed to settling the litigation. We then theorize how entrepreneurial teams will draw on the specific human capital will affect entrepreneurs' subsequent patenting decisions. We find support for most of our hypotheses, contributing to the literature on entrepreneurial crises (specifically patent lawsuits), entrepreneurial human capital, and entrepreneurial team decision-making.

HYPOTHESIS DEVELOPMENT

The Role of Human Capital in Entrepreneurial Crises and Patent Infringement Litigation

Patent infringement litigation in the U.S. occurs when patent holders sue entities for unauthorized commercialization of their inventions, serving as a key enforcement mechanism (Cohen et al., 2019). Such lawsuits have significant consequences, including knowledge spillovers (Awate & Makhija, 2022), deterrence of market entry (Onoz & Giachetti, 2023), and reduced innovation post-litigation (Cohen et al., 2019). Both NPEs and established firms use litigation strategically, often targeting resource-limited ventures (Duan, 2014; Molino, 2017). For ventures, these lawsuits can be crises, diverting resources from growth to legal defense (Welbourne Eleazar & Park, 2022) and threatening their core products. While entrepreneurship research has explored external crises like economic downturns (Williams & Shepherd, 2016), internal crises such as patent lawsuits remain underexamined.

Human capital encompasses the education, training, or and learning entrepreneurs bring to their ventures (Alomani et al., 2022). The knowledge individuals gain from their education and experience, whether general or specific, can affect their future behavior (Becker, 1962). Human capital can affect entrepreneurial performance, particularly given the uncertainty inherent to entrepreneurship that can lead to difficult decision-making and pressure (Alomani et al., 2022; Bosma et al., 2004; Davidson & Honig, 2003). The entrepreneurship crisis literature more broadly has considered the role of human capital in how entrepreneurs respond to crisis (Epure et al., 2024; Giotopoulos et al., 2017; Gruber et al., 2024; Knudsen, & Lien, 2015). That research primarily focuses on how human capital affects entrepreneurial teams' abilities to be resilient in the face of crises and continue to succeed, or how human capital is imprinted on firms by founding teams (Brymer & Rocha, 2024).

Expanding that research to the context of ventures' responses to patent infringement litigation, we theorize that specific types of human capital will determine the entrepreneurs' *risk calculus* related to the patent infringement litigation. By risk calculus, we refer to the way that the entrepreneurs perceive the different types of risks and costs associated with patent infringement litigation—both in terms of the risks and costs associated with continuing to fight the lawsuit and the risks and costs associated with a potential lawsuit. Given the ventures' limited resources, continuing to fight a costly lawsuit that involves extensive uses of resources in the form of attorney fees, court costs, expert fees, time involved in written discovery and depositions, as well as potential knowledge spillover, can be a significant risk. However, choosing to settle also comes with the risk that the venture may be targeted in future lawsuits if it is seen as a pushover that will willingly part with royalties or other settlement funds to end the lawsuit (Cohen et al., 2019).

Entrepreneurial Human Capital Risk Calculus and Litigation Duration

We theorize that the risk calculus will differ based on the specific human capital of the entrepreneurial team because the specific types of knowledge, skills, ability, and other information that the founding team possesses will shape how they view the risks and costs associated with both continuing and settling the lawsuit. We posit that entrepreneurs' decisions as to how to proceed when facing a patent infringement lawsuit are dependent on the *entrepreneurial human capital risk calculus* of the team—their cognitive balancing of the risks and costs associated with the patent infringement lawsuit based on their specific human capital endowments and the specific knowledge, skills, abilities, and other forms of information (Marvel et al., 2016) they gain from their specific human capital. We consider forms of human capital that may most likely influence the decision whether to continue to defend the lawsuit or settle. We consider venture-specific human capital in the form of prior patenting experience and the stage of the venture, reasoning that teams with prior patenting experience will emphasize the strength of the patent and resist settling, while later stage ventures may have experienced prior litigation, and thus effectively work to dismiss or settle the lawsuit early. We expect entrepreneurs' legal and STEM experience, which are closely aligned with the litigation process (legal) and subject matter (patents), will also result in shortening of the lawsuit based on the knowledge of the legal system gained from legal experience and knowledge as to alternative possible products for STEM experience. We also theorize that prior entrepreneurial experience in the form of entrepreneurs as entrepreneurs or investors will make entrepreneurial teams more likely to attempt escape the wasted resources associated with defending lawsuits, and settle early. We hypothesize:

Hypothesis 1a (H1a): Entrepreneurial teams' patents lengthen entrepreneurial ventures' patent infringement lawsuit duration

Hypothesis 1b (H1b): Entrepreneurial teams with ventures that are at a later venture stage (closer to producing revenue) shortens entrepreneurial ventures' patent infringement lawsuit duration.

Hypothesis 2 (H2): Entrepreneurial teams' legal experience shortens entrepreneurial ventures' patent infringement lawsuit duration.

Hypothesis 3 (H3): Entrepreneurial teams' STEM experience shortens entrepreneurial ventures' patent infringement lawsuit duration.

Hypothesis 4a (H4a): Entrepreneurial teams' prior experience as entrepreneurs shortens entrepreneurial ventures' patent infringement lawsuit duration.

Hypothesis 4b (H4b): Entrepreneurial teams' prior experience as investors shortens entrepreneurial ventures' patent infringement lawsuit duration.

Entrepreneurial Human Capital Risk Calculus and Subsequent Patenting

We further theorize that the entrepreneurial team's entrepreneurial human capital risk calculus will influence entrepreneurial teams' subsequent patenting decisions. We expect that human capital that arises from prior experience as a defendant in patent infringement litigation will make entrepreneurial teams more aware of the risks of being sued for patent infringement and thus be more likely to value patenting and therefore patent subsequent innovations. We expect that effect would be strengthened in teams that had patents and were able to see the value of patenting while defending against a patent infringement lawsuit. We also expect that teams in ventures at a later stage would have experienced other crises, would be more attuned to recognizing the need for patents to help prevent additional lawsuits, and would be better positioned to use their resources to obtain subsequent patenting. That experience combined with the experience defending against the patent infringement litigation, would make them even more likely to engage in subsequent patenting. Thus, we hypothesize:

Hypothesis 5 (H5): Entrepreneurial teams with experience as a defendant in patent infringement litigation are more likely to advance patent innovation.

Hypothesis 6a (H6a): Entrepreneurial teams with patent experience strengthens the positive relationship between entrepreneurial teams' experience as a defendant in patent infringement litigation and patenting innovation.

Hypothesis 6b (H6b): Entrepreneurial teams with ventures that are at a later venture stage (closer to producing revenue) strengthens the positive relationship between entrepreneurial teams' experience as a defendant in patent infringement litigation and patenting innovation.

METHOD

Sample	Ventures with VC funding from 2000-2020 in healthcare and information technology that were sued at least once within 10 years of founding
Data sources	Pitchbook, Lex Machina, Patentsview, Crunchbase, LinkedIn, venture websites
DVs	Lawsuit duration = duration in days; patent innovation = number of patents filed in the year
IVs	Venture patent experience = binary if filed for patent in the year; Venture stage = early stage (1); product development stage (2); revenue-generating stage (3); legal specific capital = number of founders with JD; STEM-specific human capital = number of founders with STEP work experience; entrepreneurial experience = number of founders with prior entrepreneurship experience; investor experience = number of founders with prior investor experience; defendant human capital = binary if venture was sued for patent infringement
Controls	Control for venture and entrepreneurial specific attributes that may affect litigation and patent innovation outcomes, as well as key years
Methods	Heckman two-stage model using probit for the first stage and linear regression with random effects for the second stage

RESULTS

Descriptive statistics indicate that there is no concern for multicollinearity. We find support for Hypothesis 1a and 1b, 3, 4a, 4b, 6a, and 6b. Robustness checks including alternative measures for *Patent experience, Venture stage,* experience variables, and using a *time trend* variable and provide generally consistent results. We further tested our hypotheses using alternative outcome variables, including legal outcomes (i.e., settlement vs. other resolutions such as pre-trial dismissal, trial, and other procedural resolutions) and patent quality (i.e., forward citations). The results remained generally consistent and aligned with our theoretical propositions.

DISCUSSION AND IMPLICATIONS

We found support for our conceptualization of *entrepreneurial specific human capital risk calculus* related to patent infringement litigation and subsequent patenting. Consistent with our theorizing, specific forms of human capital related to late venture stage, STEM experience, entrepreneurial experience, and investor experience shorten the duration of patent infringement lawsuits through dismissals or early settlement, while the specific human capital of patent experience lengthens the duration of the lawsuit. Further, human capital derived from the experience of defending the venture against a patent infringement lawsuit combined with 1) patent experience and 2) late venture stage increase subsequent patenting. Our results remain consistent throughout our robustness checks and supplemental analyses.

Theoretical Implications

We contribute to the entrepreneurship literature on crisis, such as patent litigation, human capital, and entrepreneurial team decision-making. The entrepreneurship literature has considered the importance of crises on entrepreneurs and the development of their ventures, to the extent of commenting on the poly-crisis environment in which entrepreneurs currently operate (Klyver & McMullen, 2025) but has devoted less attention toward internal crises such as patent infringement litigation. We extend the limited literature on patent infringement litigation by examining how entrepreneurial ventures respond to such litigation, focusing on the ways entrepreneurial teams' specific human capital affects their decision-making and resulting outcomes in responding to patent infringement litigation. We develop and test the concept of entrepreneurial specific human capital risk calculus related to patent infringement litigation, through which the specific knowledge, skills, and abilities entrepreneurs gain from specific experiences affect how they way and prioritize the risks and costs associated with patent litigation. As a result, that specific human capital and resulting knowledge determines whether the ventures are able to quickly resolve lawsuits through a Motion to Dismiss or settle cases early, minimizing the costs and fees associated with the lawsuit itself and potentially reducing the likelihood of a verdict against the venture that may prevent them from marketing their product altogether. This expands on existing literature related to patent infringement that has focused more on firm-level attributes or inventor-specific factors, instead of considering the role of the entrepreneurs and their knowledge, which is vital given that they are the individuals making decisions as to how to address the patent litigation in the venture.

We also contribute to the literature on human capital in entrepreneurship. Our focus on how specific human capital affects entrepreneurs' risk-calculus and decision-making related to patent litigation responds to calls for further examination of specific types of human capital on entrepreneurship (Marvel, 2013) and how specific human capital affects decision-making (Deming & Silliman, 2024). Our theory and findings suggest that specific human capital can determine what factors related to patent litigation entrepreneurial teams' cognitive emphasize in their engagement with the risk and costs involved in the litigation. Essentially, their experiences change the way they weigh the risks of moving forward or resolving the lawsuit, as well as the risks of patenting vs. not patenting innovations. Our consideration of specific forms of human capital that are not typically considered, such as legal, STEM, investor, patent, and venture stage human capital, further expands on the specific types of human capital that may affect entrepreneurs' decision-making. In addition, by considering how specific human capital affects decision-making related to patent infringement litigation, we expand on human capital literature which has not been considered extensively in relation to litigation more generally or in the case of entrepreneurial patent infringement litigation specifically.

Finally, we contribute to the literature one entrepreneurial team decision-making. The entrepreneurial team literature has considered the effect of specific and general human capital but has primarily considered broad types of outcomes such as performance (Ganotakis, 2012; Jin et al., 2017; Rauch & Rijsdijk, 2013; Shrader & Siegel, 2007) without considering how specific human capital may affect individual decision and outcomes that may have significant impacts on performance. We consider how the degree of experience in specific types of human capital affects entrepreneurial teams' decision-making related to the patent litigation, expanding on the literature on how entrepreneurial teams make decisions and the outcomes of those decisions.

Practical Implications

Our findings that specific types of human capital determine how entrepreneurial teams respond to the crisis of patent infringement litigation illustrates the importance of individual entrepreneurial team members on venture outcomes outside of those typically thought of in entrepreneurship literature such as the formation of the venture and development of the product. When entrepreneurs are deciding who to start a new venture with, they may want to think more broadly about the knowledge that each person brings to the table than just the knowledge specific to the product and venture itself. Having a lawyer on the team can provide valuable insights when lawsuits are (potentially inevitably) filed. Having someone with STEM experience can provide valuable insights into science and technology issues more broadly, such as what innovations should be patented and how to potentially avoid lawsuits. Prior entrepreneurial and investor experiences can provide knowledge related to the development of the venture itself as well as how to deal with potential challenges that may arise—and how to limit the extent to which emotions may influence decision-making related to those challenges. In addition, after the venture is formed, the specific knowledge that the entrepreneurial team gains should be used to the best of its ability in making decisions related to challenges and also how to proceed in the future, such as patenting subsequent innovations.

Expanding on the importance of this paper for entrepreneurial teams, this paper also has implications for how entrepreneurial educators teach new entrepreneurs. Our findings provide another basis for why diverse experiences can benefit ventures by helping entrepreneurs navigate the crisis of patent litigation when it occurs. Although most introductory entrepreneurial classes will discuss the value of patents to entrepreneurs, limited attention is paid to the fact that entrepreneurs may be sued for patent infringement regardless of whether they have a patent themselves, and how to respond to such lawsuits. By providing evidence as to how entrepreneurs' specific human capital affects their responses to patent litigation, we demonstrate how entrepreneurs' backgrounds can affect their cognitive responses to internal crises such as patent litigation. In realizing the importance of their backgrounds—or lack thereof—entrepreneurs can question the extent to which their response to patent litigation is based in emotions or objective knowledge based in the entrepreneurial team. For those that lack the knowledge to respond to the litigation in a relatively objective manner, they may be able to recognize sooner the need to consult with trusted advisors or lawyers. In addition, our research allows educators to examine the impact of crises such as patent litigation on subsequent decisions as to whether to innovate and obtain patents on such innovation.

Conclusion

Patent infringement lawsuits against ventures can amount to crises that put their very survival in question. Thus, it is vital to understand how entrepreneurs respond to those lawsuits and how they affect the ventures' subsequent patenting. We develop the concept of *entrepreneurial specific human capital risk calculus* in patent litigation and examine the effects of specific human capital of entrepreneurial teams on their decision-making related to the litigation and subsequent patenting. We find that venture stage, STEM, and entrepreneurial/investor experience shortens the duration of the lawsuit, while patenting experience increases the duration of the lawsuit. Defending against patent infringement litigation combined with 1) patent experience and 2) late venture stage increases subsequent patenting. This research contributes to the entrepreneurship literature on crisis, such as patent litigation, human capital, and entrepreneurial team decision-making.

CONTACT: Miranda J. Welbourne Eleazar; miranda-welbourneeleazar@uiowa.edu; University of Iowa; 108 John Pappajohn Business Building, Iowa City, IA, USA 52242-1994

SUMMARY

HYBRID ENTREPRENEURS' WELL-BEING: A COMPARATIVE CASE STUDY

Nadine Albrecht, University of Bayreuth, Germany
Matthias Baum, University of Bayreuth, Germany
Monique Boddington, University of Cambridge, United Kingdom
Pauline Reinecke, Technical University of Hamburg, Germany

Principal Topic

Hybrid entrepreneurship –the simultaneous pursuit of entrepreneurial ventures while maintaining wage employment– represents a growing and significant facet of the entrepreneurial landscape (Folta, Delmar, and Wennberg 2010). However, this hybrid approach can act as a double-edged sword: On the one hand, financial stability and security enable hybrid entrepreneurs to offset the risks of launching start-ups, potentially enhancing their well-being. On the other hand, restricted autonomy and the difficulty of balancing multiple –potentially conflicting– roles can lead to elevated stress levels, ultimately reducing their overall well-being.

Method

Drawing on arguments of energizing tensions and role conflict, we explore how hybrid entrepreneurs navigate the well-being challenges of balancing multiple roles over time. We adopted an open, theory-building methodology (Gioia, Corley, and Hamilton 2013) and conducted 15 semi-structured interviews with 8 hybrid entrepreneurs between January 2024-January 2025 (16:38:54 hours). To gain deeper insights into the lived tensions of hybrid entrepreneurs across multiple roles in their natural environment, we triangulated the interviews with additional data, including participant observations in their work settings (e.g., at their paid employment locations; 06:44:12 hours), interviews with co-founders and investors (03:32:30 hours), and their social media communication to external audiences (02:22:02 hours).

Results and Implications

Our theoretical contributions to the literature on hybrid entrepreneurship and well-being are three-fold. First, our findings reveal two main paths hybrid entrepreneurs embark on: while some achieve well-being through internal coherence, others employ external distancing strategies. These distinctions highlight how well-being tensions and coping strategies differ based on entrepreneurial aspirations, advancing our understanding of hybrid entrepreneurship's micro-level dynamics. Second, we highlight how hybrid entrepreneurs navigate role conflicts between entrepreneurship, paid employment, and family responsibilities (Carr et al., 2022; Kuske, Schulz, and Schwens 2024). For those aspiring to full-time entrepreneurship, well-being challenges arise from exiting prior ventures and navigating hybrid demands. Conversely, those pursuing hybrid careers face challenges in maintaining employment stability while growing their ventures. This challenges the view of hybrid entrepreneurship as merely transitional (Folta, Delmar, and Wennberg 2010), revealing a more intentional hybrid entrepreneurship career path. Third, while entrepreneurship literature often emphasizes the adverse effects of role conflict (Carr et al., 2022), broader management research highlights the energizing potential of tensions (Battilana et al., 2015). Specifically, our findings show how hybrid entrepreneurs transform negative tensions into positive energy (Andriopoulos et al., 2018).

CONTACT: Nadine Albrecht; nadine.albrecht@uni-bayreuth.de; University of Bayreuth; University of Bayreuth, Bayreuth, Bayern 95448, Germany

SUMMARY

CAN YOU SEE THE PIVOT? COGNITIVE FRAMEWORKS IN OPPORTUNITY AND SURVIVAL ENTREPRENEURIAL PIVOTS

Jared S. Allen, Texas Tech University, United States
Jon Carr, North Carolina State University, United States
Alanna Hirshman, Texas Tech University, United States
Ahmed Sewaid, MBSC, United Arab Emirates
Regan Stevenson, Indiana University, United States

Principal Topic

Pivoting is a critical strategy for new ventures adapting to uncertainty, shifting customer preferences, and resource constraints (Allen et al., 2024; Kirtley & O'Mahony, 2023). Despite its importance, not all entrepreneurs decide to pivot, even when it might benefit their venture (Camuffo et al., 2020; Snihur & Clarysse, 2022). Existing research often overlooks the role of cognitive frameworks—how entrepreneurs process and respond to information—in shaping pivoting decisions. Cognitive differences significantly impact entrepreneurial decision-making (e.g., Gregoire et al., 2011; Hmieleski & Baron, 2008), yet little is known about how these frameworks influence pivoting behaviors. This study explores how entrepreneurs' cognitive frameworks affect their decisions to pivot and how they interact with different types of information (opportunity- or threat-based) to influence the type of pivot chosen. Opportunity pivots and survival pivots are triggered by distinct types of information (Allen et al., 2024). We propose that entrepreneurs' regulatory focus—whether promotion-focused or prevention-focused—determines how they respond to these signals. Promotion-focused entrepreneurs, driven by the desire to pursue growth and opportunity, are more likely to pivot overall and engage in opportunity pivots. In contrast, prevention-focused entrepreneurs, motivated by the need to avoid risks and protect resources, are more inclined to pursue survival pivots in response to threat-based information.

Method

We conducted two studies: a field study using crowdfunding data and an experimental vignette study. This dual approach minimized mono-method and mono-operational bias, enhancing the robustness of our findings. Study 1 This archival study utilized data from Kickstarter, focusing on serial entrepreneurs who launched multiple campaigns. We scraped data from 182,484 campaigns and used computer-aided textual analysis (CATA) to assess entrepreneurs' regulatory focus based on their campaign descriptions. Study 2 A between-subjects experimental vignette study with 2,178 entrepreneurs tested how regulatory focus influences pivoting decisions in response to different types of information (opportunity-based or threat-based). Participants were randomly assigned to one of three conditions: opportunity-based gain, threat-based loss, or a control condition.

Results and Implications

Across both studies, we found that promotion-focused entrepreneurs were more inclined to pivot but were not significantly more likely to pivot in response to opportunity- or threat-based information. In contrast, prevention-focused entrepreneurs were less likely to pivot overall but more likely to engage in survival pivots when exposed to threat-based information. These findings highlight the role of cognitive frameworks in shaping entrepreneurial pivoting decisions.

CONTACT: Jared S. Allen; jared.allen@ttu.edu; Texas Tech University; 5937 139th st, Lubbock, Texas 79424, United States

SUMMARY

FOSTERING TRUST OR FABRICATING INTIMACY? UNPACKING THE CREDIBILITY ILLUSION OF SOCIAL MEDIA ENTREPRENEURS

Aishwarya Gayen, IE Business School, Spain
Julio De Castro, IE Business School , Madrid, Spain

Principal Topic

This paper explores social media entrepreneurship, focusing on how emotional-storytelling, authenticity, credibility, perceived expertise, and content strategies influence stakeholder engagement. Unlike traditional entrepreneurs, who are primarily recognized for their products, social media entrepreneurs simultaneously market products, influence brands, and serve as brands themselves. However, gaps remain in understanding how they cultivate trust and drive engagement. Grounded in source credibility theory and the Elaboration Likelihood Model, this research explores content types and industry-specific dynamics, including the balance between professional and personal content and strategic posting approaches. By identifying mechanisms that enhance or hinder trust and engagement, this study offers fresh insights into digital entrepreneurial success.

Method

Focusing on the beauty and lifestyle industry, this study analyses influencer-driven engagement on Instagram. A dataset of 417,957 posts from 277 publicly accessible accounts (following ethical guidelines) was collected. Profiles were identified through targeted keyword and hashtag searches (e.g., makeup artist, skincare expert, beauty influencer) and network expansion by reviewing follower connections. Data scraping adhered to Instagram's API limitations. To maintain linguistic consistency, only English-language posts were included.

Results and Implications

The findings reveal a U-shaped relationship between emotional-storytelling and stakeholder engagement. High emotional storytelling deeply resonates with stakeholders, while low emotional-storytelling provides clarity and directness that also facilitate engagement. In contrast, moderate emotional storytelling is the least effective, likely failing to evoke meaningful responses underscoring the importance of strategically balancing emotional expression to maximize stakeholder impact. The study also uncovers nuanced effects of credibility and expertise on engagement. While credibility increases likes, it has little impact on comments. Meanwhile, higher perceived expertise negatively affects both likes and comments, indicating that relatability holds greater sway over engagement than authority. Additionally, Authenticity emerges as a critical factor in building trust and engagement, as social media entrepreneurs who project genuine, transparent personas develop stronger connections with stakeholders. Content type plays a crucial role in engagement levels. Personal and leisure-oriented posts, especially those related to family, generate the highest interaction, whereas work-related content sees the least engagement. While promotion-focused content dominates social media strategies, prevention-focused posts prove more effective, in driving likes. Finally, content design with visual and linguistic elements (e.g., emojis, aspirational language) balances relatability with professionalism. Simpler language enhances interaction, while post length has mixed effects—longer posts attract more comments but fewer likes. These insights offer actionable guidelines to optimize engagement while advancing theoretical discussions on digital communication and stakeholder interaction within entrepreneurial contexts.

CONTACT: Aishwarya Gayen; gayenaishwarya@student.ie.edu; IE Business School; calle lagasca 67, Madrid, Madrid 28001, Spain

SUMMARY

NOW WHAT DO I DO? THE IMPACT OF KNOWLEDGE PROBLEMS ON ENTREPRENEURS' COPING STRATEGIES

Howard Haines, Central Michigan University, United States
Lauren Zettel, Central Michigan University, United States

Principal Topic

Entrepreneurs are faced with a variety of stressful challenges that they must cope with to establish a new venture. The coping flexibility perspective suggests that effective coping is characterized by matching the coping strategy with the appraised demands of the situation. Although a substantial body of research has examined the types of coping strategies used by entrepreneurs, we lack a theoretical understanding of how entrepreneurs' appraisals of the situation impact their choice of coping strategies. Therefore, we ask how the appraisal of the situation impacts the coping strategies deployed by entrepreneurs when they encounter problems in their ventures?

Method

To answer this question, we employ a multiple case study method using virtual, semi-structured interviews with entrepreneurs to capture situational cases common to the creation of a new venture. We asked participants to recount both the most recent and the most substantial problems they had faced in their venture over the past year, and to focus specifically on how they determined what to do in a given situation. From the interviews, we generated individual cases of the processes each entrepreneur employed to determine how to manage stressful circumstances. We then examined the data patterns across the cases before identifying higher level themes that were used to construct a theoretical framework.

Results and Implications

Entrepreneurs' descriptions of stressors revolved around a lack of control over a situation that would allow them to achieve a desirable end. This lack of control was primarily based on a lack of knowing how to reach or define a desirable end, or knowing how to access other resources required to reach the desired end. We define this individual-level perception of a knowledge problem as a knowledge deficit. Like knowledge problems, knowledge deficits took four forms: uncertainty, ambiguity, complexity, and equivocality. Furthermore, the form of the knowledge deficit perceived subsequently impacted the type of coping strategy used to mitigate the stressor. Based on these patterns, we offer a series of propositions detailing how entrepreneurs match the appraised knowledge deficit with specific coping strategies, and we describe why these coping strategies can be effective in reducing stress. Our findings inform and extend theories of coping flexibility and entrepreneurial knowledge problems and provide practical insight into how entrepreneurs cope with venture-related stressors.

CONTACT: Howard Haines; haine1h@cmich.edu; Central Michigan University; 150 E. Bellows, Mount Pleasant, Michigan 48859, United States

SUMMARY

PERCIEVED HEAVENS AND IMAGINED HELLS: EMPATHIC UNDERSTANDING AMONG COFOUNDER DYADS AND ITS EFFECT ON TEAM PERFORMANCE

Sukriti Hari, LM Thapar School of Management, India
William Gartner, Babson College, United States
Sarnish Kour, LM Thapar School of Management, India
Raja Singaram, University of Galway, Ireland

Principal Topic

Cofounders play a leading role in the establishment of new ventures and contribute substantially to their early-stage performance (Howell et al., 2022). Even though recent studies have examined issues relevant to cofounder selection (Gray et al., 2024) and exits (Preller et al., 2023), the dynamics of cofounder relationships and their effect on startup team performance is still understudied. We study the role of empathic understanding between cofounders and its effects on their relationship satisfaction and team performance (Cramer & Jowett, 2010; Klimecki, 2019; Weisz & Cikara, 2021). Drawing from Klein (2019), we hypothesize that positive empathic differential is positively related to both relationship satisfaction at the individual level and team performance at the dyadic level. Furthermore, we hypothesize that the positive relationship between empathic understanding and relationship satisfaction, as well as team performance, will be negatively mediated by the presence of relationship conflict among the founders.

Method

To study cofounder relationships, we conducted a decision-making role play experiment (Hsu et al., 2027). Taking a projective approach (Miner, 2006; Rauch, 2014), each cofounder dyad completed a series of startup activities for a new commercial technology-based startup. Dyadic data was collected in two phases from business administration students (N=60) and practicing entrepreneurs (N=27) in the technology entrepreneurship sector whose firms that are less than five years old. The two dependent variables are cofounder relationship satisfaction and team performance. Independent variables include dyadic empathic understanding measured using Barrett-Lennard Relationship Inventory (Barrett-Lennard, 2015). Empathic differential score is calculated as the difference between perceived empathy and directed empathy towards cofounders. The mediator, relationship conflict, was measured using the Intragroup Conflict Scale (Jehn, 1994). To analyze the dyadic data and the interactions, we built an Actor-Partner Interdependence model and executed multi-level analysis (Cook & Kenny, 2005; Ledermann & Kenny, 2017; Hatak & Zhou, 2021).

Results and Implications

As hypothesized, we found that positive empathic differential expressed by cofounders led to both higher relationship satisfaction ($b=0.463$, $p<.001$) and better team performance ($b=0.039$, $p<.001$). Also, we found strong support for relationship conflict negatively affecting the team performance of the cofounders ($b=0.184$, $p<.001$). Our findings reveal that overestimation of empathy from their cofounder positively affects both relationship satisfaction of cofounders and team performance (c.f., Klein, 2019; Hasan & Konig, 2019). We contribute to the literature on entrepreneurial teams through the insights on the empathic understanding between cofounders and its effects on startup team performance (Knight et al., 2020; Lazar et al., 2020).

CONTACT: Sukriti Hari; shari_phd20@thapar.edu; LM Thapar School of Management; LM Thapar School of Management, Derrabasi, Chandigarh 140507, India

SUMMARY

FIT OR MISFIT? TOWARD A THEORY OF ENTREPRENEURIAL IDENTIFICATION

James M. Hughey, University of Nebraska, United States
Dane Cannon, University of Utah, United States
Glen Kreiner, University of Utah, United States
Robert Monnot, University of Utah, United States
Devin Rapp, San Diego State University, United States

Principal Topic

Grounded in a three-year, longitudinal study of entrepreneurs, this paper aims to advance theory on entrepreneurial identity and identification. Entrepreneurial identity—the set of meanings entrepreneurs attach to their occupation—has become a core construct in entrepreneurship, revealing a myriad of role (e.g., founder, developer, inventor) and social (e.g., darwinian, communitarian, missionary) identities. However, a critical construct of entrepreneurial identity construction remains overlooked: entrepreneurial identification—i.e., the process of ascribing an entrepreneurial identity to one's self. And given identification's direct relationship with occupational entry, exit, and persistence, understanding the determinants of entrepreneurial identification is theoretically and practically critical.

Method

To build theory on identification in entrepreneurs, we draw upon a three-year field study of 70 entrepreneurs embedded in a university-sponsored accelerator. To understand the development of their entrepreneurial identities, we utilized serial interviews complemented by observation, written exercises, surveys, and archival records. Pilot data were collected during the 2021-2022 academic year. This consisted of 16 interviews with seven entrepreneurs. Our main dataset consists of 162 interviews across 63 entrepreneurs collected during the 2022-2023 and 2023-2024 academic years. Interviews were substantively rich, lasting on average 1.5 hours, with several exceeding two hours.

Results and Implications

Our findings suggest entrepreneurs rely upon mental models to guide the construction and enactment of their identities. Far from being benign construals, these mental models embody and convey important professional values, characteristics, and milestones. In addition to acting as guides throughout the identity construction process, mental models also serve as an important basis for entrepreneurs' identification (or disidentification) with their occupation. Specifically, we find that entrepreneurs traverse three distinct phases of identification—exposure, initiation, and grounding—each characterized by unique sources of information, learning motives, and referents. An important consequence of each phase concerns the mental models—stereotype, archetype, and prototype, respectively—they inspire. In the exposure phase, prospective entrepreneurs form stereotypes of the occupation via their exposure to popular figures (e.g., Mark Zuckerberg, Elizabeth Holmes). In the initiation phase, new entrepreneurs focus on achieving success and legitimacy and are focused on emulating highly successful, idealized entrepreneurs (e.g., Elon Musk)—the archetypes. Imbued with varying degrees of myth, mystery, and misinformation, we find archetypes have a pronounced effect on the identification (and disidentification) of novice entrepreneurs. Finally, in the grounding phase, experienced entrepreneurs develop an appreciation for their occupation's prototype—a holistic image of the occupation grounded in a clearer conception of the average or typical member—that attenuates the effect of the archetype on identity construction and identification.

CONTACT: James M. Hughey; jhughey2@unl.edu; University of Nebraska; 730 N 14th St, Lincoln, Nebraska 68588, United States

SUMMARY

FOR THE VENTURE OR THE SELF: DECONSTRUCTING THE ENTREPRENEURIAL FAILURE JOURNEY

Sumit Kumar Jaiswal, IIM Ahmedabad, India
K. V. Gopakumar, IIM A, India
Chitra Singla, IIM A, India

Principal Topic

This study aims to uncover how the entrepreneurial failure (EF) process unfolds over time. There is a growing consensus in the literature that the EF process consists of three stages – the causes, the failure event, and its consequences (Cardon et al., 2011; Ucbasaran et al., 2013). However, most studies address only two aspects of the EF process – the causes and the consequences, without paying much attention to the failure event (Klimas et al., 2021). Though research suggests that entrepreneurs can recognize the beginning and the end of the EF event (Klimas et al., 2021; Ucbasaran et al., 2013), little do we know how they develop the perception of a failing venture in the first place and what makes them terminate the venture, which requires tracing the pathway entrepreneurs take after establishing their ventures. By relying on a sensemaking perspective, we ask how entrepreneurs who terminated their ventures make sense of their failing experiences.

Method

Using a grounded theory method (Gioia et al., 2013), we captured the lived experiences of 28 entrepreneurs (Gehman et al., 2018), though retrospectively, as they traversed through their entrepreneurial journey until their dissociation from the venture. The duration of these interviews varied between 60 and 150 minutes. The data collection and data analysis happened simultaneously and iteratively throughout this period, which informed our theoretical model.

Results and Implications

Our findings reveal that entrepreneurs navigate three distinct yet interconnected phases on their journey toward venture termination: venture preservation, the corridor of uncertainty, and self-preservation. Entrepreneurs transition between these phases by shifting their reference points from the venture to the self, a process shaped by how they make sense of their circumstances along the failure path. In the venture preservation phase, entrepreneurs see the venture as an extension of their personal aspirations, with their sense of self deeply intertwined with its success. As they enter the corridor of uncertainty, growing doubts about the venture's viability lead to a gradual disassociation between their sense of self and the venture, marked by a loss of conviction. Finally, in the self-preservation phase, entrepreneurs clearly separate their identity from the venture, prioritizing their own well-being and self-interests over sustaining the venture. By illuminating how entrepreneurs traverse these phases, our study offers new perspectives on the dynamics and processes underpinning entrepreneurial failure.

CONTACT: Sumit Kumar Jaiswal; phd19sumitj@iima.ac.in; IIM Ahmedabad; Vikram Sarabhai Libraray, Ahmedabad, Gujarat 380015, India

SUMMARY

TOO MANY BOLD MOVES? LEARNING HOW TO OVERCOME OVERCONFIDENCE

Julian Paul Kolbe, Swiss Institute of Small Business and Entrepreneurship, Switzerland

Principal Topic

The entrepreneurial context is characterized by uncertainty, making it a challenging environment regarding information processing and decision-making. Given these characteristics, entrepreneurs tend to exhibit overconfidence, which can distort their information processing. Overprecision is a form of overconfidence that is particularly detrimental to entrepreneurial decision-making. Overprecision manifests when individuals do not recognize the limits of their knowledge. Theoretical evidence suggests that the self-regulated entrepreneurial learning (SREL) framework can reduce biased cognition, but empirical evidence is scarce.

Method

This study examines the relationship between SREL and overprecision with a cross-sectional survey (N = 198), proposing that higher SREL capacities are associated with lower levels of overprecision. SREL was measured using a 27-item scale covering forethought, performance, and reflection phases, and the dimensions cognition, metacognition and motivation. Overprecision was assessed using a probability estimation task.

Results and Implications

A linear regression analysis confirmed a significant negative relationship between SREL and overconfidence ($B = -0.10$, $p = .002$), but the effect size was small. Surprising was that while higher aggregate SREL was associated with improved knowledge calibration (lower overprecision) individual dimensions of SREL (cognition, metacognition, motivation) were not significant predictors. A post-hoc analysis found that when combined, the SREL dimensions explained a small but statistically significant variance in overconfidence (adjusted $R^2 = .038$, $p < .05$). Founding experience alone did not significantly predict overconfidence but moderated the relationship between SREL and overconfidence through a significant interaction effect ($B = 0.20$, $p = .005$). Among novice entrepreneurs, higher SREL was associated with lower overconfidence ($B = -0.24$, $p < .001$, adjusted $R^2 = 0.18$), but not for those with founding experience. In contrast to experience in founding a venture, venture phase (pre-launch, launch, post-launch) did not significantly moderate the relationship between SREL and overconfidence. The weak overall effect suggests that SREL plays a modest role in predicting overprecision. However, the stronger negative relationship observed among novice entrepreneurs in the moderation analysis indicates that SREL may be particularly valuable for those without prior founding experience, enabling them to calibrate confidence judgments and make more informed decisions. The unexpected finding that no isolated SREL dimension significantly predicted overprecision, while the combined SREL score did, suggests that SREL unfolds optimally as a cohesive construct. These findings contribute to entrepreneurial cognition research by empirically supporting theoretical claims that SREL improves knowledge calibration, offering a new theoretical lens that can be further explored by researchers, and a practical framework for entrepreneurship educators and practitioners.

CONTACT: Julian Paul Kolbe; Julian.Kolbe@unisg.ch; Swiss Institute of Small Business and Entrepreneurship; University of St. Gallen, St. Gallen, Sankt Gallen 9000, Switzerland

SUMMARY

VEERING OFF COURSE - HIDDEN INFLUENCES THAT MISLEAD ENTREPRENEURS IN EARLY-STAGE BUSINESS MODEL EXPERIMENTATION

Felix Kuhn, Technical University Munich, Germany
Jan Auernhammer, Stanford University, United States

Principal Topic

Business model experimentation plays a crucial role in early-stage ventures in the process of testing the entrepreneurs` assumptions (Andries et al., 2013; Ries, 2011; Leatherbee and Katila, 2020). Rapid iterations incorporating the feedback from these business model experiments are crucial for venture success (Ries, 2011, Felin et al., 2020). In response to negative feedback in business model experiments, entrepreneurs pivot their venture's business model (Burnell et al., 2023). Previous research indicates that the overconfidence of entrepreneurs' beliefs and their cognitive processes impact the decision-making process of entrepreneurs (Kraft et al., 2022). Furthermore, entrepreneurial methods assume entrepreneurs react to new information and environmental conditions in iterative and dynamic ways (Blank & Dorf, 2020). Although research has made important advances in studying the entrepreneur's decision-making process, yet, we still know little about which contextual influences slow down assumption-testing cycles and biasing entrepreneurs, leading necessarily to a pivot of the venture. This research on the entrepreneurial assumption-testing process is important because it reveals an underexplored yet crucial process to theoretically advance research in business model experimentation and pivoting.

Method

Our sample (N = 26) consists of early-stage entrepreneurs associated with Stanford University or acceleration programs in the Silicon Valley region. The average age of their startup is 2.3 years, and the previous entrepreneurial experience of the interviewees is heterogeneous. At the time of the interview, entrepreneurs recently had a pivot in their startup or pivoted in the months after their interview. Participants provided information about their human capital, such as educational background and entrepreneurial experience. We coded the interviews using Gioia`s methodology (Gioia et al., 2013) and tested for intercoder reliability, which was 86.40 %.

Results and Implications

This study explores the effect of contextual influences on the process of entrepreneurial assumption-testing and introduces to theory the “testing freeze”. We uncovered six key contextual factors: entrepreneurial identity, entrepreneurial mindset, customer insights, prototyping capabilities, financial constraints, and hype. These factors facilitate to slowed assumption-testing cycles and the entrepreneur’s freeze of continuously testing assumptions. Our findings advance existing theories on business model experimentation and business model pivoting by offering a nuanced understanding of how contextual factors interact with cognitive biases to influence entrepreneurial assumption-testing. We propose a process model of entrepreneurial assumption-testing, highlighting the pathways through which entrepreneurial overconfidence and contextual influences affect entrepreneurial assumption-testing.

CONTACT: Felix Kuhn; felix.kuhn@tum.de; Technical University Munich; Arcisstrasse 21, Munich, Bayern 80333, Germany

SUMMARY

THE PARADOXICAL PARTNERSHIP OF HUMILITY AND NARCISSISM IN ENTREPRENEURIAL ABSORPTIVE CAPACITY

Maya Kumar, IE University, Spain
Darren Meister, Ivey Business School, Western University, Canada

Principal Topic

Entrepreneurs operate under constant uncertainty, where their ability to learn from external information can determine success or failure. However, learning is not guaranteed, and information perceived as irrelevant may be dismissed. This study explores the tension between humility and narcissism in shaping an entrepreneur's absorptive capacity – the ability to acquire, process, and apply new knowledge. While humility fosters openness to feedback, narcissism can impede critical reflection. By examining how early-stage entrepreneurs navigate this balance, we provide new insights into cognitive mechanisms that influence entrepreneurial learning.

Method

Adopting a paradox perspective to understand the impact of these seemingly contradictory traits, we conducted a longitudinal, multiple-case study of nine accelerator-based ventures over seven months. Using 81 semi-structured interviews (76 recorded hours), we code for critical learning events, expressed narcissistic and humble narratives, and behaviors. We supplemented this with observations of meetings, pitches, events, and retail locations, and secondary data.

Results and Implications

Our results suggest that humility and narcissism present a valuable duality when entrepreneurs acquire and apply external knowledge. Cases exhibiting a balance between these behaviors progressed furthest with venture development, finding product-market fit and sales. Humility proved crucial during knowledge acquisition and integration, allowing entrepreneurs to remain open to new information and counterintuitive ideas, triggering them to admit when they are on the wrong path when feedback gathered tells a different story. In contrast, balancing narcissism and humility is key during knowledge transformation and exploitation. Narcissism provided confidence, risk-taking, and focus needed to gather and exploit resources, while humility mitigated overvaluing personal insights and insufficient feedback gathering. Excessive humility led to a cycle of feedback gathering and suspended venture progress. Narcissism was particularly valuable, however, in radical innovation contexts, protecting entrepreneurs' visions despite contradictory feedback. This study advances entrepreneurship theory by developing a model examining the interplay of humility and narcissism in entrepreneurial learning. By identifying key factors shaping absorptive capacity, we introduce humility as a crucial yet underexplored trait and offer a nuanced perspective on narcissism's role in learning. We also provide practical insights for entrepreneurs and mentors, emphasizing the importance of self-awareness in recognizing how individual traits can enhance or hinder learning from external feedback.

CONTACT: Maya Kumar; maya.kumar@ie.edu; IE University; Paseo de la Castellana, 259, Madrid, Madrid 28046, Spain

SUMMARY

EVOLUTION IN EMOTION - A PROCESS STUDY OF HOW ENTREPRENEURS PITCH IN FRONT OF VENTURE CAPITALISTS

Denis Lambert, ESCP Business School, Vietnam

Principal Topic

Entrepreneurs' emotional expressions during pitches significantly influence venture capitalists' (VC) investment decisions. While prior studies focus on discrete emotions, little is known about how emotions evolve throughout a pitch. This study examines the dynamic nature of entrepreneurs' emotional expressions and their impact on funding outcomes. By analyzing the rate of change (George and Jones, 2000) in emotions over time, we identify different patterns of emotions during pitch competitions. Conceptualizing emotions as interconnected and ever-changing states rather than isolated moments, this research contributes to the understanding of entrepreneurs' emotions and highlights the role of emotional variability in successful pitches.

Method

We analyzed 150 entrepreneurial pitch videos from major competitions, capturing 4,542 emotional state changes. Using the Body Action and Posture (BAP; Dael et al., 2012) framework and the circumplex model of affect (Russell, 1980), we coded entrepreneurs' body gestures and postures into emotional valence and arousal. We applied discrete-time Markov chains to identify transition probabilities between emotional states and used temporal bracketing to classify emotional evolution into four distinct patterns. These patterns were then mapped to funding outcomes, examining whether specific emotional trajectories correlated with higher or lower investment success.

Results and Implications

Our findings reveal four distinct emotional evolution patterns: (1) Continuous stability, where emotions remain unchanged throughout the pitch; (2) Tipped fluctuation, with an initial rise in positive emotions that declines toward the end; (3) Progressive fluctuation, characterized by increasing emotional transitions toward positive states; and (4) Absence of stability, where emotions oscillate frequently. Patterns 3 and 4 were significantly associated with higher funding success, while Patterns 1 and 2 showed lower success rates. These results suggest that emotional variability, rather than emotional stability, enhances entrepreneurs' ability to engage investors. This research advances theories of emotional dynamics in entrepreneurship and offers practical guidance for enhancing the effectiveness of pitch presentations. Entrepreneurs can leverage this understanding to refine their pitch delivery, strategically managing emotional transitions to improve persuasiveness. Investors, in turn, may develop a more nuanced perspective on interpreting entrepreneurial emotions, allowing for more informed decision-making.

CONTACT: Denis Lambert; denis.lambert@edu.escp.eu; ESCP Business School; 37 Nguyen Van Huong, Thao Dien, Ho Chi Minh City, Ho Chi Minh 700000, Vietnam

SUMMARY

DRIVERS OF ENTREPRENEURIAL PIVOTING: UNPACKING THE INTERPLAY BETWEEN AFFECT, STRATEGY, AND FUNDING

Sukanya K. Roy, Indian Institute of Management Bangalore, India
Ludvig Levasseur, Indian Institute of Management Bangalore, India

Principal Topic

In this paper, we aim to study the impact of entrepreneurial affect and cognition on the decision to pivot. A pivot represents a fundamental change in the current venture idea (Grimes, 2018) and entrepreneurs need to consistently (re)evaluate the transition from pursuing an original idea or opportunity to acting upon an alternate idea or opportunity (Martins et al., 2015; McMullen & Shepherd, 2006). Yet, the central role of Positive and Negative Affect (Baron, 2008) and cognitive factors (Grégoire et al., 2011) in pivoting decisions remain underexplored. We thus focus on how entrepreneurial PA, NA, and overconfidence impact pivot decisions and how the frequency of lean startup practices and funding moderates these relationships.

Method

Using Prolific, we collected data to form a final sample of 272 U.S. entrepreneurs in three rounds separated by two weeks. We used standard instruments to measure our independent variables including PA and NA (Watson et al., 1988) and overconfidence (Forbes, 2005). We measured pivoting using six yes/no questions in which a "yes" to any one of the questions indicated a pivot. We also replicated our main results with a final sample of 273 British entrepreneurs. We tested our hypotheses using CB-SEM and lavaan in R software.

Results and Implications

We find that entrepreneurs scoring high on PA are more likely to pivot (Hypothesis 1) whereas those scoring high on NA are less likely to pivot (Hypothesis 2). Moreover, a frequent use of lean startup practices positively moderates these relationships (Hypotheses 4a and 4b) whereas funding status has a negative moderating effect (Hypotheses 5a and 5b). Finally, we do not find support for the impact of overconfidence or related moderators (Hypotheses 4c and 5c). Beyond contributing to why some entrepreneurs do pivot while others do not, our results notably highlight that affect is a key driver in pivots, with high levels of PA and low levels of NA promoting feedback-based opportunity recognition for pivot. Moreover, while using lean practices strengthens PA's positive effect and weakens NA's negative effect on pivoting, securing funding (i.e., a source of external validation of the original opportunity) weakens PA's positive effect and strengthens NA's negative effect on pivoting. Although replication results convey that NA positively impacts pivoting and no support for the moderating effect of lean startup practices and funding status, these results increase the generalizability of our main results (Anderson et al., 2019).

CONTACT: Sukanya K. Roy; sukanya.roy20@iimb.ac.in; Indian Institute of Management Bangalore; Flat No 39106, Bengaluru, Karnataka 560087, India

SUMMARY

ENTREPRENEURIAL TEAM MEMBER EXITS: THE ROLE OF EMOTIONAL EXPRESSIONS DURING CONFLICT IN PREDICTING TEAM SATISFACTION AND INTENT TO LEAVE

Christian T. Wimmer, Technical University Munich, Germany
Theresa Treffers, Technical University Munich and Seeburg Castle University, Germany
Isabell Welpe, Technical University Munich, Germany

Principal Topic

The departure of a co-founder, even if it does not result in the complete dissolution of the venture, can significantly negatively impact the founding team and their business (Preller et al., 2023). While we know that co-founder exits are common in new venture teams (NVTs) and that their effects are severe, entrepreneurial scholars are calling for more research to better understand how team members' exit intentions and their decisions to leave the venture emerge (Patzelt et al., 2021). What literature still cannot answer is the role discrete emotions in the form of facial expressions play in conflict situations and their direct and/or indirect effect on team satisfaction and exit intentions in an entrepreneurial context such as NVTs.

Method

To test our hypotheses, we conducted a longitudinal observational study with early-stage dyadic founding teams by collecting video and audio data from an induced 15-minute conflict discussion between the co-founders. This on-site session has been adapted from the experimental framework proposed by Levenson and Gottman (1992) on freshly married couples. We capture the emotional data from the facial expressions of the founders through an emotion recognition software tool. We then analyzed the aggregated amount of discrete emotions (e.g., anger and joy) and the overall amount of positive and negative emotions expressed during the conflict while controlling for recognizability through threshold mechanisms.

Results and Implications

This study analyzed data from 60 NVTs with subsequent matching data of their exit intention and team satisfaction six months after the lab session. Our results show that joy significantly lowers co-founder exit intentions through increased team satisfaction and that anger counterintuitively had no direct or indirect effect on the exit intentions of NVTs. First, we contribute to research on team satisfaction within NVTs (Foo et al., 2006) by introducing its antecedent role in the founders' intention to exit the venture. Second, we contribute to research on NVTs by challenging the findings of Thomas et al.,(2023) by showing that expressions of joy during conflict enhance team satisfaction and reduce exit intentions, contrasting the finding that masked positivity (i.e., joy) predicts co-founder exits. Finally, we advance research on entrepreneurial team member exits (Patzelt et al., 2021) and conflict in NVTs (Ensley et al., 2002) by showing that positive interpersonal emotional exchange, especially during conflict, can have implications for success (i.e., longevity) of NVTs due to lower exit intentions of the co-founders. This suggests fostering positive emotional exchanges, even in moments of disagreement.

CONTACT: Christian T. Wimmer; christian.wimmer@tum.de; Technical University Munich; Kaulbachstraße 22, Munich, Bayern 80539, Germany

SUMMARY

EXPANDING THE LEAN STARTUP FRAMEWORK: THE ROLE OF VALIDATED LEARNING FROM TECHNICAL SPECIALISTS

Siwei Zhu, Minnesota State University Moorhead, United States
Dan Hsu, North Dakota State University, United States

Principal Topic

Entrepreneurship educators have started using the lean startup approach to teach students in higher education (Bhatia and Levina, 2020). However, the practical implementation of validated learning in entrepreneurship education reveals a critical issue: students face difficulties identifying appropriate "target" customers for validation due to a lack of technical knowledge and reliance on network-based feedback (Shepherd and Gruber, 2021). Knowledge network theory provides a valuable lens for examining this issue as it delineates the roles of different types of network members in knowledge exchange (Phelps et al., 2012). We classify the networks of student entrepreneurs into those with business generalists and those with technical specialists. Our work further suggests that failure analysis is also aided by the student entrepreneurs' own engagement (Dahlin et al., 2018), especially through the individual networks they turn to for support.

Method

We conducted a between-subjects and randomized field experiment using a longitudinal design with 291 students from an MBA entrepreneurship course at a major university in the Midwest, USA. The participants were tasked with seeking feedback through three assigned conditions: network with technical specialists, network with business generalists, or no network at all. Data collection occurred at three time points. Time 1 Survey was collected for demographic information once the participants were invited. Four months later, we surveyed the same participants and measured the amount of interviewing time and self-reported failure analysis as Time 2 Survey. Two weeks later, the participants presented their MVPs, which were graded by 5 judges, in a competition as Time 3 Survey.

Results and Implications

We examined the mediating model and results show a significant and positive total effect of the technical specialist condition with the control condition for success of MVPs; a significant and positive total effect of the business generalist condition with the control condition for success of MVPs; failure analysis is a significant predictor of success of MVPs and the indirect effect on success of MVPs through failure analysis is significant, meaning that compared to those in the business generalist condition, student entrepreneurs in the technical specialist condition are more likely to conduct failure analysis, which leads to success of MVPs. We highlight that student entrepreneurs—who often struggle to implement the lean startup approach—can develop superior MVPs by integrating validated learning from both technical specialists and business generalists. This study underscores the pivotal role of an individual's knowledge network as a precursor to their active engagement in failure analysis.

CONTACT: Siwei Zhu; siwei.zhu@mnstate.edu; Minnesota State University Moorhead; 1104 7th Avenue South, Center for Business, Q207, Moorhead, Minnesota 56563, United States

INTERACTIVE PAPER

INTUITIVE CAPABILITIES AND THE IMPACT ON ORGANIZATIONAL LEARNING

Marc Dal-Canton, Rhein Main University of Applied Science, Germany
Gunnar Lang, Rhein Main University of Applied Sciences, Germany
Mette Søgaard Nielsen, University of Southern Denmark, Denmark

Principal Topic

Organizational learning – change that occurs as organizations acquire experience (Argote & Miron-Spektor, 2011) – is essential for entrepreneurial performance and long-term success (Zhao et al., 2011; Zhang et al., 2022). The 4I-Framework (Crossan et al., 1999) posits that individual intuition – affectively-charged judgments that arise through rapid, nonconscious, holistic association (Dane & Pratt, 2007) – is the basis of organizational learning. Indeed, intuition is referred to as the "seed of any entrepreneurial action" (Dutta & Crossan, 2005), yet research on entrepreneurial intuition is limited, mostly relying on self-reports such as interviews or questionnaires (e.g., Baldacchino, 2016).

Method

To test our hypothesis that an entrepreneur's intuitive capabilities relate to organizational learning, we draw on neuroscientific research to measure these capabilities. Studies indicate that individuals develop intuitive cues (somatic markers (SM)) via the sympathetic nervous system (SNS), guiding decisions before they consciously understand the respective cognitive tasks (Bechara et al., 1997; Bierman 2005). The SNS typically reacts to danger (fight-or-flight response) (Jansen et al., 1995), suggesting that before individuals consciously understood the task, the SNS adequately responds, i.e., upcoming incorrect decisions are perceived as danger. Research also shows that individuals with strong SM perform better in cognitive tasks (Merchán-Clavellino et al., 2019), indicating that SM provide a way to evaluate intuition: the higher the SNS activity before wrong decisions, the better the intuitive capability. In our study, entrepreneurs participated in implicit tests while SNS before incorrect decisions were measured via Skin Conductance Response Analysis. Implicit tests are defined by individuals' inability to articulate the rules but still allow for the acquisition of implicit knowledge (Shanks 2005), leading to correct decisions without being able to explain why. Given the inherent limitations of measuring intuition, in addition to the SNS pre-decision response, we also assessed entrepreneurs' intuition through implicit test results which assess their ability to unconsciously acquire and apply knowledge (Kaufman et al., 2010), aligning with intuition's definition. To measure organizational learning, entrepreneurs answered a questionnaire weekly for 6 weeks about changes in practices, product, and performance (Argote & Miron-Spektor, 2011).

Results and Implications

Our current results from 18 entrepreneurs did not show a significant relation between entrepreneurs' intuition (measured via SNS and implicit tests) and organizational learning. This contradicts the 4I-Framework and previous empirical research in entrepreneurship, which investigated intuition's influence based on self-reports. Our research therefore currently questions whether intuition truly is the "seed of any entrepreneurial action".

CONTACT: Marc Dal-Canton; marc.canton@hs-rm.de; Rhein Main University of Applied Science; König-Adolf-Straße, Wiesbaden, Hessen 65191, Germany

INTERACTIVE PAPER

STRESS AND ASSESS: ENTREPRENEURIAL OPPORTUNITY EVALUATION IN A CONDITION OF INDUCED STRESS

Gustav R. Hjerting, Technical University of Denmark, Denmark
Carina Lomberg, Technical University of Denmark, Denmark

Principal Topic

Entrepreneurs frequently make decisions in high-stress environments, yet the specific impact of acute stress on opportunity evaluation remains underexplored. Stress can impair decision-making, increasing cognitive errors, reliance on stereotypes, and disregard for contextual factors, although previous research suggests some exposure to stress can improve cognitive performance in the short term. Our study examines how stress influences entrepreneurial opportunity evaluation, considering both first and third-person perspectives. We hypothesize that stress affects cognitive flexibility and the depth of information processing, potentially leading to the overvaluation or undervaluation of opportunities. Additionally, the timing of stress exposure—before or after processing opportunity-related information—may differentially affect opportunity evaluation. By studying nascent as well as experienced entrepreneurs, this research advances the understanding of cognitive processes in entrepreneurial decision-making under stress.

Method

We conducted a between-subjects experiment with 60 entrepreneurs, both nascent and experienced, assigned to one of four conditions: two experimental groups (stress induced before or after opportunity presentation) and corresponding control groups for each. Stress induction followed the "method to induce stress in human subjects in online research environments" by Almazrouei et al., involving timed arithmetic and general knowledge questions with performance monitoring and negative feedback. The control group answered questions without time pressure or feedback. Participants evaluated a graphene-based technology opportunity under these conditions. Key variables included first and third-person opportunity evaluation, information recall, and physiological stress responses measured via Empatica EmbracePlus wristbands. Controls included entrepreneurial experience, market knowledge, age, and gender.

Results and Implications

Preliminary findings indicate that acute stress significantly influences opportunity evaluation, though not always negatively. Understanding when stress enhances versus impairs decision-making is critical. This study contributes to entrepreneurial psychology by highlighting how stress impacts cognitive processes during opportunity assessment. It also aligns with broader discussions on entrepreneurial mental health, emphasizing the role of stress in decision-making. Recognizing the impact of stress on opportunity evaluation could inform interventions that enhance entrepreneurial resilience and business success. By identifying when stress management strategies are most pertinent, our research provides practical implications for supporting entrepreneurs in high-pressure environments.

CONTACT: Gustav R. Hjerting; grahj@dtu.dk; Technical University of Denmark (DTU); DTU Entrepreneurship, Kongens Lyngby, Hovedstaden 2800, Denmark

INTERACTIVE PAPER

THE IMPACT OF WORKPLACE IMPOSTOR THOUGHTS ON ENTREPRENEURIAL HUSTLE: EXAMINING THE MODERATING ROLE OF FELT OBLIGATION

Pia Leyendecker, RWTH Aachen University, Germany

Principal Topic

The impostor phenomenon, marked by self-doubt and fear of being exposed as a fraud, affects nearly two-thirds of individuals (Gravois, 2007). In the work environment, research primarily examines its negative effects, such as reduced satisfaction and increased stress (Bravata et al., 2020). However, recent work conceptualizes workplace impostor thoughts (WIT) as self-doubt stemming from the perception of being overestimated by others at work , revealing positive effects such as enhanced interpersonal effectiveness (Tewfik, 2022). While its impact on workplace behavior is gaining recognition, its role in entrepreneurship remains unexplored (Guedes, 2023). Our study examines how WIT influence entrepreneurial hustle, a driver for venture performance, comprising urgency (speed of action) and unorthodoxy (unconventional actions) (Burnell et al., 2023). Drawing on Social Cognitive Theory (Bandura, 1986), we propose that WIT relate positively to unorthodoxy and negatively to urgency, with felt obligation (i.e., a sense of personal duty toward the venture) moderating both relationships.

Method

Our study uses a cross-sectional design, surveying 358 entrepreneurs in Germany, Austria, and Switzerland. We used validated multi-item scales, with responses recorded on a 7-point Likert scale. Entrepreneurial hustle, our dependent variable, was measured using Burnell et al.'s (2021) scale and WIT were assessed following Tewfik (2022). Our moderator felt obligation was measured using Eisenberger et al.'s (2001) scale. To ensure robustness, we controlled for firm- and individual-level factors. Hypotheses were tested using Structural Equation Modeling in AMOS. Additionally, six semi-structured interviews complemented the quantitative analysis.

Results and Implications

Our findings reveal a dual effect of WIT on entrepreneurial hustle. While they negatively impact urgency, causing hesitation and overpreparation, they positively influence unorthodoxy, fostering creative, non-traditional approaches. Furthermore, felt obligation strengthens the positive relationship between WIT and unorthodoxy. However, our hypothesis that felt obligation weakens the negative impact on urgency was not supported. Our study contributes to research on the impostor phenomenon in three ways. First, we contextualize WIT within entrepreneurship, exploring their impact beyond traditional organizations. Second, we challenge the prevailing views that focus primarily on negative outcomes by demonstrating that WIT can lead to more unorthodox actions, which have been shown to positively impact venture performance. Third, we introduce felt obligation as a moderator, illustrating how entrepreneurs' internalized sense of duty toward their venture can amplify the positive impact of WIT on unorthodox behavior. Despite our findings, further research is needed to identify mechanisms that can reduce hesitation to support the overarching behavior of entrepreneurial hustle.

CONTACT: Pia Leyendecker; leyendecker@time.rwth-aachen.de; RWTH Aachen University; Erftstr. 8, Cologne, Nordrhein-Westfalen 50672, Germany

INTERACTIVE PAPER

AH YOU ARE ONE OF THOSE ENTREPRENEURS: AN INVESTIGATION INTO INVESTOR EVALUATIONS AND SUPPORT OF TYPECASTED ENTREPRENEURS

Jonathan C. Preedom, Louisiana State University, United States
Chad Navis, Clemson University, United States

Principal Topic

Investors evaluate numerous ventures under time constraints, relying on heuristics (Huang & Pearce, 2015) and observable cues (Glynn & Navis, 2013) amid uncertainty (Fisher & Neubert, 2023). This sensemaking process (Weick, 1995) leads to typecasting entrepreneurs as CEOs, Salesmen, Technicians, Humanitarians, or Novices, shaping resource allocation. Financial, social, and human capital decisions follow distinct logics—profit maximization, identity/status, and interpersonal alignment—affecting investor perceptions and the resources entrepreneurs receive (Kalvapalle, Phillips, & Cornelissen, 2024).

Method

This research employs a mixed-methods approach (Creswell & Poth, 2016). First, a qualitative study analyzed investor evaluations of entrepreneurial pitches, identifying key cues used for entrepreneurial typecasting. Study two examined 777 investor decisions from 62 live pitches, integrating qualitative coding and hierarchical multiple regression to assess how such typecasting impacts financial, social, and human capital support.

Results and Implications

The first study revealed investors consistently identified and interpreted specific cues to aid in their decision-making. The primary types emerging from the data include: 1. Technician: Identified by technical jargon, weak communication skills, and visible discomfort. 2. Novice: Characterized by enthusiasm disrupting their ability to deliver a smooth pitch, often stumbling through a memorized script. 3. CEO: Delivers an engaging, confident pitch while appearing professional and drawing audience attention. 4. Humanitarian: Deeply committed to a personal cause, focusing their pitch on the significance of that cause, rather than the business. 5. Salesman: Seen as overly polished, attention-seeking, through a flashy presentation. Entrepreneurial typecasting strongly influences support decisions. CEOs and Salesmen are more likely to receive financial backing, while Humanitarians, Technicians, and Novices attract social capital due to their perceived social value. Human capital support remains unaffected by typecasting, suggesting interpersonal factors like homophily play a greater role. These findings highlight how investors use heuristics to categorize entrepreneurs, shaping resource allocation. While typecasting helps investors navigate uncertainty, it also determines the types of capital entrepreneurs receive. This study contributes to entrepreneurship and investment research by providing empirical evidence on the role of typecasting in resource distribution, emphasizing distinct decision logics for financial, social, and human capital.

CONTACT: Jonathan C. Preedom; jcpreedom@lsu.edu; Louisiana State University; 18926 Lake Harbour Ave, Baton Rouge, Louisiana 70816, United States

INTERACTIVE PAPER

REVISITING NARCISSISM: FORMS OF GRANDIOSE NARCISSISM AND THEIR EFFECTS ON PERSPECTIVE TAKING AND NEW VENTURE IDEATION

Aakash Sapru, Iona University, United States
Brandon Mueller, Iowa State University, United States

Principal Topic

High levels of narcissism in entrepreneurs coupled with uncertainty in the environment may lead these entrepreneurs to overvalue the correctness, suitability, and underlying rationale of the ideas they develop, and undervalue the perspective of others who disagree with their assessment. During new venture ideation, this self-centered perspective can create challenges and lead the entrepreneur to evaluate the originality, usefulness, and user alignment of their new venture ideas based on their own viewpoint rather than that of the potential customers or users. Counterintuitively, however, a recent meta-analysis exploring the impact of narcissism on entrepreneurial performance found that narcissism has an overall positive effect on new venture performance. To reconcile such apparent anomalies, we explore the complex nature and evolving conceptualization and operationalization of the narcissism construct. We argue that narcissism is not a monolithic construct (as conceptualized in prior studies) and that disparate forms of narcissism have different (i.e., positive and negative) impacts on perspective-taking and the quality of new venture ideas. We integrate the agency-communion model of narcissism with the narcissistic admiration and rivalry concept and underscore how communal narcissism, narcissistic admiration, and narcissistic rivalry influence new venture idea quality through their differential associations with perspective-taking.

Method

The study comprised 206 undergraduate students enrolled in an entrepreneurial class at a midwestern university. Validated measures were used to assess communal narcissism, narcissistic admiration and rivalry, and perspective-taking. We provided participants with a description of facial recognition software and asked them to generate new venture ideas and applications. Two independent judges familiar with the local context and experienced in assessing business ideas assessed the ideas using the consensual assessment technique. We controlled for age, gender, openness to experience, number of words written, time spent in generating the idea, number of entrepreneurship classes taken, and whether the students had started a business. Hierarchical regression analysis was used to test our hypotheses. To test our mediation hypotheses, we employed the bootstrapping resampling technique.

Results and Implications

We found that communal narcissism and narcissistic rivalry explained a significant amount of variance in perspective-taking, which (i.e., perspective-taking) explained a significant amount of variance in NVI quality. Similarly, communal narcissism and narcissistic rivalry explained a significant amount of variance in NVI quality. Furthermore, we found that perspective-taking fully (not partially, as hypothesized) mediates the association between communal narcissism and idea quality. Results for perspective-taking partially mediating the association between communal narcissism and idea quality were supported as hypothesized.

CONTACT: Aakash Sapru; asapru@iona.edu; Iona University; 387 Huguenot Street, New Rochelle, New York 10801, United States

INTERACTIVE PAPER

SURVIVING THE CHAOS: WHEN DO ENTREPRENEURS ENGAGE IN DISINHIBITED BEHAVIOR?

Jun Zhou, ESCP Business School, France
Daniel Lerner, IE University, Spain

Principal Topic

Disinhibited behavior, defined as unrestrained, impulsive actions that arise from reduced self-regulation or inhibition, is a meaningful action logic of decision-making and opportunity pursuit for entrepreneurs navigating uncertainty (Lerner et al., 2018). While prior research has explored what this behavior is and why it is important to new venture creation, we lack the knowledge about "when" entrepreneurs engage in disinhibited behavior. Traditionally, scholars treat neurodiverse traits (e.g., ADHD) as the main antecedents of disinhibited behavior. While traits like impulsivity or ADHD-related tendencies provided an important starting point, this dispositional perspective fails to account for the situational nature of disinhibition. Under high uncertainty, entrepreneurs, frequently experiencing tensions and dissonance, are motivated to engage in discrepancy reduction via different behaviors (Festinger, 1957). Thus, we integrate tension literature and Cognitive Dissonance Theory (CDT) to explore when entrepreneurs, beyond dispositional-based predispositions, engage in disinhibited behavior.

Method

This study adopts a qualitative design centered on in-depth interviews, capturing entrepreneurial behaviors during the new venture creation process (Jiang & Tornikoski, 2019). The research was conducted among 18 start-ups in the biotech and health sectors in France. The unique combination of high uncertainty, complexity, and innovation-driven challenges, makes these sectors ideal for examining the interplay between tension and entrepreneurial behavior (Pisano, 2006). We performed two steps in data analysis. First, we coded internal (dissonance) and external tension, the underlying action logic, and new venture outcomes (Hinojosa et al., 2017; Lerner et al., 2018; Putnam & Mumby, 2014; Wood et al., 2012). This analysis generated 316 tension events. Second, we translated our qualitative coding into a quantitative database to perform a qualitative comparative analysis (QCA) (Ragin, 2008).

Results and Implications

Our results found that disinhibited behavior is strongly associated with the presence of cognitive dissonance and absence of external tension. In contrast, reasoned behavior aligns with the absence of cognitive dissonance and the presence of external tension. We offer three contributions. First, we reveal that regardless of dispositional trait predispositions, entrepreneurs engage in disinhibited behavior, particularly when experiencing cognitive dissonance. Second, we refine the construct of disinhibited behavior by shifting from a dispositional-based view to a situational perspective, showing how cognitive, emotional, and environmental factors shape it. This challenges the notion that disinhibition is simply dispositional and highlights its adaptive role in decision-making. Third, we refine action logic framework, showing that entrepreneurs alternate between disinhibited and reasoned behavior depending on tension type. This advances the understanding of entrepreneurial action logic and practice.

CONTACT: Jun Zhou; jun.zhou@edu.escp.eu; ESCP Business School; 5 Rue Valentin Hauy, Paris, Ile-de-France 75015, France

THE PARADOX OF PASSION: THE INTERPLAY BETWEEN ENTREPRENEURIAL PASSION AND FEAR OF FAILURE IN RESPONSE TO NEGATIVE FEEDBACK

Md Kamrul Hasan, Washington State University, United States
Benjamin J. Warnick, Washington State University, United States

ABSTRACT

This paper explores the simultaneous experience of passion and fear in entrepreneurs, highlighting the interplay of their emotional dynamics. Passion and fear are studied separately and positioned as antithetical: passion drives dedication, while fear of failure can deter action. We bridge this gap by examining how passion moderates the impact of negative feedback on fear and subsequent decisions to pivot or persist. An experiment with nascent entrepreneurs reveals that passion increases fear of failure in response to negative feedback but also strengthens the intention to pivot. We offer an important bridge between research on entrepreneurial passion and fear of failure and provide a deeper understanding of the emotional mechanisms in entrepreneurial decision-making.

INTRODUCTION

Passion drives entrepreneurs to persist in their ventures even in the face of significant adversity. But is that always the case? Entrepreneurial passion is an intense positive emotion for venture-related opportunities, tasks, and activities (Baum & Locke, 2004; Cardon et al., 2009). Entrepreneurial passion has a motivational effect that stimulates entrepreneurs to overcome obstacles and remain engaged (Cardon et al., 2009). Such obstacles are usually faced during difficult times for the venture when entrepreneurs must deal with negative emotions, such as fear of failure— "an emotional reaction emerging from the cognitive appraisal of experiences that represent a threat to one's ability to accomplish a personally meaningful goal" (Cacciotti et al., 2020, p. 3).

Recent research suggests that fear of failure can explain why entrepreneurs might pivot during difficult times, making "a change in a firm's strategy that reorients the firm's strategic direction through a reallocation or restructuring of activities, resources, and attention" (Kirtley & O'Mahony, 2020, p. 199). In contrast, entrepreneurial passion enhances persistence, defined as continual repeated effortful actions despite failures or threats (Gimeno et al., 1997) or sustained entrepreneurial actions (Cardon & Kirk, 2015). When faced with challenges or adversity, entrepreneurs must decide how to proceed, including whether to pivot for the survival of the venture.

Entrepreneurship scholars have not yet investigated the complexities of pivoting decisions made by entrepreneurs who simultaneously experience both passion for their venture and fear that it might fail. Therefore, we aim to unravel the complex interplay between passion and fear. By examining how passion shapes the experience of fear and how entrepreneurs respond to fear, we provide a deeper understanding of the emotional mechanisms at play in entrepreneurial decision-making and offer an important bridge between research on entrepreneurial passion and fear of failure.

HYPOTHESIS DEVELOPMENT

The Moderating Effect of Entrepreneurial Passion on Fear of Failure

Entrepreneurs continuously seek feedback from stakeholders as they navigate uncertainty and strive for venture success (Pryor et al., 2016; Kaffka et al., 2021). Negative feedback challenges entrepreneurial ideas for products and services and helps them assess their potential (Davila et al., 2003; Shepherd et al.,

2009; Clausen, 2020; Domurath et al., 2020). In the face of negative feedback, entrepreneurs often experience fear of failure if that feedback indicates a potential threat to venture survival (Cacciotti et al., 2016, 2020). Building upon this, we theorize that entrepreneurs' passion for their venture's product or service, or *product passion* (Warnick et al., 2018), plays a role in increasing entrepreneurs' fear of failure.

The positive, personal identification of entrepreneurs with their products and services plays a critical role in connecting entrepreneurial passion with fear of failure in the presence of negative feedback. Humans naturally worry about losing beloved people or objects in our lives when faced with uncertainty and risk (Cacciotti et al., 2020; Conroy, 2001). Recognizing this, we hypothesize that entrepreneurs' passion for their product or service (i.e., product passion; Cardon et al., 2017; Warnick et al., 2018) increases their experience of fear of failure in response to negative feedback. Entrepreneurs who are passionate about their ventures form a deeply personal connection with their product or service ideas (Cardon et al., 2009). We theorize that this passion leads to heightened emotional reactivity when faced with criticism or setbacks. For entrepreneurs with a high degree of product passion, negative feedback is perceived not just as a critique of the venture's potential, but as a personal affront, amplifying the entrepreneur's fear of failure. The stakes feel inherently higher because the entrepreneur's self-worth and identity are closely tied to the success of their venture.

Hypothesis 1a: Negative feedback about the potential of their business idea increases entrepreneurs' fear of failure regarding the potential of their business idea.

Hypothesis 1b: Entrepreneurs' product passion strengthens the effect of negative feedback on fear of failure.

Fear of Failure, Pivoting Intentions, and Persistence

Entrepreneurs' fear of failure impacts their cognition and behavior in both beneficial and detrimental ways (Cacciotti & Hayton, 2015). Fear can either motivate or inhibit entrepreneurial behavior (Cacciotti et al., 2016). Previous research has predominantly viewed fear as a psychological barrier to entrepreneurship, inhibiting entrepreneurial action (Shinnar et al., 2012; Cacciotti & Hayton, 2015). However, fear of failure significantly influences entrepreneurial cognition, emotions, and behaviors, particularly when failure is anticipated (Heckhausen, 1975).

First, fear impacts entrepreneurs' pivoting decisions. Entrepreneurs experiencing a high degree of fear of failure during periods of uncertainty and challenge exhibit a greater inclination to learn and adapt novel solutions for venture survival, making them more likely to pivot. Negative emotions enhance goal-directed behaviors (Carver & Scheier, 1990; De Dreu et al., 2008), and as a result, fear drives entrepreneurs to seek innovative solutions (McGrath, 2001) and focus more on learning (Shepherd et al., 2011). However, fear of failure can also affect entrepreneurial intentions (Arenius & Minniti, 2005) and their decisions to persist. Negative emotions can activate demotivation and an avoidance orientation, leading individuals to withdraw from their goals or abandon them altogether (Bandura, 1991; Gabriel et al., 2022), suggesting fear reduces entrepreneurial persistence. Consequently, when venture survival necessitates changes in strategic direction or in the products and services offered, entrepreneurs are more likely to pivot and less likely to persist, driven by their fear of failure.

Hypothesis 2a: Entrepreneurs' fear of failure is positively related to their intention to pivot from the business idea.

Hypothesis 2b: Entrepreneurs' fear of failure is negatively related to persistence with their business idea.

The Moderating Effect of Entrepreneurial Passion on Pivoting Intentions and Persistence

We focus on entrepreneurs' passion for the venture's product and services (Warnick et al., 2018), where product passion is conceptualized by consciously accessible, intense positive feelings for a product or service that is central to an entrepreneur's identity (Cardon et al., 2017; Warnick et al., 2018). Entrepreneurs who are passionate about their venture's product/service are even more likely to persist when faced with uncertainty (Winnen, 2005; McMullen & Shepherd, 2006; Stenholm & Renko, 2016; Murnieks et al., 2020). Passion plays a crucial role in shaping how entrepreneurs respond to fear. Research on emotional ambivalence explains that experiencing positive (e.g., passion) and negative emotions (e.g., fear) together encourages individuals to think critically and reevaluate their goal-directed cognition (i.e., flexibility in strategy development), eventually leaning more toward persistence in self-regulatory efforts (Gabriel et al., 2022). Rather than succumbing to the pressure to pivot or abandon their efforts altogether (cf. Kirtley & O'Mahony, 2020; Wood et al., 2019), passionate entrepreneurs may exhibit a stronger resolve to continue, driven by their positive identification with their product or service. Experiencing positive and negative emotions at the same time encourages individuals to invest more energy toward achieving a desired goal (Lord et al., 2010; Gabriel et al., 2022). This suggests that entrepreneurs who experience both fear and passion are particularly likely to persist with their venture idea compared to those experiencing only one (or neither) of these emotions. Passionate entrepreneurs persist in difficult situations (Bird, 1989; Cardon & Kirk, 2015), experience high enthusiasm to pursue further, and tend to invest significant effort even when the venture performs poorly (Meyer & Zucker, 1989; McGrath, 1999; Cardon et al., 2005). Passionate entrepreneurs often persist despite significant obstacles, but some are willing to pivot for the sake of higher market value and growth of the venture (Cardon et al., 2009).

In contrast with entrepreneurial persistence, pivoting involves changing product features or aspects of the business model (Andries & Debackere, 2006), incorporating a completely new concept (Bajwa et al., 2017), business model innovation (Koen, et al., 2011) or strategic change (Greve, 1998) to better align with the market demands based on feedback to promote firm survival (Kirtley & O'Mahony, 2020). In that passion involves strong identification with the object of passion (Cardon et al., 2009), identity theory provides a useful lens to understand entrepreneurial behavior by explaining why entrepreneurs behave in a certain way that they believe appropriate in certain circumstances (Gruber & MacMilan, 2017). Thus, this theory creates a link between entrepreneurial passion and their behavior, actions, and decisions (Gruber & MacMilan, 2017) and creates an interesting tension for entrepreneurs to resolve—whether, when, and how to pivot in response to negative market feedback. Market influences often pressure changes that conflict with entrepreneurs' passion, such that entrepreneurs who respond to market pressures might find that they are "pivoting away their passion" (McMullen, 2017). Passion limits entrepreneurs from exploring alternatives (Cardon et al., 2009), which eventually limits the process of creative change (e.g., pivoting; Grimes, 2018), suggesting that entrepreneurs with high passion are less likely to pivot in response to negative feedback. Indeed, passion exerts a motivational effect that helps entrepreneurs remain engaged and deal with obstacles during uncertain and difficult times of the venture (Cardon et al., 2009).

Therefore, we theorize that product passion attenuates the positive relationship between fear of failure and pivoting intentions. Product passion fuels steadfastness by virtue of the deep emotional commitment it entails, holding potential to transform fear into a motivating force rather than a paralyzing one. We further expect that passion significantly moderates the countervailing links between negative feedback, fear of failure, and reduced persistence. Product passion not only increases entrepreneurs' propensity to experience failure in response to negative feedback, it further increases their inclination toward persistence in response to fear of failure.

Hypothesis 3a: Entrepreneurs' product passion attenuates the positive relationship between fear of failure and intention to pivot from their business idea.

Hypothesis 3b: Entrepreneurs' product passion attenuates the negative relationship between fear of failure and persistence with their business idea.

METHOD

We test our hypotheses via a two-wave online experiment with a sample of 158 nascent entrepreneurs. At Time 1, participants described their business idea, which included detailing its features, value proposition, target market, and a comprehensive plan for value creation. Passion was assessed by capturing both of its dimensions, including identity centrality and the intense positive feelings for the proposed product/service. At Time 2 (one week later), participants were randomly assigned to receive either positive or negative feedback regarding the potential of their business idea. A manipulation check validated the effectiveness of this feedback intervention. Subsequently, we measured participants' fear of failure and intentions to pivot and persist.

RESULTS

Negative feedback increased the entrepreneurs' fear of failure regarding the potential of the business idea ($b = 1.46$, $p = .000$), supporting Hypothesis 1a. In turn, fear of failure increased intentions to pivot ($b = 0.50$, $p = .000$), supporting Hypothesis 2a. We found evidence for first-stage and second-stage moderated mediation. First, consistent with our theorizing, product passion moderated the effect of negative feedback on fear of failure, such that the positive effect of negative feedback on fear of failure was greater for higher levels of product passion, supporting Hypothesis 1b (Figure 1; $b = 0.07$, $p = .014$). The conditional effect of negative feedback on fear of failure was strongest at higher levels (+1*SD*) of product passion ($b = 2.10$, $SE = .36$, $p < .001$), moderate at mean levels ($b = 1.46$, SE = .26, $p < .001$), and weakest at low (-1*SD*) levels ($b = .82$, SE = .36, $p < .05$).

Product passion also moderated the effect of fear of failure on pivoting intentions. Counter to our expected attenuation effect, the positive effect of fear of failure on pivoting intentions was greater for higher levels of product passion (Figure 2; $b = 0.015$, $p = .045$). We thus fail to find support for Hypothesis 3a. The conditional effect of fear of failure on pivoting intentions was strongest at high levels (+1*SD*) of product passion ($b = .64$, SE = .09, $p < .001$), moderate at mean levels ($b = .50$, SE = .07, $p < .001$), and weakest at low levels (-1*SD*, $b = .37$, SE = .10, $p < .001$). Moreover, the conditional indirect effect of negative feedback on pivoting intentions through fear of failure was significant at low (-1*SD*, 95% CI [.04, .63]), mean (95% CI [.45, 1.08]) and high levels (+1*SD*, 95% CI [.86, 1.96]) of product passion.

Finally, we found that fear of failure significantly predicted persistence ($b = -.378$, $p = .000$), supporting Hypothesis 2b. The two-way interaction of entrepreneurial product passion and fear of failure did not significantly predict persistence with the venture idea, such that Hypothesis 3b is not supported ($b = .004$, $p = .438$). However, the conditional indirect effect of negative feedback on persistence through fear of failure was significant at low (-1*SD*, 95% CI [-.78, -.04]), mean (95% CI [-.89, -.30]) and high levels (+1*SD*, 95% CI [-1.26, -.25]) of product passion.

DISCUSSION AND IMPLICATIONS

Our theorizing and findings offer important contributions to extant entrepreneurship research. First, we contribute to the entrepreneurial passion and fear of failure literature by demonstrating that when entrepreneurs receive negative feedback, their passion not only amplifies their experience of fear but also increases their likelihood of pivoting. In doing so, we bridge research streams on passion and fear in entrepreneurship literature, theorizing and finding that passion, as an intense emotional investment, heightens entrepreneurs' susceptibility to experience fear of failure in response to negative feedback. Understanding these intertwined emotional experiences is crucial, as it not only aids in comprehensively addressing the emotional challenges faced by entrepreneurs but also enhances scholarly insight into how these emotions influence entrepreneurial behavior and outcomes.

Second, we contribute an emotion-based perspective on entrepreneurs' decisions to pivot or persist by highlighting responses to emotional ambivalence in which positive and negative emotions are experienced in concert (cf. Gabriel et al., 2022). The literature on emotional ambivalence suggests that individuals exhibit greater flexibility and persistence when experiencing high levels of both positive and negative emotions simultaneously, as this activates a more advantageous motivational process (Gabriel et al., 2022). Our research indicates that when experiencing fear, entrepreneurs with high product passion become more adaptable and demonstrate increased pivoting intentions. Positive emotions, such as passion, enhance confidence and the likelihood of idea revision, thereby fostering flexibility and improved performance through elevated cognitive functioning (Chawla et al., 2019; Fredrickson, 2001; Isen, 2001). Conversely, negative emotions tend to demotivate individuals and increase withdrawal tendencies (Bandura, 1991). Thus, our findings complement previous research by elucidating the conditions under which entrepreneurial passion encourages pivoting, encouraging entrepreneurs to adapt and refine their ideas with greater receptivity to negative feedback.

Third, we enrich broader research on negativity bias in judgment and decision-making (e.g., Baumeister et al., 2001; Lerner et al., 2023) by demonstrating that passion, given the strong personal identification it entails, enhances this negativity bias to drive entrepreneurial pivoting intentions in the face of fear. Our study reveals the intricate relationship between passion and fear in driving pivoting intentions among entrepreneurs when they receive negative feedback on their business ideas. High levels of positive emotions, such as passion, coupled with low levels of negative emotions, such as fear, tend to trigger avoidance-oriented behaviors and goal abandonment. Furthermore, when coupled with high levels of negative emotions, it enhances flexibility, ultimately fostering greater pivoting intentions (Gabriel et al., 2022). Our theorizing and findings demonstrate product passion makes entrepreneurs more likely to experience fear upon receiving negative feedback, which subsequently leads to increased pivoting. We address previously unanswered questions in the literature, such as why some entrepreneurs choose to pivot while others persist despite significant uncertainties and challenges. By doing so, we advance a more comprehensive understanding of the interaction between passion and fear in entrepreneurial decision-making.

CONTACT: Md Kamrul Hasan; mdkamrul.hasan1@wsu.edu; Washington State University, Carson College of Business, 300 NE College Ave, Pullman, WA, USA, 99163

Figure 1. Interaction of Negative Feedback and Passion on Fear of Failure

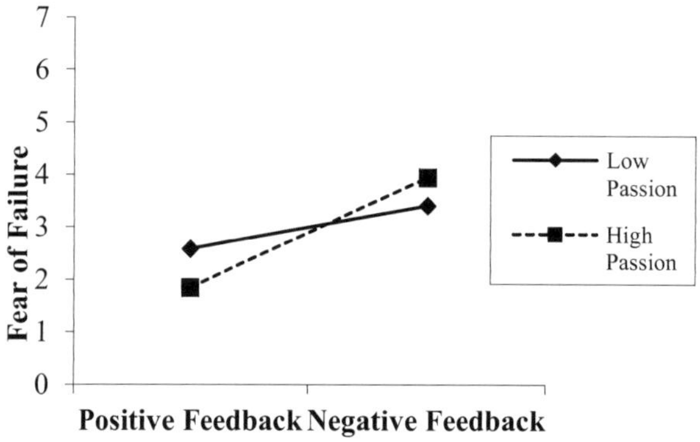

Figure 2. Interaction of Fear of Failure and Passion on Pivoting Intentions

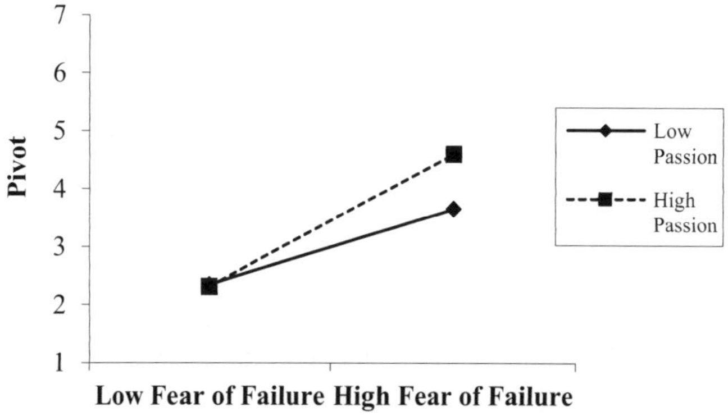

HOW DO UNCONVENTIONAL ENTREPRENEURS OVERCOME THEIR MULTIPLE LIMINALITIES? ENTER THE "GRANMAPRENEURS"

Hayfaa Tlaiss, ADA University, Azerbaijan
Dmitry Khanin, University of New Mexico, United States

ABSTRACT

Prior research has examined how unconventional entrepreneurs operating in contested contexts create new ventures to alleviate their compatriots' sufferings in the wake of a natural disaster or to survive in a hostile social environment (e.g., refugee or migrant entrepreneurs). Such necessity entrepreneurs may encounter liminality, or a transitional state between the past from which they had been separated and the future that yet needs to be secured. Based on intersectionality research and in-depth interviews, we show that new business founders may wrestle with multiple liminalities and apply *intervention agency* to convince themselves and others in their entrepreneurial credentials and competence. Our participants are what we describe as *granmapreneurs*: older women, with children and grandchildren, who gain courage and commitment to start a business while facing overwhelming odds. We establish that such a life change from housewife to granmapreneur involves overcoming the *cataclysmic liminality* of an environment devastated by compound crisis; the *social liminality* of older women in a deeply patriarchal society denying them employment or participation in business activities; and the *specialized liminality* of individuals whose socio-economic status prevented them from receiving education and work experience.

INTRODUCTION

The concept of liminality (from the Latin word *limen*, or threshold) was introduced by the French anthropologist Charles-Arnold Kurr van Gennep in 1909. Van Gennep studied rites of passage in tribal societies, often observed in special ceremonies such as celebrations of harvest, new year, or marriage (Van Gennep, 1909;1960). The scholar suggested that liminality can be broken down into three stages: (1) preliminal, separation stage, which emphasizes that the previous life stage is completed; (2) liminal, or transitional stage, characterized by trials and uncertainty as the past is gone and entry into a future state has yet to be earned; and (3) postliminal. reincorporation stage that embraces and celebrates the achievement of a new state or status. In the last thirty years, scholars increasingly have used the concept of liminality to address the positive aspects of collective and individual transitions viewing them as liberating, transforming endeavors (Ibarra & Obodaru; 2011; Johnson & Sorensen, 2015). Thus, in the area of entrepreneurship, researchers have related liminality to (1) abandoning previous employment or occupation (Kelly & McAdam, 2023), (2) experiencing a period of soul searching and uncertainty as individuals test their fit for entrepreneurship (Garcia-Lorenzo et al., 2020), and (3) acquiring and advancing new occupational and personal identities (Frost et al., 2024).

Other scholars have emphasized that such entrepreneurial transitions are not painless as they represent "ambivalent places of becoming" (Doussard et al., 2023:938), involve internal struggles, and may entail identity change and balancing an older identity with a new identity (Garcia-Lorenzo et al., 2020). Importantly, unconventional entrepreneurs, for example, entrepreneurs who are older, women entrepreneurs, refugee and migrant entrepreneurs, entrepreneurs with disabilities and entrepreneurs from marginalized communities (Bakker & McMullen, 2022) may experience the negative aspects liminality more sharply compared to conventional entrepreneurs. Thus, older entrepreneurs may go through identity interruptions and polarization (Garcia-Lorenzo et al., 2020). Women abandoning their previous employment to embrace a new career as digital entrepreneurs may also suffer from self-doubts and existential precarity as they are not completely accepted by their male colleagues and must learn to navigate the digital liminal spaces in which they find themselves through identity work and identity play (Kelly & McAdam, 2023). Applying the intersectionality lens (Collins, 2019; Crenshaw, 1991), unconventional entrepreneurs may be simultaneously exposed to distinct yet overlapping and mutually reinforcing kinds of marginalization, such as race, gender, age, and sexual orientation.

Advancing these ideas, we inquire how unconventional entrepreneurs overcome their liminality and the intersectional obstacles to entrepreneurship they face as they seek to create new ventures in contested contexts (Simarasl et al., 2024). For this purpose, we conducted a qualitative study examining the entrepreneurial transitions of *granmapreneurs* we define as older women, with children and grandchildren, who create new businesses while facing overwhelming odds. Our participants come from modest socioeconomic backgrounds. They live in a deeply patriarchal society that denies women agency and regards them merely as caretakers. Furthermore, this society is shaken by a severe compound crisis: economic, financial, political and social. We study how granmapreneurs overcome the multiple liminalities they face to complete their entrepreneurial transitions.

Based on semi-structured interviews with Lebanese granmapreneurs, we identify three types of liminalities that represent substantial obstacles to entrepreneurship. First is *cataclysmic liminality* caused by natural and/or societal disturbances that can both stifle economic activities as businesses close and invite new economic entrants. These new economic entrants are necessity entrepreneurs (Dencker et al., 2023; Gruber et al., 2024) that create new, informal businesses for the purpose of assisting their extended families suffering from a deepening compound crisis. Second is *social liminality* that arises due to ingrained inequities in society. It puts entire categories of people (e.g., women, migrants, racial, ethnic and sexual minorities, older people, people with disabilities) into a liminal state. Such social liminality stems from discriminatory social norms and beliefs that define women and/or older people as unsuitable for occupations requiring a strong physical effort or strong analytical and creative abilities. Third is *specialized liminality* that emerges from social inequities that hinder individuals' access to socially favored occupations because of a lack of general education, professional education, work or business experience, capital, and other prerequisites for obtaining employment or starting a business. Such specialized liminality causes personal doubts regarding one's competence and low self-esteem. It is often linked to lower economic status relegating individuals and collectivities to subsistence and crisis contexts.

HYPOTHESES DEVELOPMENT

This is a qualitative study that does not require explicit hypotheses formulation. Nevertheless, we did specify the research questions that guided our empirical examination of the participants' narratives of venture creation. Prior research on liminality (Van Gennep, 1960; Turner, 1982), contrasted its hierarchical, spatial, situational, and temporary types (Soderlund & Borg, 2018) and identified various subjects of liminality, from individuals to national states (Thomassen, 2015). Developing these ideas, we proposed that liminalities can arise due to various disturbances (e.g., natural disasters or economic/financial crises) as well as social inequities (e.g., biases related to gender, race, age, sexuality, and/or socio-economic status). We also argued that although subjects may be exposed to multiple liminalities, they can use their *intervention agency* to claim authority and facilitate the unraveling of their liminal status. Respectively, we identify several liminality types that may thwart entrepreneurship. These are related to macro or micro disturbances or inequities: (1) cataclysmic liminality caused by extreme negative events (e.g., a devastating earthquake or hurricane, migration, Civil war and crises); (2) social liminality (e.g., the humiliating situation of individuals and collectivities regarded by their sexist, elitist, racist, ethnocentric, or agist society as lacking in some positive characteristics); and (3) specialized liminality (e.g., lacking education or work experience as a result of a low socio-economic status, or migration, disease and incarceration). Hence, our research questions can be formulated as follows: (1) What is the effect of multiple liminalities on business creation? (2) How do the subjects of liminality overcome the liminal states related to environmental disturbances, social disadvantages, and social inequities aggravated by life circumstances? 3) Which strategies do the subjects of liminality utilize to overcome multiple liminalities that hamper new business creation?

METHOD

We used face-to-face, in-depth interviews that allowed us to delve deeply into the participants' perspectives (Treanor and Marlow, 2019) and unearth the "interplay between historical conditions, local situations and individual agency" (Foss, 2004: 82). The interviews were semi-structured and two-partite. The first part of the interviews was comprised of questions about the participants' personal demographics and used to create a friendly atmosphere conducive to building trust and sharing personal information. The second part was comprised of open-ended questions about the participants' experience of opening a business, with a special focus on how the decision to start a business was made, the obstacles faced afterwards, and the strategies used to overcome these obstacles.

We applied a purposive sampling strategy (Patton, 1990) to identify a group of participants who founded their own business and operated it during or after the advent of the compound crisis and asked what made them decide to start a business. Considering that starting a business by women in a patriarchal culture during a national crisis was very challenging, the participants were more than willing to share their highly emotional experiences. Such readiness to talk was previously observed in other studies (e.g., Tlaiss and Khanin, 2025). The logic of purposeful sampling lies in selecting information-rich cases (Patton, 1990) that would contribute to examining our research questions, advance knowledge (Corley & Gioia, 2004) and help generate constructive data.

We analyzed the data using Gioia et al.,'s (2013) thematic analysis protocol. As our primary methodology, we used a combination of abduction and induction. Abduction is defined in research on methods as generation of new ideas that could help better understand a phenomenon of interest, or an anomaly and evaluation of such insights (Satre & Van de Ven, 2021). Scholars pointed out that abduction has some advantages over both deduction and induction due to the tight connection between theory and data (Behfar & Okhuysen, 2018). Combining abduction and induction is especially useful for identifying obstacles to change and strategies used by actors to overcome them (Golden-Biddle, 2020). The analysis was conducted in three steps. In the first step, we began with thoroughly familiarizing ourselves with the interviews through iterative readings of the transcripts (Tlaiss and Khanin, 2025). This process involved open coding of the data to allow for the identification of the initial concepts and then grouping them together to form *a priori* categories. In the second stage, we engaged in axial coding (Gioia & Corley, 2004). Thus, we searched and identified the relationship between and among the categories, and subsequently, collapsed them into higher, second order themes. Afterwards, we compared the codes across the interviews and developed a list of codes grounded in the participants' characteristic expressions (Gioia & Corley, 2004). Finally, we listed all the open codes to identify the possible connections between them and group them under larger subcategories. In the third stage, we focused on exploring the relationship between the second-order themes and collapsed them into the several overarching categories (Gioia and Corley, 2004).

RESULTS

Cataclysmic liminality and strategies to overcome it

As the participants related their entrepreneurship narratives, i.e., stories about the life events that have led them to the decision to start a business, they inevitably started with the description of the compound crisis as a startling catastrophe. Throughout their lives, the participants thought of themselves as being safe and financially secure although hardly wealthy. The compound crisis brought about a drastic deterioration in their financial situation and standards of living pushing them to extreme deprivation. The participants reminisced nostalgically about their relatively comfortable lives before the crisis erupted and expressed their shock and disbelief at the abrupt collapse of their lives.

Importantly, the crisis has led to closing of businesses throughout Lebanon. Thus, the participants' relatives who had jobs were laid off as the businesses stopped operating. This created feelings among the participants (and possibly other Lebanese people) that they have entered a new reality, separated from the past, yet unable to advance into the future. Consequently, we described the situation caused by large-scale disturbances including natural disasters and/or economic, social, and political crises/calamities as cataclysmic liminality. Cataclysmic liminality's negative effect lies in precluding action. Thus, the participants' family and community members spent much time talking about the horrors of the present and reminisced nostalgically about the "good" life they had in the past while doing absolutely nothing to improve their situation. Although the participants also contrasted the turbulent present with the more secure past, they took a more proactive, interventionist position by asking themselves what they could do, in practical terms, to help their families. Prior research coined the term negotiated agency as "the urge to maintain institutionally valued goals in the face of crisis" (Venugopal & Viswanathan, 2021: 337) as opposed to completely replacing previous institutional goals. In our study, the participants adopted a more active, interventionist position we described as *intervention agency* as they expressed overt criticism of others who languished in their recollections of the past instead of resolutely acting. Capitalizing on their *intervention agency,* the participants decided to start a business to support their families and convinced themselves and others in their entrepreneurial credentials and competence.

Social liminality and strategies to overcome it

Despite their resolution to act as change agents, the participants could not participate in economic activities (e.g., get a job or start a business) because they were "women of age." In the patriarchal society of Lebanon, women are expected to stay home and start a family. Their main duty is caregiving. Previously, the participants did not have a problem with this established gender role and order of things. However, with the arrival of a national compound crisis this acquiescence was gone. The participants decided that starting a business was the only way they could help their families stuck in cataclysmic liminality. As they considered these initial plans for business creation, the participants understood, however, that they were unrealizable. Hence, we described the situation they encountered as social liminality since it prevented their entrepreneurial transition. The participants knew of no woman in their community who started a business. Although the participants were separated from the past by activating their agency, they could not realize their plans because of their social liminality that precluded action. While cataclysmic liminality acted on a macro level by subjugating people to lethargy and thus creating external barriers to entrepreneurship, social liminality acted on a micro level by creating internal barriers, that is, making the participants doubt their abilities and chances to succeed as entrepreneurs. Furthermore, based on intersectionality theory (Crenshaw, 1991), it was not just gender that hindered the participants' entrepreneurial transition. It was also age. In a patriarchal community neither women nor older individuals could get a job as they were deemed incapable of exerting the physical and mental effort it required. The participants used a strategy of self-legitimation to overcome these internal barriers to entrepreneurship. They used intervention agency and claimed that there is no sacrilege in women seeking to own a business, especially in an industry traditionally regarded in society as "feminine."

Specialized liminality and strategies to overcome it

The participants had no college education or any kind of business education. The reason for that was that they were low-middle-class women expected to drop out of school to get married at an early age. Furthermore, they had zero work experience in professional organizations because of gender roles and patriarchal social expectations of women to work solely in their homes. This created specialized liminality precluding the participants' entrepreneurial transition. While being ready to start a new venture, the participants were hesitant to take this step because they questioned their own ability to run a new venture considering their lack of business education and business experience. The participants overcame this specialized liminality by applying their intervention agency. The participants stated that they already had the competencies needed to run a business as they were recognized in the community for their mastery of a certain skill, such as knitting. In addition, the participants claimed that they would be able to pick up the lacking skills along the way. Thus, we described this strategy of overcoming specialized liminality as practical self-empowerment. Although the participants realized their professional limitations, they thought that such limitations should not stop them as they already had many practical skills and could learn more in the process, experientially, as running a business represents a hands-on activity of constant learning.

DISCUSSION AND IMPLICATIONS

Prior research has applied the concept of liminality introduced and developed in the field of anthropology (Van Gennep, 1960; Thomassen, 2015) to shed light on the nature of entrepreneurial transitions (Garcia-Lorenzo et al., 2020; Kelly & McAdam, 2022; 2023). Advancing this research, we establish that unconventional entrepreneurs (Bakken & McMullen, 2022) operating in contested contexts (Simarasl et al., 2022; 2024) may encounter multiple liminalities that they must learn to overcome by applying appropriate strategies geared to each liminality type. Specifically, unconventional entrepreneurs, such as women, refugees, migrants, and ethnic and sexual minorities may find themselves in the situation of a natural disaster or compound crisis. The ensuing cataclysmic liminality could make it seem like individual action is impossible in the face of mass-scale disturbances causing many institutions to collapse or become inoperable. Thus, banks in Lebanon refused to cash their depositors' checks, beyond minimal amounts, making them useless for running a business. While cataclysmic liminality creates lethargy as entrepreneurial action appears to be pointless, unconventional entrepreneurs may use intervention agency to set realizable, practical goals that can be accomplished even in the situation of the devastating crisis. Our participants can be regarded as necessity entrepreneurs (Dencker et al., 2021; Gruber et al., 2024) striving to create small, informal businesses to return to pre-crisis consumption levels. However, to realize these goals they needed to exercise intervention agency to undermine traditional norms and institutions that prevent marginalized individuals from action.

Before the participants could begin creating new, informal businesses, they needed to convince themselves and others that they are entitled to act as independent economic actors. Patriarchal society refuses to grant older women the right to be employed or run a business. We described this obstacle to entrepreneurship as social liminality precluding entrepreneurial activities on the part of women and older individuals discounted by society as weak both physically and cognitively. The participants used intervention agency to convince themselves and others that the existing institutional barriers to entrepreneurship were not reasonable and could be reconsidered. Although the participants clearly believed that such barriers were nonsensical, in principle, they chose to only lower the bar by claiming that women should be allowed to create businesses in those areas that are traditionally considered as feminine. Thus, the participants used intervention agency and self-legitimation strategy to overcome social liminality and secure entrepreneurial transition.

Finally, the participants also found an appropriate strategy for wrestling with specialized liminality. While recognizing their deficiencies in terms of education and business experience, the participants used their intervention agency to emphasize the importance of experiential learning, claiming that they would be able to develop any missing competencies in the process. Hence, we described this strategy as practical self-empowerment. Overall, this study contributes to research on unconventional entrepreneurs by showing that, to overcome multiple liminalities that prevent their entrepreneurial transitions, unconventional entrepreneurs, such as the granmapreneurs whose narratives we analyzed, need to use intervention agency, complemented by strategies that help overcome these challenges.

CONTACT: Dmitry Khanin; dkhanin2023@unm.edu; University of New Mexico, USA

FALLING TO RISE: DOWNWARD MOBILITY AND ENTREPRENEURSHIP

Qiong Wu, IE Business School, Spain
Julio De Castro, IE Business School, Spain

ABSTRACT

Prior research has extensively examined the role of entrepreneurship in facilitating social mobility, but less is known about how intergenerational downward mobility shapes entrepreneurial decisions. In the context of slowing economic growth, labor market shifts, and credential devaluation, individuals today face increasing difficulty not only in surpassing but even in maintaining their parents' social status. While existing research suggests that moving down the social ladder influences individuals' psychology and behaviors, its implications for economic decision-making remain largely unexplored. In this study, we conceptualize intergenerational downward mobility as status loss, defined as a decline in an individual's occupational status relative to that of their parents. Drawing on research on status discordance, we argue that the misalignment between childhood status expectations and adulthood outcomes triggers a psychological need for status restoration, which in turn drives individuals toward entrepreneurship as a compensatory mechanism. Specifically, individuals experiencing downward mobility seek to regain lost status through entrepreneurial endeavors, leveraging financial success, autonomy, and social recognition. Moreover, we explore how this relationship varies by gender and rural/urban environment as structural constraints and social expectations may moderate the impact of status loss on entrepreneurial entry. To test our hypotheses, we use nationally representative Chinese data from 2010 to 2018. The results support our theoretical predictions, demonstrating a link between downward mobility and entrepreneurial engagement. Our study has both theoretical and practical implications.

INTRODUCTION

The popularity of the "from rags to riches" narrative reflects a widespread societal belief in the transformative potential of entrepreneurship, reinforcing the perception that individuals can leverage entrepreneurial endeavors to improve their social standings (Quadrini, 2000). This belief aligns with empirical findings that entrepreneurship can facilitate economic advancement, enhance social recognition, and ultimately foster upward mobility (Lippmann et al., 2005). However, while previous studies have established that entrepreneurship can be an important precursor to social mobility (Alvarez & Barney, 2014; Bruton et al., 2013; Rider et al., 2023), the impact of social mobility on entrepreneurial activities remains relatively understudied. Social mobility, broadly defined as the relative change in one's position within the social structure, implying corresponding shifts in the features of that position, such as status, economic means, and social identity (Friedman, 2016). Such transformations, according to prior research, have substantial psychological and behavioral implications (Breen & Jonsson, 2005; Phillips, Martin, et al., 2020; Phillips, Stephens, et al., 2020). Nonetheless, how these mobility-induced psychological shifts translate into entrepreneurial activity remains an open question.

In this study, we focus on the increasing prevalence of downward mobility, a phenomenon with profound implications for social cohesion, economic stability and individual well-being, positioning it as a critical issue in contemporary social and political research (Heath & Li, 2023). This trend is particularly evident in cases of intergenerational downward mobility, where an individual's occupational status falls below that of their parents (Black & Devereux, 2011). Contrary to the common perception that downward mobility is primarily attributed to individual failures in meritocratic selection processes or extreme life events such as divorce or illness, recent research highlights the growing role of structural and economic forces (Sobel et al., 1985). These forces include transformations in occupational hierarchies, shifts in labor market demand, and credential inflation (Autor & Dorn, 2013; Goos et al., 2009; Oesch, 2014; Moss-Pech et al., 2021), all of which hinder individuals from attaining the same socio-economic standing as previous generations. Indeed, it is reported that when individuals compare their own job status with that of their fathers at age 16, an increasing proportion perceive their father's job as having higher status, rising from 16% to 21% on average across OECD countries over the past two decades (OECD, 2018). Given the prevalence and significance of this phenomenon, it is thus crucial to examine how intergenerational downward mobility shapes individuals' psychological responses and career strategies (Bean et al., 1973). Specifically, we aim to investigate whether individuals turn to entrepreneurship as a means of responding to downward mobility.

To explore this question, we draw on the concept of status discordance (Kurer & Staalduinen, 2022). In their work, Kurer and Staalduinen conceptualize social mobility as status discordance—defined as the discrepancy between the status expectations shaped by one's childhood social class and the outcomes realized in adulthood, or their current social class. As such, intergenerational downward mobility represents a negative status discordance where individuals' achieved status falls short of the expectations shaped by their childhood social environment (i.e., their parents' status) (Hope, 1975). Prior research suggests that such status loss can trigger status anxiety, identity conflict, and a heightened need for status restoration (Bean et al., 1973; Brown, 2016; Simandan, 2018). As a result, when facing this psychological distress, individuals are likely to seek proactive ways to regain social recognition and reestablish a coherent self-concept (Gest et al., 2017; Gidron & Hall, 2017; Hope, 1975). Building on this framework and literature on entrepreneurship (Carsrud & Brännback, 2011), we argue that entrepreneurship serves as a compelling response to status loss because it offers opportunities for both social recognition and meets individual psychological needs. Specifically, individuals experiencing downward mobility may turn to entrepreneurship as a compensatory means of reclaiming autonomy, restoring lost status, and reinforcing their self-identity (Bourdieu, 1984; Wiklund et al., 2019).

Consistent with the findings that mobility is a gendered phenomenon (Damaske, 2020), we also explore how men and women may respond differently to experiences of status loss in their decisions to pursue entrepreneurship. Specifically, we propose that gender uniquely shapes these decisions, such that women who experience downward mobility are more likely than men to pursue entrepreneurship as a pathway to status recovery. Additionally, we explore the moderating role of contextual environments, proposing that urban settings attenuate the positive relationship between downward mobility and entrepreneurial entry. Urban environments, with their broader array of employment opportunities (Acs & Armington, 2004), may reduce the necessity or aspiration for individuals to engage in entrepreneurship as a response to status loss.

To test these hypotheses, we use a nationally representative sample of 8597 individuals from 25 provinces in China. The dataset spans the years 2010 to 2018, providing comprehensive information on the family dynamics and economic activities of Chinese residents. China provides an ideal context for this study due to its significant transformation in intergenerational occupational mobility over the past three decades following the reform and opening-up period (Bian, 2002). Our empirical analysis provides support for our hypotheses.

Our study makes several contributions to the existing literature. First, we offer new explanatory insights to the entrepreneurship literature by demonstrating that intergenerational downward mobility serves as a driver of entrepreneurial engagement. While existing research primarily considers social mobility as an outcome of entrepreneurship, our findings suggest that social mobility—particularly downward mobility—can also function as an antecedent of entrepreneurial activity. While prior studies tend to emphasize the economic constraints associated with downward mobility (Iversen et al., 2019), we extend this perspective by highlighting status loss as a psychological motivator that compels individuals to pursue entrepreneurship as a means of status restoration. By incorporating this psychological mechanism, our study provides a more comprehensive understanding of why individuals engage in entrepreneurship beyond purely economic considerations.

We also contribute to the literature on social mobility by demonstrating the psychological and behavioral impacts of downward mobility, extending its impact to economic outcomes. Responding to the call to examine the consequence of social mobility (Phillips et al., 2020), our study reveals that the experience of downward mobility not only affects financial well-being but also shapes individual agency and career decision-making. Specifically, we show that status loss alters individuals' perceived agency, increases their motivation to regain lost social standing, and heightens their propensity to engage in entrepreneurial behavior as a coping mechanism. These insights broaden our understanding of social mobility by demonstrating how status-related experiences influence occupational pathways and decision-making beyond financial constraints.

Third, by integrating perspectives from entrepreneurship and social mobility research, our study provides a more comprehensive view of the relationship between entrepreneurship and social mobility, highlighting how social dynamics shape entrepreneurial behavior. These insights deepen our understanding of the motivations underlying entrepreneurial decisions and offer potential implications for policies aimed at fostering entrepreneurship in contexts of economic and social instability, where downwardly mobile individuals may increasingly turn to entrepreneurship. Understanding this relationship can inform policies aimed at supporting entrepreneurship as a response to social and economic displacement, ensuring that individuals facing downward mobility have the necessary resources and institutional support to succeed in entrepreneurial endeavors.

HYPOTHESIS DEVELOPMENT

In this study, we theorize that mobility-induced status loss creates a misalignment between expected and achieved social standing, leading to psychological tension and identity reconstruction needs (Brown, 2016; Simandan, 2018; Tumin, 1957). This negative status discordance imposes socio-psychological strain, compelling individuals to actively seek pathways for status restoration. Among various coping strategies, we argue that entrepreneurship provides a unique and viable response to status loss, as it simultaneously addresses both external societal perceptions and internal individual psychological needs, thus providing a pathway to reclaim agency and redefine their social identity (Carsrud & Brännback, 2011).

First, successful entrepreneurs are frequently portrayed as exemplary "self-starters" who overcome challenges and uncertainties (Claire, 2012; Pilotta, 2016). Research has long revealed that, in contemporary society, entrepreneurship is inherently associated with high risk, innovation, and resilience—qualities that align with cultural ideals of individualism and societal progress (Hamilton, 1957). This image is reinforced by the narratives surrounding celebrity entrepreneurs, whose stories of perseverance and groundbreaking innovation have become symbols of entrepreneurial excellence (Kenney, 2000). Media portrayal amplifies these "from rags to riches" narratives, positioning entrepreneurship as a pathway to break through social barriers and achieve upward mobility, even for individuals starting from disadvantaged positions (Ge et al., 2022).

Second, entrepreneurship experiences are found to be associated with autonomy, control, and a sense of achievement, all of which are critical sources of individual well-being (Wiklund et al., 2019). The entrepreneurial process enables individuals to design their own work paths, set personal goals, and exert control over their careers, fostering a sense of agency and empowerment (Shepherd et al., 2015). This autonomy is particularly valuable for individuals experiencing status loss, as it allows them to reclaim control over their circumstances and redefine their social trajectory. Furthermore, entrepreneurship provides a platform for personal achievement and meaning-making. By tackling challenges, launching innovative ideas, and solving real-world problems, entrepreneurs experience a profound sense of purpose and fulfillment (Newman et al., 2019). These positive emotional experiences together not only alleviate identity crises caused by status discordance but also reinforce psychological resilience, enabling individuals to persist through setbacks and rebuild their self-identity (Farmer et al., 2011). On the whole, the intrinsic rewards of entrepreneurship—autonomy, achievement, and purpose—serve as critical mechanisms for psychological recovery and identity reconstruction.

Third, prior research shows that entrepreneurial success may lead to expanded social networks (Van Praag & Versloot, 2007), both of which play a critical role in restoring individuals' social standing and adapting to class changes. Entrepreneurs build relationships with investors, clients, collaborators, and other stakeholders, gaining access to valuable information, resources, and opportunities (De Carolis & Saparito, 2006). These networks not only facilitate business growth but also enhance the entrepreneur's integration into higher-status social groups, further solidifying their restored social position. In this sense, entrepreneurship is expected to not only rebuild an individual's status through increased wealth and social capital, but also achieve a renewed sense of belonging and purpose, further reinforcing their psychological and social recovery.

Last, prior research suggests that entrepreneurship contributes directly to society and creates social value (Acs, 2006). At the societal level, entrepreneurship drives economic development, generates employment opportunities, and fosters innovation, which are critical to sustaining economic growth (Audretsch, 2018). Furthermore, entrepreneurship is increasingly recognized as a mechanism for addressing grand societal challenges, such as poverty alleviation, regional economic revitalization, and environmental sustainability (Bruton et al, 2013). The resulting support national policies and cultural narratives in society further enhanced societal recognition of entrepreneurs. These societal contributions provide entrepreneurs with a sense of purpose and fulfillment, reinforcing their self-worth and identity as agents of positive change (Brieger, Clercq, & Meynhardt, 2020). In this way, entrepreneurship bridges the gap between societal recognition and personal psychological needs, making it a compelling pathway for status restoration.

Based on these arguments, we propose:

Hypothesis 1: Intergenerational downward mobility is positively associated with subsequent entrepreneurial entry.

Hypothesis 2: The positive relationship between intergenerational downward mobility and subsequent entrepreneurial entry is more pronounced for women than for men.

Hypothesis 3: The positive relationship between intergenerational downward mobility and subsequent entrepreneurial entry is less pronounced for urban residents compared to rural residents.

METHOD

Sample and Data Data for this study are drawn from the China Family Panel Studies (CFPS), a nationally representative longitudinal survey of Chinese households and individuals. CFPS provides comprehensive data on family dynamics, economic activities, and other key aspects of Chinese residents' lives. The survey, initiated by the Institute of Social Science Survey at Peking University in collaboration with the Survey Research Center at the University of Michigan, has been conducted biennially since its inception. CFPS is widely regarded as a reliable source for empirical research in sociology and entrepreneurship (e.g., Li & Tong, 2023; Xie et al., 2023; Yu & Cheng, 2024), offering rich insights into household behavior and individual economic outcomes.

China provides a compelling context for examining the relationship between intergenerational mobility and entrepreneurship due to its unique economic and social transformations. First, the structural transformations driven by industrialization and urbanization (Lippit, 2018) have fundamentally reshaped societal hierarchies, generating both new opportunities and significant challenges for upward and downward mobility. Second, over the past decades, China has witnessed a surge in entrepreneurial activities and the proliferation of privately-owned businesses, contributing to its evolving role as a key player in the global entrepreneurial landscape (He et al., 2019). This dual phenomenon of significant mobility and entrepreneurial expansion offers a fertile ground for studying how structural and individual factors interact to influence entrepreneurial behavior.

Our analysis focuses on individuals who were of working age (16 to 60 years old) in 2010 (Feng et al., 2019). Intergenerational mobility is measured by comparing individuals' occupational status from 2010 with their parents' occupational status collected in 2012. Given the stability of parental background information over time, combining the 2012 data with the 2010 baseline survey, which includes the largest number of respondents, is methodologically appropriate and ensures robust measurement of intergenerational mobility. To comprehensively capture entrepreneurial activities, we examine entrepreneurship information across all subsequent waves (2014, 2016, and 2018), using the 2010 wave as the baseline. To ensure robustness, the sample is restricted to individuals who participated in all five waves of the survey and who were not engaged in entrepreneurial activities in 2010 or 2012. After excluding observations with missing values for key variables, the final sample comprises 8,597 individuals.

Dependent variable The dependent variable Entrepreneurship is a dummy variable that takes the value 1 for individuals who were not engaged in entrepreneurial activities in 2010 but reported being self-employed in at least one of the subsequent waves (2012, 2014, 2016, or 2018). Excluding individuals engaged in entrepreneurial activities in 2010 ensures that the observed self-employment status in subsequent waves reflects a shift into entrepreneurship after experiencing downward mobility. These two waves also provide the baseline data for constructing the independent variable—whether respondents experienced downward mobility prior to 2014. This design ensures a clear temporal sequence between the independent variable and the dependent variable, thereby reducing concerns about reverse causality. A question in the survey explicitly asks respondents whether they work for themselves, i.e., self-employment. Following prior studies (Li & Tong, 2023; Sørensen & Fassiotto, 2011; Sørensen & Sharkey, 2014), we operationalize entrepreneurship as self-employment. This operationalization aligns closely with the study's focus on intergenerational mobility and its impact on entrepreneurial activities. Self-employment serves as a low-barrier form of entrepreneurship (Parker, 2004), making it particularly relevant for individuals experiencing downward mobility, who may face resource constraints. Additionally, in the Chinese context, self-employment captures a broad spectrum of entrepreneurial activities, reflecting the unique socioeconomic dynamics shaped by rapid transitions and structural inequalities.

Independent variable Intergenerational Downward mobility is measured by changes in occupational prestige between individuals and their parents, coded as 1 if an individual's occupational prestige is lower than their parents' when the individual was 14, and 0 otherwise. Parental occupational prestige was collected in the 2012 follow-up survey, as it provided more complete and reliable data on parental background compared to the baseline 2010 wave. Individual occupational prestige, in contrast, was measured in 2010. The combination of these two datasets ensures robust and accurate measurement of intergenerational mobility. This prestige-based measure aligns with our theoretical framework on social class and provides a reliable and consistent approach for assessing mobility, as commonly used in sociology (Treiman, 1977; Lersch et al., 2020). The use of occupational prestige offers the advantage of cross-generational stability and comparability, ensuring a robust measure of intergenerational mobility. Additionally, its applicability for international comparisons has been validated in prior research (Lin & Xie, 1988), making it particularly suited for our study.

Moderators There are two moderators in our model. The first moderator, Gender, is a binary variable coded as 1 for females and 0 for males. The second moderator, Urban, is also a binary variable, coded as 1 if the individual resides in an urban area and 0 if in a rural area. Urban and rural areas are defined based on the respondent's current place of residence, rather than their household registration (hukou). This approach reflects the increasing mobility in China due to the relaxation of the hukou system and allows us to capture the influence of the living environment on individual behaviors and opportunities.

Controls We include several control variables to account for factors that may influence the likelihood of entrepreneurship, including age, marital status, migration status, party membership, number of siblings, and county-level variables, such as per capita GDP and employment rate.

RESULTS

The descriptive statistics and correlations for the key variables in this study suggest a preliminary association consistent with our study's hypothesis. Probit regression analysis is employed to test the hypotheses. Hypothesis 1 (H1) posits that individuals who experience downward mobility are more likely to engage in entrepreneurship. Model 2 tests this main effect, and the results provide support for H1, as the marginal effect of downward mobility is positive and statistically significant (0.028, z=2.96, p<0.01). Hypothesis 2 (H2) posits that gender moderates the relationship between downward mobility and entrepreneurship. Consistent with our expectations, Model 3 reveals that the positive effect of downward mobility is stronger for women, as evidenced by the positive and significant coefficient of the interaction term between downward mobility and gender ($\beta = 0.217$, $z = 7.6$, $p < 0.05$). This result suggests that women are more likely to use entrepreneurship as a response to downward mobility, providing support for H2. Hypothesis 3 (H3) proposes that the effect of downward mobility on entrepreneurship is moderated by the urban context. However, Model 4 does not find statistically significant evidence to support H3.

DISCUSSION AND IMPLICATIONS

This study investigates how intergenerational downward mobility influences entrepreneurship, addressing a critical yet underexplored phenomenon in the context of modern economic and social changes. Building on research on status discordance (Kurer & Staalduinen, 2022), we conceptualize intergenerational downward mobility as negative status discordance, defined as the discrepancy between status expectations formed by one's childhood social class and the outcomes realized in adulthood. This status discordance creates unique psychological and social challenges, as individuals face a misalignment between their early internalized class-based aspirations and their current social reality. Drawing on Bourdieu's (1984) theory of habitus, we argue that individuals experiencing downward mobility encounter conflicts between the values, norms, and behaviors embedded during their upbringing and those required to adapt to their new, lower social class. This psychological tension, we argue, motivates individuals to pursue entrepreneurship as a strategy to restore their lost status. Our analysis of CFPS data provides support to our hypotheses.

We contribute to the literature on status loss by extending its application to the domain of entrepreneurship, particularly through the lens of intergenerational mobility and occupational change (Audretsch, et al, 2013; Quadrini, 2000; Ge et al., 2022). While much of the existing research in organizational studies examines the outcomes of status in terms of race, language, or hierarchical position, our study introduces a novel perspective by focusing on intergenerational comparisons (Clemens, 2006; Neeley, 2013; Marr & Thau, 2014). By examining how status changes between parents and their children influence an important organizational outcome—entrepreneurship—our research bridges the gap between mobility studies and entrepreneurial behavior. This intergenerational perspective not only broadens the scope of status loss research but also highlights the long-term implications of downward mobility for entrepreneurial decision-making.

We also make contribution to the entrepreneurship literature by highlighting status attainment as an important and underexplored motivation for entrepreneurship (Cuervo, 2005). The inquiry of what motivates individuals to become entrepreneurs has remained a persistent focus of interest (Walter & Heinriches, 2015). Much of the prior research on entrepreneurial motivations has focused on factors such as economic success, autonomy, achievement, and flexibility. While these are critical drivers, our study suggests that the desire to restore or enhance social status plays a pivotal role, particularly for individuals experiencing downward mobility. By integrating downward mobility into the entrepreneurial framework, we demonstrate how individual-level psychological factors and class-level environmental constraints interact to shape entrepreneurial decisions. This integrated approach provides a more holistic understanding of entrepreneurial motivations and offers a theoretical foundation for future studies to explore the dynamic interplay between social mobility, status, and entrepreneurship.

CONTACT: Qiong Wu; Email: qiongwu@student.ie.edu; IE Business School; Alvarez de Baena 4, Bajo, 28006 Madrid, Spain.

SUMMARY

RESILIENCE OF ENTREPRENEURIAL TEAMS: IMPACT ON VENTURE OUTCOMES

Shafagh Kashef, Université Libre de Bruxelles, Belgium
Judith Behrens, Université libre de Bruxelles, Belgium
Matthias Weiß, Ruhr-Universität Bochum, Germany

Principal Topic

Resilience—the ability to recover from adversity through positive adaptation—is crucial for entrepreneurial teams facing uncertainty (Blatt, 2009; Fleming & Ledogar, 2008). While research on resilience is growing, most studies focus on individual resilience, neglecting the unique challenges entrepreneurial teams encounter, such as rapid decision-making, resource constraints, and external fluctuations (Baroncelli, Caputo, Santini, & Theodoraki, 2024; Blatt, 2009; Conz, Magnani, Zucchella, & De Massis, 2023; Fu, Zhu, Liu, Jiao, & Li, 2023). This study examines how team traits and dynamics contribute to resilience throughout the entrepreneurial process, from ideation to growth. Additionally, it explores how factors like team composition changes and external shocks influence resilience. By addressing gaps in the literature, this research provides insights into the role of team resilience in entrepreneurial success.

Method

Building on existing frameworks in entrepreneurship, this study investigates the factors and dynamics that support team resilience. Specifically, it explores: What factors contribute to resilience in entrepreneurial teams? How do team dynamics and individual traits shape resilience? To answer these questions, we conducted a systematic literature review and employed bibliometric analysis alongside NVivo-supported content analysis. We interviewed 100 individuals from 78 entrepreneurial teams in Belgium, examining both successful ventures and failed teams to identify key resilience factors. The interviews covered individual and team-level resilience, and findings were synthesized into a conceptual framework. Thematic analysis, supported by double-coding, minimized subjectivity and ensured rigor in qualitative interpretation.

Results and Implications

Entrepreneurial team resilience is shaped by the interplay between individual backgrounds and collective team dynamics and gradually evolves through shared experiences and the lessons learned from overcoming challenges. Individual and collective factors influence team resilience. Personal factors include optimism, self-efficacy, and emotional regulation, which help members cope with stress (Vakilzadeh & Haase, 2021). Collective factors are shaped by team culture and communication patterns, fostering a supportive environment where members rely on each other during adversity (Alliger, Cerasoli, Tannenbaum, & Vessey, 2015). The literature suggests that resilient teams are more likely to achieve positive outcomes (Fisher, Maritz, & Lobo, 2016). This study identifies gaps, including longitudinal research needs. It builds on existing frameworks by identifying specific factors and dynamics within entrepreneurial teams and emphasizes the intersection of entrepreneurship and team dynamics. Qualitative interviews offer valuable insights into how team resilience is crucial for the survival and growth of entrepreneurial teams, providing rich data that theoretical models may overlook. The findings present essential guidelines for founders, VCs, and business incubators to enhance team resilience.

CONTACT: Shafagh Kashef; shafagh.kashef@ulb.be; Université Libre de Bruxelles; box 3, No 66, Rue Dodonée, Brussels, Brussels Hoofdstedelijk Gewest 1180, Belgium

SUMMARY

CULTIVATING HOPE TO EMPOWER SUSTAINABLE ENTREPRENEURIAL ACTION

Carmen Mayer, Technical University Munich, Germany
Anne Tryba, Technical University Munich, Germany

Principal Topic

Sustainable entrepreneurship education is pivotal for equipping young people with the entrepreneurial competencies needed to drive societal change. However, many hesitate to take action, feeling powerless and frustrated in the face of complex challenges. Scholars thus call for redesigning entrepreneurship education to integrate hope - a cognitive-motivational construct characterized by agency and pathway thinking. Hope enhances well-being, counteracts despair, and fosters venture creation. While research highlights hope's importance in entrepreneurship education, little is known about how pedagogies of hope empower students' sustainable entrepreneurial action. Therefore, we ask: How can integrating pedagogies of hope in sustainable entrepreneurship education empower students to take sustainable entrepreneurial action?

Method

Following an inductive qualitative research approach, we developed an interdisciplinary curricular sustainable entrepreneurship course at the business school of a leading European technical university. We integrated pedagogies of hope using dedicated hope-focused exercises, learning goals, reflections, and theoretical insights. During two course iterations, we collected 984 individual reflections, rich course observations, and 84 semi-structured post-course interviews. In line with a process research approach, we triangulated and analyzed the data using a bottom-up grounded theory approach.

Results and Implications

This data informed the development of our "Hope-Induced Empowerment for Societal Change Model", which illustrates how students' hope evolved throughout the course and empowered sustainable entrepreneurial actions post-course. We identified three distinct hope trajectories: (1) "Hope as an obligation," (2) "Hope as a balancing act," and (3) "Hope as a gradual ascent." These trajectories reflect students' volatile, wave-like hope fluctuations when facing societal challenges, shaped by high-hope moments, like shared group passion, or low-hope moments, such as engagement with systems thinking. Post-course, students following different trajectories experienced five drivers of sustainable entrepreneurial action, including gaining confidence or learning to break down societal problems. Hence, first, we add to the literature on sustainable entrepreneurship education by revealing how course participation shapes students' hope trajectories. Second, we contribute to research on pedagogies of hope in higher education by offering a nuanced understanding of how different pedagogical elements evoke high- and low-hope experiences among students. Third, we extend the literature on empowerment for entrepreneurship and sustainability action by showing how students' distinct hope experiences during a course empower them differently for sustainable entrepreneurial action.

CONTACT: Carmen Mayer; carmen.mayer@tum.de; Technical University Munich; Schulstraße, Hochdorf-Assenheim, Rheinland-Pfalz 67126, Germany

SUMMARY

ENTREPRENEURIAL U-TURNING: THE ANTECEDENT CONDITIONS AND MULTI-LEVEL IMPACTS OF EMPLOYEE-TO-ENTREPRENEUR-TO-EMPLOYEE MIGRATIONS

Avery Ortiz-Hunt, Apex Center for Entrepreneurs, United States
Parul Manocha, University of Alabama - Birmingham, United States
Joseph Simpson, Virginia Tech, United States

Principal Topic

Although career migrations from entrepreneur-status to employee-status have been thinly researched, they are quite a common occurrence. Since, on average, new founders have well over a decade of professional experience prior to starting a business, most entrepreneurs have been previously employed. In these cases, the move from founder-status to employee-status constitutes an "entrepreneurial U-Turn," wherein individuals leave employee-status to become an entrepreneur, only to later return to the role of employee, once again working for other people (Employee-to-Entrepreneur-to-Employee – "E-to-E-to-E"). This leads us to pose several related questions: Why do individuals make "entrepreneurial U-Turns"? Are they purely a function of necessity or are there other discernible drivers, such as burnout, aspirational shifts, or issues concerning employee-firm fit? Reciprocally, from the perspective of hiring firms, what is the impact of hiring U-turners? Our study examines these questions through the lens of attraction-selection-attrition (ASA) theory in order to ascertain the decision-action-outcome dynamics embedded in the complex utility functions of "E-to-E-to-E" individuals.

Method

Emulating the research design developed by Hunt (2013), we used data from highly regulated service sectors in healthcare, hazardous materials, and construction, as well as selected, specialized plastics manufacturing. Firms in these industries are required by law to report exhaustive, publicly available data involving employees, products, processes, projects, customers, suppliers, and firms. Through formal requests to state-level departments of health, we obtained data on 1,107 entrepreneurial firms, 72% of which were spinouts, and 57% of which were entrepreneurial U-turners. Using a matched-pair design, we compared the pool of U-turner with non-U-turners, using the publicly available data to analyze comparative odds ratios through a logistic regression.

Results and Implications

Interestingly, our analysis suggests that successful and unsuccessful entrepreneurs return to employee-status in approximately equal proportions, supporting the Gimeno et al., (1997) premise that entrepreneurial utility functions cannot be boiled down to strictly financial considerations. We find that age, gender, marital status, years of prior professional experience, and previous managerial experience are all significant predictors of "U-Turning," but not venture performance.

CONTACT: Avery Ortiz-Hunt; rahunt@stratwa.com; Apex Center for Entrepreneurs; 1345 S Pitkin Ave, Superior, Colorado 80027, United States

SUMMARY

HYBRID ENTREPRENEURS' TRANSITIONS TO FULL-TIME ENTREPRENEURSHIP: THE ROLE OF WORKPLACE INCIVILITY

Nicholas A. Smith, Northern Illinois University, United States
Jon Carr, North Carolina State University, United States
Alexander McKelvie, Syracuse University, United States
Tim Michaelis, Northern Illinois University, United States
Jeffrey Pollack, North Carolina State University, United States

Principal Topic

Hybrid entrepreneurship, the processes by which individuals maintain their wage-based employment during their transition to entrepreneurship, is a rapidly growing area of focus in entrepreneurship literature. To date, research on hybrid entrepreneurs mainly focuses on individual aspects that pull people to engage in hybrid versus full-time entrepreneurship, or the transition from hybrid to full-time entrepreneurship (e.g., age, core self-evaluations, risk aversion, Raffie & Feng, 2014; Thorgren et al., 2016). While these contributions are critical to understanding positive aspects of entrepreneurship that attract hybrid entrepreneurs, we know precious little about the negative situational factors that push individuals away from their employee roles toward their ventures. We propose that the negative interpersonal experiences (i.e., workplace incivility) hybrid entrepreneurs have in their wage role will motivate them to leave their wage job (i.e., turnover intentions) and engage in entrepreneurial action (i.e., venture startup effort).

Method

We test our framework with an autobiographical recall experiment (Study 1, $N = 479$), as well as a time-lagged mediation study (Study 2, $N = 523$). We used best practices for reducing method bias and improving online-panel data quality. OLS regression and Structural Equation Modeling (SEM) were the primary analytical techniques used.

Results and Implications

Our empirical results support our theoretical model. Hybrid entrepreneurs who recalled (and wrote about) a recent incident of workplace incivility in their wage job indicated higher state anger (manipulation check; $\beta = .42$, $p < .001$), higher turnover intentions ($\beta = .14$, $p < .001$) and startup effort ($\beta = .11$, $p = .021$) as compared to the control group. The mediation model for Study 2 highlighted work-to-venture conflict as a key mechanism linking workplace incivility to hybrids' intentions to leave their wage job (ACME = .13, 95% CI = [.07, .20], $p < .001$) and higher levels of startup effort (ACME = .02, 95% CI = [.001, .05], $p = .046$). Venture optimism positively moderated the direct effect of workplace incivility on work-to-venture conflict ($\beta = .63$, $t = 2.88$, $p = .004$), and positively moderates the indirect effect of workplace incivility on turnover intentions via work-to-venture conflict (indirect effect = .05, CI = [.01, .11], $p = .031$), but not startup effort. This study provides a new theoretical lens for understanding hybrid entrepreneur role transitions and highlights opportunities for future research to apply workplace mistreatment constructs as drivers of entrepreneurs' behaviors.

CONTACT: Nicholas A. Smith; nsmith6@niu.edu; Northern Illinois University; 6330 Collina Springs Ct., Houston, Texas 77041, United States

SUMMARY

BETWEEN MOTIVATION AND AMBIGUITY: HOW SUPERVISOR–STUDENT INTERACTIONS SHAPE DOCTORAL ENGAGEMENT IN ACADEMIC ENTREPRENEURSHIP

Laurin Waldmann, Technical University Munich, Germany
Anne Tryba, Technical University of Munich, Germany

Principal Topic

Academic entrepreneurship (AE) is vital for economic, social, and ecological progress (Block et al., 2017). Yet, despite growing support for AE, science-based ventures remain scarce and are even declining (Ding et al., 2023). Beyond various macro-level influences on AE (e.g., limited early R&D funding or inefficiencies in technology transfer offices), researchers are increasingly recognizing micro-level factors, particularly the complex role of individual scientists, such as doctoral students, in driving AE (Hmieleski & Powell, 2018). Science doctoral students conduct a substantial portion of university research and play a crucial role in advancing technology-based AE (Hayter et al., 2017). However, their involvement in AE remains largely understudied (Boh et al., 2016), especially regarding how interactions with their doctoral supervisors shape their motivation and engagement.

Method

Using an exploratory qualitative approach, we collected data at two leading German research universities, conducting 43 semi-structured interviews with life science doctoral students, faculty members, and incubation program leaders, complemented by field observations and archival data (Corbin & Strauss, 2008). We conducted open and axial coding, then structured the data into first-order concepts, and grouped them into second-order themes, which were aggregated into overall dimensions. Finally, we present a model illustrating our findings.

Results and Implications

We found that the interaction between science doctoral students and their supervisors can both foster and inhibit student engagement in AE. On the one hand, supervisors can motivate their students by creating awareness for entrepreneurship as an achievable career path, by shifting their stereotypical image of an entrepreneur, and by demonstrating that their research competencies overlap with those needed for AE. On the other hand, we found that there is a role ambiguity between doctoral students and their supervisors, which is particularly evident in what we categorized as "motivation", "competencies", and "leadership" for initiating AE. This ambiguity stems from doctoral students attributing an illusory competence to their supervisors and from expectation mismatches between the two, which potentially hinder their active involvement in AE. By presenting a framework of expectation mismatches between doctoral students and supervisors, we add previously underexplored micro-level barriers to the AE literature. We contribute to the AE motivation literature by presenting motivational factors specific to the overlooked group of doctoral students. Finally, we extend the role episode model. As a practical implication, university policies and entrepreneurship education should mitigate role ambiguity through targeted programs that enhance role clarity and promote effective mentoring strategies.

CONTACT: Laurin Waldmann; laurin.waldmann@tum.de; Technical University Munich; Lichtenbergstrasse 6, Garching, Bayern 85748, Germany

INTERACTIVE PAPER

SHIFTING ENTREPRENEURIAL ATTENTION AND ENTREPRENEURIAL ENTRY AND EXIT CYCLES

Aleksios Gotsopoulos, Sungkyunkwan University, Republic of Korea
José Lejarraga, IE Business School, IE University, Spain
Maud Pindard-Lejarraga, IE Business School, IE University, Spain

Principal Topic

Despite its central role in entrepreneurship research, the study of new industry emergence has largely focused on economic and institutional factors, with little attention to how entrepreneurs interpret environmental signals when making entry and exit decisions. Using an attention-based view of entrepreneurial action, we propose that boundedly rational entrepreneurs selectively direct their attention to concrete, easy-to-process signals (Valliere, 2013, Srivastava et al., 2021) to decide upon entry and exit from the industry, and that these signals may not always align with economic indicators. We hypothesize that entrepreneurs direct their attention to mass media signals and to the behavior of their peers, potentially leading to excessive entry and unsustainable competition levels. The novel status of an industry and its visibility in public discourse may generate overoptimistic expectations, leading entrepreneurs to enter based on perceived popularity rather than resource availability. Conversely, as novelty fades and expectations are unmet, rapid exits may follow—even for economically viable firms—contributing to boom-and-bust cycles weakly connected to economic fundamentals. We investigate these dynamics using data on small-scale Greek stock brokerage firms.

Method

We examine Greek stock brokerage firms in the Athens metropolitan area from 1996 to 2005, alongside media coverage of the industry. These firms resembled major brokerage firms but were smaller, restricted from portfolio management, and operated with low capital requirements, lowering barriers to entry. To measure public attention, we tracked newspaper articles on stock markets and brokerage firms in a major Greek daily. To capture economic resources, we used changes in trade volume (millions of Euros) at the Athens Stock Exchange (ATHEX). We modeled organizational failure using an event history model and estimated entry through a negative binomial model.

Results and Implications

We find that economic resource availability, as measured by trade volume, does not affect entry rates. Instead, prior entries positively influence contemporaneous entry, suggesting entrepreneurs rely on peer behavior as a heuristic for opportunity evaluation. However, excessive prior entries reduce subsequent entry, as overcrowding lowers success prospects. Media coverage significantly influences entry, with increased attention leading to higher entry rates. Similarly, exits are not tied to economic resources but are strongly influenced by media coverage. These results suggest that industry entry is shaped by social and cognitive dynamics rather than rational economic assessments. Our study extends the attention-based view of the firm to nascent entrepreneurial ventures, showing how media-driven expectations can transform industry evolution into a fad-like movement, decoupling entry and exit from economic fundamentals.

CONTACT: Aleksios Gotsopoulos; agotsopo@skku.edu; Sungkyunkwan University; 25-2, Sungkyunkwan-ro, Jongro-gu, Seoul, Seoul-t'ukpyolsi 110-745, Republic of Korea

FROM FAITH TO FORTUNE? RELIGIOUS ENTREPRENEURS' RESPONSES TO INSTITUTIONAL VOIDS

Todd W. Moss, University of Oklahoma, United States
Ana Cristina Dahik, IPADE Business School, Mexico
Eduardo Melendez, University of Oklahoma, United States
Antonio Casanueva, IPADE Business School, Mexico

Religion influences billions, yet its intersection with entrepreneurship remains underexplored. This abductive study examines how religious entrepreneurs respond to institutional voids, shaping positive or negative outcomes. We show how individual religiosity transfers to enterprises in visible or less visible ways, influencing responses to institutional voids. Visible manifestations lead to confrontation, purposeful evasion, or remorseful acceptance, while less visible manifestations favor rational evasion and pragmatic acceptance, avoiding confrontation. Our study contributes by contextualizing entrepreneurial action, expanding outcomes at the organizational, community, and institutional levels, and developing a related process model.

INTRODUCTION

Religion is fundamental to billions of people (Tracey et al., 2014), and its integration into entrepreneurship research remains nascent (Busenitz & Lichtenstein, 2019). Entrepreneurial action and religious expression are shaped by institutional environments, particularly in contexts with institutional voids where individuals face tensions between incompatible practices and beliefs (Greenwood et al., 2011). These voids, prevalent in developing economies, create market failures that businesses must address (Khanna & Palepu, 1997). While prior research examines entrepreneurs' responses to formal voids through informal institutions (Mair & Marti, 2009), little is known about how religious beliefs influence such responses. This study explores how religious entrepreneurs navigate institutional voids and their implications at organizational, community, and institutional levels. Our contributions link entrepreneurs' religiosity to enterprise religiosity manifestations, highlighting contextual influences on entrepreneurial mechanisms (Brown et al., 2022; Welter, 2011). We examine how institutional voids in developing economies magnify social and economic inequalities, positioning religious entrepreneurs as both alleviators and exacerbators of these ills (Webb et al., 2020). We propose a process model illustrating how religious entrepreneurs reconcile tensions between beliefs, institutional voids, and decision-making.

LITERATURE REVIEW

Religion is a "system of meanings embodied in a pattern of life" (Schmidt 1999, 10) that shapes individual and group decisions (Dodd & Seaman, 1998). As an informal institution, religion manifests through practices—repeated activities infused with shared meanings (Furnari, 2014), which embody beliefs that establish normative frameworks (Schwitzgebel, 2011) and provide guidelines for decision-making (Edgell, 2012). Religious practices constrain and enable action within institutional environments (Dodd & Seaman, 1998), yet individuals often navigate tensions between their religious values and other societal norms (DiMaggio & Powell, 1983). Religion, as an institutional order, interacts with other institutions and entrepreneurship.

Institutions are shared rules and norms shaping behavior (Barley & Tolbert, 1997). Institutional complexity emerges when individuals face incompatible practices and beliefs (Greenwood et al., 2011), such as in contexts with institutional voids (Khanna & Palepu, 1997). These voids, prevalent in developing economies (Mair & Martí, 2009), create tensions for individuals attempting to reconcile conflicting norms (Martin et al., 2017). Despite extensive research, little attention has been given to informal institutional voids—the failure of norms, values, and beliefs to enable stability (Webb et al., 2020). Institutions are key drivers of entrepreneurship (Baumol, 1996), and voids impede market growth (Webb et al., 2020). Entrepreneurs in developing economies mitigate formal voids through informal networks and trust-based relationships (Puffer et al., 2010), with religion generating trust and reciprocity (Neubert et al., 2017). While entrepreneurs often rely on family ties to navigate institutional gaps (Ge et al., 2019), research on religion's role in filling voids remains scarce. We extend this discussion to explore how religion and institutional voids interact in mitigating and exacerbating tensions within institutional complexity.

METHODS

Our literature review suggests that religion and institutional voids in entrepreneurship remain under-theorized, requiring a conceptualization beyond a contextual boundary condition (George & Jones, 2000). We adopt an abductive approach and an interpretivist lens to theorize religious entrepreneurs' responses to institutional voids.

Research Context: Religious Entrepreneurs in Mexico

Our research examines entrepreneurs in Jalisco, Mexico, an extreme case (Eisenhardt & Graebner, 2007) due to its strong Catholic identity and distinct institutional environment. Religion is a top life priority and influences private and family enterprises, fostering a sense of belonging (Cortés Guardado, 2019). Catholicism—deeply rooted since the Spanish conquest and reinforced by the Cristero War—shapes regional development and workplace values, with religious symbols like the Virgin of Guadalupe playing a significant role (Renée, 2019). Jalisco faces institutional voids such as weak governance, poor education, crime, and food insecurity. For instance, 42% of the population lacks quality housing (CONEVAL, 2024), contrasting with developed economies where such voids are rare.

Data Collection and Analysis

Our study uses primary and secondary data, including four rounds of data collection with 56 in-depth interviews, four focus groups, and three company site visits, totaling 76 informants. Additional data came from non-participant observations of entrepreneur interactions, WhatsApp messages, emails, and phone conversations, capturing spontaneous religious expressions in business settings. We began with purposeful sampling (Lincoln & Guba, 1985) and refined our open-ended questions as theoretical relationships emerged. After developing a rough model, we conducted deeper interviews focusing on religion and institutional voids, reaching theoretical saturation by the fourth round. The first two rounds explored entrepreneurs' perceptions of business culture and social investment. The third round examined relationships with stakeholders and community impact efforts. Religion and institutional voids emerged as key themes, prompting the fourth round to focus on their tangible and intangible manifestations, entrepreneurs' experiences with institutional voids, and their strategies to mitigate resulting tensions.

We used Kreiner's (2016) twin slate approach, an abductive grounded theory method that integrates theory and data (Murphy et al.,2017). Unlike purely inductive methods, twin slate balances theory engagement with grounded analysis, making it suitable for both micro- and macro-level studies (Smith et al., 2010). We conducted detailed coding, starting with in vivo codes and progressively incorporating theoretical codes on religion, entrepreneurship, and institutional voids. We independently coded transcripts before refining them in joint meetings, developing 1st- and 2nd-order codes (Kreiner et al., 2017), and identifying patterns and theoretical mechanisms in entrepreneurs' responses to institutional voids and religious practices.

FINDINGS

All informants acknowledged the role of personal religious beliefs and practices in their lives, but their willingness to integrate them into the workplace varied. One group actively used their businesses to foster religious engagement, while the other refrained from openly manifesting religious beliefs within their organizations. These differences shaped their interactions with institutional voids and influenced outcomes at the institutional, community, and organizational levels. Our process model (Figure 1) highlights shared and distinct mechanisms through which informants bring religion into the workplace, explaining variations in outcomes across levels.

Individual religiosity

Religiosity—a person's religious beliefs and practices (Holdcroft, 2006)—can be experienced individually or collectively, with individual religiosity consisting of personal religious disciplines and practices (Zhang et al., 2019). Building on Allport and Ross (1967), we found that all informants described their personal relationship with God and exhibited intrinsic religiosity, internalizing core beliefs beyond attending religious services. Informants referenced their relationship with God, religious practices, family legacy, and interactions with others. Said one informant: "I try to thank God every day, every single day... I am a strong believer in the blessings, I try to bless every day and every time my girls. I'm a strong believer that I have a purpose in life because of God" (Inv24). Another informant referred to the legacy of ancestors having a direct influence in everyday actions and the relationship with God: "I always remember to be with God, because when I'm tempted to do bad things, I always remember grandma. She passed away a long time ago, and I remember my father-in-law, and I remember my aunt, and I think, what would they think of me, if I do bad things?" (Inv19) We observed six dimensions of how individual religiosity was revealed in our informants: their personal relationship with God, sacraments, practices and doctrines, vocation and purpose, helping others, and legacy.

Corporate religiosity

Beyond individual religiosity, we observed a corporate religiosity in the enterprises that occurs in group settings through shared religious practices, worship, and guidance (Zhang et al., 2019). This form of religiosity varies in openness, particularly regarding entrepreneurs' willingness to use their businesses as platforms for religious expression and their connection to a religious institution (Holdcroft, 2006). One group of informants openly embraced their Catholic identity, integrating religious rituals into their enterprises. They displayed religious symbols in offices and factories, initiated workdays with company-wide prayer, celebrated Catholic holidays, and invited priests to hold mass and confessionals at the workplace. We refer to this group of informants as those exhibiting *visible/high manifestations of corporate religiosity with high denominational connection*: "We believe that work and our vocation is an occasion to glorify God. So let us put the right things in place to glorify God and His greatness through the finished work. I have every confidence in God that we will come through with the commitments we have with the client... The most important thing is to touch our clients, collaborators, allies" (Vis12). A particularly common theme among informants was the visibility and access they provided to employees for Catholic imagery, sacraments, and traditions: "On Ash Wednesday there is imposition of ashes inside the plant and a mass. Machines are stopped. It doesn't matter if someone tells me that there is an urgent order, the machines are stopped" (Vis17). Informants in this group also emphasized implicit tangible demonstrations—physical expressions of their beliefs that do not explicitly reference their denominational connections.

The second group of informants we identified expressed opposite intentions regarding the purpose of businesses and religion compared to the first group. These informants mentioned that it is not necessary to be outspoken about religion in the workplace: "I'm not that devout... However, I believe that with the right actions and the right intentions in everyday work, you do that without having to be very outspoken about it" (Inv22). We call this second group of informants as those exhibiting *invisible/low manifestations of corporate religiosity with low denominational connection* as the group of informants with personal religiosity but who do not show group manifestations of religion in the workplace. When asked about religious expression in their organizations, informants preferred discretion, avoiding references to denominational connections. Instead, they manifested religiosity through routines and daily behaviors rather than explicit religious practices. These implicit intangible demonstrations included discreet actions and anonymous deeds linked to their corporate religiosity: "You're reinforcing the way of thinking and coming to religion, even though we're Catholics and we have a strong sense of religion, it's more the way of how you behave" (Inv23).

We identified two mechanisms linking individual religiosity to corporate religiosity. The key difference lay in how each group connected religious manifestations to their business purpose. Visible/high corporate religiosity informants emphasized a business logic for transcendence, while invisible/low corporate religiosity informants prioritized bottom-line strategies for survival. We term these mechanisms *aiming for transcendence* and *aiming for permanence*: "So, my relationship with God has been very close. When we do transmissions of religious content, people are happy. I said to the priest, 'Father, I'll put the means' but the message of God must be transmitted as a strategic concept of the business" (Vis20). Another key distinction between the two groups is how they viewed their organizations as *bearers of values*. While both saw religion as shaping values, those with visible/high corporate religiosity openly linked their values to Catholic principles. In contrast, those with invisible/low corporate religiosity referred to common values without connecting them to their personal denomination: "It's values of the Catholic social doctrine... The most important is that you must be highly productive, with fully human and socially responsible companies. Make money...but be responsible with people" (Vis13).

Interactions of institutional voids with manifestations of corporate religiosity

A key distinction between the two groups was how they identified and responded to formal and informal institutional voids affecting entrepreneurial growth. Commonly cited voids included corruption, crime, gender inequality, lack of social safety nets, business illegalities, and poverty. While informants varied in their perceptions and responses, all recognized that voids increase uncertainty and risk. For instance, one informant linked poor education to a shortage of skilled workers, directly affecting business growth. Another in agribusiness highlighted the challenge of crime in the supply chain, noting that criminal groups control 75% of businesses, making confrontation risky. "The study programs in government are fragmented, in decline... The business community sees this in the graduates who arrive, especially from public universities, who are very poorly prepared." (FG2)

Responses to institutional voids

Our finding reveal that religion shapes entrepreneurial action in response to institutional voids in three ways: *confronting*, *evading*, and *accepting*. These responses differ based on corporate religiosity levels. Informants with visible/high corporate religiosity actively sought to *confront* institutional voids through forming discussion groups with the government, involving priests in private-sector talks, and using religion as a moral weapon against illegalities. One informant considered holding public office to challenge voids from within, creating pressure against corruption: "Moral awareness is the most lethal weapon you have. For example, it just happened to me in the real estate business where certain offices are not reporting transactions. My partner asked me, 'What do you think?' I told him, 'I would love for them to audit me, because that would give them an idea of who I am.' Words convince, actions compel" (Vis 19).

A second type of response, *evasion*, involved avoiding situations that reinforced institutional voids. Both groups engaged in evasion, but those with visible/high corporate religiosity sought meaningful consequences through their actions through *purposeful evasion*. For instance, one informant negotiated with the government to redirect bribe money toward a community project. "Something we aim to avoid promoting in the company is "mordidas" (mordidas is the local language used to refer to bribes) We spoke with the municipal presidents and told them not to ask us for them. If you ask me for "mordidas", I won't give you anything. Let me help you with your community—what do you need? They would say: a multipurpose court, and a walkway. Our response was: very well, I'll make your courts and your walkway" (Vis19). In contrast to purposeful evasion, we observed how informants with invisible/low corporate religiosity were often reluctant to accept the void situation through a more practical thought process we term *rational evasion*. For example, one informant mentioned how even though there is a possibility to help alleviate poverty, is it more useful to make money through the rich population: "When first starting my business, I wanted to make anything that makes money. Now I don't need more money, I am established. Money is okay. I like to help the people making business... I think that if I want to help the poor people, maybe I can. But it is very difficult. Difficult to receive a remuneration correctly. If I help the rich people, the money is easier, the money works. My projects are with the rich people." (Inv32)

A third response, *acceptance*, involved accepting institutional voids as unavoidable. Many informants felt they had to pay bribes or face significant business obstacles. Acceptance was seen as necessary to sustain entrepreneurial goals but varied in justification. Those with visible/high corporate religiosity experienced *remorseful acceptance*, expressing guilt over their compliance: "When there is no guilt, decisions can be made quickly. But when I need to fire someone, you never fire them for the reason they are leaving, never. We talk about all their qualities, all their virtues, we praise the person" (Vis21.). In contrast, those with invisible/low corporate religiosity viewed it as a *pragmatic acceptance*, acknowledging voids without shame: "I do believe that people prioritize their current earthly plane before the spiritual one. I sincerely have done it, and I have had to give up many of my beliefs because I say that if specifically, where

I worked before because I say that if I don't do it, I will be left out and at home I have a girl and I have rent to pay" (Inv29).

Outcomes resulting from responses to institutional voids

Our study identified three levels of outcomes from responses to institutional voids: institutional, community, and organizational. Some entrepreneurs aimed to transform societal rules and norms driving social tensions. An informant with visible/high corporate religiosity exemplified this by advocating for a public agenda to actively fill voids in their locality: "I made a revolution and I brought together all the political sectors, the productive sectors, the farmers' association, the pig farmers' association, the cattle breeders' association, the engineers, the architects, the lawyers, the unions and I lined them up in a project and I sat them together and I said, let's see what model of our region we want, let's create an agenda" (Vis20). We group these outcomes in relation to the *institutional-level viability,* meaning an attempt to cause improvements in how the norms and rules in the context where entrepreneurs and their enterprises operate. This level of outcomes might not yet have achieved significant change; however, informants' responses reflected effects of responses at a level above their communities and their organizations.

A second level of outcomes had to do with the well-being of those surrounding the founders and their enterprises: relatives, proximate community members, and other individuals in vulnerable conditions. We call this *community-level prosperity*, and it entails outcomes related to the well-being, inclusion and opportunities of the population: "The goal of our businesses is for vulnerable women to do well. In Mexico, there's a whole issue with women working away from home, the absence of the mother, and all of that. This business model aims to ensure that there is a mother who is present, working on her own time and at her own pace" (Vis21).

We also observed how informants aimed to achieve *organizational-level stability,* related to outcomes affecting the ongoing existence of businesses. This group of outcomes referred to positive or negative effects of their responses to institutional voids in terms of what the enterprises gained or sacrificed regarding survival: "There are [tensions] when you're in a highly competitive environment, and you think that the extra three or four pesos I pay to provide my people with all the benefits put me at a disadvantage compared to those who don't offer them" (Inv26).

Varying responses to institutional voids led to both positive and negative outcomes across levels. Confrontation and purposeful evasion were linked to institutional and community benefits, while pragmatic acceptance sometimes harmed long-term institutional viability. Some outcomes were intentional, while others were unintended. For instance, paying a bribe might help an organization survive but perpetuate corruption. Multi-level trade-offs emerged: visible/high corporate religiosity informants faced organizational setbacks (e.g., slower progress due to legal hurdles or lost government contracts) but contributed to institutional change by challenging voids. In contrast, invisible/low corporate religiosity informants benefited organizationally (e.g., faster business growth, employee support) but risked long-term institutional harm by normalizing voids.

DISCUSSION AND IMPLICATIONS

Our study examines how entrepreneurs' religious beliefs and practices interact with institutional voids, shaping outcomes at organizational, community, and institutional levels. We make three key contributions. First, we link entrepreneurs' religiosity to differences in corporate religious manifestations, addressing calls for more research on religion and institutional demands (Tracey et al., 2014). We show how institutional voids alter business mechanisms based on entrepreneurs' differing approaches (Brown et al., 2022; Welter, 2011). Second, we extend research on institutional voids beyond market transactions (Webb et al., 2020) to explore broader societal impacts. Entrepreneurs with visible religiosity often confront voids at the cost of financial performance, while those with lower religiosity tend to perpetuate voids through pragmatic acceptance, benefiting financially. Third, we present a process model illustrating how corporate religiosity and denominational connection shape responses to institutional voids and subsequent outcomes. Entrepreneurs with similar personal religiosity split into two groups, each employing distinct mechanisms that influence organizational decisions.

Our findings reveal how high corporate religiosity is expressed through practices like Catholic traditions and group prayers, while lower corporate religiosity manifests through commitment to employees and shared values. Entrepreneurs with high religiosity actively challenge institutional voids, while others use rational evasion or acceptance to normalize constraints. In developing economies, religious entrepreneurs play dual roles—either mitigating or exacerbating societal inequalities. Our study complements prior research in developed contexts where institutional voids are minimal (Jones et al., 2017) and in vulnerable settings, expanding theories on family business structures and social embeddedness (Eze et al., 2021).

CONTACT: Todd Moss; tmoss@ou.edu; University of Oklahoma; 307 West Brooks Street, Norman, Oklahoma, 73019, United States

FIGURE 1. Conceptual model

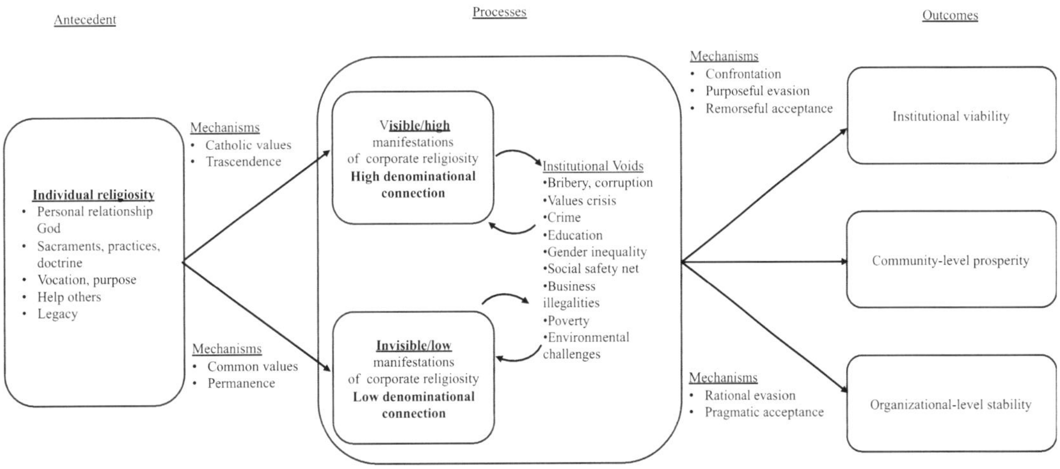

THE ONES WHO WALK AWAY FROM BOSTON: TOWARDS A MODEL OF ENGAGING AND DISENGAGING FROM ENTREPRENEURIAL ECOSYSTEMS

Ben Spigel, Babson College, United States
Eliana Crosina, Babson College, United States
Angela Randolph, Babson College, United States

ABSTRACT

Our paper explores how and why entrepreneurs engage with and disengage from their local ecosystems. We argue that traditional top-down approaches overlook entrepreneurs' lived experiences of their ecosystem. In particular, drawing on 45 interviews with Boston-based founders, and using mental mapping techniques, our study examines how emotions influence entrepreneurs' networking practices, and importantly, their willingness to engage with the ecosystem. Our findings reveal different forms of disengagement and highlight the importance of understanding entrepreneurs' emotional experiences to create more inclusive and supportive ecosystems.

INTRODUCTION

Research on entrepreneurial ecosystems has traditionally focused on top-down approaches that seek to identify the key actors, factors, structures, and processes linking localized contexts with high-growth entrepreneurship (Spigel, Kitagawa and Mason 2020). This has led to a large literature around measuring and mapping ecosystems to identify key resources, actors, and networks of a place to inform policy decisions and support entrepreneurs (Andrews et al., 2021; Bergman 2024; Hruskova, Mason and Herzog 2022). While this has contributed to a vibrant debate with numerous methodological innovations (Hess Wahl and Johnson 2025; Bennett Vedula Araki 2025; von Bloh et al., 2019), it has also led to tranchet critiques for lacking an understanding of the micro foundations that link an entrepreneur's presence in a well-functioning ecosystem with superior growth performance (Wurth, Stam and Spigel, 2023). As a result, while we have strong definitions of ecosystems, identified their key elements and actors, and have developed new approaches to measure and differentiate these, we still only have a rough understanding of how these factors contribute to the ability of entrepreneurs to scale their firms.

This top-down approach has had a particularly pernicious effect: ecosystems research frequently ignores entrepreneurs' own agency. This is to say, researchers have focused on understanding the structures of ecosystems rather than how entrepreneurs act within them. This is ironic as foundational ecosystem literature focused on the importance of understanding entrepreneurs as the center of their ecosystems (Isenberg, 2010; Feld, 2014). As a result, we still have a poor understanding of entrepreneurs' lived experiences of their ecosystem, including how they understand their ecosystem's role in the development of their ventures, and how they strategically navigate situated opportunities and tensions. This is most notable when considering research on ecosystem use. While there are exceptions (e.g., Korber Swail and Krishanasamy 2022; Huggins et al., 2024), research has traditionally implicitly assumed that entrepreneurs have equal access to the resources of an ecosystem. Or, if some entrepreneurs do not have equal access, that this is purely due to structural exclusion rather than more complex social dynamics (Motoyama et al., 2021; Lassalle and Johnston 2018).

Thus, we need to better understand how entrepreneurs *experience* their ecosystem and how these experiences impact if and how they engage with the ecosystem, as well as if and how they can access the localized resources they need to scale their firms. This is particularly important for understanding how entrepreneurs from under-represented minorities utilize their ecosystems. To simply say that they face discrimination and exclusion (Motoyama Golatt and Etienne 2024) robs them of their agency to develop alternatives and workarounds within the local ecosystem (Ozkazanc-Pan and Muntean 2021; Sheares 2024).

THEORY DEVELOPEMENT

We are primarily interested in how entrepreneurs interpret, experience, and use their ecosystem. To sensitize our understanding of these dynamics, we borrow from research on emotions and entrepreneurship (Ashkanasy Humphery and Huy 2017; Goss 2005). This work sees emotions as a broad concept encompassing various subjective feelings an individual has towards other people or phenomena (Hatfield, Cacioppo, and Rapson 1994) as part of the entrepreneurship process (Cardon et al., 2012). Drawing on theories from psychology, our focus is on how events and conditions create positive or negative emotions for entrepreneurs, and how these emotions subsequently affect entrepreneurial attitudes, judgements, and practices (Goss, 2008).

Existing research suggests that the main process connecting emotions with entrepreneurs' experience of their ecosystem is networking. The entrepreneurial networking literature, in particular, focuses on the practices underlying the creation, maintaining, and use of personal and business social connections in service of entrepreneurial activities (Elfring and Hulsink 2007; Zheng Ashan and DeNoble 2019). This scholarship suggests that the emotional energy connected with developing and maintaining social ties — which varies based on entrepreneurs' personalities and goals — can impact entrepreneurs' willingness to engage in networking activities (Nielson and Klyver 2020). That is, entrepreneurs who associate networking with positive emotions will be more likely to engage in networking activities than those who associate it with negative emotions such as anxiety or fear.

Based on these insights, we zoom in on two oppositional phenomena: how and why entrepreneurs engage with and experience their ecosystem. and, more importantly, how, and why entrepreneurs *do not* engage with their ecosystem. These research questions provide an agency-centric perspective on how ecosystems operate that foregrounds the emotional experiences of founders rather than the structural characteristics of their local context.

METHOD

We employ mental mapping methods, which focuses on how actors experience the spaces that they inhabit (Lynch, 1960). Mental mapping methods involve asking respondents to draw maps or sketches of particular places in order to capture their understanding of those places (Gieseking 2013; Guelton 2023). In our case, rather than asking them to draw a physical space we asked them to draw their entrepreneurial ecosystem. This creates a novel perspective of ecosystems that identifies the resources, networks, and support that entrepreneurs perceive and use rather than those that are merely available to them. Mental mapping approaches explicitly focus on the emotions and meanings actors attach to different places, people, and organizations to understand why they may or may not seek to engage with different entrepreneurial communities.

To do this, we conducted qualitative interviews with a purposeful sample of high-growth entrepreneurs in Boston, Massachusetts in 2024-2025. We define high-growth as the entrepreneur developing a firm with a scalable business model. The authors drew on their own personal contacts with accelerators — which specifically curate high-growth firms — as well as used business intelligence databases such as Pitchbook to identify firms. We oversampled founders from under-represented groups (e.g., women and racialized minorities) and outside of technology sectors. The interviews focused on how respondents engaged with their ecosystem. We framed how they overcome common entrepreneurial, managerial, and growth challenges, probing them about who provided them the resources, knowledge, or support to address these problems. In addition to enquiring about the nature of these relationships, we also asked about the emotional saliency of their interactions within the ecosystem.

Towards the end of the interview, we asked respondents to draw their entrepreneurial ecosystem. The directions were intentionally vague to allow the interviewee to decide what they would draw and how they would draw it. Respondents typically spent about 3 to 5 minutes drawing and the rest of the interview focused on discussing what they created, emphasizing the specific people and organizations (if any) that they included, the nature of the relationships, and any connections between actors on the map.

We employed a two-part analysis. The first part involved a close read of the interview transcripts to identify themes and patterns. We were focused on entrepreneurs' perceptions of their ecosystem, the emotions associated with their engagement in different parts of the ecosystem, and the types of resources they drew on to solve the problems they faced. We were particularly interested in the experiences of informants who did not draw on the ecosystem and the underlying logic for their choice. Second, we performed a quantitative analysis of respondents' maps. This involved creating a categorization of the types of actors included in maps and recording the relationships between them. We created a five-part categorization based on the nature of what was listed. [A] Ego: the respondent representing themselves; [B] Individuals: Specifically named people such as a family member or mentors; [C] Groups: organizations or groups of people that are expected to have close, familiar relationships. This might include a small startup group, an accelerator cohort, or a group of friends; [D] Collectives: broader, more diffuse collections of people with similar roles who do not have strong relationships with each other. Examples include things such as "venture capitalists" or "other founders"; [E] Organizations: named organizations such as accelerators, support programs, or universities. In addition to categorizing each person or organization listed, we also quantified the connections drawn between them (normally indicated by arrows or inscribed circles) and if they were located within the Boston ecosystem.

To date we have completed interviews with 45 entrepreneurs. The majority (55%) represent technology companies with 28% of the sample involved in consumer goods ventures, 13% in life science and 5% in manufacturing organizations. Fifty-two percent of the respondents were male and 48% were female. Thirty-seven percent of the respondents identified as having a non-white ethnic background. On average, their firms had about 17 employees and had raised 4 million USD in investment.

RESULTS

Of the 45 interviews conducted to date, we have performed in-depth analysis of 19. The results show a striking pattern: more than half of the interviewees *do not* presently engage with their Boston ecosystem (e.g., attending local events, participating in training programs or having informal conversations with other founders) but have disengaged from the ecosystem. Founders discussed several "on-ramps" that connected them initially to the ecosystem, as well as emotional and business-related "off-ramps" that led some of them to disengage. Through a qualitative analysis of the interviews and maps we gathered, we have identified three forms of ecosystem disengagement, what we called "transcending," "opting out" and "pushed out," in addition to those who exhibited "continued engagement" in the ecosystem. So far, we have identified two founders who transcended the ecosystem, eight who retreated from it, three who were pushed out and six who exhibit continued ecosystem engagement (see Table 1).

Transcending the ecosystem refers to founders who, despite being based in Boston, do not see the local ecosystem as having the resources they need to grow their business. As a result, these founders tend to look beyond Boston for growth resources. Many of the founders who transcend the Boston ecosystem have been through accelerators in other places, such as Y Combinator in San Francisco, or have otherwise built global networks that they actively maintain and frequently use. These entrepreneurs are 'promiscuous' networkers: they enjoy building ties with other people and see it as a key entrepreneurial function. For example, one technology entrepreneur discussed how he gathered knowledge to enter a new market: "we ended up making that list where we called heads of commercial excellence … [and then I asked people] we knew in an adjacent space and just saying, hey, I went through your LinkedIn, can you introduce me to these like four people?" (I01). Despite being based in Boston, these founders do not see themselves as being part of the Boston ecosystem. Rather, they see themselves as transnational entrepreneurs.

Founders who *opted out of the Boston ecosystem* found limited enjoyment or value when engaging with it. In the words of one founder, "I don't find that the community... is very helpful to me. It's not alive for me. Right? It's kind of very flat for me." (I03). These entrepreneurs typically saw practices associated with engaging with the ecosystem, such as attending events and building connects as "more of a distraction" (I11). One discussed how she had to "force myself to go" to events, where they "sit in the corner all night" (I12). The most common reason for disengagement was that they failed to find peers they could learn from. One cleantech founder based in a popular startup coworking space found that his conversation at events had "not been as deep as I would like" (I08) while another found that when he tried to engage... "at networking events there were either like, people trying to, like get money or like people trying to like, make weird artificial connections." (I07).

Finally, founders who were *pushed out of the ecosystem* were those who felt excluded, despite their best efforts to engage. These founders generally came from under-represented groups: in one case from a racialized minority, and in the other from working-class background. These founders pointed to numerous examples where they felt excluded from the local entrepreneurial community. For example, a founder making software for the beauty industry said that "So there's definitely an expectation coming from blue collar and maybe blue collar combined with a beauty industry where you think people are just, you know, gossiping and chewing gum all day" (I10). These feelings led them to spend less time trying to access the "mainstream" Boston ecosystem and pushed them to create alternative communities elsewhere.

By contrast, entrepreneurs who *continued to engage in the ecosystem* generally found it both valuable and enjoyable. These founders frequently cited early mentors or advisors who introduced them to the ecosystem, giving them an 'on-ramp' to Boston. Such 'concierges' by which we mean early advisors, often associated with an accelerator, curated critical introductions aligned with founders' industries, business models, and visions. Such concierges were pivotal "entry points" (I02), and particularly valuable for first time entrepreneurs with limited pre-existing connections.

The differences between these groups are evidenced in their maps (see Figure 1 and Table 1). Entrepreneurs who felt pushed out of the Boston ecosystem drew complex diagrams of their communities. Their communities comprise an array of individual people — family members, friends, mentors, and fellow founders — rather than broader collectives or groups. They drew comparatively many more people and organizations on their maps and indicated more connections between them than the founders in the other two groups.

Table 1: Statistics from Founder Mental Maps

Ecosystem Engagement Category	Number	Avg. Individuals	Avg. Group	Avg. Collective	Avg. Org	Avg. Node Number	Avg. Node Degree	Proportion of Local Connections
Continued Engagement	6	1.0	2.5	4.5	3.3	11.8	.52	.59
Transcend	2	0.5	3.0	2.0	3.5	9.5	1.3	.43
Opt Out	8	1.75	1.75	1.75	0.13	6.0	.52	.53
Pushed Out	3	8.0	1.67	1.33	3.0	14.67	1.74	.76

These preliminary findings suggest that founders pushed out of the Boston ecosystem have worked to build an alternative ecosystem, likely more welcoming to them than the 'mainstream' ecosystem.

Entrepreneurs who opted out of the ecosystem drew the sparsest maps, with few people and almost no organizations. Their maps typically had few connections linking the various elements in them. Conversely, the maps of those who transcended the Boston ecosystem tended to portray organizations and more diffuse/generalized collectives and exhibited few local elements.

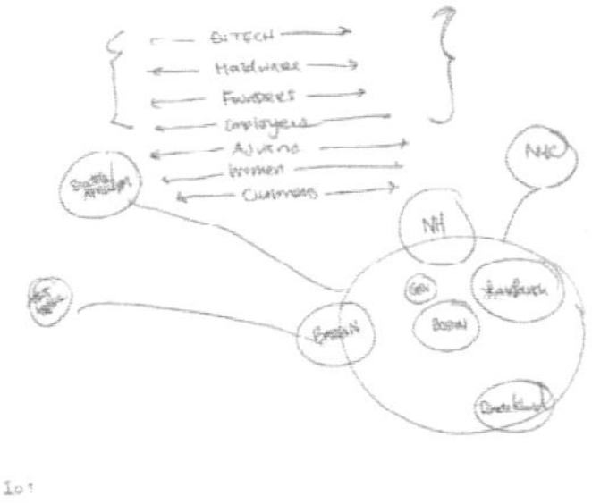

Continued Engagement (I09)

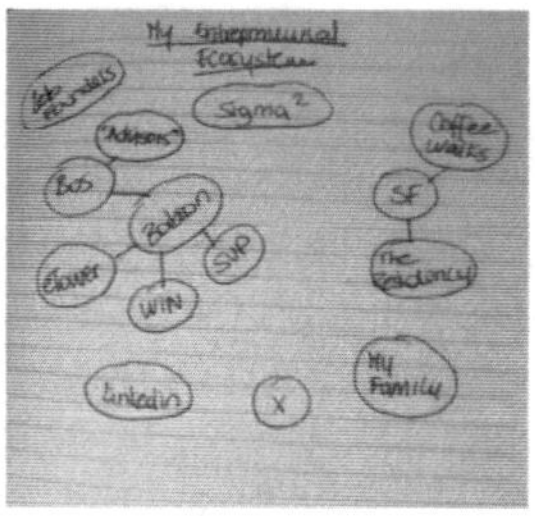

Transcending Ecosystem (I17)

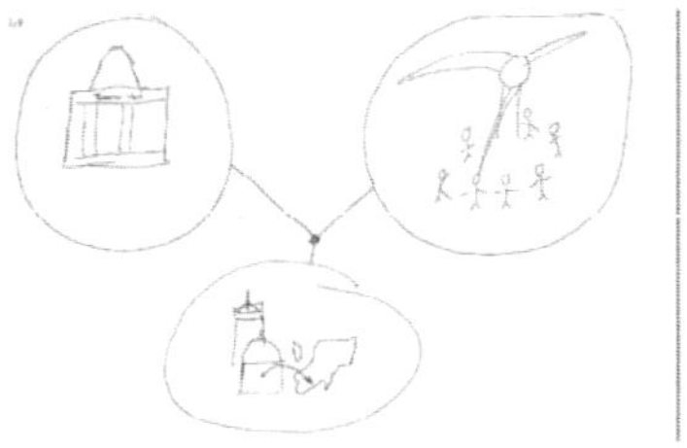

Opting Out (I07)

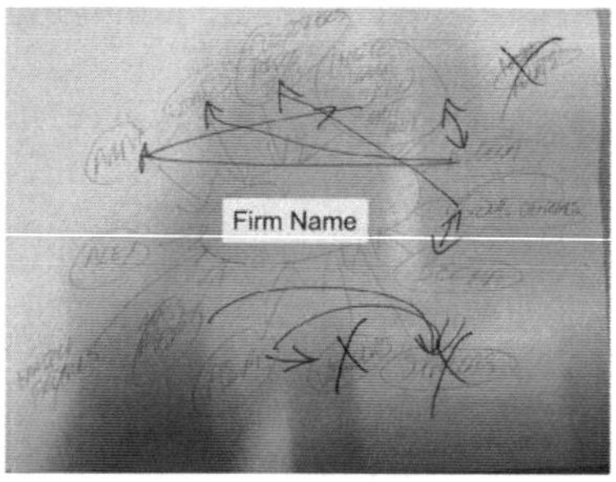

Pushed Out (I14)

Figure 1: Representative Mental Map Drawings

DISUCSSION AND IMPLICATIONS

Our analyses so far suggest that entrepreneurs do not engage equally in their ecosystem. There are multiple reasons for this, from the belief that Boston is too small for their ambitions to feelings of exclusion. However, in all cases of engagement and disengagement, entrepreneurs' relationship to their ecosystem was based on their underlying feelings, emotions, and attitudes towards the value of the ecosystem and the effort required to gather the resources and support they need from it. The maps show complex relationships between entrepreneurs and different elements of their ecosystem, allowing us to better understand how entrepreneurs dynamically create and recreate their ecosystem – in their minds and hearts.

The visual methodology we employed allows for a more nuanced perspective of ecosystem that privileges entrepreneurs' experiences rather than structural characteristics. This is a critical perspective as the role of formal entrepreneurship support organizations and state-sponsored networking groups depends on the willingness of founders to engage with them. Ultimately, understanding founders' attitudes towards their ecosystem is ultimately important for developing more inclusive and equitable ecosystems.

CONTACT: Ben Spigel; bspigel@babson.edu; 231 Forest Street Babson Park, MA 02457; Boston, Massachusetts

ACCELERATORS IN FOCUS: UNPACKING THE NECESSARY ACCELERATOR DESIGN FEATURES TO ENABLE EARLY-STAGE FINANCING

Maud Thys, KU Leuven, Belgium
Marjana Subotić, KU Leuven, Belgium

ABSTRACT

This exploratory study investigates the critical role of accelerator design in enhancing early-stage ventures' access to external equity funding. Drawing on necessity logic, we explore the key necessary conditions in accelerator program design that influence post-acceleration equity investment outcomes. Using necessary condition analysis (NCA) and qualitative comparative analysis (QCA) to examine necessity, we analyze the open-source data from Entrepreneurship Database Program (GALI Initiative) across North America, Latin America, and South Asia. Our results suggest that offering accelerator grants is a necessary condition for securing high post-acceleration equity investments. A structured curriculum emerges as a secondary bottleneck. Region-specific factors, such as demo days in North America and an impact focus in South Asia, are necessary to attract any equity investments.

INTRODUCTION

Accelerators are crucial components of entrepreneurial ecosystems, drawing attention from scholars, practitioners, and entrepreneurs (Assenova & Amit, 2024; Cohen et al., 2019). Their impact on new ventures is widely debated, serving as identifiers and enhancers of high-growth startups (González-Uribe & Reyes, 2021; Venâncio & Jorge, 2022), and with Y Combinator serving as a prominent example of success in fostering unicorn ventures (PitchBook, 2023). The number of accelerators has grown significantly worldwide (Drori & Wright, 2018), yet their overall effectiveness remains insufficiently understood (Bergman & McMullen, 2022). Variations in accelerator programs contribute to differences in new venture performance (Chan, Patel, & Phan, 2020), raising the question: What conditions within accelerator programs are necessary for fostering ventures' growth in securing substantial early-stage funding?

Early-stage entrepreneurs face challenges such as liability of newness (Bruderl & Schussler, 1990), uncertainty (McKelvie, et al., 2011), and resource scarcity (Gupta, et al., 2006). Accelerators aim to mitigate these issues by providing critical resources (Cohen, 2013). The intermediation theory suggests accelerators connect new ventures with investors and ecosystem agents (Vandeweghe et al., 2019), with empirical evidence supporting their role in reducing information asymmetry and improving startup financing (González-Uribe & Reyes, 2021; Venâncio & Jorge, 2022). However, the effectiveness of accelerator programs remains inconsistent, with some studies indicating potential negative effects (Browne et al., 2024; Moroz Sierra & Anderson, 2024). The diversity of accelerator structures highlights the need to examine key design features (Cohen et al., 2019; Hallen et al., 2020). While structured curricula, accelerator grants, and demo days are recognized as beneficial (Assenova & Amit, 2024), their relative importance remains debated. Some argue that learning and mentorship drive high-growth potential (González-Uribe & Reyes, 2021), whereas others find accelerator seed funding to be the strongest determinant of growth (Assenova & Amit, 2024). Additionally, the importance of these design features appears to be region-specific (Uhm, et al., 2018), with no clear consensus on which elements are universally critical, thus, offering no clear guidance for various stakeholders

This exploratory study addresses this gap by applying necessity logic through Necessary Condition Analysis (NCA) (Dul, 2016, 2020) and complementing it with fuzzy-set Qualitative Comparative Analysis (fsQCA) (Ragin, 2008). By integrating necessity logic, this study shifts the focus from identifying factors that enhance investment outcomes to pinpointing those without which high levels of equity investment would not occur. This distinction is particularly relevant in understanding accelerator effectiveness, as prior research shows mixed findings on their impact (Browne et al., 2024; Moroz Sierra & Anderson, 2024). Additionally, by focusing on high-equity investment cases, examining unusual but impactful funding patterns (Ruef & Birkhead, 2024), we aim to provide fresh insights on the topic. Through this novel approach, we contribute to both theory and practice by clarifying the fundamental conditions that enable accelerators to support startups in securing substantial early-stage financing.

THEORY

While accelerators are widely recognized as distinct organizational forms (Browne et al., 2024), no universal definition exists (Assenova & Amit, 2024; Drori & Wright, 2018). They are commonly understood as: "support organizations designed to foster the growth of early-stage ventures through structured programs, which often include funding, mentorship, and educational content." (Cohen et al., 2019, p. 1782). Despite general agreement on their core components, significant variation exists in accelerator design, leading to diverse post-acceleration outcomes (Bergman & McMullen, 2022; Hallen et al., 2020). Given exploratory nature of this study, outlined presumed relationships are guiding our empirical analysis, without explicitly stating hypothesis (similarly Van Der Valk et al., (2016)).

Program Structural Design Elements

Program package features address key challenges faced by early-stage entrepreneurs, such as liability of newness, uncertainty, and resource constraints (Bruderl & Schussler, 1990; Gupta, Smith, & Shalley, 2006). First, accelerator grants address early-stage funding gaps, allowing startups' focus on development and scaling (Cohen et al., 2019). Offering grants may serve as signal of quality, reducing investor uncertainty and increasing the likelihood of follow-on equity investments (Cohen et al., 2019). Specifically, accelerators' willingness to provide seed capital signals "skin in the game", demonstrating their commitment and belief in potential of selected startups (Frid et al., 2015). In mentor-based programs, grants—often in exchange for equity—raise the stakes for mentors and founders, driving higher-quality feedback (Assenova & Amit, 2024). Second, structured curriculum provides formal education, equipping entrepreneurs with strategic knowledge to improve decision-making, refine business models, and enhance innovation, ultimately strengthening startup market readiness and investment potential (Cohen et al., 2019; Gonzalez-Uribe & Leatherbee, 2018). Third, demo days play a crucial role in increasing startup visibility and creating direct networking opportunities with investors, serving as credibility signals that reinforce investor confidence and allow entrepreneurs to refine their pitches (Assenova & Amit, 2024; Uribe & Leatherbee, 2017). In developed ecosystems (e.g., North America), with intense funding competition, demo days may be particularly valuable (Uhm, Sung, & Park, 2018). Together, these program elements—structured curricula, demo days, and grants—help startups overcome early-stage barriers, enhancing their ability to secure equity investment.

Program setting features shape broader accelerator environment and influence program effectiveness. Accelerators' focus provides specialized mentorship, tailored resources, and industry-specific network (Assenova & Amit, 2024). Some studies suggest that accelerator credibility varies by industry, influencing investor perceptions (Browne et al., 2024). Given that
venture capital firms can specialize or diversify (Gompers et al., 2009), accelerator's industry focus may affect its signaling power. Longer programs allow startups to refine strategies, iterate on product development, and deepen investor relationships (Assenova & Amit, 2024; Cohen et al., 2019b). Extended support periods may enhance startup resilience, increasing likelihood of securing external funding. Finally, impact focus may be particularly relevant in ecosystems with limited access to traditional equity financing. Impact-driven accelerators connect startups with mission-driven investors, bridging funding gaps for

ventures addressing socio-economic challenges (Roberts & Lall, 2019; Uhm, et al., 2018).

Regional Differences

While Y Combinator has inspired accelerator models worldwide (Hallen et al., 2020), regional factors significantly influence accelerator effectiveness (Chowdhury & Audretsch, 2024). The economic landscape, investor activity, and ecosystem maturity shape early-stage financing opportunities (Fehder & Hochberg, 2014). In established startup hubs like Silicon Valley and London, accelerators benefit from deep investor networks, facilitating access to equity financing (Caccamo & Beckman, 2022). In contrast, accelerators in emerging markets face structural challenges, such as weaker investor infrastructure and limited funding availability (Fischer et al., 2022). Consequently, program priorities differ—while North American accelerators emphasize investment readiness and networking, Latin American and South Asian programs focus on foundational support, such as business model validation and early market access (Crişan et al., 2021).

METHOD

Necessary Condition Analysis (NCA) is a research method used to identify variables that are necessary but not sufficient for an outcome to occur (Dul, 2016). In NCA, the focus is on determining whether absence of specific conditions guarantees the absence of the desired outcome. Unlike additive models where variables can compensate for each other, necessity logic implies that a necessary condition cannot be offset by higher values in other variables. Importantly, the presence of a necessary condition does not automatically ensure the outcome; it merely sets a prerequisite. Although not the main goal, fuzzy-set Qualitative Comparative Analysis (fsQCA) can also identify necessary conditions. Both NCA and fsQCA detect necessity 'in kind' which indicates the presence (or absence) of X is necessary for the presence (or absence) of Y, with NCA additionally providing a more detailed necessity analysis by providing necessity 'in degree', which means certain level of X is necessary for a certain level of Y.

Our analysis draws on open-source data from the Entrepreneurship Database Program at Emory University, part of the Global Accelerator Learning Initiative (GALI). The dataset comprises information on 23,364 early-stage ventures that applied to 408 accelerator programs worldwide between 2013 and 2019. We applied strict sampling criteria to ensure that only for-profit, first-time accelerator participants and first-time early-stage investment (e.g., business angels/venture capital) receivers are included. Ventures with less than seven years of age at the time of application were retained, and observations with missing values for key variables were removed. To capture regional differences, we focused on three distinct geographic areas, resulting in a final sample of 70 accelerators in North America[1], (2) 79 accelerators from Latin America[2], and (3) 22 accelerators from South Asia[3]. Data are aggregated at the accelerator level, ensuring that the performance measure reflects the accelerator's overall ability to facilitate high equity investments. Outcome variable (Y) is the maximum amount of external equity investments the best early-stage venture in the accelerator's portfolio received in the year after accelerator participation from a business angel or venture capital in dollars, enabling us to learn from unusual data points (Ruef & Birkhead, 2024). Independent variables (X), the accelerator program design features, are operationalized primarily as dummy variables.

---

[1] *USA and Canada.*

[2] *Antigua and Barbuda, Argentina, Barbados, Bolivia, Brazil, Chile, Colombia, Costa Rica, Curaçao, Dominica, Dominican Republic, Ecuador, El Salvador, Grenada, Guatemala, Guyana, Haiti, Honduras, Mexico, Nicaragua, Panama, Paraguay, Peru, Puerto Rico, Saint Kitts and Nevis, Saint Lucia, Saint Vincent and the Grenadines, Uruguay, Venezuela.*

[3] *India, Bangladesh, Nepal*

RESULTS

NCA Results

Column (1) in Table 1 shows the results for the North American accelerator design features. While all effect sizes exceed 0.1, only accelerator grants are statistically significant, with an effect size of approximately 0.86 ($p < 0.05$). This indicates that accelerator grant opportunities are necessary for the top 15% of accelerators to achieve higher BA or VC equity investments in the sample. A structured curriculum also appears necessary, though significant only at the 10% level ($p < 0.1$) for the top 9% of accelerators. Additional robustness tests, which included maximum equity amounts from all sources and shifted the analysis from the accelerator to the firm level, confirmed the consistency of these findings. An analysis of demo days initially showed a small effect size due to a single extreme outlier; after removing the outlier, the effect size increased to 0.951. However, statistical significance for demo days remained elusive because of a skewed distribution—only a few accelerators in North America do not offer demo days, and thus statistical testing for necessity is not possible (Dul, 2024).

" See Table 1 in the Appendix"

Having a structured curriculum and opportunity for accelerator grants are the most important conditions in the Latin American sample, similar to the North American sample. As shown in *Table 1*, *column (2),* the effect sizes for curriculum (0.86) and accelerator grant (0.79) are the largest. However, we only find a significant p-value for accelerator grants ($p<0.05$). Thus, providing opportunities for accelerator grants becomes a bottleneck preventing accelerated firms from achieving higher BA or VC equity investments and belonging to the top 4 percent performing accelerators. Similar to North America, the accelerator grant feature is a necessary condition at the higher levels of BA and VC investments. However, for lower amounts of equity investment, there are no necessary conditions. Regarding the extensions, these results appear to be robust when changing to a firm-level set-up.

Column (3) in Table 1 shows the results for South Asian accelerators. Here, impact focus is the dominant necessary condition. The NCA results yield an effect size of 1 for impact focus, indicating that every accelerator receiving BA/VC equity investments had an impact focus, indicating necessity in kind. Although the effect size has the maximum value, due to the small sample and the extremely skewed distribution, a statistically significant necessity cannot be found (Dul, 2024). When the analysis is expanded to the firm level and additionally the inclusion of other equity types, the effect size of impact focus becomes gradually significant due to an increase in sample size.

FsQCA Results

We calibrate the "fully-out" point at 0 and the crossover at 0.01, signaling that any amount above 0 dollars represents the presence of a condition. To fully capture the significance of large equity investments, the "fully-in" threshold is set at the highest observed investment value. These calibrations are consistent with NCA thresholds, ensuring methodological comparability.

Table 2 presents the fsQCA necessity analysis for accelerator program features influencing the highest BA/VC equity investments per accelerator. fsQCA does not identify accelerator grants or structured curricula as necessary, aligning with NCA's finding that these conditions are necessary only in degree at higher investment levels. In North America, demo days emerge as the only necessary condition in kind, with a consistency of 0.93. This suggests that accelerators must include demo days for their startups to secure equity funding in the subsequent year. As fsQCA is more tolerant of outliers, it confirms the additional NCA analysis, validating demo days as necessary in kind for equity investments of any amount. For Latin America, fsQCA does not identify any necessary conditions in kind, consistent with NCA's finding that accelerator grants are necessary only in degree for securing high BA/VC equity investments. The absence of strong effect sizes for other program features suggests no hidden necessary conditions in kind. In South Asia, fsQCA identifies impact focus as a necessary condition in kind, with a consistency of 0.98. Though statistical limitations prevent NCA from establishing significance, both methods strongly indicate that South Asian accelerators must have an impact focus for their startups to obtain BA/VC investments post-acceleration.

" See Table 2 in the Appendix"

DISCUSSION AND IMPLICATIONS

The primary contribution lies in identifying the design features necessary for achieving the highest equity investments. Our NCA findings reveal that offering grants is a necessary condition for securing high equity investments, particularly among the top 15% of accelerators in mature ecosystems like North America. While less pronounced in emerging markets, grants also emerge as necessary in Latin America, suggesting broad applicability across regions. This aligns with prior studies highlighting grants as key signals that enhance investment prospects (Assenova & Amit, 2024; Cohen et al., 2019). At a lower significance level, structured curricula are necessary for top-performing North American accelerators, supporting research suggesting formal education fosters high-quality ventures (Gonzalez-Uribe & Leatherbee, 2018; Hallen et al., 2020). Unlike sufficiency-based approaches (e.g., Assenova & Amit, 2024), our findings reveal that necessity logic offers essential insights into why certain features, such as accelerator grants, are crucial for achieving exceptional equity outcomes. This methodological contribution addresses calls for more nuanced analytical frameworks (Dul, 2020) and provides insights into program elements that cannot be offset by other factors. By shifting the focus from general accelerator effectiveness to the unique design elements that differentiate top-performing startups, our NCA findings help to fill gaps in the existing literature regarding the variability of post-acceleration outcomes (Bergman & McMullen, 2022).

In our second contribution, we highlight complementarity insights of necessity in degree (high investment levels) from necessity in kind (any investment level) from NCA and fsQCA approaches. NCA indicates necessity in degree for accelerator grants and structured curricula —relevant only for securing the highest equity amounts. FsQCA, on the other hand, identifies necessity in kind for demo days in North America and impact focus in South Asia —essential for obtaining any investment. Our third contribution challenges the "one-size-fits-all" approach to accelerator design. While programs often imitate successful models like Y Combinator (Hallen et al., 2020), direct replication may not be the most effective strategy. In mature ecosystems such as North America, demo days enhance visibility, making them crucial for securing equity financing in competitive markets (Cohen, 2013). Accelerators serve as credible intermediaries, and pitching events act as valuable signals to investors (Uhm et al., 2018). In South Asia, where entrepreneurship is often socially driven, an impact-focused accelerator is essential for attracting mission-driven investors (Roberts & Lall, 2019; Uhm et al., 2018). These findings reinforce the need for accelerator designs tailored to local ecosystem demands, supporting the view that program heterogeneity is essential (Bergman & McMullen, 2022; Crișan et al., 2021).

Finally, this study provides actionable insights for policymakers, accelerator managers, and entrepreneurs by highlighting that accelerators are not homogeneous—certain designs are better suited to attract equity investments and nurture growth of startups. Necessary design features vary by region and investment level, influencing accelerated startup funding success. A strategic, tailored approach is needed, as securing any investment differs from obtaining top-tier funding. Identifying necessary conditions for higher equity investments enables more effective, regionally adapted accelerator strategies.

CONTACT: Maud Thys; maud.thys@kuleuven.be; KU Leuven; Naamsestraat 61, 3000 Leuven, Belgium

TABLES

Table 1: Effect sizes and p-values for max equity investment (BA/VC) on accelerator level across regions

	North America (1)		Latin America (2)		South Asia (3)	
	d	P	d	p	d	p
Impact focus	0.250	0.335	0	1	1	0.14
Sector focus	0.250	0.310	0	1	0.778	0.239
Program duration	0.368	0.137	0.393	0.534	0	1
Accelerator grant	0.862**	0.016	0.786**	0.038	0.818	0.11
Curriculum	0.783*	0.083	0.857	0.465	0.556	0.495
Demo days	0.131	0.915	0	1	0.667	0.744
Sample size	70		79		22	

Note: d effect size, p p-value. ***p < .01; **p < .05; *p < .10

Table 2: Necessary analysis of program features for max BA/VC equity on accelerator level in fsQCA

	North America		Latin America		South Asia	
	Consistency	Coverage	Consistency	Coverage	Consistency	Coverage
Impact focus	0.745	0.310	0.478	0.182	*0.980*	0.384
Sector focus	0.685	0.299	0.392	0.165	0.811	0.378
Program duration	0.741	0.409	0.685	0.196	0.564	0.395
Accelerator grant	0.792	0.281	0.415	0.187	0.882	0.411
Curriculum	0.732	0.250	0.820	0.151	0.658	0.445
Demo days	*0.902*	0.240	0.685	0.196	0.811	0.318

SUMMARY

DISRUPTING THE STATUS QUO: HOW STRATEGIC ISSUE CREATION ENGAGES STAKEHOLDERS

Janelle Bremer, University of Tennessee, Knoxville, United States

Principal Topic

By examining pre-existing environmental conditions that influence the ability of entrepreneurs to enroll key stakeholders, this study moves beyond the prevailing agent-centric focus of previous literature on stakeholder enrollment (Ramoglou et al., 2023). I draw on institutional entrepreneurship theories that explain how actors initiate change in highly institutionalized environments (Battilana et al., 2009; Micelotta et al., 2017) to understand how and why stakeholders engage with a new venture that disrupts the status quo. I do this by examining the stakeholder enrollment process in an environment dominated by a powerful incumbent and where stakeholders have vested interests in maintaining the existing order.

Method

I use a historical case study design (Wadhwani et al., 2020) to examine the golf industry from 1890-1974. Qualitative data sources include historical and contemporary books, autobiographies of key actors, and media accounts from newspapers and trade journals. Quantitative data contextualizing changes in power and status over time was gathered from PGA Tour Media Guides. Case data was analyzed using process analytical methods (Langley, 1999), including case narratives, chronologies, and visual mapping. Findings emerged through iterating between data collection, analysis, and theory in an abductive process (Locke et al., 2008).

Results and Implications

Preliminary findings suggest that stakeholder decision-making processes vary in the two phases of stakeholder enrollment. In the first phase, entrepreneurs leverage environmental conditions to organize a critical mass of pre-commitments (Sarasvathy, 2001) from self-identifying stakeholders. In the second phase, when the entrepreneurial endeavor is introduced to external stakeholders, they must be forced into a "decision to decide" through strategic issue creation. This process transforms an external event or development into a decision event (Dutton & Ashford, 1993). This study makes three contributions. First, my process model connects previously disparate literature focused on either early (Alvarez & Sachs, 2023) or later-stage (Burns et al., 2016) stakeholder enrollment but not the relationship between the two. Second, the findings suggest that stakeholders are initially reluctant to engage with a new, potentially disruptive venture and readily dismiss it as disrupting the status quo. Stakeholders move to evaluate the opportunity only after an issue is created through urgency, positive framing, and a powerful coalition. Third, these findings suggest that entrepreneurs can tailor their efforts at stakeholder enrollment toward the specific decision-making processes of stakeholders at each stage.

CONTACT: Janelle Bremer; jbremer@vols.utk.edu; University of Tennessee, Knoxville; 3613 Sevier Heights Rd, Knoxville, Tennessee 37920, United States

SUMMARY

EMPOWERMENT OR ENTRAPMENT? THE RESOURCE DEPENDENCE DILEMMA IN ENTREPRENEURIAL SUPPORT ORGANIZATIONS

Ashley Gomez, University of Pittsburgh, United States
Banu Ozkazanc-Pan, Brown University, United States

Principal Topic

Entrepreneurial support organizations (ESOs) are vital to the entrepreneurial ecosystem, providing resources, networks, and guidance to founders (Bergman & McMullen, 2022). For marginalized entrepreneurs—such as women and racial/ethnic minorities—ESOs help mitigate inequities in access to financing, market entry, networks, sustainability, and knowledge (Brush et al., 2019; Motoyama et al., 2023; Özkazanç-Pan & Muntean, 2021). However, ESOs navigate a dual challenge: empowering marginalized founders while contending with resource constraints, institutional limitations, and biases. Drawing on resource dependence theory (RDT) (Pfeffer & Salancik, 1978, 2015), this study explores how ESOs balance their organizational sustainability with fostering agency among the marginalized entrepreneurs they support.

Method

This study employs a constructivist grounded theory (CGT) approach, providing a systematic methodology for developing theory from qualitative data (Charmaz, 2006, 2014). To gather insights, semi-structured interviews were conducted with ESO staff members who primarily serve minority entrepreneurs, as well as entrepreneurs who received support from at least one of these organizations. To enhance data triangulation, both participant and non-participant observations were utilized as secondary data collection methods. Observations took place at ESO locations, including staff meetings, daily interactions, and various entrepreneurial events throughout the region.

Results and Implications

ESOs often struggle to balance their own sustainability with the support they provide to marginalized entrepreneurs, leading to unintended consequences. While aiming to foster agency, they may inadvertently create dependency, straining resources and limiting entrepreneurs' capacity for independent growth. To navigate this, ESOs adopt a continuum of strategies, including immediate resource provision and transitional scaffolding to gradually reduce reliance. At the ecosystem level, ESOs engage in advocacy and structural interventions to foster broader conditions for entrepreneurial independence. This interdependent model suggests that rather than full autonomy, marginalized founders often benefit from evolving, adaptive support structures that shift over time. This study challenges the traditional RDT assumption that organizations should minimize dependence, instead highlighting the role of strategic interdependence between ESOs and founders. ESO-led resource enactment is shaped by both internal constraints (e.g., funding limitations, mission trade-offs) and external ecosystem conditions (e.g., policy shifts, institutional support). Rather than solely mitigating dependence, ESOs operate within a dynamic system where interdependence emerges as a viable and necessary strategy. This multi-level, temporal perspective reframes how support organizations manage founder dependency, sustainability, and systemic change within entrepreneurial ecosystems.

CONTACT: Ashley Gomez; agomez@katz.pitt.edu; University of Pittsburgh; 5850 Centre Ave Apt 213, Pittsburgh, Pennsylvania 15206, United States

SUMMARY

FRACKING WARS: EXPLORING HOW THE MICRO-DYNAMICS OF FRAMING CONTESTS AROUND NEW PRACTICES GENERATE SETTLEMENTS

Kevin McSweeney, Oklahoma State University, United States
Jordan McSweeney, Suffolk University, United States

Principal Topic

The purpose of this study is to advance understanding on how the micro-dynamics of framing contests around new practices generate settlements, with a particular focus on the sustainable development context. Entrepreneurs are often thought of as being integral to finding solutions to sustainable development concerns because of their ability to create new practices (Hall et al., 2010; Shantz et al., 2024). Despite prior research acknowledging the importance of framing contests for shaping settlements around new practices, the tendency for research to focus on how consensus and support for alternative frames in a framing contest are generated has left an important gap in our understanding. Consequently, recent calls from scholars have emphasized the need for greater attention to the "micro-level framing strategies" employed on both sides of a discursive contest and how these produce "common ground and temporary consensus" (Baba, Sasaki, Vaara, 2021: 1961). We respond to these calls by exploring the micro-dynamics of a framing contest around a new practice in the field of sustainable development to understand how the contest unfolded and enabled a settlement to be achieved.

Method

To explore our research question, we study the commercialization of hydraulic fracturing (fracking) in the United States. Following exemplars such as Maguire and Hardy (2013) or Hardy and Maguire (2010), we conduct a longitudinal discourse analysis to capture how contestants use the logics underpinning the anti and pro-fracking framings to contest the practice of fracking, shape a shared meaning, and achieve a settlement. Our final sample consisted of 29 sources and 3,074 documents.

Results and Implications

Our findings show how contestants invalidate their opponents' framing through three counterframing tactics which we label temporalizing, visualizing and radicalizing. Common to all three, they expose the internal incoherence in the opposing framing. In short, they weaponize the logics underpinning their opponents framing to advance their own cause. We then integrate the three counterframing tactics into a process model. We found a progressive filtering of concerns through the three increasingly aggressive counterframing tactics of temporalizing, visualizing and radicalizing which, in turn, surfaced areas of shared concerns that ultimately enabled a field settlement. Taken together, our study provides an important step forward in advancing understanding of how the micro-dynamics of framing contests around new practices generate field settlements around important issues like sustainable development.

CONTACT: Kevin McSweeney; kevin.mcsweeney@okstate.edu; Oklahoma State University; 920 S Murphy Street, Stillwater, Oklahoma 74074, United States

SUMMARY

THE DEVELOPMENT OF ENTREPRENEURIAL ECOSYSTEMS AND THE ROLE OF EXTERNAL ENABLERS: THE CASE OF THE NORDIC SPACE RACE

Sølvi Solvoll, Nord University Business School, Norway
Andrew Corbett, Babson College, United States
Marta Lindvert, Nord University Business School, Norway
Iselin Kristine Mauseth Steira, Nord University Business School, Norway

Principal Topic

Entrepreneurial ecosystems are comprised of interdependent factors governed in a way that enables productive entrepreneurship within a particular territory (Stam, 2015; Stam and Spigel, 2018). A lesser-explored aspect of ecosystem research is external enablers – defined as "distinct, external circumstance[s]" that have "the potential of playing an essential role in eliciting and/or enabling a variety of entrepreneurial endeavors…" (Davidsson, 2015, p. 683) (Juma et al., 2023). This paper examines the role of externals enablers in entrepreneurial ecosystem development by exploring the emerging new space economy, the third evolutionary phase of the industry. The dominance of entrepreneurs like Elon Musk's SpaceX, which owns approximately 70% of satellites in Low Earth Orbit, exemplifies this shift. As competitors worldwide pursue entrepreneurial opportunities in the space industry, we ask: What actors and factors influence the development of a new entrepreneurial ecosystem?

Method

Taking a qualitative approach, we follow two burgeoning entrepreneurial ecosystems in Norway and Sweden, competing for their first successful low orbit launch. We use multi-method data collection to understand the development of these ecosystems in real time, through in-depth interviews, observations, field notes from events, and transcribed podcasts with relevant actors. The rich data set is analyzed using Nvivo.

Results and Implications

Preliminary findings indicate that external enablers are critically important in terms of technological development such as the reduction of production costs and the enablement of innovative technologies. Finance knowledge and physical infrastructure have been essential in building up the entrepreneurial ecosystem in the new space industry. Separately, we also examine the efforts on both countries. In Norway, we found entrepreneurial ecosystem champions (Johnstone-Louis & Love 2021; O'Connor et al., 2018) to be crucial, where a few individuals in central positions managed to carry and implement their visions for the space sector to grow. In Sweden, we found that collaboration between different actors, at distinct levels within the ecosystem, has played a more significant role for the development of the sector. Furthermore, the role of the military industry has been significantly different in the two countries. We also develop contributions to theory, while simultaneously providing practical implications for entrepreneurship policy in general and for the space industry in the Nordic region more specifically.

CONTACT: Sølvi Solvoll; solvi.solvoll@nord.no; Nord University Business School; Bukken Bruse 12, Bodø, Nordland 8028, Norway

SUMMARY

(DON'T) JUDGE A VENTURE BY ITS COLOR: THE IMPACT OF GREENWASHING ACCUSATIONS ON SUSTAINABLE NEW VENTURES' EMPLOYER ATTRACTIVENESS

Johanna Isabell Supe, University of Bayreuth, Germany
Matthias Baum, University of Bayreuth, Germany
Mirjam Nicole Streeb, University of Bayreuth, Germany

Principal Topic

Sustainable ventures often walk a fine line between visionary ambition and overpromising, making them vulnerable to accusations of greenwashing, referring to overstating sustainability efforts without substantive actions to support these claims (Schultz et al., 2013; Siano et al., 2017). Such accusations can be very harmful for firms in general (Walker & Wan, 2012) but might be particularly detrimental for ventures with a sustainable vision that aims to attract job seekers. So, what can sustainable ventures do to remain attractive to job seekers when confronted with greenwashing accusations? Is it viable to conduct costly actions that serve as potential strong signals (Kleinert & Hildebrand, 2024) of the authentic green vision enactment? Or does the venture require endorsements, particularly from reputable third-party organizations (Mochkabadi et al., 2024)? Building on signaling and cognition literature, we argue that the interplay of such ambiguous signals triggers and modulates different cognitive attributions, such as the assumption of a genuine motivation for the new venture and uncertainty perceptions (Küsshauer & Baum, 2023) that predict applicants' job-pursuit intentions.

Method

We conducted a randomized between-subjects experiment (vignette-study) with 342 U.S. participants, employing a 2x2x2 design that manipulated greenwashing accusations, the venture's green vision enactment, and external endorsement (all: weak vs. strong). We included perceived value-driven and strategic motives (Vlachos et al.,2009), uncertainty perceptions (Küsshauer & Baum, 2023) as mediators, and job-pursuit intentions as outcomes. Our sample comprised job seekers with higher education and at least three years of experience.

Results and Implications

We use lavaan in R to analyze moderated-mediation models. Our preliminary analyses yield models with a decent fit (e.g., RMSEA=0.036) and indicate that greenwashing accusations particularly influence job-pursuit intentions when green vision enactment of the venture is on low levels ($\beta = -0.15$, $p = 0.011$). Moreover, perceived value-driven motives and uncertainty are both affected by greenwashing accusations and strongly predict job pursuit intentions (value-driven motives: $\beta = 0.58$, $p < 0.001$; uncertainty: $\beta = -0.29$, $p < 0.001$). However, these effects seem to be conditional upon green vision enactment and external endorsements. Green vision enactment significantly reduces the negative effect of greenwashing accusations on perceived value-driven motives of the venture ($\beta = 0.26$, $p = 0.016$), while external endorsements can mitigate the effect of such allegations on perceived uncertainty ($\beta = 0.21$, $p = 0.060$). This suggests a complex interplay between internal and external signals influencing the diverse cognitions that greenwashing accusations stimulate.

CONTACT: Johanna Isabell Supe; johanna.supe@uni-bayreuth.de; University of Bayreuth; Hattsteiner Straße 20, Frankfurt, Hessen 60489, Germany

INTERACTIVE PAPER

"HOW DOES THE HEISENBERG COMPENSATOR WORK? VERY WELL, THANK YOU!" A QUANTUM MECHANICS PERSPECTIVE TO ENTREPRENEURIAL UNCERTAINTY

Daniel R. Clark, Ivey Business School, Canada

Principal Topic

Quantum theory addresses behaviors at the smallest discrete unit of a phenomenon. For physics, where the theory originated, this involves the study of subatomic particles. For management, quantum theory is of interest among scholars who seek better understanding of markets and the general nature of environmental uncertainty. The hope is that a quantum approach can lend novel insights, and breakthrough explanations of phenomena currently treated as error terms in most analytical work. Clark and Hunt (2024), suggest employing the Heisenberg Uncertainty Principle to better understand entrepreneurial action under uncertainty. They propose a canonically conjoined variable approach of modeling market uncertainty, employing price elasticity and competitive intensity. Our study formally tests this model by collecting computational and experimental data concerning inter-related dimensions that have two key properties: (i.) identifiability of the central need or want that provides actor attraction – that is, the source of opportunity; and (ii.) variability in attraction strengths – that is, the opportunity's relevance varies individual-to-individual. In a fashion faithful to quantum principles, we operationalize these key dimensions through two canonical conjugated variables – price elasticity and competitive intensity – each ultimately unknowable yet completely responsive to intervention.

Method

Operationalizing superposition and entanglement in entrepreneurial decision making is a formidable challenge. Fortunately, entrepreneurship has dynamic, canonically conjoined variables that are amenable to experimentation -- price elasticity and competitive intensity – which we used in two separate research designs: (a) an agent-based model and (b) an experiment conducted with two subject pools: (i.) college students, and (ii.) entrepreneurs and entrepreneur-aspirants. For the experiment, subjects were exposed to various scenarios in which information was provided concerning either, both, or none of the conjoined variables. Subjects were then instructed to make predictions about the remaining variable and/or the overall level of uncertainty. Partial or complete knowledge of one variable should influence the level of uncertainty for the conjoined variable. Using ANOVA analyses of between and within effects, we then compared critical thresholds to results from the computational model.

Results and Implications

Experimental results suggest that the application of the Heisenberg Uncertainty Principle has exciting potential and that entrepreneurs can constructively employ use of canonically conjoined variables in better understanding – and partially mitigating – environmental uncertainty. Validation of the Clark-Hunt model (2024) represents a critical step in demonstrating the manner in which quantum principles can be usefully applied to dynamically complex and causally convoluted decision environments, such as those commonly arising in entrepreneurship.

CONTACT: Daniel R. Clark; dclark@ivey.ca; Ivey Business School; 1255 Western Rd, London, Ontario N6G 0N1, Canada

INTERACTIVE PAPER

THE ROLE OF FOUNDERS' POLITICAL IDEOLOGIES FOR GREEN VENTURES

David Flore, TU Dortmund University, Germany
Steffen Strese, TU Dortmund University, Germany

Principal Topic

Green ventures can act as change agents (Hällerstrand et al., 2023) to combat human-made climate change. Recently, researchers call to investigate sustainability-related outcomes of green ventures (Anand et al., 2021), but the measurement of such outcomes needs clarification (Fichter et al., 2023). Drawing on upper echelons perspective (UEP) (Hambrick & Mason, 1984) and literature on CEOs, leaders' characteristics are important drivers of environmental innovations (e.g., Arena et al., 2018). Particularly, scholars find leaders' political ideologies to influence sustainability-related decisions (e.g., Lesage et al., 2024) based on differences between liberal and conservative individuals' climate change beliefs (Ziegler, 2017). However, we lack theoretical and empirical knowledge about the role of founders' political ideologies for green ventures' efforts towards environmental innovations. This gap is intriguing because founders' characteristics influence the creation of a venture since the first day (Shepherd et al., 2021) and environmental entrepreneurship research seeks insights on the ecological part of green ventures' dual mission (Schaltegger & Wagner, 2011).

Method

Our sample is drawn from Crunchbase and complemented with US election and patent data. We empirically evaluate our hypotheses with a unique dataset of 875 US-based green ventures founded between 2012 and 2021 utilizing a zero-inflated negative binomial model. Ventures are considered green when they are tagged to have "economically-sound processes that minimize negative environmental impacts while conserving energy and natural resources" (Crunchbase, 2025).

Results and Implications

We reveal that founder liberalism promotes environmental innovations in green ventures. Specifically, green ventures with founders who have more liberal values develop more green patents. This relationship seems unaffected by regional-level conservatism. This result diverges from insights on CEOs (e.g., Fuchs et al., 2024) which could be explained by the more intense coupling of founders' identities with their ventures' goals (York et al., 2016). Moreover, this relation is enhanced by higher shares of females in founding teams, showing that green ventures' founders integrate their stakeholders' interests into their decision-making regarding environmental innovations. Our study contributes to environmental entrepreneurship literature and UEP. First, we introduce a new antecedent for environmental innovations in green ventures by showing the importance of founders' liberal political values. Second, we extend research on leaders' political ideologies to the new context of green ventures. Third, we introduce new boundary conditions for green ventures. Green ventures' founders seem unaffected by diverging, dominant political ideologies in their region but are influenced in their decisions towards driving environmental innovations by characteristics of their direct stakeholders such as co-founders.

CONTACT: David Flore; david.flore@tu-dortmund.de; TU Dortmund University; Bernkasteler Str. 55, Cologne, Nordrhein-Westfalen 50969, Germany

INTERACTIVE PAPER

LOCATION MATTERS: THE INFLUENCE OF URBAN DYNAMICS ON ENTREPRENEURIAL VENTURE SUSTAINABILITY

Michael D. Hudecheck, University of St. Gallen Institute of Responsible Innovation, Switzerland

Principal Topic

Entrepreneurial ventures are predominantly located in urban environments. Modern urban development patterns, such as the proliferation of car-centric urban features, predispose entrepreneurial ventures to implicitly participate in unsustainable practices with detrimental ecological and economic repercussions. Despite growing interest among entrepreneurship scholars in exploring how entrepreneurial ventures engage in sustainability practices, little research has been done to examine the impact of entrepreneurial venture environmental embeddedness on venture sustainability. Our study, grounded in the sustainability transition literature, explores whether entrepreneurial venture proximity to "green" urban features influences the likelihood that those ventures engage in sustainable practices. In doing so, we contribute to the literature on entrepreneurial venture sustainability by showing that environmental embeddedness within green urban ecosystems is positively associated with venture sustainability practices.

Method

We obtained aerial imagery composites of San Francisco from the San Francisco Planning Department to algorithmically calculate vegetation coverage levels from 2019 to 2024. We then geolocated approximately four thousand six hundred extant entrepreneurial ventures headquartered in San Francisco, along with various green urban spatial features from Google Places and OpenStreetMap (OSM). We calculated functional distance estimates between each known venture location, each binned location with high levels of vegetation coverage, and each known geolocated urban spatial feature. This resulted in our independent variable, Environmental Embeddedness. We used our imagery composites, along with a trained state-of-the-art YOLOv11 image detection algorithm, to create annual counts of the number of open parking spaces within a three-minute distance of each venture address. Using the COVID-19 lock downs in 2020 as a baseline, we estimated levels of pre- and post-pandemic office-space utilization, or Return to Work, which we use to operationalize sustainability. We include control variables on entrepreneurial venture activity. We test our theorizing with a pooled ordinary least squares model and conduct comprehension and robustness checks.

Results and Implications

Our results suggest a positive relationship between entrepreneurial venture environmental embeddedness and levels of estimated sustainability. More explicitly, our results suggest that venture embeddedness in gray urban landscapes increases the likelihood that an entrepreneurial venture will engage in less sustainable practices, which we attribute to ecosystem proximity effects. In doing so, our results contribute to the literature on entrepreneurial venture sustainability by showing a relationship between entrepreneurial venture sustainability practices and their surrounding environmental ecosystems and providing practical insights on how office location can positively influence venture sustainability practices.

CONTACT: Michael D. Hudecheck; michael.hudecheck@unisg.ch; University of St. Gallen Institute of Responsible Innovation; Dufourstrasse 40a, St. Gallen, Sankt Gallen 9000, Switzerland

INTERACTIVE PAPER

INFORMAL INVESTORS, RULE OF LAW, AND HOUSEHOLD INCOME IN NEW VENTURE INTERNATIONALIZATION

Victor Martin-Sanchez, University of Southern Denmark, Denmark
Cecilia Alvarez Vargas, Universitat Autònoma de Barcelona, Spain
Joan-Lluís Capelleras, Universitat Autònoma de Barcelona, Spain

Principal Topic

The internationalization of early-stage entrepreneurial firms sits at the intersection of international business and entrepreneurship research (Knight & Liesch, 2016; Oviatt & McDougall, 1994). A key focus in this domain is the expansion of new ventures (Shane & Venkataraman, 2000), with substantial research exploring their internationalization (Jones et al., 2011; Schwens et al., 2018). Despite progress, a fundamental question remains: what drives new venture internationalization? Informal investors play a crucial role, providing early-stage financing and actively engaging in firm creation (Qin et al., 2022; Korosteleva & Mickiewicz, 2011). However, their specific impact on internationalization remains underexplored. Additionally, home-country institutional conditions, particularly the rule of law, can shape this relationship by reducing transaction costs and offering regulatory support for international expansion (Kenneth-Southworth et al., 2018). Beyond country-level factors, individual-level determinants are also critical. Entrepreneurs from high-income households not only possess greater financial resources but also better access to quality opportunities (Arenius & De Clercq, 2005). This paper examines three key aspects: (1) the impact of informal investors on new venture internationalization, (2) the moderating role of the rule of law, and (3) how entrepreneurs' household income influences these relationships. Our framework integrates insights from entrepreneurial finance (Denis, 2004), the resource-based view (Wernerfelt, 1984), and institutional theory (North, 1990).

Method

Empirically, we employ a dataset combining individual-level observations from the Global Entrepreneurship Monitor (GEM) with country-level data for income classification from the World Bank (World Development Indicators, WDI) and the Worldwide Governance Indicators (WGI). Our dataset encompasses 78,349 early-stage entrepreneurs (those initiating firms within the past 42 months) representing 108 countries from 2005 to 2020. We utilize a multilevel logistic regression model to predict the likelihood of new venture internationalization.

Results and Implications

Our research highlights three key findings. First, a higher proportion of informal investors in a country positively influences new venture internationalization, as they provide crucial resources like financial capital, industry knowledge, networks, and mentorship. Second, a strong rule of law weakens this effect, as entrepreneurs may prefer formal financing options like bank loans or venture capital, which they perceive as more reliable. A robust legal framework also enhances contract enforcement and intellectual property protection, reducing reliance on informal investors. Third, in strong legal systems, entrepreneurs from lower-income backgrounds benefit less from informal investors, likely due to regulatory complexities and limited resources to navigate them, making informal investment less attractive for internationalization.

CONTACT: Victor Martin-Sanchez; vms@sam.sdu.dk; University of Southern Denmark; Campusvej 55, Odense, Syddanmark 5230, Denmark

INTERACTIVE PAPER

ONE CITY, TWO ECOSYSTEMS: DUAL ECOSYSTEM ADAPTATION OF FOREIGN ENTREPRENEURS AS A RESPONSE TO ECOSYSTEM CONSTRAINTS

Maximilian Scheu, University of Hohenheim, Germany
Leif Brändle, University of Hohenheim, Germany
Andreas Kuckertz, University of Hohenheim, Germany

Principal Topic

While the ecosystem literature has provided valuable insights into the beneficial role of foreign entrepreneurs in entrepreneurial ecosystems (Harima et al., 2021, 2024), it has largely neglected the study of the liabilities foreign entrepreneurs face in accessing the ecosystem's resources and institutions. Entrepreneurial ecosystem theories primarily focus on the availability of institutional arrangements and resources that foster entrepreneurial activities in a region (Stam & Van de Ven, 2021) but neglect variance in the accessibility of these ecosystem elements for different entrepreneurs (Scheidgen, 2021). Our study investigates how foreign entrepreneurs respond to liabilities of foreignness in entrepreneurial ecosystems.

Method

We closely followed foreign entrepreneurs in Shanghai's entrepreneurial ecosystem from 2019 to 2024 and conducted over 50 interviews with 32 informants. We also performed 1.5 years of an on-site observational study that produced over 50 field notes and accessed more than 100 archival data sources. We employed purposive and snowball sampling to identify knowledgeable informants who could provide a diverse perspective (Eisenhardt, 1989). The study sample included foreign entrepreneurs, support providers, investors, and company and government representatives. The interviews lasted between 30 minutes and two hours. Moreover, the lead researcher resided in Shanghai during the initial data collection phase and extended his stay until early 2021, providing the research team with first-hand experience within the ecosystem. We iteratively analyzed the interview data to build a data structure using an open coding scheme (Gioia et al., 2013).

Results and Implications

Our results indicate that liabilities of foreignness, including sociocultural and formal constraints, exclude foreign entrepreneurs from accessing crucial elements of the entrepreneurial ecosystem. However, two mutually dependent response strategies emerged from our data. First, foreign entrepreneurs counter their exclusion from the main ecosystem through local adaptation. Second, facing the limits of the local adaptation strategy, foreign entrepreneurs participate in a foreign entrepreneurial sub-ecosystem. Our longitudinal data collection allowed us to witness multiple escalations in ecosystem exclusion, affecting how foreign entrepreneurs reacted to these increasingly hostile conditions by adapting their response strategies or withdrawing from the ecosystem. This study highlights the function, boundaries, and interactions of a foreign entrepreneurial sub-ecosystem, contributing to the growing discussion on the sub-structures of entrepreneurial ecosystems (Scheidgen, 2021; Cloutier & Messeghem, 2022; Theodoraki & Catanzaro, 2022). Since attracting foreign entrepreneurial talents is critical for the economic impact of entrepreneurial ecosystems, the practical implications of our study inform policy-makers aiming to design globally oriented and inclusive entrepreneurial ecosystems.

CONTACT: Maximilian Scheu; Maximilian.Scheu@uni-Hohenheim.de; University of Hohenheim; Wollgrasweg 49, Stuttgart, Baden-Wurttemberg 70190, Germany

INTERACTIVE PAPER

ASSET OF NEWNESS IN CONTEXTS OF INSTITUTION FAILURES: THE CASE OF JACKSON, MISSISSIPPI

Sharon Simmons, Jackson State University, United States
Sumaia Alsharari, Jackson State University, United States
Miles Anthony, Analytix Research Group LLC, United States
Hailie Webb, Jackson State University, United States

Principal Topic

This study investigates the "liabilities of newness," which posits that newly established businesses incur higher operational costs and face greater risks of failure due to their lack of legitimacy among customers, suppliers, and financiers (Pfeffer & Salancik, 1978). Aldrich and Fiol (1994) expand this view by attributing these liabilities to the inherent uncertainty and limited resources of new firms, making them particularly susceptible to external shocks. Focusing on Jackson, Mississippi—a city burdened by aging infrastructure, economic disparities, and recent water crises—this study explores how institutional failures, particularly water contamination, exacerbate these challenges. Additionally, the paper introduces a new concept, the "asset of newness," which suggests that young firms may benefit from greater agility and adaptability during widespread crises.

Method

Using a case study approach, the analysis examines the impacts of Jackson's 2020 Pearl River flood and the 2022 water treatment plant failure on local businesses. Data were collected from the Mississippi Secretary of State's corporation database and refined to focus on over 50,000 companies within Jackson. The study employs logistic regression analysis, incorporating variables for formation year, entity type and 2-digit NAICS industry codes. The dependent variable captures the dissolution status of firms. Special attention is given to firms formed in recent years to assess how these variables influence survival rates during institutional crises.

Results and Implications

The findings reveal that while young firms typically experience the "liabilities of newness," during a systemic crisis that disrupts essential services like water, these liabilities may be counterbalanced by what we term the "asset of newness." The result is that these young firms had a lower probability of dissolution. Established firms with rigid routines and legacy systems also encounter legitimacy challenges, higher operational costs, and diminished agility. In contrast, young firms, unburdened by entrenched processes, can rapidly adapt and innovate to navigate disruptions. This agility enables them to adjust to fluctuating resource availability and shifting market dynamics more effectively than their older counterparts. Consequently, during city-wide crises affecting water infrastructure, the survival advantage may shift toward younger, more flexible enterprises. The study underscores the importance of investing in resilient infrastructure and rethinking traditional notions of firm vulnerability. Future research should further explore how crisis-specific characteristics interact with firm age and structure, and refine the "asset of newness" theory to better understand how flexibility in young firms can serve as a strategic advantage in turbulent environments.

CONTACT: Sharon Simmons; sharon.simmons@jsums.edu; Jackson State University; 123 N Wyoming Ave, South Orange, New Jersey 07079, United States

BERTARELLI INSTITUTE FOR FAMILY ENTREPRENEURSHIP AWARD (BABSON COLLEGE) FOR THE BEST PAPER ON FAMILY ENTREPRENEURSHIP

MANAGING DUAL ROLES: COPING WITH WORK-FAMILY CONFLICT AND ENHANCING WELL-BEING OF FAMILY BUSINESS MEMBERS

Moonum Zehra, University of Siegen, Germany
Arndt Werner, University of Siegen, Germany
Meike Stephan, University of Siegen, Germany
Sepideh Ghahremani Tehrani, University of Siegen, Germany

ABSTRACT

This study explores how time-based work-to-family interference (WIF) impacts satisfaction with work-life balance (SWLB), focusing on the moderating roles of business family members (BFMs) status and boundary management. While WIF is typically seen as detrimental, we argue that BFMs may perceive work demands as contributing to the family legacy. Drawing on identity theory and institutional logics, we find that BFMs experience less time-based WIF than non-BFMs. Moreover, strong boundary management further buffers negative effects. These findings highlight the context-dependent meaning of WIF and call for a rethinking of boundary theory in integrated work-family domains.

INTRODUCTION

The boundary between work and family has long been portrayed as a site of tension—a zero-sum game where time spent in one domain comes at the cost of the other. Time-based work-to-family interference (WIF) is widely recognized as a key threat to individual well-being, undermining satisfaction with work-life balance (SWLB) and overall role functioning (Greenhaus & Beutell, 1985; Allen et al., 2000). Such views assume that work and family are inherently separate and often in conflict—but what if they *aren't*?

In family firms, business is not external to family rather embedded within it. In this study, we challenge conventional models of work-family conflict by exploring how business family members (BFMs) interpret and experience time-based WIF. Drawing on identity theory (Stryker & Burke, 2000) and institutional logics (Thornton et al., 2012), We argue that BFMs may view work demands not as interference, but as meaningful investments in the family's shared legacy. We further theorize that the strength of this reinterpretation depends on individuals' boundary management skills—the capacity to integrate or segment roles with strategic intent.

Drawing on longitudinal panel data from Germany (Pairfam), we demonstrate that the negative relationship between time-based WIF and SWLB is significantly attenuated for BFMs, and that this buffering effect is amplified when boundary management is high. These findings challenge dominant theories of WIF and reveal that under specific cultural and identity conditions, interference can be reimagined not as burden—but as belonging or investment.

HYPOTHESIS DEVELOPMENT

WIF is commonly framed as a conflict between incompatible roles, resulting in lower life satisfaction and perceived work-life balance (Greenhaus & Beutell, 1985; Adisa et al., 2016; Kalliath & Brough, 2008). According to the traditional work-family conflict perspective, WIF occurs when the demands of the work domain undermine one's ability to engage in family responsibilities. Frone et al., (1992) further expanded this perspective by distinguishing between time, strain, and behavior-based conflicts—where limited time, emotional exhaustion, or incompatible behaviors from the work domain negatively affect family life.

On the other hand, work-life balance, though variably defined, is commonly understood as the individual's perception of how well they can meet both work and non-work role demands (Kalliath & Brough, 2008). SWLB is not merely about equal time spent in each domain but about subjective alignment between role expectations and individual preferences (Greenhaus et al., 2003). The depletion of time resources caused by work overload leaves individuals with less discretionary time and flexibility, resulting in lower perceived balance (Voydanoff, 2005; Allen et al., 2000).

From a theoretical standpoint, the Conservation of Resources (COR) theory (Hobfoll, 1989) provides a useful framework for understanding how demands in the work domain consume resources such as time and energy, thereby contributing to negative spillover effects and dissatisfaction with work-life balance. Empirical evidence consistently identifies time-based WIF as an important predictor of negative outcomes in work and life domains. For instance, long working hours and perceived time pressures strongly correlate with reduced satisfaction across various life domains, including job and overall life satisfaction (Allen et al., 2000). Similarly, Byron's (2005) meta-analysis demonstrates a robust negative relationship between time-based WIF and family satisfaction, further supporting the assumption that time-based interference negatively impacts satisfaction with work-life balance. Thus, drawing on theoretical insights and empirical findings, the following hypothesis is proposed:

H1. Time-based WIF has negative influence on SWLB.

In the context of family businesses, researchers argue that intertwined roles, time pressures, and blurred boundaries may intensify WIF (Poelmans, 2005; Zahra & Sharma, 2004). However, this argument relies on the assumption that work, and family are inherently separate and competing spheres—an assumption that may not fully apply to BFMs. Embedded in family businesses where work and family roles are deeply interwoven, BFMs operate within a unique socio-cultural context that reshapes the meaning and impact of WIF. Identity theory (Stryker & Burke, 2000) suggests that conflict arises when incompatible roles demand competing behaviours or values. Yet for BFMs, work and family roles are often aligned rather than at odds. The family business is not merely a workplace—it is a shared project, an identity, and a long-term legacy. This alignment alters the emotional interpretation of time-based WIF. Instead of perceiving long hours as sacrificial or burdensome, BFMs may experience them as meaningful investments in family continuity and success.

This reinterpretation can be further supported by institutional logics theory (Thornton et. al., 2012) which holds that people interpret actions through prevailing cultural logics. In BFMs, a "familial logic" legitimizes work demands as acts of collective responsibility and emotional contribution. As a result, while BFMs may still encounter time-based WIF, they are more likely to cognitively reframe it in ways that buffer its negative impact on work–life balance. This challenges conventional WIF models and positions BFMs as resilient—not because they experience less interference, but because they make sense of it differently. Therefore, we hypothesis that,

H2. The negative relationship between time-based WIF and SWLB is weaker for BFMs than for non-BFMs.

Boundary theory adds another layer to the story, suggesting that individuals manage these boundaries through segmentation (keeping roles distinct) or integration (blending roles) (Ashforth et al., 2000). The effectiveness of these strategies often depends on personal preferences and contextual demands. Considering the complicated nature of roles, BFMs often cannot—nor always wish to—clearly separate these roles. Instead, they may pursue strategic integration, selectively managing when and how boundaries are crossed. For example, setting shared family work hours or co-creating routines can help sustain work life balance even under high WIF. Those who consciously develop boundary management skills—like time-blocking, role negotiation, or psychological detachment—may transform role conflict into role synergy. Research indicates that a misalignment between an individual's preferred boundary management style and their actual work-family interface can exacerbate work-family conflict. For instance, when there's a poor fit between desired and enacted boundary management, individuals may experience increased time- and strain-based conflicts, negatively impacting family satisfaction (Chen et al., 2009). Conversely, aligning boundary management strategies with personal preferences can reduce such conflicts and enhance overall well-being. For BFMs, who are frequently exposed to role integration, the development of boundary management skills may be especially critical in mitigating the negative effects of WIF. Thus, boundary management amplifies the buffering potential of BFM status.

H3. The moderating effect of BFM status on the relationship between time-based WIF and SWLB is stronger when boundary management is high, such that BFMs with high boundary management skills are least negatively affected by time-based WIF.

METHOD

This study uses data from the German Family Panel (PAIRFAM), initiated in 2008 with 12,402 randomly selected individuals from cohorts born 1991-93, 1981-83, and 1971-73, expanded in 2018-19 with individuals from the 2001-03 cohort. We specifically analyze waves 6, 8, 10, and 12 (2013–2020), which include key measures of time-based WIF. The sample comprises 1,104 BFMs in Germany, operationalized following Stamm et al., (2023).

The dependent variable is SWLB, and the independent variable is time-based WIF. Boundary-setting construct, introduced in wave 6 based on Kossek et al., (2012), serve as a moderator and were reverse-coded for consistency. Controls include age, gender, household and individual income, weekly working hours (respondent and partner), household size, years of education, financial satisfaction, number of children, and partner's employment status). All variables are standardized and mean-centered. Table of operationalization of variables is provided upon request.

Hypotheses are tested via hierarchical regression analysis (Tabachnick et al., 2013; Cohen et al., 2003; Field, 2013). A Hausman test indicated a preference for fixed-effects (FE) modeling (χ^2=104.67, p<0.001), suggesting significant correlation between regressors and individual-specific effects. Thus, all analyses apply FE specifications with robust standard errors.

RESULTS

According to the descriptive statistics, respondents reported moderate to high SWLB (M = 6.26, SD = 2.04) and moderate levels of time-based WIF (M = 2.51, SD = 1.10), suggesting the presence of conflict without it being overly pronounced. About 7.1% of the sample are BFMs, and participants generally show high boundary management (M = 4.23, SD = 1.17). The average respondent is 35 years old (SD = 8.27), lives in a three-person household (SD = 1.31), and has nearly 13 years of education (SD = 3.04). Net individual income varies widely (M = €1,796, SD = €1,449), as does household income (M = €3,479, SD = €2,383). Respondents work 36 hours per week on average (SD = 12.70), with partners working 38 hours (SD = 11.69). Financial satisfaction is relatively high (M = 6.89, SD = 2.18). On average, respondents have one child, 46.4% are married, and 38% of partners are employed. The sample is gender-balanced (51% female).

Pearson correlations show that SWLB is significantly and negatively correlated with WIF (r = –0.621, p < .01), supporting the expected link between time conflict and reduced balance (Greenhaus & Beutell, 1985; Allen et al., 2020). BFM status is not significantly correlated with either SWLB (r = 0.003) or WIF (r = 0.035). Boundary management, however, is positively associated with SWLB (r = 0.114, p < .05) and negatively with WIF (r = –0.259, p < .01), consistent with prior findings (Clark, 2000; Kossek et al., 2012). Among controls, longer work hours negatively relate to SWLB (r = –0.264, p < .01) and positively to WIF (r = 0.385, p < .01), while financial satisfaction is positively linked to SWLB (r = 0.186, p < .01), and individual income is positively related to WIF ($r = 0.163$, $p < .01$), possibly reflecting role strain (Shockley & Allen, 2007). The table of descriptive statistics as well as matrix of correlation is provided upon request.

Multicollinearity was examined through variance inflation factors (VIFs) and found to be non-problematic (Hair et al., 2010). Although some controls—e.g., number of children and household members ($r = 0.796$), and marital status and partner employment ($r = 0.864$)—are strongly correlated, they are theoretically justified. All VIFs remained below the accepted threshold of 10.

Regression results are shown in Table 1. Model 1 includes control variables, Model 2 adds the main predictors, and Model 3 includes interaction terms. Across models, WIF is negatively associated with SWLB (Model 2: $\beta = -.5078$, $p < .001$; Model 3: $\beta = -.5139$, $p < .001$), supporting H1.

While main effects of BFM status and boundary management are not significant, Model 3 reveals a significant interaction between BFM status and WIF ($\beta = .1774$, $p < .001$), indicating that BFMs are less negatively affected by WIF—supporting H2.

Furthermore, the three-way interaction between BFM status, WIF, and boundary management is significant ($\beta = .0837$, $p < .05$), supporting H3. Figure 1 shows that BFMs with strong boundary skills experience minimal decline in SWLB under WIF, while those with weak skills show a steep drop. Among non-BFMs, boundary management has little effect. Figure 2 further illustrates that for BFMs, the negative impact of WIF diminishes with stronger boundary management and even becomes negligible or slightly positive at high levels—emphasizing its amplifying role.

Among control variables, most results align with expectations: longer working hours are linked to lower SWLB (Model 3: $\beta = -.125$, $p < .001$), while financial satisfaction and education levels are positively related. A notable but counterintuitive finding is the negative relationship between individual income and SWLB in Model 1 ($\beta = -.03$, $p < .05$), which disappears in later models—possibly reflecting hidden costs of high-income roles. Other demographics such as number of children, household size, and marital status do not show consistent or significant effects.

DISCUSSION AND IMPLICATIONS

This study shows that time-based WIF persistently undermines SWLB—but not for everyone, and not in the same way. While prior research confirms the negative effect of WIF (Byron, 2005; Allen et al., 2000; Greenhaus & Beutell, 1985) and the buffering role of boundary management (Kossek et al., 2012; Allen et al., 2014; Kreiner et al., 2009), this study contributes a novel perspective by introducing BFM er status as a socio-culturally embedded identity that reshapes the very experience of work-family conflict. In these contexts, work is not an outsider to family life rather part of its fabric. This lens complicates traditional assumptions in boundary theory and highlights the conditions under which interference may be perceived as investment, not intrusion.

For BFMs, the experience of WIF is meaningfully different. They are embedded in a unique socio-cultural and organizational context (Zellweger et al., 2012; Chrisman et al., 2005; Sharma & Irving, 2005; Dyer, 2003; Habbershon & Pistrui, 2002) where work and family roles are often aligned, rather than competing. This alignment fundamentally reshapes how time pressure from work is appraised and managed (Greenhaus & Powell, 2006; Clark, 2000).

At the center of our findings is a simple but powerful insight: when work is family, interference is not always conflict—it can be contribution. While WIF typically signals strain, guilt, and imbalance (Greenhaus & Beutell, 1985; Michel et al., 2011), BFMs often interpret the same time demands as meaningful investments in a collective legacy. This cognitive reframing, supported by identity theory (Stryker & Burke, 2000) and institutional logics theory (Thornton, 2004), helps explain why BFMs report higher resilience to time-based WIF. The business is not just work—it's "ours," and it matters emotionally, not just economically (Zellweger et al., 2013).

This study also clarifies the role of boundary management. In traditional work-family research, boundary management is often portrayed as a defensive tool that is used to protect one role from the encroachment of the other (Ashforth et al., 2000). However, our findings suggest a different logic for BFMs: boundary management isn't about building walls—it's about blending strategically. BFMs who consciously manage their time, expectations, and transitions between roles can amplify the buffering effects of their identity. In short, they turn natural role overlap into functional role synergy. This aligns with the findings of Kossek and Lautsch (2012), that strategic alignment between preferred and enacted boundary styles leads to reduced conflict and higher well-being.

These insights lead to two key theoretical implications. First, they challenge the assumption that work-family conflict operates uniformly across all contexts. Role identity and cultural logic, clearly matter (Ashforth & Mael, 1989; Rothbard & Edwards, 2003). Second, they call for a more nuanced application of boundary theory in contexts of high role integration. Strategic boundary management remains essential—but not in the form of separation. For BFMs, the "skill" lies in harmonizing overlapping roles, not in keeping them apart.

From a practical perspective, these findings offer valuable guidance for family businesses, HR professionals, and policymakers. Rather than viewing WIF as something to eliminate at all costs, organizations might instead help family members reframe and recontextualize work demands in ways that align with family values. Simultaneously, training in boundary management—especially for younger or newer BFMs—can provide the tools to navigate demanding environments more sustainably (Kossek et al., 2012).

Finally, one surprising control variable deserves attention: individual net income was negatively associated with SWLB in the base model. This counterintuitive result may reflect the hidden costs of high-earning roles—where income comes bundled with long hours, responsibility pressure, or reduced flexibility (Perlow, 1999; Voydanoff, 2005). It's a reminder that financial success does not always equate to subjective balance.

This study faces several limitations. First, key constructs were measured with short scales—boundary management via a single item and time-based WIF with two items. While this restricts certain psychometric evaluations (e.g., internal consistency and CFA), prior research supports the validity of such measures when constructs are clearly defined and unidimensional (Kossek et al., 2012; Kreiner, 2006; Allen et al., 2014). Reliability for the two-item scale was acceptable ($\alpha = 0.80$; $r = 0.67$, $p < .001$), aligning with existing recommendations (Eisinga et al., 2013; Fuchs & Diamantopoulos, 2009). Second, the study focused solely on time-based WIF, excluding strain- and behavior-based dimensions, which may interact in meaningful ways (Michel et al., 2011). Future research should adopt a multidimensional approach (Greenhaus & Beutell, 1985; Byron, 2005). Third, the limited sample size of BFMs (7.1%) reduced statistical power to detect nuanced interaction effects. While representative, this small subgroup warrants targeted oversampling in future studies (Aguinis & Edwards, 2014). Boundary management was assessed cross-sectionally as a stable trait, though it's likely dynamic over time (Kossek et al., 2012; Allen et al., 2014). Longitudinal or experience-sampling designs could better capture this variation. Lastly, while the sample included multiple age cohorts (1971–2003), generational differences in boundary preferences and digital integration were not modeled—an important avenue for future research (Wayne et al., 2006; Derks et al., 2016).

In sum, despite limitations, this study offers a more textured understanding of how work-to-family interference is experienced and mitigated. By foregrounding the identities and strategies of BFMs, it moves beyond the binary of "conflict versus balance" and shows that under the right conditions, even time pressure can become part of the family fabric.

CONTACT: Moonum Zehra; moonum.zehra@uni-siegen.de; University of Siegen, Lehrstuhl für KMU-Management und Entrepreneurship, Campus Unteres Schloss, Kohlbettstr. 17, 57072, Siegen, Germany

APPENDIX

Table 1. Regression: dependent variable - SWLB

	Model 1	Model 2	Model 3
Time-based WIF		-.5078***	-.5139***
		(.0165)	(.0169)
BFM		.0962	.1096
		(.0669)	(.0781)
Boundary management		-.0138	-.0097
		(.0147)	(.0158)
Time-based WIF x Boundary management			-.002
			(.0131)
BFM x Time-based WIF			.1774***
			(.0567)
BFM x Boundary management			-.0188
			(.0412)
BFM x Time-based WIF x Boundary management			.0837**
			(.0327)
Constant	.0476	.0129	.0108
	(.0829)	(.0694)	(.0693)
Observations	9747	9747	9747
R-squared	.0593	.2519	.2538

Robust standard errors are in parentheses
**** p<.01, ** p<.05, * p<.1*

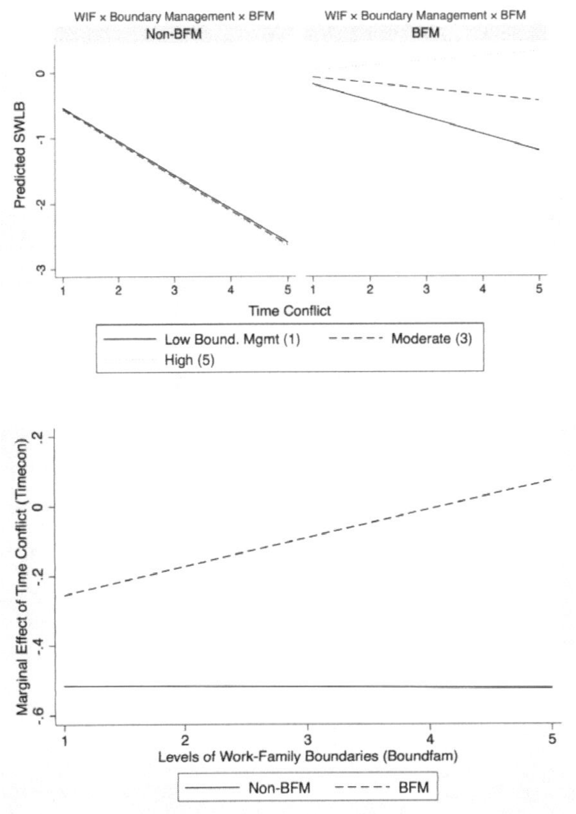

Figure 2. Marginal effect of WIF on SWLB by boundary management - BFM status

SUMMARY

FAMILY FIRMS: BAD EMPLOYERS OR GOOD SAMARITANS?

Mateja Andric, University of St. Gallen & Stockholm School of Economics, Switzerland
Mohamed Genedy, Stockholm School of Economics & Jönköping International Business School, Sweden
Lucia Naldi, Jönköping International Business School, Sweden
Mattias Nordqvist, Stockholm School of Economics, Sweden

Principal Topic

Research suggests that family firms offer less favorable employment conditions to employees than non-family firms (Neckebrouck, Schulze, and Zellweger, 2018). However, perspectives of family firms as "bad employers" neglect that lower compensation levels in family firms may in part be attributable to the sorting of low-wage employees into family firms (Bassanini, Breda, Caroli, & Rebérioux, 2013). Building on this sorting perspective, we introduce the argument that family firms may act as "good Samaritans" providing employment opportunities to individuals that otherwise face structural barriers in the labor market. We thereby argue that family firms, due to their sense of responsibility toward their community, hire and retain individuals with otherwise unfavorable opportunity structures in the labor market to a greater extent than non-family firms.

Method

We draw on longitudinal register data from Statistics Sweden covering all Swedish residents and firms from 1990 until 2020. We match employee and employer data to construct a firm-level dataset with aggregated measures of firms' employee composition. We identify family firms as firms in which at least two members of the same family are working; thereof, at least one family member assuming a managerial role and declaring ownership in the tax declaration. We adopt a conservative two-way fixed effects design and exploit changes from family-firm to non-family-firm status over time as a within-firm source of variation in family influence, examining how the demographic profiles of joining, staying, and leaving employees change over time. Using a stacked difference-in-differences regression design, we test how the proportion of structurally disadvantaged population groups (women, older-aged, immigrants) among joiners, stayers, and leavers change as firms transition from being a family firm to a non-family firm.

Results and Implications

Our preliminary findings suggest that once a family firm transitions into a non-family firm: (1) less women join, more women leave, and the overall share of women among staying employees decreases; and (2) the share of older-aged workers (age 55+) increases among leaving employees and decreases among staying employees. Our study contributes to the literature on employment conditions in family businesses (Neckebrouck et al., 2018), introducing sorting of disadvantaged population groups into family firms as additional explanation for low compensation levels in family firms. We also contribute to CSR literature (Mariani et al., 2023) by examining the hiring and retainment of disadvantaged employee groups as a potential hidden form of social responsibility exercised by family firms.

CONTACT: Mateja Andric; mateja.andric@unisg.ch; University of St. Gallen & Stockholm School of Economics; Pionierstrasse 24, Winterthur, Zurich 8400, Switzerland

SUMMARY

THE IMPACT OF PARENTS' WORK-LIFE (IM)BALANCE ON FAMILY FIRMS' SUCCESSION AND CONTINUITY

Juliana Binhote, University of Louisville, United States
Kristen Lucas, University of Louisville, United States

Principal Topic

Succession is a major challenge for family firms (FFs), requiring strategic planning, leadership grooming, and a gradual transition beyond ownership transfer (Le Breton-Miller et al., 2004). Despite its importance, many FFs fail to survive beyond the first generation (Kets de Vries, 1993). Successor readiness, family dynamics, and industry sector influence outcomes (De Massis et al., 2015; Sharma et al., 2003), yet the role of work-life balance (WLB) remains underexplored. While some studies address WLB in family businesses (Michael-Tsabari et al., 2020), little is known about how intergenerational WLB expectations shape succession. WLB research in organizational behavior (OB) mainly focuses on employees (Greenhaus & Beutell, 1985; Casper et al., 2018), defining it through role conflict, enrichment, or fit. More recent perspectives emphasize goal-setting and strategic regulation (Hirschi et al., 2019). While these insights apply to entrepreneurs and family businesses (FBs), the FB context introduces complexities. Successor decisions may be shaped by previous generations' work-family imbalance, influencing firm continuity.

Method

We conducted in-depth interviews in two phases with family business owners, next-generation members, and their spouses (N = 23). Interviews were transcribed and analyzed using qualitative software. Following an abductive approach (Dubois & Gadde, 2002, 2014), we iteratively refined our theoretical framework. Through "thick description", we identified three key succession pathways for managing work-life imbalances.

Results and Implications

Our interviews reveal that FB owners face high work intensity and low WLB. Founders, particularly in the early years, rarely reflect on balance, often working extensively and bringing their young children to the business—either to wait or contribute, sometimes without pay. This early exposure shaped the perspectives of the next generation in distinct ways. One successor mirrored their parent's high workload (emulation path), though their spouse was unhappy with their lack of household involvement. Others became highly aware of the need for work-family boundaries, making conscious efforts to avoid repeating their parents' patterns (divergent and departure paths). These findings reveal key insights. Generational differences in WLB exist, as founders prioritized firm survival over balance, only achieving it after stepping down. Vicarious learning from next-generation members' exposure to work-family dynamics is often overlooked in succession models. Self-awareness and spousal dynamics uncover unresolved conflicts that may not be captured in surveys. Finally, spousal support—as highlighted in OB research—is crucial for FB continuity. Overall, FB owners experience and perceive WLB and work-family conflict (WFC) differently from what is commonly stated in OB literature, with direct implications for business survival and continuity.

CONTACT: Juliana Binhote; juliana.binhote@louisville.edu; University of Louisville; 416 Benjamin Ln, Louisville, Kentucky 40222, United States

SUMMARY

FAMILY AT THE HEART: AN EXPLORATION OF SOCIAL FOUNDATIONS IN ENTREPRENEURSHIP

Jose A. Cerecedo Lopez, Elon University, United States

Principal Topic

This study challenges the traditional separation of family and business by examining family as a critical socio-cultural enabler in entrepreneurship. Drawing on the Family Embeddedness Perspective (Aldrich & Cliff, 2003) and External Enablers of Entrepreneurship (Davidsson et al., 2020), the research argues that families fundamentally shape entrepreneurial processes. The study broadens the traditional view of entrepreneurship by exploring how family dynamics influence entrepreneurial activity.

Method

The research employed a qualitative, exploratory approach using the Gioia methodology (Gioia et al., 2013) to analyze 80 in-depth interviews from the "How I Built This" podcast. The dataset comprised over 90 hours of audio and 1,035,433 words, representing diverse entrepreneurial experiences. The qualitative coding process identified family-related keywords, around parental influence, family support and collaboration. The study integrated artificial intelligence to cross-validate thematic analysis, enhancing the consistency and reliability of findings through a comparative approach of human interpretation and AI pattern recognition.

Results and Implications

The thematic analysis revealed six ways in which family serves as a socio-cultural enabler of entrepreneurship: (1) Family Role Models—Entrepreneurs often cite family members as role models who demonstrated the entrepreneurial mindset, work ethic, and lifestyle, inspiring their own ventures. (2) Family Values—Entrepreneurs are often shaped by values instilled in them from a young age. These family-taught principles guide decision-making and persistence in the entrepreneurial journey. (3) Entrepreneurial Skills—Many entrepreneurs acquire essential business skills through early exposure to family-owned enterprises or through direct involvement in family-led entrepreneurial initiatives. (4) Family Resources—Families frequently provide critical financial, human, and social capital. (5) Family Social Support—Entrepreneurs benefit from the psychological and emotional support that families provide. This social safety net enables them to take risks, endure failures, and be resilient to navigate the uncertain path of entrepreneurship. (6) Family as a Community of Practice—Families often function as incubators for entrepreneurial learning, where experimentation, problem-solving, and collaborative innovation take place. This study advances the field by integrating the Family Embeddedness Perspective with the External Enablers framework. While previous research focused on external factors like technological and economic changes (Shane & Venkataraman, 2000), this research empirically demonstrates family's role as a powerful socio-cultural enabler. By detailing specific mechanisms of family influence into entrepreneurial activities, the study advocates for a holistic understanding of entrepreneurship that recognizes family as a critical analytical unit.

CONTACT: Jose A. Cerecedo Lopez; jcerecedolopez@elon.edu; Elon University; 910 Lakecrest Ave, High Point, North Carolina 27265, United States

SUMMARY

ENTREPRENURSHIP THROUGH ACQUISITION: AN ENTREPRENEURIAL OPTION FOR THE NEXT GENERATION MEMBERS OF A FAMILY FIRM

John Trimble, University of Louisville, United States
Isabel Botero, University of Louisville, United States
Tomasz Fediuk, Family Business Strong LLC, United States

Principal Topic

Entrepreneurship through acquisition (ETA) describes the acquisition of an existing SME (i.e., firm revenue under $50 million) by an entrepreneur for the purpose of expanding and enhancing the business through transformational strategies that reshape market processes. Most of the literature about family business treats ETA as an exit opportunity for family business owners who do not have a successor. In this paper we argue that ETA can also be an opportunity for family firms in at least three ways. First, it provides family businesses a way to continue aspects of their legacy beyond the family. Second, ETA offers family businesses a way to diversify and grow into a family enterprise (i.e., a family that owns more than one business). Finally, ETAs can an offer next generation family members an alternative to working inside their family business.

Method

This project was designed to qualitatively explore how ETA provides next generation family members the opportunity to maintain the entrepreneurial spirit of the family while working in a business outside of their family venture. Based on 20 in-depth interviews with searchers that have engaged in ETA or are in the process of doing so, we explore why next generation members choose to work outside the family business, what role does the family play in the acquisition, and the level of support that they receive from their business families in this process.

Results and Implications

With the prevalence of SME owners who do not have a transition plan or are too small to attract a Private Equity buyer, the potential socioeconomic impact of this shortfall is significant. Information from our interviews suggest that next generation members engage in ETA however, they do so independent of the business family. Individuals that engage in ETA buy SMEs to drive improvement, enabling revitalization and growth. Our study has several contributions. First, our study begins to connect ETA as a resource and an opportunity for family businesses. Our findings provide insights into one of the ways that ETA and family business intersect. Second, our findings help us understand how ETAs can become an opportunity for next generation family members to continue the entrepreneurial legacy of the family without directly working in the family business. In our study we explore the motivations of the next generation, and believe our findings provide insights into how business families can support their next generation members even when they work outside of the business.

CONTACT: John Trimble; john.trimble@louisville.edu; University of Louisville; 1265 Enterprise Road, Lawrenceburg, Kentucky 40342, United States

INTERACTIVE PAPER

UNCOVERING THE PATHWAYS OF PROFESSIONALIZATION - AN EXPLORATION OF FAMILY VENTURES

Kristin Sabel, Jönköping International Business School, Sweden
Karin Hellerstedt, Jönköping International Business School (JIBS), Jönköping University, Sweden

Principal Topic

Professionalization is a crucial process for expanding ventures, involving changes in goals, management, functional roles, and operations. However, ventures do not follow a uniform professionalization process and influential milestones of the venture's development likely influence how roles and responsibilities are changed and negotiated over time. In family ventures, professionalization processes are particularly complex and multidimensional, as there are several challenges due to the overlap of family and business systems. However, there is a lack of understanding of how and why family ventures develop these dimensions over time and how certain milestones influence the professionalization process. Consequently, in this paper, we address the question of how and why family ventures professionalize their business.

Method

To gain a deeper understanding of how and why family ventures professionalize over time, we rely on longitudinal case studies. We conducted 20 semi-structured interviews with family venture owners in 2019 and followed up with 30 interviews in 2024/25. The collected material has been systematically coded and analyzed using established tools and techniques, furthermore, interpreted with support from theory and previous literature to extend existing theory and generate new conceptual insights.

Results and Implications

The findings show that as family ventures grow, the family still remains in control. Interestingly, we see that later generations take on a more operational role than their predecessors, also during venture growth. This results in family owners/managers taking on several additional roles to remain in control. Additional roles are also appointed to employees, as the family venture grows and expands. Furthermore, several family ventures remove middle management as they grow, to maintain operational control. Contrary to previous research that suggests professionalizing by hiring external CEOs or structuring HR functions during growth, our findings indicate that family owners/managers prefer to stay in hands-on roles even while pursuing business expansion. Even if earlier generations have had more formalized strategic roles, later generations choose to become more operational in their leadership. The results also show that internal as well as external shocks may force the family ventures into new ways of operating. The findings raise the question of whether professionalization activities in family ventures in fact lead to professionalization. This study brings new knowledge about family-owned ventures at different stages of the professionalization process, exploring underlying aims and decisions guiding this process.

CONTACT: Kristin Sabel; kristin.sabel@ju.se; Jönköping International Business School; Gjuterigatan 5, Jönköping, Jonkopings Lan 55318, Sweden

THE HYPE AROUND VENTURE CAPITAL AND ITS IMPLICATIONS FOR THE STANDARD MODEL OF ENTREPRENEURSHIP

Eduardo Boada, IE University, Spain
Pablo Muñoz, Durham University, United Kingdom
Daniel A. Lerner, IE University, Spain

ABSTRACT

This study investigates the evolving discourse surrounding venture capital (VC) and its implications for a dominant form of high-growth, innovation-driven entrepreneurship. Conceptualizing both VC and this entrepreneurial model as nested sociotechnical systems, we analyze two decades of media coverage as well as expert interviews conducted to trace cycles of sentiment and expectation, or hype. Our findings reveal recurring patterns of heightened enthusiasm followed by disillusionment, reflecting ongoing struggles for legitimacy within financial markets and entrepreneurial ecosystems. Despite its global influence and symbolic status, this particular form of entrepreneurship remains contested, with persistent critiques regarding its effectiveness, societal contribution, and structural limitations. Our analysis advances understanding of how public expectations shape and reflect the evolving position of VC and the model of entrepreneurship it co-evolved with.

INTRODUCTION

Venture Capital is one of the main sources of monetary resources for new ventures (Gompers et al., 2020), and it has enabled the emergence of a distinct entrepreneurial approach known as the "Silicon Valley Model" (Aldrich & Ruef, 2018; Herrmann, 2019; Pahnke & Welter, 2019). This model is characterized by its focus on growth, innovation, and creativity, along with high capitalization primarily through VC investment (Aldrich & Ruef, 2018). It can be conceptualized as a sociotechnical system (Geels, 2002), encompassing technologies and artifacts, institutions and regulations, cultural and symbolic meanings, ideological underpinnings, and financial infrastructure. From a static, contemporary vantage point, this configuration may seem definitive—a dominant paradigm fully adopted and accepted (Welter et al., 2019). Upon closer examination, however, the dominance of this model and its constituent elements appear less uncontested. Criticism surrounding the practices and consequences of this highly financialized, disruption-oriented, and fast-growing entrepreneurial model is found both in academia (Audretsch, 2021; Welter et al., 2019; Zankl & Grimes, 2024) and the popular press (e.g., Savage, 2021; Waikar, 2018).

Using terminology from sociotechnical systems and transformation research (Geels, 2002; 2004), we could say that the sociotechnical regime guiding entrepreneurial activity does not appear static, with the elements that compose what we perceive as the "standard model" of contemporary entrepreneurship fully and undisputably integrated. Rather, these elements we collectively call a sociotechnical system seem engaged in an ongoing process of integration, encountering simultaneous support and opposition from actors and networks with vested interests in its outcomes. These struggles have been described as "technological dramas," leaving evidence of their existence and content in public discourse (Kriechbaum et al., 2021).

Studying the dynamics of public discourse surrounding the standard entrepreneurial model and its components is thus critical for improving our understanding of its trajectory, societal role, and future potential. Therefore, this study takes a first step toward this goal by tracing historical sentiment and expectations regarding VC throughout the past two decades, given VC's role as a main pillar supporting entrepreneurship as practiced under the model of Silicon Valley. We leverage the concept of hype (Borup et al., 2006) to theoretically and empirically ground these sentiments and expectations. Hype is broadly defined as heightened expectation and excitement (Logue & Grimes, 2022; van Lente et al., 2013), typically unfolding in cycles of excitement followed by disillusionment (Linden & Fenn, 2003). Recent research has applied this framework to study the trajectory of new technological systems as they transition from niche creation toward mainstream adoption (Kriechbaum et al., 2018; 2021).

Our empirical strategy aligns with previous research in science, technology, and innovation (STI), particularly studies examining hype cycles (Kriechbaum et al., 2018; Melton et al., 2016; van Lente et al., 2013). We conducted sentiment analysis (Pang & Lee, 2008) of VC-related articles from a leading media outlet over the past two decades, complemented by inductive analysis of statements within these articles (Gioia et al., 2013). Additionally, we carried out expert interviews to enhance interpretation of media content and provide supplementary data for triangulation. Findings indicate that despite VC's widespread popularity and apparent success, it remains subject to hype cycles reflecting dynamic shifts in its positioning and that of its associated entrepreneurial model.

This study contributes to scholarly debates challenging the standard entrepreneurial model and its treatment as synonymous with entrepreneurship more broadly (Aldrich & Ruef, 2018; Audretsch et al., 2021; Pahnke & Welter, 2019). We propose a theoretical framework examining this model as a distinct sociotechnical system within the entrepreneurial spectrum, following its developmental trajectory toward maturation (Perez, 2010). Findings underscore that the model's dominance should not be uncritically accepted, highlighting limitations inherent in treating it as prototypical for understanding entrepreneurial opportunity discovery and exploitation (Shane & Venkataraman, 2000). Additionally, we address a gap in VC literature by explicitly analyzing hype and expectations surrounding the VC industry itself, where existing research primarily focused on industry-level hype and bubbles (Valliere & Peterson, 2004). Finally, we contribute to emerging literature on entrepreneurial hype (Heupel et al., 2024; Logue & Grimes, 2022; Wood et al., 2024). While previous studies predominantly examined hype from strategic (entrepreneur-focused) or institutional perspectives (Wadhwani & Lubinski, 2025), our emphasis on the hype around entrepreneurial activity itself offers an additional dimension, enhancing understanding of its broader implications.

THEORY

To understand the position of the standard model of entrepreneurship, we adopt a perspective rooted in the sociology of technology (Geels, 2002; Rip & Kemp, 1998). This framework conceptualizes technologies as "configurations that work," embedded within social environments enabling them to fulfill societal functions through arrangements of heterogeneous elements—collectively termed sociotechnical systems (Geels, 2002). The constellation of elements recognized today as the standard model of entrepreneurship (embodied by the Silicon Valley ecosystem where it originated) represents such a structure, comprising its own culture, rules, infrastructure, and interrelated components aimed at enabling the discovery and exploitation of entrepreneurial opportunities (Shane & Venkataraman, 2000). Within this system, VC represents a crucial element, forming the core of its financial infrastructure. Additionally, VC itself comprises a nested system with its own constituting elements, integral to its existence and functioning, similar to how road infrastructure is indispensable to contemporary transportation.

Beyond simply characterizing these entities as sociotechnical systems, we inquire into their placement within technological transition processes through which new systems move from emergence to full adoption (Geels & Schot, 2007). This process has been described using a multi-level perspective, comprising niches (micro-level), regimes (meso-level), and the landscape (macro-level) (Geels, 2002). Innovations (new systems) first appear in niches acting as incubating spaces, providing protection from selective forces. When pressures from the landscape or instability within regimes emerge, new systems can break through, replacing older systems and reshaping the regime (Geels, 2004). Where, then, can VC and its accompanying entrepreneurial model be situated?

The system recognized as the standard entrepreneurial model is clearly no longer confined within the Silicon Valley niche where it originated (Baker & Welter, 2024). However, it does not yet enjoy the uncontested status expected of a fully integrated system. Despite its apparent ubiquity and success, the model has recently criticism, raising public doubts about its practices (Savage, 2021; Waikar, 2018). Academic critiques have focused on its underlying rent-seeking ideology, tendencies toward unproductive entrepreneurship, and externalities often ignored by proponents (Zankl & Grimes, 2024). Scholars also question the model's ability to address contemporary societal challenges, such as inequality and inclusion (Audretsch, 2021). These conflicting perspectives highlight complexity beneath the apparent universal acceptance and success of this system for creating and scaling new companies.

Recent perspectives on VC, much like those on the entrepreneurial system it supports, vary widely. Some portray venture capitalists as reckless speculators fixated on "unicorns," flooding unsustainable business models with excessive capital and distorting the broader economy (Cassidy, 2020). Others positively depict VC as a vital driver of innovation and economic growth (Rist, 2021). Academic scholarship similarly captures this duality, highlighting VC's contributions to economic growth (Gornall & Strebulaev, 2021; Kaplan & Lerner, 2010), while simultaneously acknowledging drawbacks such as behavioral biases (Franke et al., 2006; Valliere & Peterson, 2004; Zacharakis & Shepherd, 2001), conflicts of interest and moral hazard (Bradley et al., 2015; Coakley et al., 2009), and uneven capital flows into the industry (Gompers & Lerner, 2001).

To assess the extent to which Silicon Valley entrepreneurship and VC as nested sociotechnical systems have become established within the regime, we leverage recent research on technological transitions, viewing expectations and hype as reflections shaping the journey of an emerging sociotechnical system toward widespread adoption (Kriechbaum et al., 2018; 2021; Sovacool & Hess, 2017). In STI studies, hype is characterized as "waves of high rising expectations" (van Lente et al., 2013, p. 1615), shaping innovation dynamics by guiding agendas and conferring legitimacy, thereby attracting resources and actors to emerging technologies (van Lente & Bakker, 2010). In the context of sociotechnical transitions, expectations and hype play a critical role, with positive expectations both promoting niche protection and mobilizing resources (Geels & Raven, 2006; Kriechbaum et al., 2021; Schot & Geels, 2008). Additionally, expectations around new sociotechnical systems represent narratives associated with power struggles between niches and regimes (Geels, 2010; Kriechbaum et al., 2021). Under this perspective, the shape of a hype cycle mirrors the broader trajectory of discursive struggles between niche and regime actors, indicating how successfully niche technologies have gained a foothold within the dominant regime.

Applying this perspective to the sociotechnical system of Silicon Valley entrepreneurship enables the use of hype and discourse to better understand niche-regime struggles it may have recently experienced, as well as providing new insights into its current societal status. However, directly studying hype around the abstract concept of the standard entrepreneurial model through traditional public discourse analysis methods presents challenges, as explicit references primarily appear in academic literature, limiting observable evidence in broader societal contexts. Conversely, VC represents a more tangible phenomenon, and given its inseparable link with the overall entrepreneurial system, analyzing VC hype can serve as a valuable proxy to understand struggles and the transition trajectory of the entire sociotechnical system.

METHOD

Our analytical approach involved two complementary strategies. First, we followed prior research that studied hype and collective expectations in other fields (Kriechbaum et al., 2018; Kriechbaum et al., 2021; van Lente et al., 2013) and analyzed public discourse on VC in mass media to investigate hype fluctuations over the past two decades and the themes accompanying these variations. Specifically, we selected articles from The Financial Times, a prominent financial media outlet, as our primary data source. Our analysis period spanned from early 2005 to the end of 2024, resulting in a dataset of 110 articles. We manually identified statements explicitly reflecting market hype, focusing either on future-oriented expectations or current assessments of the VC industry. Following identification (n = 783 statements), we conducted open coding, aggregating codes inductively into categories and subsequently broader themes through iterative revision (Gioia et al., 2013). In parallel, we conducted sentiment analysis by categorizing statements as positive (1), neutral (0), or negative (-1) based on semantic meaning. Combining content and sentiment analyses, we explored how the four identified themes evolved over time, both in frequency and sentiment.

Our second approach involved conducting 34 semi-structured expert interviews with active professionals in fields related to new venture financing, including VC, angel investing, and private equity. Interviews centered around three core topics: participants' views of the VC market, the impact of hype cycles, and current market conditions. These discussions provided valuable context and also served as a complementary data source, allowing triangulation with findings from media analysis and enhancing the overall robustness of our results. Moreover, interview insights highlighted the causes, immediate effects, and long-term implications of hype in the VC market.

To construct a coherent narrative from our data, we employed a temporal bracketing strategy (Langley, 1999). Given the cyclical nature of hype, this approach was well-suited for interpreting emerging themes. After reviewing both datasets and analysis results, we identified three discrete periods of VC market hype. Each period exhibited distinct sentiment patterns, corroborated by interview accounts and by VentureBeat data on annual VC investment totals, which showed notable shifts around the identified cutoff points.

RESULTS

Our temporal bracketing approach led us to divide the twenty-year period into three discrete phases, corresponding closely to stages of the archetypal hype cycle as conceptualized by Gartner (Linden & Fenn, 2003). This cycle is also evident in our sentiment analysis. The first phase (2005–2013) revealed low media coverage and low overall optimism toward the industry, suggesting a period of disillusionment following an earlier cycle of VC hype (Ruef & Markard, 2010). During the second period (2014–2019), we observed an increase in both metrics, indicating the beginning of a hype phase (Ruef & Markard, 2010; van Lente et al., 2013). The third phase (2020–2024) presents inconsistencies between these metrics, complicating interpretation yet providing an intriguing analytical opportunity. During this period, the average number of expectation statements per year increased, but their sentiment value markedly decreased. Notably, we often found contradictions within themes and categories during the same period, with some categories and themes containing significant numbers of both positive and negative expectation statements (Figure 1). This result echoes prior research on hype dynamics in other fields (e.g., van Lente et al., 2013), reinforcing the characterization of hype cycles as "technological dramas" (Pfaffenberger, 1992) reflecting competition between optimistic and pessimistic views on the technology (Kriechbaum et al., 2021, p. 4). Our inductive analysis identified four overarching themes. The first and broadest in terms of categories is "Challenges of VC." The second theme is "Role and Impact of VC," followed by a third theme titled "Global Dynamics and Geopolitics." The fourth theme identified was "Evolving Dynamics in VC Investing."

Figure 1: Themes by Period

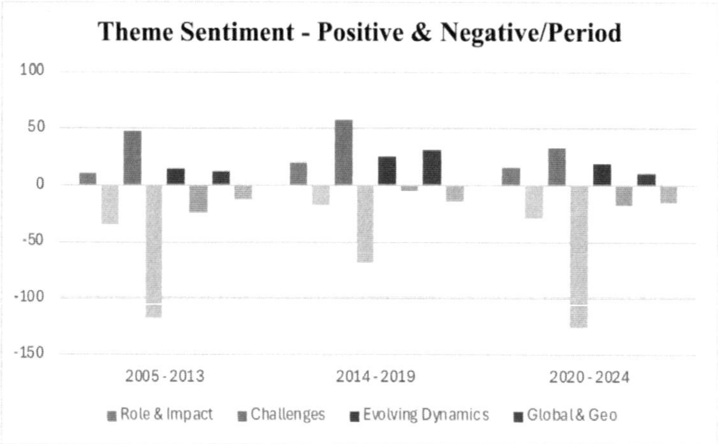

For the most part, our inductively derived themes followed a consistent pattern: their frequency remained relatively stable across the three periods, displaying net-negative sentiment during the troughs of the cycle (2005–2013 and 2020–2024) and net-positive sentiment during the peak (2014–2019). One notable exception was the *Challenges of VC* theme, which consistently contained the highest number of data points across all themes and captured inward-looking aspects of the VC industry, such as technical capabilities. Sentiment within this theme remained net-negative even during the peak of VC hype and was contentious throughout the observed timeframe, characterized by a significant number of both negative and positive statements across all three phases. This pattern was also evident in specific topics within the theme, including industry resilience in both short- and long-term horizons, persistent concerns about market bubbles, and the impact of external factors—all of which remained consistently negative in aggregate across the entire period. Other problematic topics, such as market volatility and VC's efficiency in capital allocation, emerged predominantly during the troughs. Certain topics, however, exhibited different dynamics: discussions of excesses in the industry became prominent during the peak period and persisted into the downturn, while positively portrayed individual events increased notably during the same period and remained frequent thereafter. Investment strategy changes, captured under the *Evolving Dynamics* theme, were viewed negatively in the first period but shifted positively in subsequent periods. Alternative funding options for startups outside of VC were primarily discussed in the first and third periods when overall sentiment was lower.

Within the *Role and Impact of VC* theme, only two topics retained relevance across multiple periods, underscoring their centrality in broader discussions about VC's societal implications. These were VC's capacity to drive technological and economic growth and its ability to deliver adequate returns to limited partners (LPs). While the industry's role as a growth driver was contested, it remained net-positive in aggregate during the last two periods. In contrast, views regarding VC's ability to deliver sufficient returns to LPs remained consistently negative across all three periods. Meanwhile, discourse within the *Global Dynamics and Geopolitics* theme consistently revolved around nations' capacity to cultivate entrepreneurial ecosystems capable of producing and retaining unicorns and market-leading companies. This discussion frequently involved the role of governments, as well as international movements of financial and human capital.

DISCUSSION

Our analysis revealed specific themes within the hype cycle, which we interpret as "manifestations of the discursive struggles associated with periods in which niche technologies challenge and compete with existing socio-technical systems" (Kriechbaum et al., 2021, p. 14). Tracking the hype patterns for VC and their accompanying discourse, we identified struggles of VC to establish itself more securely within the broader socio-technical regime from two distinct but related perspectives. From an entrepreneurial standpoint, as a key component of the standard model of entrepreneurship, VC shares in its efforts to incorporate as the predominant approach for creating and scaling new companies, particularly highly innovative or technology-driven ventures. From a financial market perspective, VC competes as an asset class, vying for resources from institutional investors, wealthy individuals, and other limited partners. On both fronts, our results reveal discourse that shifts somewhat in content and sentiment over the cycle yet maintains certain persistent features (see Cross-period Analysis).

Considering both our data sources, VC in its current form—and by extension the system it supports—appears as a flawed socio-technical configuration whose place vis-à-vis the regime remains contested. The trajectory of expectations surrounding VC indicates ongoing struggles with varying degrees of success over the observed period. Our study highlights that VC has faced criticism for its incentive structure, its susceptibility to biases (particularly social ones), and lapses in discipline and prudence during periods of exuberance. These characteristics create volatility seen as threatening its long-term viability. Additionally, its effectiveness as an intermediary between capital sources and entrepreneurial ventures has been debated, especially concerning efficient fund allocation and support for impactful entrepreneurship. Upstream, it faces skepticism as an asset class from LPs, alongside fundamental questions about whether generated returns reflect genuine expertise or mere luck. Downstream, we found doubts about VC's value-added beyond funding and its suitability for expansion beyond the niche of high-growth, consumer-oriented technology ventures. Nonetheless, we also found strong recognition that, despite its flaws and apparent wastefulness, VC and the entrepreneurial system it is tied to have successfully fostered innovation and created market-leading companies. Even in periods of pronounced skepticism, high-profile events such as large fund establishments or successful venture financings spark excitement and optimism, underscoring the model's continued relevance.

Contributions

We advance theory by conceptualizing VC and the so-called standard model of entrepreneurship as nested socio-technical configurations embedded within the broader sociotechnical system of entrepreneurship (Geels, 2002; 2004). This framing offers analytical benefits, particularly by allowing us to leverage the concept of hype as an indicator of socio-technical transitions (Kriechbaum et al., 2021). Thus, we provide a clearer understanding of the discourse surrounding these entities, revealing their perceived strengths and pitfalls, their current social roles, and the expectations that persist around them. More broadly, our framework establishes a theoretical foundation from which these structures can be clearly distinguished from entrepreneurship as a whole, enabling more precise study. We also make an empirical contribution by applying methods from other fields that study hype cycles to measure and analyze the expectations surrounding VC for the first time. Our empirical strategy enabled us to quantify waves of excitement and disillusionment around VC over the past two decades, identifying a clear pattern consistent with a hype cycle. These findings validate our framing of VC (and, by extension, the model of entrepreneurship it is embedded in) as socio-technical innovations. Additionally, we contribute to the growing literature on hype in entrepreneurship (e.g., Heupel et al., 2024; Logue & Grimes, 2022; Wood et al., 2024), which primarily examines hype from strategic perspectives (how entrepreneurs leverage hype) and, more recently, institutional perspectives (Wadhwani & Lubinski, 2025). Prior research has mostly focused on hype surrounding specific industries, technologies, or individuals (e.g., Fisher et al., 2024; Heupel et al., 2024; Garud et al., 2021). By stepping back to examine hype around entrepreneurial activity itself (specifically one type), our study introduces a novel dimension, enhancing our understanding of hype's broader implications within entrepreneurship.

Limitations & Future Research

One limitation of our study is using venture capital hype as a proxy for the broader entrepreneurial sociotechnical system. Although theoretically justified—given VC's integral role within this system (Aldrich & Ruef, 2018; Baker & Welter, 2024)—future research could explore alternative operationalizations specific to the entire model. Another limitation is our reliance on a single media source (Financial Times); while this aligns with prior studies (Kriechbaum et al., 2021; Melton et al., 2016), it potentially limits generalizability. However, our data included diverse industry perspectives and was triangulated through expert interviews. Finally, future studies could deepen analyses by applying additional concepts from sociology of expectations (van Lente et al., 2013; Borup et al., 2006) and sociotechnical transitions literature (Geels, 2002; 2004) to explore system emergence and destabilization dynamics.

CONTACT: Eduardo Boada; eduardo.boada@student.ie.edu; Calle María de Molina 13, Madrid, Spain, 28006

GOOD BUSINESS BUT BAD MORALS? HOW INVESTORS REACT TO EARLY-STAGE ENTREPRENEURS' ETHICAL BREACHES

Sara Luisa Kappelhoff, Technical University of Munich, Germany
Nadja Mirjam Born, Technical University of Munich, Germany

ABSTRACT

Early-stage entrepreneurs can be tempted to prioritize profitability over ethicality to increase competitiveness and attract investors. Yet, investors' reactions to profitable but unethical behavior of entrepreneurs remain largely unknown, even though investors rate personal characteristics as the most important criterion when selecting early-stage investments (Gompers et al., 2020). We investigated investors' reactions to early-stage entrepreneurs' profitable but unethical behavior in two experiments (total $N = 603$). Our findings suggest that unethical entrepreneurs are evaluated less positively, receive less funding, and are less frequently broadcasted by investors. Reduced funding is mainly driven by investors' negative warmth evaluation of unethical entrepreneurs. Surprisingly, we find that female entrepreneurs do not face more negative consequences for unethical behavior than their male counterparts. We contribute to the literature on ethics in entrepreneurship by demonstrating that entrepreneurs' profitable but unethical behavior is penalized by investors.

INTRODUCTION

Early-stage entrepreneurs compete in highly complex environments where they must effectively navigate uncertainty while attracting potential investors (McKelvie, Haynie, & Gustavsson, 2011). In such competitive contexts, they are often drawn to make business decisions that seem profitable but ethically questionable, challenging moral norms or standards (Fassin, 2005; Lewis, 1985; Vallaster et al., 2019). Technological advances, such as in artificial intelligence, amplify these ethical challenges for entrepreneurs by promoting efficiency and growth without providing ethical guidelines (Hall & Rosson, 2006; Obschonka & Audretsch, 2020). Entrepreneurs might even perceive ethical compromises for business purposes as necessary to demonstrate business acumen to potential investors and enable growth (Uriarte et al., 2024). However, investors' reactions to entrepreneurs' unethical behavior, even if not explicitly illegal, remain largely unknown. Existing management research suggests that perceived moral character strongly impacts evaluations of other people (Goodwin et al., 2014). Specifically, unethical behavior is associated with lower perceived competence and warmth, the two universal dimensions for evaluating others (Fiske et al., 2007; Stellar & Willer, 2018). However, it has not been examined whether this holds for entrepreneurs operating in a different environment with additional ethical challenges compared to managers. Accordingly, investors could evaluate profitable but unethical business decisions as a sign of poor moral judgment or as evidence of strong business acumen of entrepreneurs. Closing this research gap seems essential, given that investors rate personal evaluations and management team characteristics as the most important criteria when selecting investments, especially in the early stages (Gompers, et al., 2020; Huang & Pearce, 2015). As research points to a double standard for evaluating the ethicality of actions performed by men and women (e.g., Kennedy et al., 2016; Klein & Shtudiner, 2021; Mai et al., 2020; Montgomery & Cowen, 2020), we expect entrepreneur gender to moderate the relationship between unethical behavior and its consequences.

Our research addresses two fundamental questions with a pilot study and a main study. First, we investigate whether early-stage entrepreneurs' unethical behavior is rewarded or punished by investors and explore the underlying mechanisms. Second, we inquire whether female entrepreneurs face more negative consequences for unethical behavior than male entrepreneurs. We examine investors' investment behavior, investment broadcasting decisions, and evaluations of the entrepreneur using a pilot study with manipulation checks and an incentive-compatible online experiment.

HYPOTHESIS DEVELOPMENT

Unethical Behavior in Entrepreneurship

In recent decades, interest in research connecting ethics and entrepreneurship has grown significantly (Uriarte et al., 2024; Zhu & Chang, 2013). Prior work on ethics in entrepreneurship outlines that entrepreneurs face additional ethical challenges compared to managers in organizations (Solymossy & Masters, 2002; Vallaster et al., 2019). Ethical decision-making is difficult for entrepreneurs, as they need to navigate the uncertainty and the psychological and financial pressures in their environment (Hall & Rosson, 2006; Vallaster et al., 2019). What is perceived as unethical behavior is dependent on an individual's evaluation: the same behavior can be perceived as unethical by one party and as usual business practice by another (Collewaert & Fassin, 2013). Prior research defines behavior as unethical when it is "either illegal or morally unacceptable to the larger community" (Jones, 1991: 367). Importantly, this paper does not focus on illegal behavior but on areas "where the law ends" (Crane & Matten, 2004: 359). Following Collewaert & Fassin (2013), we define unethical behavior as the perception of a stakeholder party that an action violates moral norms or standards. Unethical behavior can manifest in various forms, including deceptive communication, prioritization of self-interest over collective goals, breach of agreement, or abuse of authority (Crane & Matten, 2004).

Initial research on the consequences of unethical behavior in entrepreneurship investigated the unethical behavior of different stakeholders. For example, scholars propose that in the later stages of the investor-entrepreneur relationship, unethical behavior from either the investor or the entrepreneur harms the existing relationship, leads to conflict, and negatively influences willingness to collaborate again in the future (Collewaert & Fassin, 2013). Research from China demonstrates how an entrepreneur's unethical behavior negatively impacts the company's public image (Zhu & Chang, 2013). Last, an investor's ethical reputation seems to influence entrepreneurs' willingness to collaborate when choosing funding partners (Drover et al., 2014).

We draw upon attribution theory to investigate the potential consequences of entrepreneurs' unethical behavior, even when such behavior appears profitable. It outlines how actions or events are interpreted by individuals (Weiner, 1985). According to attribution theory, observers attribute individuals' actions either to internal factors, such as their personal character, or to external factors, such as situational characteristics (Weiner, 1985). When early-stage entrepreneurs exhibit unethical behavior, investors likely attribute it to the entrepreneur's personal character or values, reflected in their evaluation of the entrepreneur. While the profitable aspects may seem appealing, unethical behavior likely reduces perceptions of the entrepreneur's warmth and competence (Stellar & Willer, 2018), the predominant factors of social perception (Fiske et al., 2007). Specifically, warmth comprises the perception of social traits, while competence comprises the perception of intellectual traits (Fiske et al., 2007). In management research, unethical actions are shown to reduce perceived competence as they are attributed to a person's low social intelligence and poor moral judgment (Stellar & Willer, 2018).

Early-stage investors often lack financial or performance data of new ventures and mainly depend on their gut feeling of entrepreneurs (Huang & Pearce, 2015). We expect that unethical actions strongly impact their social perception of an entrepreneur. Even if unethical behavior benefits the business, investors may attribute it to a character flaw that damages the overall perception of entrepreneurs and their ventures. Therefore, we expect that showing unethical behavior leads to a negative competence and warmth evaluation of an entrepreneur, which is formulated in our first hypothesis (H1). Drawing upon attribution theory (Weiner, 1985), we expect that investors penalize early-stage entrepreneurs for unethical behavior despite its profitability. Even if investors decide to invest, they might choose smaller investment amounts to account for the increased risks. Investors might also fear reputational damage from investing in or broadcasting entrepreneurs who behave unethically and, therefore, minimize reputational harm by hiding these kinds of investments. Accordingly, we hypothesize that when entrepreneurs show unethical behavior, investors are less willing to invest (H2) in their ventures and to broadcast their investment (H3).

H1: Entrepreneurs who display unethical behavior are less positively evaluated than those who do not display unethical behavior.

H2: Entrepreneurs who display unethical behavior receive less funding than those who do not display unethical behavior.

H3: Entrepreneurs who display unethical behavior are less frequently broadcasted than those who do not display unethical behavior.

While we did not preregister a hypothesis about the relationship between unethical behavior, personal evaluations of entrepreneurs, and funding outcomes, we aim to explore the underlying mechanism. Specifically, we investigate how investors' warmth and competence evaluation of an entrepreneur following an unethical action drives investors' investment decisions.

Entrepreneur Gender

Previous research points to a moral double standard in evaluating the unethical behavior of men and women (Kennedy et al., 2016; Klein & Shtudiner, 2021). Therefore, we investigate whether the interpersonal consequences of unethical behavior differ by entrepreneur gender, potentially adding to the difficulties female entrepreneurs face in the entrepreneurial landscape (Pistilli et al., 2023). Prior research shows that women receive less funding and different types of questions from investors (Brooks et al., 2014; Kanze et al., 2018). Moreover, they are evaluated based on different criteria and held to higher performance standards (Lee & Huang, 2018; Pistilli et al., 2023). Stronger penalties for unethical behavior might increase the difficulties for women in the entrepreneurial landscape and make them reluctant to enhance profitability through ethical compromises, such as in the context of new technologies. Unethical behavior deviates strongly from the prevailing stereotype of women to be warm and communal (Eckes, 2002; Rudman & Fairchild, 2004; Rudman & Glick, 2001).

As role congruity theory (Eagly & Karau, 2002) posits, women are subject to gender bias when they engage in behaviors that conflict with societal norms and role expectations. Prevailing stereotypes portray women as warm and relationship-oriented and men as dominant, competent, and risk-taking (Eckes, 2002; Rudman & Fairchild, 2004; Rudman & Glick, 2001). Hence, unethical behavior, or violating moral norms to drive profitability, is incongruent and counter-stereotypical for women. Accordingly, management literature suggests that women experience stronger punishment for unethical behavior at work (e.g., Kennedy et al., 2016; Mai et al., 2020; Montgomery & Cowen, 2020). Therefore, we expect that when female entrepreneurs engage in profitable but unethical behavior, they are evaluated disproportionately negatively and must face harsher consequences. In contrast, the same behavior performed by male entrepreneurs might align more closely with their societal role to be dominant, risk-taking, and less communal (Eckes, 2002). Taken together, we expect female entrepreneurs to experience stronger punishment for unethical behavior, as formulated in our last hypothesis (H4).

H4: Entrepreneur gender moderates the relationship between unethical behavior and its consequences such that female entrepreneurs face more negative consequences (evaluations, funding, broadcasting) for unethical behavior than their male counterparts.

METHOD

To empirically test our hypotheses, we conducted a between-subject pilot study and a subsequent preregistered incentive-compatible online experiment.

Sample: US and UK-based individuals with prior investment experience recruited via prolific Pilot Study $N = 169$, online experiment Main Study $N = 434$, preregistered incentive-compatible online experiment

Dependent Variables = investment behavior, broadcasting behavior, personal evaluations of entrepreneur (warmth and competence)	
Independent Variables = ethicality, entrepreneur gender (only main study)	
Controls (not significant) = venture investment knowledge, cancer treatment knowledge	
Statistical Analysis = independent samples t-tests, chi-square tests, mediation analysis	

RESULTS

Pilot Study

To examine the effect of unethical behavior on personal evaluations, and investment behavior, we conducted independent samples t-tests (using Bonferroni correction). Consistent with H1, participants in the unethical condition perceived the entrepreneur as significantly less warm ($M = 3.21$, $SD = 1.74$) than in the neutral condition ($M = 6.96$, $SD = 1.35$; $t(159.64) = -15.72$, $p < .001$, $d = -2.41$). On average, unethical entrepreneurs were also perceived as less competent ($M = 6.67$, $SD = 1.58$) than those who did not display unethical behavior ($M = 7.76$, $SD = 1.05$, $t(148.48) = -5.31$, $p < .001$, $d = -0.81$). Moreover, participants reduced their investments significantly in the unethical condition ($M = 16.87$, $SD = 24.12$) compared to the control condition ($M = 44.20$, $SD = 29.78$, $t(157.70) = -6.54$, $p < .001$, $d = -1.01$). To investigate the investment broadcasting decisions of investors (H3), we conducted a chi-square test with only those participants who decided to invest. The test results show a significant difference in participants' broadcasting behavior for the two groups $\chi^2(2) = 8.75$, $p = .01$ with a small to moderate effect size (*Cramer's* $V = .26$, $p = .01$). 39.3% of participants from the unethical condition chose not to broadcast the entrepreneur in a personal format, while only 16.7% from the neutral condition made the same choice. Importantly, all these participants chose to invest before.

Main Study

We conducted the analyses for the first three hypotheses on the full sample, as subsequent results revealed no significant differences between the stimulus groups based on entrepreneur gender (female, male, or unspecified). Participants evaluated entrepreneurs displaying unethical behavior lower in the warmth dimension ($M = 4.11$, $SD = 2.16$) than those who do not display such behavior ($M = 7.32$, $SD = 1.24$, on a 9-point scale, $t(346.76) = -19.05$, $p < .001$, $d = -1.83$) Participants in our sample assessed unethical entrepreneurs also as less competent ($M = 6.68$, $SD = 1.76$) than their neutral counterparts ($M = 8.00$, $SD = 1.09$, $t(362.21) = -9.46$, $p < .001$, $d = -0.91$). Moreover, participants invested a significantly lower amount in entrepreneurs displaying unethical behavior ($M = 34.11$, $SD = 31.90$) than in entrepreneurs who do not display such behavior ($M = 53.38$, $SD = 26.97$, $t(421.57) = -6.80$, $p < .001$, $d = -0.65$). Last, there was a significant association between entrepreneurs' type of behavior and investors' broadcasting decisions ($\chi^2(3) = 12.00$, $p = .01$) with a small to moderate effect size (*Cramer's* $V = .18$, $p = .01$). Together, these results show a significant negative effect of entrepreneurs' unethical behavior on investors' investment decisions, broadcasting behavior and evaluations of the entrepreneur, providing support for H1-H3.

Additionally, we were interested in the mechanism of how unethical behavior influences behavioral investment decisions. Therefore, we conducted a mediation analysis with Hayes' PROCESS macro model 4 (Hayes, 2018), using unethical behavior as the independent variable (X), investment size as the dependent variable (Y), and the warmth and competence dimensions as mediators (two main factors of social perception, M1 and M2). We used bootstrapping with 5,000 samples to estimate indirect effects. Results from the mediation analysis show a total significant effect of unethical behavior on investment size ($b = -19.27$, $SE = 2.84$, $t = -6.79$, $p < .001$). However, the direct effect of unethical behavior on investment amount becomes insignificant when including the mediators ($b = -0.78$, $SE = 3.58$, $t = -0.22$, $p = .83$). The indirect effects demonstrate that both the warmth and the competence dimension take on significant mediating roles, suggesting full mediation with warmth being the primary driver.

Last, we investigated whether female entrepreneurs displaying profitable unethical behavior experienced more negative consequences than their male counterparts (H4) by comparing two groups (male/female entrepreneur) in the unethical condition. We did not find support for H4, as our results reveal that for the personal evaluation variables, unethical female entrepreneurs were not evaluated differently from male entrepreneurs. We did also not find significant differences in investment decisions. Unethical female entrepreneurs (M = 34.56, SD = 29.47) received even slightly more funding than unethical male entrepreneurs (M = 33.07, SD = 34.69), but the difference was not significant ($t(143) = -.28$, $p = .78$, $d = 0.05$). The same pattern emerged for the broadcasting decision, in which we included the participants who chose to invest. Again, the differences were not significant ($\chi^2(3) = 1.05$, $p = .79$) with a small effect size (*Cramer's V* = .10, $p = .80$).

DISCUSSION AND IMPLICATIONS

With the recent advancements in new technologies, there has been increasing interest in the role of ethics in entrepreneurship and the ethical challenges entrepreneurs face (Uriarte et al., 2024). While much attention has been devoted to understanding the antecedents of unethical behavior of entrepreneurs (Baron et al., 2015; Tacke et al., 2023), less is known about the consequences of such behavior. Closing this research gap is important because it is unclear whether investors perceive such behavior as evidence of poor moral judgment and punish it or whether they perceive it as a sign of strong business acumen in the highly uncertain environment of early-stage entrepreneurship and accept or even reward it. Across both studies, we find support that entrepreneurs who prioritize profitability over ethicality are evaluated more negatively by investors, receive less funding, and are broadcasted less frequently than those who do not engage in such behavior. Specifically, we find that investors evaluate unethical entrepreneurs as significantly less warm and competent. Investors provide them with less funding, even in an incentive-compatible setting. Last, investors who decide to invest are less willing to broadcast their investment, especially in a post focused on the entrepreneur. We find that the effect of entrepreneurs' unethical behavior on investment amount is fully mediated by the investors' lower warmth (primary driver) and competence (secondary driver) evaluation of unethical entrepreneurs. We also find the warmth evaluation to be the primary driver of the likelihood of broadcasting the investment (competence not significant). However, we do not observe differences in the consequences of unethical behavior based on entrepreneur gender.

We contribute to the literature in several ways. First, we contribute to the literature on ethics in entrepreneurship (e.g., Uriarte et al., 2024; Vallaster et al., 2019) by showing investors' negative reactions to entrepreneurs' profitable but unethical behavior. We use innovative measures to assess the investors' cognitive and behavioral reactions to entrepreneurs' profitable but unethical behavior. While entrepreneurs might be tempted to make ethical compromises to benefit their ventures, we show that investors are reluctant to invest in unethical entrepreneurs' ventures and are less likely to publicly associate with unethical entrepreneurs, even in incentive-compatible settings. The usage of multiple investor reaction measures across two studies with non-overlapping samples increases our confidence in the effects. Second, we extend the literature on business ethics (e.g., Crane & Matten, 2004; Motro & Sullivan, 2022; Stellar & Willer, 2018) by exploring the mechanisms of how unethical behavior leads to investors' behavioral outcomes. Specifically, we show that unethical behavior reduces the warmth and competence perception of an entrepreneur, fully mediating the negative effect of unethical behavior on investors' investment behavior. Last, we contribute to the literature on women in entrepreneurship suggesting that female entrepreneurs are disfavored in the entrepreneurial landscape (e.g., Huang et al., 2021; Kanze et al., 2018; Pistilli et al., 2023). Our work adds to this literature by using role congruity theory (Eagly & Karau, 2002) to examine gender bias in investors' reactions to entrepreneurs' unethical behavior. Surprisingly, our findings suggest that female and male entrepreneurs face equal negative consequences when engaging in unethical behavior, which is mostly inconsistent with the literature on gender bias in judging unethical behavior (Kennedy et al., 2016; Klein & Shtudiner, 2021; Montgomery & Cowen, 2020) or the literature on women in entrepreneurship (Huang et al., 2021; Kanze et al., 2018; Lee & Huang, 2018). A possible explanation for this finding is that the unethical act might be more important for evaluations and behavioral outcomes than the gender of the entrepreneur, as some previous studies did not find evaluators to hold different standards for men and women across different scenarios of unethical behavior (Maher & Bailey, 1999; McNichols & Zimmerer, 1985).

We aim to inform entrepreneurs and investors that both male and female entrepreneurs are punished for profitable but unethical behavior. The severity of the punishment is demonstrated through the large magnitude of the effects on investment behavior, broadcasting, and personal evaluations. Even though ethical compromises might be tempting, early-stage entrepreneurs should be aware that prioritizing profitability over ethicality can represent high obstacles when attracting potential investors. Our findings underline the importance of ethical decision-making for early-stage entrepreneurs, especially in uncertain situations where information for investors is scarce and decisions are mainly based on personal impressions. Moreover, we inform investors by shedding light on the significant influence of entrepreneurs' warmth and competence evaluations on investors' early-stage investment decisions.

Our work is not without limitations. First, the experiments were conducted online with participants with investment experience from Prolific. While Prolific represents a state-of-the-art platform for conducting experiments (Palan & Schitter, 2018), future research should recruit venture capitalists and angel investors to increase the ecological validity. Second, future scholars are encouraged to compare samples from different cultures as opposed to the traditional focus on Western cultures. This is relevant because prior research has shown differences in gender and moral perceptions across cultures (Roxas & Stoneback, 2004). Third, we incorporated incentive compatibility in our main study to increase the stakes of the investment decision. However, future research should replicate the study in a field setting, where the stakes, especially the risk of losing the investment, are even higher. Overall, our work emphasizes the negative impact of entrepreneurs' profitable but unethical behavior on investors' reactions and calls for future scholars to further investigate trade-offs between ethicality and profitability, contextual moderators, and gender effects.

CONTACT: Sara Luisa Kappelhoff, sara@kappelhoff-ol.de, Technical University of Munich, Arcisstraße 21, 80333 Munich.

ACKNOWLEDGEMENTS: The authors acknowledge the financial support by the Federal Ministry of Education and Research of Germany in the program "Innovative Frauen im Fokus"

THE IMPACT OF GOVERNMENTAL AND PRIVATE VENTURE CAPITAL ON CORPORATE INNOVATION

Norah Kennis, Katholieke Universiteit Leuven, Belgium
Steven Vanhaverbeke, Katholieke Universiteit Leuven, Belgium

ABSTRACT

This paper investigates the impact of governmental venture capital (GVC) and private venture capital (PVC) on corporate innovation in Europe. Using patent-based indicators, we assess the value of innovation, distinguishing between exploitative and explorative innovation. Employing a staggered difference-in-differences design, we compare the innovation outcomes of firms financed by GVC and PVC. Our findings reveal that while PVC-backed firms produce more patents, GVC-backed firms generate patents with higher citation counts. This suggests that GVC-backed firms tend to produce more influential, particularly explorative, innovations. We contribute by highlighting the distinct objectives and incentives of GVCs and PVCs in fostering innovation.

INTRODUCTION

Start-ups play a crucial role in driving innovation and economic development, yet they often encounter financial constraints that hinder their growth and potential. These constraints arise from market failures, such as information asymmetries and risk aversion, which can prevent private investors from fully recognizing the value of innovative ventures. To address these financial constraints, policymakers worldwide have increasingly implemented governmental venture capital (GVC) funds as a means of support. The primary objectives of government investments in start-ups are to foster economic growth, generate positive R&D spillovers, support innovation, and stimulate regional development (Bertoni & Tykvová, 2015; Da Rin et al., 2013). Unlike private venture capital (PVC) funds, which primarily seek financial returns, GVCs are driven by broader socio-economic goals, reflecting governmental interests in promoting societal welfare through entrepreneurial support (Abrardi et al., 2019; Leleux & Surlemont, 2003; Lerner, 2002).

There is a substantial body of literature that examines the differences between governmental venture capital (GVC) and private venture capital (PVC), with the general tendency being that PVC outperforms GVC in achieving desired outcomes. Research has shown that GVCs often fall short of their goals, such as stimulating growth, supporting innovation, and compensating for the scarcity of private venture capital investments in start-ups (Alperovych et al., 2015; Bertoni & Tykvová, 2012, 2015; Cumming et al., 2017; Grilli & Murtinu, 2014; Munari & Toschi, 2015; Pierrakis & Saridakis, 2017; Shane, 2009). For example, Bertoni and Tykvová (2015) find that while PVC funding positively impacts innovation, GVC funding has no significant effect on innovation in young biotech firms in Europe. This calls into question the need for such policy interventions and suggests that resources could potentially be better allocated elsewhere.

However, while private venture capital is often considered more effective in stimulating innovation, this prevailing view largely arises from the use of traditional output metrics that may undervalue the unique contributions of governmental venture capital. Previous studies often rely on measures like patent counts, which might fail to capture GVCs' emphasis on high-risk, early-stage, and transformative innovations. By exploring alternative metrics that reflect the broader socio-economic goals and distinct innovation pathways supported by GVCs, this paper aims to provide a more comprehensive understanding of their role and potential in complementing PVC efforts.

HYPOTHESIS DEVELOPMENT

We posit that there are intrinsic differences in the types of innovation pursued by GVC-backed versus PVC-backed firms. Drawing on March's (1991) distinction between exploration and exploitation in organizational learning, we lay out a conceptual framework linking venture capital type to innovation strategy through two primary mechanisms: investment time horizons and strategic objectives. While PVCs' shorter investment horizons and emphasis on financial returns may drive higher rates of incremental innovation, GVCs' longer time horizons and broader social objectives may foster more experimental, exploratory innovation. Noteworthy, we follow the reasoning of Fitzgerald et al., (2021) to establish a distinction between the explorative versus exploitative value of innovation. Explorative innovation is defined as innovation based on novel technologies and approaches that are distant from the firm's current knowledge and capabilities. While exploitative innovation refers to incremental innovation based on a firm's existing competencies, for instance by refining current patented knowledge and capabilities (Fitzgerald et al., 2021).

On one hand, private venture capital operates under relatively short investment horizons, typically seeking to exit investments within 5-7 years to provide returns to limited partners (Gompers & Lerner, 1999; Sahlman, 1990). This temporal constraint creates pressure for quick commercial success, as PVCs must demonstrate strong performance to raise subsequent funds. The compensation structure of PVC fund managers, which ties carried interest to successful exits, further reinforces this orientation toward near-term results (Gompers & Lerner, 1999). These institutional features create what Manso (2011) terms "short-termism" - an emphasis on immediate performance that can discourage experimentation and exploration. Furthermore, private venture capital funds place significant emphasis on building strong patent portfolios, as these directly impact exit valuations through IPOs or acquisitions (Hsu & Ziedonis, 2013; Mann & Sager, 2007). This emphasis on portfolio building creates incentives for strategic patenting - the systematic accumulation of patents primarily for their defensive and bargaining value rather than their technological merit (Hall & Ziedonis, 2001). In conclusion, we argue that PVC-backed firms will generate more exploitative innovations.

On the other hand, government venture capital operates under broader strategic objectives that extend beyond maximizing financial returns. GVCs are held accountable by the public authorities that established them and accordingly have to adhere to set out economic and political policy objectives. Subsequently, GVCs have the incentive to create positive R&D spillovers and foster regional development (Bertoni & Tykvová, 2015; Da Rin, Hellmann & Puri, 2013). Additionally, without the pressure to provide regular returns to limited partners, GVCs can maintain longer investment horizons that better accommodate the uncertainty and experimentation inherent in exploratory innovation (Lerner, 2002). Theoretical and empirical work on innovation in principal-agent settings – like the venture capital and investee firm context - suggests that successful exploration requires tolerance for early failure and longer evaluation periods (Azoulay et al., 2011; Ederer & Manso, 2013; Manso, 2011). Thus, we argue that GVC-backed firms will generate more explorative innovations.

METHOD

Data Sources & Proxies to Measure the Value of Innovation

Venture capital data is collected from Bureau van Dijk–Orbis Historical Ownership Database. Since 2007, this database has provided information on ownership structures between subsidiaries and their shareholders (i.e., parents, owners). Based on this information, we classified VC funds as GVCs if government involvement was apparent in their shareholder structure; otherwise, they were classified as PVC. Next, we identified firms in which venture capital funds held an equity stake, establishing a firm-year-level dataset that records the type of VC financing each firm received and the corresponding year. Finally, control variables were also retrieved from Orbis.

Further, citation data about patents available in PATSTAT is used to construct our measures of explorative and exploitive innovation. In this study, we propose a methodological shift that leverages alternative measures of innovation beyond patent counts. Building on the work of Trajtenberg et al., (1997), Hall et al., (2005), Balsmeier et al., (2017), as well as Fitzgerald et al., (2021), we create measures that differentiate between exploitative and explorative innovation. Specifically, we proxy exploitative innovation using self-citations, uncited patents, and blocked patents, while explorative innovation is measured through non-self-citations, top-cited patents, and novel patents.

First, self-citations refer to citations made to patents owned by the same assignee as the patent in question, i.e., subsequent patents developed "in-house" by the same firm. These citations signify further advancements of the original invention, acting as pathways for appropriating returns. A higher proportion of self-citations indicates a greater degree of internalization, allowing firms to capture more private benefits from their own R&D investments. Furthermore, self-citations possess a greater private value compared to external non-self-citations. For the reason that the latter could potentially signify spillover effects that benefit others, including competitors. As such, self-citations serve as a proxy for exploitative innovation. Conversely, non-self-citations, i.e., citations made to patents owned by other firms, indicate that the patent draws upon diverse sources of prior knowledge, suggesting a more explorative innovation strategy. Non-self-citations also highlight the extent to which an innovation contributes to knowledge spillovers, promoting cross-fertilization of ideas across firms and industries. Therefore, non-self-citations serve as a proxy for explorative innovation.

Second, uncited patents are those that receive no citations, while top-cited patents are those ranked in the top 10% of most-cited patents. Generally, a higher citation count signifies greater technological importance, quality, or impact of a patent (Czarnitzki & Lopes-Bento, 2014). Fundamental innovations tend to garner many citations due to their broad impact, while incremental innovations yield fewer or no citations. Accordingly, we argue that GVC-backed firms are likely to produce more top-cited patents, whereas PVC-backed firms may generate a higher proportion of uncited patents.

Lastly, blocked patents are those rejected due to citations that challenge their novelty. Because of a certain citation, the claimed invention in the patent cannot be considered novel or cannot be considered to involve an inventive step, leading to the patent being blocked. By contrast, novel patents are those successfully granted without such impediments. GVC-backed firms, driven by broader socio-economic goals and a willingness to explore uncharted territory, are expected to create more novel patents. Conversely, PVC-backed firms, constrained by shorter-term objectives, may produce more blocked patents.

Sample

The study examines data spanning from 2007 to 2020, comprising 44,257 firm-year observations corresponding to 4,733 unique European firms in total. Approximately half of these firms received VC financing at some point during the sample period. For each treated firm, a matched control firm that did not receive VC funding was added to the sample. Furthermore, descriptive statistics in Table 1 show that respectively, 27.42% of firm-year observations receive PVC financing, and 9.71% GVC financing. This distribution is expected, as traditional private venture capital funds are more common and have a longer history of market presence. Other databases providing VC data also reflect a similar pattern, with significantly more PVC-backed firms and relatively few GVC-backed firms (VICO | RISIS Docs, 2024).

Research Design

To rigorously assess the influence of venture capital on firm innovation performance, we apply a staggered difference-in-differences research model. The regression model contains firm and country-industry-year fixed effects and heteroskedasticity-robust standard errors. The dependent variables are the various patent citation measures. The independent variables of interest are PVC funding and GVC funding, dummies that equal 1 once a firm-year observation receives PVC or GVC financing, respectively. Hence, the dummies switch from 0 to 1 in the year the firm receives its first VC investment and remain at 1 in all subsequent years. The two dummies are not mutually exclusive as sometimes a firm-year observation can receive both PVC and GVC funding. The control variables are ROA, innovative intensity, and firm age.

RESULTS

Table 2 reports the regression results of the main analysis. In the first column, the influence of private and governmental venture capital on the number of patents is investigated. The variables PVC funding and GVC funding both have a significant positive influence on patent count. When comparing the significance and size of their coefficients, we conclude that PVCs have a more pronounced positive influence on patent count. PVC-backed firms spur more patents than GVC-backed firms, a finding consistent with previous research (Berger et al., 2024; Bertoni & Tykvová, 2015). Based on this finding, previous research would conclude that PVCs perform better than GVCs on corporate innovation. However, we expand the analyses to several citation-related metrics to measure the value of innovation. These metrics give us a more meaningful insight into the various ways in which venture capital can influence innovation. Most notably, column two indicates that GVC financing leads to a significant increase in forward citations, in stark contrast to PVC financing, which is associated with a decline in patent citations for investee firms. Hence, these results provide a first indication that although GVC-backed firms might patent less, they do seem to have patents of greater importance (evidenced by the higher number of citations). Untabulated results confirm this insight, as only GVCs have a positive significant effect on top-cited patents, whereas only PVCs have a positive significant effect on uncited patents. Again, this proofs that GVC-backed firms spur patents that especially produce many citations, though this is clearly not the innovation strategy followed by PVC-backed firms.

Next, consider the regression results when dividing citations in self- versus non-self-citations. Non-self-citations reflect the broader impact and explorative value of an innovation, indicating knowledge spillovers and cross-fertilization of ideas across different firms. These citations suggest that the innovation draws upon and contributes to a diverse range of external knowledge sources, ultimately fostering broader societal benefits. Consistent with this reasoning we find in column four that GVCs have more non-self-forward citations than PVCs. Interestingly, a comparison across columns two to four—covering total, self-, and non-self-citations—reveals that the citation advantage of GVC-backed firms is largely driven by non-self-forward citations. The coefficient for total forward citations closely mirrors that of non-self-forward citations, suggesting that the observed citation premium for GVCs stems predominantly from their broader external impact rather than internal reuse. On the other hand, self-citations are related to the exploitative value of innovation. The more self-citations, the easier firms succeed in reaping the benefits of their prior R&D investments and the easier firms can appropriate private exploitative returns from innovation. The regression results in column three partially align with our expectations regarding self-forward citations. Contrary to what we anticipated, PVC-backed firms do not exhibit higher levels of self-citation compared to GVC-backed firms. However, it is worth noting that the negative effect of PVC financing on self-citations is relatively modest compared to its stronger negative effect on total citations. This suggests that while PVC-backed firms produce patents that receive fewer citations overall, the decline in self-citations is less pronounced—indicating a relatively stronger retention of internal, exploitative value.

Lastly, consistent with our hypotheses, untabulated results show that patents from PVC-backed firms are more frequently blocked due to impeding citations, whereas GVC-backed firms are more likely to produce novel, unblocked patents. This finding further supports the view that PVCs tend to emphasize incremental, exploitative innovation, while GVCs are more oriented toward explorative, groundbreaking innovation.

DISCUSSION AND IMPLICATIONS

This study contributes to the entrepreneurship literature by offering new insights into how venture capital funding—both governmental and private—shapes the innovation outcomes of start-ups. The main insight of the paper is that governmental venture capital has a positive influence on explorative innovation. In this way, the paper shows that governmental venture capital is not necessarily less effective compared to private venture capital. Substantiating the reasoning that PVCs and GVCs exhibit different objectives and incentive structures: while PVCs primarily focus on private returns, GVCs have a dual purpose and focus on private and social returns.

Our findings have implications for both policymakers and entrepreneurs. For policymakers, the results suggest that government venture capital serves a unique and valuable role by supporting more experimental, exploratory innovation that private venture capital tends to avoid. Rather than viewing GVCs as underperforming alternatives to private VCs, policymakers should recognize them as complementary institutions that fill critical gaps in innovation financing. For entrepreneurs, the choice between GVC and PVC funding should align with their innovation strategy and time horizons. Those pursuing more experimental technologies or requiring longer development periods may find better alignment with GVC investors, while those focused on commercializing established technologies or seeking faster market entry may benefit more from PVC backing. Additionally, entrepreneurs should recognize that PVC funding may create pressure to shift toward more exploitative innovation strategies after investment.

Table 1

DESCRIPTIVE STATISTICS

Panel A: Independent variables						
Variable	N	mean	Sd	p1	p50	p99
PVC funding	44,257	0.274	0.446	0.000	0.000	1.000
GVC funding	44,257	0.097	0.296	0.000	0.000	1.000
Panel B: Dependent variables						
Variable	N	mean	Sd	p1	p50	p99
Patent Count	44,257	5.529	18.581	0.000	0.000	126.000
Forward Cites	44,257	10.136	41.884	0.000	0.000	300.000
Self Forward Cites	44,257	1.740	8.273	0.000	0.000	61.000
Non-self Forward Cites	44,257	7.906	32.259	0.000	0.000	231.000
Backward Cites	44,257	30.013	117.088	0.000	0.000	836.000
Self Backward Cites	44,257	2.992	13.409	0.000	0.000	96.000
Non-self Backward Cites	44,257	26.686	103.635	0.000	0.000	742.000
Top 10 Cited Patents	44,257	0.319	1.308	0.000	0.000	9.000
Uncited Patents	44,257	3.732	12.433	0.000	0.000	83.000
Novel Patents	44,257	1.681	5.973	0.000	0.000	41.000
Blocked Patents	44,257	1.610	5.437	0.000	0.000	37.000
Panel C: Control variables						
Variable	N	mean	Sd	p1	p50	p99
ROA	44,257	-0.148	0.482	-1.954	0.002	0.659
Innovative Intensity	44,257	2.963	15.089	0.000	0.187	62.851
Log(Age)	44,257	2.518	0.865	0.693	2.485	4.625

Notes: This table presents the sample descriptive statistics. All dependent and control variables are winsorized.

Table 2

REGRESSION RESULTS

Outcome:	Patent Count	Forward Cites	Self Forward Cites	Non-self Forward Cites
	(1)	(2)	(3)	(4)
PVC funding	0.686***	-1.357*	-0.341**	-0.925
	(2.929)	(-1.876)	(-2.525)	(-1.595)
GVC funding	0.548**	2.302***	0.314**	1.770***
	(2.288)	(3.115)	(2.323)	(2.885)
ROA_{t-1}	0.265*	-0.318	0.049	-0.392
	(1.743)	(-0.725)	(0.530)	(-1.131)
Innovative $Intensity_{t-1}$	-0.004	-0.006	-0.002	-0.008
	(-0.742)	(-0.536)	(-0.560)	(-0.843)
log(Age)	0.841***	6.465***	0.848***	5.318***
	(2.694)	(6.454)	(4.503)	(6.694)
Firm FE	Yes	Yes	Yes	Yes
Country-Industry-Year FE	Yes	Yes	Yes	Yes
Observations	44,257	44,257	44,257	44,257
R-squared	0.823	0.697	0.717	0.678

Notes: This table presents OLS regressions on innovation. Values between brackets represent t-values. *, **, and *** denote significance at the 10 percent, 5 percent, and 1 percent level (two-tailed), respectively.

CONTACT: Norah Kennis; norah.kennis@kuleuven.be; Katholieke Universiteit Leuven; Naamsestraat 69 - bus 3525, 3000 Leuven, Belgium

The authors gratefully acknowledge helpful comments from Jeroen Mahieu, Dirk Czarnitzki, Mark Giebel, Maikel Pellens, and seminar participants at KU Leuven. Norah Kennis gratefully acknowledges financial support from the Flemish Science Foundation.

MOBILIZING SUPPORT FOR INDIGENOUS VENTURES: THE ROLE OF CO-CULTURAL STRATEGIES AND ALLYSHIPS

Annaleena Parhankangas, Iowa State University, United States
Rick Colbourne, Carleton University, Canada

ABSTRACT

We set out to explore how entrepreneurs' expressions of their Indigenous identity affect their crowdfunding success. Based on the Social Identity Model of Collective Action and Co-Cultural Theory, we predict that entrepreneurs' references to Indigeneity boost their fundraising efforts. Furthermore, this effect is expected to be stronger for ally-led as opposed to Indigenous-led campaigns. We tested these hypotheses on 521 Indigenous Kickstarter campaigns and found general support for these predictions.

INTRODUCTION

Entrepreneurship can be seen as a pathway to emancipation and empowerment for disadvantaged and marginalized individuals and groups (Rindova, Barry, & Ketchen Jr, 2009). However, to grow their ventures, disadvantaged entrepreneurs often need to mobilize support from a wide variety of resource-holders, also from the socially dominant groups. While a growing body of literature has documented evidence of discrimination against women, minority and geographically remote entrepreneurs by the traditional providers of entrepreneurial finance (Aernoudt & de San José, 2020; Sorenson, Assenova, Li, Boada, & Fleming, 2016; Younkin & Kuppuswamy, 2018), less is known on the strategies deployed by the disadvantaged groups to communicate their marginalized identity and how these strategies affect the support mobilized from important stakeholders.

In this study, we focus on Indigenous entrepreneurs who are involved in a process of drawing value from community-based resources (people, land, capabilities, culture) and contributing value back that is responsive to a community's particular set of socioeconomic conditions (Colbourne, 2017; Jack & Anderson, 2002). *Indigenous entrepreneurship* is a means by which *Indigenous peoples* exercise and sustain their rights to design, develop and maintain political, economic and social systems or institutions that secure their own means of subsistence and development and enables community members to engage in traditional, cultural and/or economic activities occurring on or near their traditional territories (Peredo, Anderson, Galbraith, & Honig, 2004; UN, 2009). As Indigeneity plays a key role in these entrepreneurs' individual identities and their companies' mission, it becomes crucial for Indigenous entrepreneurs targeting the mainstream markets to find a way to communicate their Indigenous identity in a way that resonates with non-Indigenous audiences. In this context, co-cultural theory (Orbe, 1998, 2005; Orbe & Roberts, 2012) provides a framework to identify co-cultural communication practices that underrepresented groups may use to communicate with dominant social structures.

In addition to co-cultural theory, we rely on an emerging body of work on intergroup prosociality (Louis et al., 2019; Stürmer & Snyder, 2009; Van Leeuwen & Zagefka, 2017) that recognizes that individuals from various social groups may seek to reduce perceived injustice and help disadvantaged groups regardless of whether their own group is negatively impacted by the status quo. In the context of reward-based crowdfunding, we claim that funders are more likely to engage in collective action and support ventures promoting Indigenous products, services or causes, if they feel psychological closeness and identification with the Indigenous peoples and their aspirations (Craig, Badaan, & Brown, 2020; Klavina & van Zomeren, 2020; Tropp & Barlow, 2018). Therefore, we predict that those Indigenous crowdfunding campaigns that use co-cultural practices involving culturally embedded cues, such as depictions of ancestral heritage; culture-specific assimilation experiences; and other cultural-bound traditions, practices, and symbols will be more favorably viewed by the backers on crowdfunding platforms.

Furthermore, we predict that the impact of co-cultural practices used to express Indigeneity differs depending on whether they are utilized by Indigenous-led as opposed to ally-led campaigns. Previous research on intergroup prosociality maintains that positive cross-group contact – such as Indigenous entrepreneurs working together with their allies- increases the mainstream society's willingness to help the members of the marginalized group (Glasford, 2013; Louis et al., 2019; Tropp & Barlow, 2018). In a similar vein, previous work has demonstrated that racial justice messaging is perceived to be more convincing and less threatening by the members of the dominant group, if this message is conveyed by the member of their own group (Lane, Coles, & Saleem, 2019; Roden, 2022). Therefore, we predict that the use of co-cultural practices, such as highlighting one's own cultural uniqueness, experienced injustices and quest for greater empowerment, are more effective in terms of raising funding when utilized by ally-led campaigns.

We tested our hypotheses on a sample of 521 Indigenous Kickstarter campaigns, i.e., Indigenous-led or ally-led campaigns that promoted Indigenous culture or causes. Our results lend general support to our hypotheses, showing that co-cultural strategies of using Indigenous symbols, references to and expressing pride in Indigeneity increase the likelihood of crowdfunding success. Furthermore, the impact of many of these communication strategies was strengthened when used by ally-led campaigns.

Our study makes several important contributions. We contribute to the entrepreneurial finance literature by discussing how the expression of marginalized identity impacts entrepreneurs' ability to mobilize support from important stakeholders. We extend this literature by examining how the use of various co-cultural communication strategies appeals to the dominant group members. In a similar vein, we highlight the critical role of allyships – largely ignored in the prior entrepreneurial finance literature- in enabling marginalized entrepreneurs to expand beyond co-ethnic markets.

HYPOTHESES DEVELOPMENT

The study of intergroup prosociality and collective action has illuminated many ways with which people may change the world positively for others, from making charitable donations to engaging in political allyships (Craig et al., 2020; Louis et al., 2019; Stürmer & Snyder, 2009). Intergroup prosociality and collective action can include both public (e.g., participating in a protest) and private (e.g., signing an online petition) behaviors (Radke, Kutlaca, Siem, Wright, & Becker, 2020) and may take the form of transfer of resources (tangible or intangible) to individuals in need (Van Zomeren, 2013). Recent years have marked an increase in collective action on online communities (Agostini & van Zomeren, 2021; Monik & Parzuchowski, 2023), including crowdfunding platforms that allow entrepreneurs to mobilize financial resources from a community of non-professional resource providers during their venture's earliest and most uncertain stages (Murray, Kotha, & Fisher, 2020). This community-based resource-mobilization is largely motivated by non-financial incentives, such as taking part in something that helps nascent ventures get off the ground, promoting innovation and bringing new products to the market (Belleflamme, Lambert, & Schwienbacher, 2012; Gerber, Hui, & Kuo, 2012; Ordanini, Miceli, Pizzetti, & Parasuraman, 2011), especially by entrepreneurs who backers relate to or who they see in need of support (Greenberg & Mollick, 2017).

Although the Social Identity Model of Collective Action (SIMCA) originally focused on the motivations of *disadvantaged* groups to engage in collective action to improve their situation (Klavina & van Zomeren, 2020), it has been successfully extended to the members of advantaged groups (Cakal, Hewstone, Schwär, & Heath, 2011; Van Zomeren, Postmes, Spears, & Bettache, 2011), and allied minority groups (Dixon et al., 2015). We extend the Social Identity Model of Collective Action by considering communication strategies available for marginalized groups wishing to evoke support from advantaged group members in the crowdfunding setting. In here, co-cultural communication theory proves helpful (Orbe, 1996, 1998) in analyzing the interactions between underrepresented and dominant group members and providing insight into how marginalized groups choose to negotiate their ''cultural differentness'' with others (Orbe & Roberts, 2012).

In particular, the co-cultural theory has identified several communication practices available for equity seeking groups shaped by their desired degree of assimilation to the mainstream culture (Orbe, 1996, 1998). At the assimilation end of the continuum, Indigenous entrepreneurs attempt to fit in with the dominant cultural norms by eliminating cultural differences and minimizing distinctions between groups. At the other extreme, the separation end of the continuum, Indigenous entrepreneurs distance themselves from the mainstream society by creating spaces that are reflective of their own values, mores, and norms. Somewhere in the middle fall accommodation strategies, where minority entrepreneurs work within dominant societal structures so that they reflect co-cultural experiences (Orbe, 1998).

Entrepreneurs promoting Indigenous products, culture or causes are likely to gravitate towards accommodation and separation when it comes to adopting co-cultural practices. For instance, Indigenous entrepreneurs may choose to maintain co-cultural presence with dominant structures by increasing their visibility through *references to Indigeneity*. Furthermore, they may choose to *educate* others by taking the role of a teacher in co-cultural interactions enlightening dominant group members of Indigenous norms, values and history. Finally, Indigenous entrepreneurs may choose to *express pride* by promoting the recognition of co-cultural group strengths, past accomplishments, and contributions to society. In the following, we discuss how the use of various co-cultural practices impacts the fundraising success of Indigenous crowdfunding campaigns.

An emerging body of research on intergroup prosociality (Louis et al., 2018; Sturmer & Snyder, 2010; van Leeuwen & Zagefka, 2017) highlights that individuals from diverse social groups may strive to address perceived injustices and support disadvantaged groups, regardless of whether their own group is adversely affected by the existing social order. In the realm of reward-based crowdfunding, we propose that funders are more likely to participate in collective action and back ventures advocating for Indigenous products, services, or causes when they experience psychological closeness and identification (Craig et al., 2020; Klavina & van Zomeren, 2020; Tropp & Barlow, 2018) with Indigenous peoples and their aspirations. Indigenous entrepreneurs can foster this connection by embracing co-cultural practices that emphasize uniqueness and shared cultural experiences. Consequently, we anticipate that Indigenous crowdfunding campaigns incorporating co-cultural practices—featuring culturally embedded cues such as portrayals of ancestral heritage and other culturally grounded traditions, practices, and symbols (Williams, 1995)—will be viewed more favorably by backers on crowdfunding platforms.

Hypothesis 1: Culturally embedded cues increase the likelihood of fundraising success for Indigenous ventures.

Hypothesis 1a: References to Indigeneity increase the likelihood of fundraising success for Indigenous ventures.

Hypothesis 1b: Use of Indigenous symbols increases the likelihood of fundraising success for Indigenous ventures.

Hypothesis 1c: Expressing pride in Indigeneity increases the likelihood of success for Indigenous ventures.

Hypothesis 1d: Educating the crowd about Indigenous peoples and their experiences increases the increases the likelihood of success for Indigenous ventures.

In our final set of hypotheses, we predict that the impact of co-cultural practices used to express Indigeneity differs depending on whether they are utilized by Indigenous-led as opposed to ally-led campaigns. Previous research on intergroup prosociality maintains that positive cross-group contact – such as Indigenous entrepreneurs working together with their allies- increases backers' willingness to help the members of the equity-seeking group (Louis et al., 2018; Glasford, 2013; Tropp & Barlow, 2018).

This positive moderating effect is likely to occur via two mechanisms: First, positive cross-group contact through allyship is likely to lend credence to the cause of the equity seeking groups group and elicit greater acceptance (Louis et al., 2019; Radke et al., 2020). In a similar vein, previous work has demonstrated that racial justice messaging is perceived more convincing and less threatening by the members of the dominant group, if this message is conveyed by the member of their own group (Lane et al., 2019; Roden et al, 2022). Second, positive cross-group contact through allyship is likely to evoke greater empathy and create a greater sense of closeness with the stigmatized group (Craig et al., 2020). Therefore, we predict that the use of co-cultural practices, such as highlighting one's own cultural uniqueness, their experiences of resilience, resistance, continued pursuit of justice and quest for greater empowerment is more effective in terms of raising funding when utilized by ally-led campaigns.

Hypothesis 2: Culturally embedded cues are more effective when used by ally-led rather than Indigenous-led campaigns.

Hypothesis 2a: References to Indigeneity are more effective when used by ally-led rather than Indigenous-led campaigns.

Hypothesis 2b: Use of Indigenous symbols is more effective when used by ally-led rather than Indigenous-led campaigns.

Hypothesis 2c: Expressing pride in Indigeneity is more effective when used by used by ally-led rather than Indigenous-led campaigns.

Hypothesis 2d: Educating the crowd about Indigenous peoples and their experience is more effective when used by used by ally-led rather than Indigenous-led campaigns.

METHOD

Sample

We tested our hypotheses on a sample of 521 Indigenous Kickstarter campaigns, i.e., Indigenous-led or ally-led campaigns that promoted Indigenous culture or causes. We used the search words "Native American," "American Indian," "Indigenous," "Aboriginal," "Alaska Native," "Inuit," "M´etis," "First Nation," and "Reservation" to extract Indigenous peoples' campaigns. Only campaigns that were completed by June 2020 were included.

It is important to note that our sample only includes campaigns that identify themselves as Indigenous. Not all entrepreneurs self-identify when using crowdfunding platforms. Therefore, we do not claim our sample to be representative of all crowdfunding campaigns on these platforms initiated by Indigenous individuals or organizations. Instead, we focus on campaigns that chose to self-identify and/or refer to Indigenous peoples, communities, and entrepreneurs as well as their traditions, values, beliefs, culture, or particular Indigenous-centric issues when promoting the campaigns.

Our major sources of information were the campaign pages posted on crowdfunding platforms, containing written campaign descriptions, videos, and campaign background information. We carefully read through the campaign descriptions to determine whether the campaign was Indigenous-led or ally-led.

Measures

We measured our dependent variable - crowdfunding success - to a degree to which entrepreneurs were able to reach their campaign goals. To operationalize this variable, we divided the amount raised by funding goal set for the campaign (Cumming, Leboeuf, & Schwienbacher, 2020; Gafni, Marom, & Sade, 2019).

We extracted the co-cultural communication strategies by analyzing the campaign pitches of Indigenous entrepreneurs. We started by identifying instances where Indigenous entrepreneurs referred to their Indigenous heritage, culture and relationship with the mainstream society. The communication practices used to express their Indigeneity were coded by using the co-cultural communication theoretical model as a point of departure (Orbe, 1998; 2005). As a result, four major co-cultural strategies emerged from our textual analysis: *use of Indigenous symbols*, *references to Indigeneity*, *expressing pride in Indigeneity*, and *educating others about Indigeneity*.

The use of Indigenous symbols was operationalized as a number of Indigenous symbols displayed in the crowdfunding campaign description. References to Indigeneity was measured as the number of sentences describing Indigenous culture, issues or traditions. Expressing pride in Indigeneity captures the number of sentences praising the achievements and culture of Indigenous peoples. Educating others about Indigeneity refers to the number of sentences informing the audience about Indigenous culture, history and traditions.

In addition to coding the co-cultural strategies used, we also analyzed whether the crowdfunding campaign was ally-led (as opposed to Indigenous-led). This binary variable was coded 1, if the campaign was launch by a non-Indigenous individual or organization.

Following prior crowdfunding studies, we controlled several factors contributing to crowdfunding success, such as the number of comments, updates, the emotional tone of the campaign description (Chan & Parhankangas, 2017; Kuppuswamy & Bayus, 2013). The emotional tone of the campaign description was operationalized using the LIWC variables positive and negative emotions as a proxy (Bednar, 2012; Pennebaker, Mehl, & Niederhoffer, 2003; Tausczik & Pennebaker, 2010). We also controlled for the location of the campaign (rural vs urban), whether there were any women in the entrepreneurial team and whether the campaign was commercial (vs. social) in nature.

RESULTS

Our sample consists of 521 campaigns launched on Kickstarter during the years 2010 and 2020. Most campaign initiators came from the United States (347), followed by Latin America (40), Canada (66) and Australia (23). In most cases, the campaigns (78 percent) were involved in the production of Indigenous culture through films, fiction, visual arts, photography or documentaries. Of all campaigns, 18 percent were based on commercializing Indigenous products or services, such as apparel, crafts, food, fashion or games. On average, the campaigns raised $5,356 from 75 funders.1 Overall, 43% of our campaigns received some funding. In 80 percent of the cases, campaign creators came from urban locations. Half of the campaigns were ally-led (as opposed to Indigenous-led).

We first ran a baseline model that included the control variables. In the second and third steps, we examined whether use of culturally embedded cues were positively associated with crowdfunding success, and whether their impact was contingent on whether the campaign was Indigenous or ally-led.

According to Hypothesis 1, culturally embedded cues boost the fundraising success of Indigenous crowdfunding campaigns. The positive, and statistically significant beta coefficients for the use of Indigenous symbols, references to Indigeneity, and expressing pride in Indigeneity lend support for Hypotheses 1a, 1b and 1c. However, educating the crowd about Indigeneity was negatively associated with crowdfunding success, thus refuting Hypothesis 1d.

Hypothesis 2 proposes that culturally embedded cues are more effective when used by ally-led rather than Indigenous-led campaigns. We detected a positive, statistically significant interaction effect between a crowdfunding campaign being ally-led and references to Indigeneity and expressing pride in Indigeneity, thus supporting Hypotheses 2b and 2c. Contrary to our Hypothesis 2d, the interaction term ally-led campaign *educating the crowd about Indigeneity is negative and statistically significant, whereas the interaction term ally-led campaign * use of Indigenous symbols remains non-significant. As a result, Hypotheses 2a and 2d were rejected.

DISCUSSION AND IMPLICATIONS

We set out to explore how entrepreneurs' expressions of their Indigenous identity affect their ability to mobilize support from the crowdfunding community. We found that entrepreneurs that used various co-cultural strategies to highlight Indigeneity were favored by the crowd. Furthermore, many of these co-cultural practices were more effective when utilized by ally-led rather than Indigenous- led campaigns.

Our paper makes several theoretical contributions. First, our knowledge of how expressions of marginalized identities affect entrepreneurs' resource mobilization efforts remains limited. In this regard, our study sheds light on the role of co-cultural communication practices and allyships in mobilizing support that extends beyond one's ethnic ties. Second, we contribute to the crowdfunding literature by examining how entrepreneurs can best tap on intergroup prosociality and mobilize collective action. While the crowdfunding literature has focused on the motivations of the crowd to support individual entrepreneurs – both financial and non-financial (Belleflamme et al., 2012; Gerber et al., 2012; Ordanini et al., 2011)- less is known on what makes the crowd to support the causes of equity seeking groups on crowdfunding platforms, especially when their messaging might provoke guilt and discomfort among the dominant group members. Furthermore, management scholars have recently expressed a growing interest in the role of allies and allyships in management and organizations (Creary, 2024; Kossek et al., 2021). However, our knowledge on how allyships can be used to facilitate resource acquisition efforts by disadvantaged groups remains limited. We contribute to this stream of research by demonstrating how the efficacy of co-cultural communication strategies depends on whether marginalized entrepreneurs have joined their forces with dominant group allies.

Besides adding to the conversations in entrepreneurial finance and resource mobilization, we also contribute to co-cultural theory and literature on intergroup prosociality. While co-cultural theory has identified myriads of ways in which disadvantaged groups communicate their marginalized identities, less is known on the effectiveness of these communication strategies, especially in terms of entrepreneurial resource mobilization. In a similar vein, the literature on intergroup prosociality has rarely considered the financial implications of collective action.

Finally, the results of our study have many important practical implications for entrepreneurs wishing to mobilize support from important stakeholders- especially for those entrepreneurs whose ventures cannot be separated from their marginalized identities. Such entrepreneurs often cannot avoid bringing up topics such as resilience, resistance and continued pursuit of justice that may be perceived as threatening or polarizing by dominant group members (DiAngelo, 2022; Ford et al., 2022). Our study suggests that such entrepreneurs will benefit from explicitly communicating their ethnic heritage and – if possible- using dominant group allies when seeking to mobilize support on crowdfunding platforms.

CONTACT: Annaleena Parhankangas; leena@iastate.edu; Iowa State University, 2167 Union Drive, Ames, IA 50011, USA.

I'M SMARTER, BUT AM I BETTER OFF? USING AI TOOLS TO GENERATE MORE ACCURATE VENTURE CAPITAL VALUATIONS

Mohammad Rady, Virginia Tech, United States
David Townsend, Virginia Tech, United States
Richard Hunt, Virginia Tech, United States

ABSTRACT

This study explores the integration of AI into venture capital decision-making and its impact of the bidding behaviors of VCs in competitive early-stage capital markets. Our central premise is that while AI models can yield more accurate predictions of enterprise value and future investment returns than traditional human judgment, our research reveals a paradox: strict reliance on AI could prevent firms from winning competitive bids. We illuminate this paradox and identify several important contingencies through a comprehensive agent-based model that demonstrates critical trade-offs between algorithmic precision and human intuition in VC decision-making. Ultimately, the study raises important questions about the limitations and challenges of enhancing technical accuracy with AI to improve overall investment outcomes in a manner that inadvertently hinders a VC's ability to secure investments in competitive environments.

INTRODUCTION

By its very nature, the process of seeking to generate supra-ordinary returns through early-stage investments in fledging companies and emergent sectors means that venture capital (VC) firms inherently navigate a decisional environment that is characterized by what McGrath (1999) referred to as "*a priori* irreducible uncertainty," and Knight (1921) called "pure indeterminism." Since the very first firms -- American Research and Development Corporation and J.H. Whitney – were launched in 1946, the battle to successfully pick winners amidst uncertainty has driven VC to search for more and better tools to improve decisional outcomes by gathering and integrating financial data, market analyses, technological trajectories, and even personal intuition into increasingly sophisticated models (Ewens et al., 2018). For much of the past eighty years, human judgment has been the key element in this framework, especially when it comes to early-stage venture investment decisions (Huang & Pearce, 2015).

However, with the development of Artificial Intelligence (AI) models, AI is now enabling firms to process vast amounts of data and to forecast with a higher level of accuracy than what human capability has traditionally been able to achieve (Jarrahi, 2018). In light of this, the VC decision-making process is experiencing significant change. AI-integrated decision-making processes raise questions about how well these fit within the existing frameworks. AI systems may be able to provide an objective valuation of startups and their potential, yet they can fail to consider nuanced qualitative factors such as market sentiment, entrepreneur characteristics, and competitive pressures (Rettberg, 2022).

Recent years have witnessed an explosion in algorithmic goods and services, built on machine learning-enabled systems and complex computational systems that are capable of exceeding human performance across a myriad of specified tasks (Townsend et al., 2023). In particular, generative AI (Gen AI) has spawned large language models that have the capacity to process massive amounts of structured and unstructured data that is highly germane to decision-making. Paralleling this surge in new tools, there has been an explosion in scholarly literature on the roles and impacts of AI, including work devoted to better understanding how AI affects VC investing (Arroyo et al., 2019). However, while this stream is quite extensive, extant scholarship is odiously narrow, tending to emphasize the benefits of using AI in enhancing decision-making processes such as computation speed and accuracy (Fei et al., 2024). In relatively uncritical fashion. Although AI can reduce certain aspects of uncertainty given its data-driven approach, this data-driven approach can eliminate biases such as overconfidence or herd behavior (Zacharakis & Shepherd, 2016; Zhang et al., 2021).

Yet, while AI may enhance the technical side of decision-making, venture capital is not purely a numbers game. It involves assessing intangible factors such as the quality and experience of the startup founders (Gompers et al., 2020). These factors are difficult, if not impossible to quantify and so are not captured by the AI models. Furthermore, VCs frequently operate in competitive environments where other investors' behaviors can influence their decisions. For example, overbidding in competitive rounds can drive up valuations beyond what might be considered rational by AI models. If AI systems consistently undervalue startups due to their emphasis on hard data, VCs could find themselves losing out on key deals to competitors. As a result, the introduction of AI into venture capital decision-making raises questions about whether these systems might hinder the ability of VCs to make holistic, well-rounded decisions that incorporate both data and experience-based intuition. This leads to the central research question of this study: *When does the use of AI hinder or help investment decision-making among VC firms in competitive markets?*

To explore this research question, the purpose of this research is to explore the trade-offs and contingencies that arise when AI is introduced to the decision-making process which had traditionally been human-driven. The question challenges the assumption that more data and more sophisticated analytical tools will lead to better investment outcomes by highlighting some of the limitations of deploying AI tools in competitive environments. Our central premise is that in highly uncertain environments the use of AI-powered analytical tools creates a series of paradoxical trade-offs: On one hand, due to their abilities to process and analyze large amounts of data, AI tools will provide more precise long-term investment bids based on accurate valuation of an early-stage startup. At the same time, however, despite the gains in bid accuracy, these AI tools increase the risk VCs will *underbid* for different deals thereby increasing the risk they will miss out on key deals in highly uncertain early-stage capital markets. In this sense, the growing use of AI will make VC funds smarter but perhaps less capable of competing in early-stage markets with humans that are more biased and optimistic about a venture's future prospects.

HYPOTHESIS DEVELOPMENT

Venture capital (VC) investment decisions are inherently characterized by significant uncertainty and high stakes, often described by scholars as situations of "pure indeterminism" or "a priori irreducible uncertainty." Historically, VC firms have relied heavily on human intuition, judgment, and experience, complemented by increasingly sophisticated analytical frameworks. The integration of artificial intelligence (AI) into these decision-making processes represents a significant shift, enabling firms to systematically analyze vast amounts of structured and unstructured data to generate precise valuations and investment predictions. While AI has the potential to significantly enhance technical accuracy, reduce cognitive biases such as overconfidence and herding, and streamline analytical processes, it simultaneously introduces critical challenges. Particularly, AI-driven valuation methods can overlook qualitative, intangible factors including entrepreneurial characteristics, market sentiment, and competitive behaviors, which are difficult to quantify but essential to holistic investment decisions.

This research specifically explores the paradox emerging from the enhanced precision provided by AI tools. Theoretically, while AI can offer more accurate and less biased valuations, strict adherence to these algorithmically determined valuations may inadvertently disadvantage VCs in competitive bidding environments, where market dynamics often reward optimistic and aggressive bidding strategies. Hence, while AI-driven precision theoretically reduces investment risks by ensuring bids closely match rational valuations, it simultaneously creates a practical disadvantage in competitive markets where human investors regularly overbid based on optimism, intuition, or strategic market positioning.

METHODS

To systematically investigate this paradox, we developed a comprehensive agent-based simulation designed to emulate the complexities of early-stage venture capital markets. The simulation incorporates detailed agent attributes reflecting both human and AI-driven venture capitalists, each characterized by unique behavioral parameters such as risk aversion, bargaining power, overconfidence, information advantage, anchoring biases, and herding tendencies.

The simulation environment spans a 10-year investment period, incorporating a broad array of factors influencing market dynamics, including sector-specific volatility, growth rates, failure probabilities, and varying exit multiples. Three distinct sectors (AI, energy, and consumer) were modeled to reflect varying risk and return profiles that are grounded in data analyzed from Crunchbase. Additionally, the model includes realistic VC operational constraints such as fund size limitations, syndication patterns in funding rounds (seed, series A, series B), and adaptive behaviors that allow agents to update biases based on historical performance outcomes we estimated using available Crunchbase data.

Agents representing AI-driven VCs are equipped with enhanced analytical capabilities, demonstrated through lower valuation noise and significantly reduced cognitive biases. Conversely, human-driven VC agents exhibit higher valuation variability, reflecting intuitive and judgment-driven decisions, including significant susceptibility to market sentiment and competitive pressures. The simulation also incorporates contagion effects, reflecting real-world scenarios where market sentiment, previous deal activity, and competitive dynamics influence both the number and characteristics of emerging startups and investment behaviors.

To evaluate the robustness of these dynamics, we conducted a comprehensive sensitivity analysis focused particularly on market sentiment, examining how variations in sentiment impact both AI-driven and human-driven VCs in terms of bidding behaviors and investment outcomes.

RESULTS

The results of the simulation underscore a critical tension in VC decision-making enhanced by AI. AI-driven venture capitalists consistently produce more precise and objective valuations due to their algorithmic capability to process large datasets and reduce biases. However, this precision significantly limits their bidding flexibility, particularly in competitive and sentiment-driven market environments. AI-driven VCs consistently place bids close to their accurate valuations, resulting in lower overall winning bid frequencies when compared to human-driven VCs, who more frequently adapt bids upward based on intuitive optimism and competitive dynamics.

Human-driven VCs demonstrate significant variability in their bidding behaviors, frequently leveraging intuition and sentiment to place strategically higher bids, thus enhancing their competitive advantage in securing key deals. The sensitivity analysis specifically highlighted how market sentiment strongly influences human VC outcomes, with positive sentiment substantially increasing human VC bidding levels and resultant market participation. Conversely, AI-driven VCs demonstrate negligible responsiveness to market sentiment fluctuations, indicating algorithmic rigidity in bidding strategies.

In terms of investment outcomes, the precision of AI-driven bids translated into relatively stable but often conservative returns, whereas human-driven VCs exhibited greater volatility in returns due to the variability and optimism embedded in their decision-making processes. Over multiple simulation runs, this translated into scenarios where AI-driven VCs, despite their superior analytical rigor, regularly missed out on key investment opportunities due to systematic underbidding in competitively intense environments.

DISCUSSION AND IMPLICATIONS

Overall, our study illuminates an important set of insights for research on the integration of AI in venture capital and entrepreneur decision-making. The integration of AI in VC decision-making, while valuable for reducing certain biases and enhancing valuation precision, must be carefully balanced with human intuitive judgment, particularly in environments characterized by high levels of uncertainty, market volatility, and competitive bidding pressures. Hybrid decision-making approaches that combine AI-driven analytical insights with human intuition appear necessary for enhancing venture capital investment outcomes.

Our research also highlights the complexities of adopting AI-driven methods in competitive environments. The ability of AI systems to evaluate the objective valuation of startups must be complemented by human judgment to adequately identify and account for dynamic changes in the environment. VCs that successfully integrate AI analytics with human intuition are likely to achieve superior investment results by capturing the benefits of data-driven valuation assessments with human intuition.

At the same time, our research also highlights the critical challenges 'competitive recursion' will increasingly play in AI-influenced markets. In a recent study, Townsend and colleagues (2024a) argue that the growing use of AI tools in entrepreneurship can exacerbate the problems of '*competitive recursion*' in decision environments. Our study sheds further light on these problems by illuminating how AI limits the bidding behaviors of AI VCs in a manner that limits their ability to successfully invest in new startups. In fundamental ways, the use of AI tools can enhance the sophistication and accuracy of VC investment behaviors but at what cost? Policymakers and regulators must consider these dynamics as AI becomes increasingly prevalent in financial markets. Understanding the limitations and strengths of AI-driven decision-making can guide appropriate regulations and standards, ensuring AI adoption enhances market efficiency without inadvertently diminishing competitive dynamics or introducing systemic risks.

Lastly, the findings contribute significantly to the theoretical understanding of decision-making under conditions of Knightian uncertainty, highlighting the necessity of balance between analytical rigor and strategic intuition. Future research should continue to explore optimal integration strategies, the psychological dimensions of human-AI collaboration, and longitudinal impacts of AI integration on market outcomes. Hybrid approaches that integrate human intelligence and AI will provide useful complementary benefits that enhance the competitiveness of VCs and entrepreneurs to contend with the problems of Knightian uncertainty.

CONTACT: Mohammad Rady; mrady@vt.edu; Virginia Tech; 880 West Campus Drive, Blacksburg, VA 24061, United States

SUMMARY

WHEN STORIES DO NOT MATTER: AN EXAMINATION OF LEGITIMATE DISTINCTIVENESS IN POOLING EQUILIBRIA

Jorge Arteaga-Fonseca, Mississippi State University, United States
Matthew Rutherford, Oklahoma State University, United States

Principal Topic

If investors cannot perceive differences in quality among ventures, how can they decide which ventures to invest in? The fact that signals create a separating equilibrium is critical, yet the existence of different types of equilibria has not been fully appreciated by entrepreneurship scholars (Bergh et al., 2014). By marrying signaling theory with the legitimate distinctiveness literature, we argue that a pooling equilibrium—a context wherein investors cannot discriminate higher quality signalers from lower quality signalers—will limit the benefits of differentiation as a resource acquisition strategy. The well-established legitimate distinctiveness paradigm (Navis & Glynn, 2011) outlines how new ventures need to conform (for legitimacy), but also must differentiate (for competitive advantage). This results in entrepreneurs transmitting conformity signals that align their venture with established norms, values, beliefs; while also transmitting differentiation signals to separate the venture from those established norms, values, beliefs. Therefore, our baseline hypothesis is that legitimate distinctiveness (i.e., conformity and distinctiveness claims) will increase the likelihood of private-equity investments. However, we argue that the benefits accrued from legitimate distinctiveness will hold in a separating equilibrium where high-quality ventures stand out; but in a pooling equilibrium different dynamics are at play. In pooling equilibrium, investors cannot make sense of, nor perceive, the value of differentiation signals. This is because differentiation signals have no informational value to separate legitimately distinctive ventures from others. Thus, all signalers are pooled despite some having superior quality.

Method

We employ an experimental vignette design. Participants are placed in the screening stage of an investment decision, and 4-items measure the likelihood of moving to evaluation stages. We test the moderation of signaling equilibria (pooling vs. separating equilibria) on high/low levels of conformity and distinctiveness, and funding decisions. We test this on 321 financial professionals from Prolific.co (e.g., Heupel et al., 2024). We find a significant positive relationship between distinctiveness and conformity on private equity funding. Although we theorized a negative relationship between distinctiveness and access to resources in a pooling equilibrium, this is not significant. However, the positive relationship between conformity and access to resources is significant in a pooling equilibrium.

Results and Implications

Our logic runs counter to legitimate distinctiveness prescriptions. This study introduces pooling equilibria into the entrepreneurial process to "explain a broader range of social and organizational phenomena" (Connelly et al., 2011: p. 56). Our study also relaxes the tension in the conformity versus distinctiveness paradox contributing to the integration of institutional theory and strategic management.

CONTACT: Jorge Arteaga-Fonseca; jorge.arteaga@msstate.edu; Mississippi State University; 100 Maroon Dr., Starkville, Mississippi 39759, United States

SUMMARY

INVESTIGATING HEURISTIC DECISION-MAKING STRATEGIES OF INDIVIDUAL GOVERNMENT VENTURE CAPITAL INVESTORS IN EARLY-STAGE START-UP INVESTMENTS

Gillian Barry, The Technological University of the Shannon, Ireland
Patrick Lynch, The South East Technological University, Ireland
William O'Gorman, The South East Technological University, Ireland

Principal Topic

Government Venture Capital (GVC) funds play a crucial role in addressing funding gaps for early-stage startups that struggle to attract private investment due to high uncertainty and commercialisation challenges. Unlike venture capitalists and other private investors, who primarily seek financial returns, GVC Investors must balance commercial viability with broader public policy objectives, such as fostering innovation, job creation, and economic development. Despite their significance, limited research has examined how individual GVC Investors make decisions. This study explored their heuristic decision-making strategies - cognitive processes that simplify decision-making under uncertainty - focusing on the cues that trigger heuristics, the types of heuristics employed, and how they support investment decisions.

Method

This study employed a qualitative research design using Verbal Protocol Analysis (VPA) and in-depth interviews to investigate the decision-making processes of GVC Investors. Twelve experienced GVC Investors participated, each with over five years of experience assessing early-stage startups for government-backed programmes. VPA captured real-time cognitive processes as investors evaluated anonymised business plans, identifying key cues and heuristics. Semi-structured interviews followed to explore decision-making themes further. In total, the study generated nearly 21 hours of recordings, which were transcribed and analysed using thematic analysis to classify heuristic applications, focusing on search, stopping, and decision rules within the adaptive toolbox framework.

Results and Implications

Findings reveal that GVC Investors consistently rely on heuristics to streamline decision-making. Fast and Frugal Decision Trees (FFTs) and satisficing strategies enabled investors to prioritise key cues - such as the entrepreneur's capabilities, value proposition, financials, and socio-economic alignment - while streamlining their evaluation process. These cues were interconnected rather than assessed in isolation. For example, financial projections often triggered further scrutiny of the entrepreneur's ability to execute their plan. Investors used satisficing to set minimum acceptable thresholds for key cues. Once these thresholds were met, decisions advanced; however, if a threshold was not met, this could trigger a review of other cues or influence the decision to reject the proposal. These strategies enabled efficient, nuanced evaluations. This study challenges the assumption that GVC investment decisions rely solely on structured models, emphasising heuristics as adaptive tools that improve decision efficiency. Practical implications include enhancing decision environments, integrating fast and frugal heuristics into investor training, and offering clearer guidance for entrepreneurs. By recognising how GVC Investors apply heuristics, policymakers can enhance investment effectiveness, while entrepreneurs can tailor their applications to align with investor decision-making strategies, ultimately improving funding outcomes.

CONTACT: Gillian Barry; Gillian.Barry@TUS.ie; The Technological University of the Shannon; Tígrua, Kilmore, Clare V94D9P3, Ireland

SUMMARY

IN PURSUIT OF JUSTICE: POTENTIAL INVESTOR RESPONSES TO THE UNDERDOG NARRATIVE IN NEW VENTURES

Mark T. Bolinger, Appalachian State University, United States
Katrina Brownell, Virginia Tech University, United States

Principal Topic

Early ventures often rely on entrepreneurial narratives to attract resources and investors. While positive narratives highlighting a venture's potential are most common, some ventures use "underdog" narratives, portraying themselves as disadvantaged yet striving against the odds. Such narratives are counterintuitive to both intuition and established theories. However, research across a variety of fields has demonstrated a human tendency to support perceived underdogs. We aim to explore whether underdog narratives enhance resource provision in the entrepreneurial context, and specifically explore characteristics of resource providers that may lead them to support these ventures.

Method

Across a pilot (N = 185) and two full studies (N = 250, 124), we conducted a series of randomized experiments in which we manipulated perceptions (via their narrative) of a nascent venture as either a "favorite" (i.e., a venture with resource abundance) or an "underdog" (i.e., a venture with resource deficiencies), then measured respondents' investment intentions. We varied respondent's personal risk, and considered their views of the justice of the world. In our follow-up studies, we manipulated their perceptions of the justice of the venture's situation and feelings of schadenfreude (i.e., feelings of pleasure at another's misfortune) toward a venture.

Results and Implications

We found a negative main effect of the underdog narrative on investment intensions, but importantly, also identified boundary conditions based on potential investor beliefs. Personal risk to potential investors negatively moderated intended investment, while potential investor justice beliefs acted as a positive moderator. Specifically, investors who considered the world to be generally unfair were more likely to invest in underdog ventures while those who perceived general fairness were more likely to invest in favorite ventures. A subsequent study confirmed that intended support corresponded to justice perceptions, with respondents apparently wishing to restore justice. Intended investment was increased for favorite ventures that deserved their status compared to those that didn't, and for underdog ventures that didn't deserve their status (i.e., were unfairly disadvantaged) compared to those that did. Furthermore, there was no difference between deserved favorites and undeserved underdogs, and intended investment was even marginally higher for undeserved underdogs than undeserved favorites. This suggests underdog narratives may be superior to favorite narratives in some situations. Finally, underdog ventures experienced less schadenfreude than favorites, but this had no effect on intended investment. Our studies contribute to a rapidly expanding area of the literature examining how entrepreneurs might best facilitate resource acquisition through manipulation of entrepreneurial narratives and audience perceptions.

CONTACT: Mark T. Bolinger; bolingermt@appstate.edu; Appalachian State University; 416 Howard St., Boone, North Carolina 28607, United States

SUMMARY

AN EXPLAINABLE MACHINE LEARNING APPROACH TO EVALUATING FOUNDER ATTRIBUTES AND THEIR IMPACT ON FUNDING SUCCESS ACROSS STARTUP GROWTH STAGES

Joyce Ann Clarize Galang, Technical University Munich, Germany
Michael Mansfeld, Technical University of Munich, Germany

Principal Topic

Founders are frequently viewed as pivotal to a startup's trajectory, shaping product strategy, financial decisions, and market expansion. Recent ML-based studies (e.g., Maurer et al., 2024; McCarthy et al., 2023) have primarily focused on predictive accuracy for startup outcomes, often overlooking how different founder attributes—such as social capital, industry expertise, and personality—vary in importance across funding stages. This study addresses that gap by examining a dataset of 183,410 solo founders, thereby quantifying which characteristics contribute most to startup success across different venture funding stages.

Method

Gradient Boosting Models serve as the primary predictive tool, operating as binary classifiers that uncover complex feature interactions in predicting funding success across multiple investment stages. "SHapley Additive exPlanations (SHAP)" are then used to interpret these predictions, offering insights into the relative importance of each founder attribute at distinct points in a venture's lifecycle. By pairing structured commercial data with text-derived features from Large Language Models, the study constructs a robust pipeline that highlights how each set of attributes evolves from the early seed rounds to late-stage funding events.

Results and Implications

Our analysis reveals that social capital and adaptability matter most in early-stage fundraising, reflecting the importance of trust-building and resourcefulness among nascent ventures (e.g., Pakura et al., 2020; Park & Kim, 2023). By contrast, execution capacity and specialized expertise increasingly drive success in later rounds, aligning with prior evidence that organizational elements and innovation capacity gain prominence over time (Argaw & Liu, 2024). By integrating qualitative insights from textual data, the "Founder-Stage-Fit Model" also demonstrates enhanced predictive accuracy, thereby supporting investors who seek granular, data-driven insights rather than relying on subjective judgments. From a theoretical standpoint, these findings extend current entrepreneurship research by confirming that founder attributes do not retain a static level of influence across the venture lifecycle. Instead, they interact with stage-specific funding criteria, shaping both investor decisions and the strategic paths founders pursue. In practice, venture capitalists may deploy this framework to strengthen their due diligence processes and to fine-tune deal sourcing, while entrepreneurs can use the resulting insights to cultivate or emphasize those attributes most critical for achieving their next funding milestone. Overall, the study promotes a more transparent and evidence-based approach to founder evaluations and investment decision-making.

CONTACT: Joyce Ann Clarize Galang; joyce.acdg@gmail.com; Technical University Munich; 142 Herzogstrasse, Munich, Bayern 80797, Germany

SUMMARY

THE RIGHT LANGUAGE AT THE RIGHT TIME: ENTREPRENEURS' USE OF CODESWITCHING TO MOBILIZE FINANCIAL RESOURCES

Sheila Guo, University of Oklahoma, United States
Matthew Wood, University of Oklahoma, United States

Principal Topic

A key mechanism for entrepreneurs to garner investor support is through narratives (Lockwood et al., 2019). While existing research has primarily focused on identifying types of narratives, it sidelines the audience and leaves narrative interactions underexamined (Martens et al., 2007). Drawing on codeswitching (Nilep, 2006) and audience design theory (Bell, 1984), we propose that alignment between the entrepreneur's use of narrative codes and investor's analytic or intuitive cognitive preference is positively related to funding outcomes. Specifically, the quicker the entrepreneur switches to match the investor's cognitive style, the more likely funding is secured. However, we argue that excessive codeswitching will reduce that likelihood.

Method

We test our hypotheses using data from Shark Tank (247 entrepreneurs and 1,235 investment decisions). The unit of analysis is entrepreneur-investor narrative interactions. The independent variable, narrative code alignment, is content-analyzed using Linguistic Inquiry and Word Count's "cognitive process" and "affect" categories to capture frequency of analytical and intuitive narratives. Code (mis)alignment is calculated as the absolute difference between the entrepreneur's and investor's scores. We also analyze entrepreneur's response to each individual investor question, constructing entrepreneur's narrative sequences ("I I A I A I"). From this, we derive speed of first switch and frequency of switches. The dependent variable is a binary measure of funding outcome. We employ multilevel logistic regression, controlling for prototype stage, industry, and revenue.

Results and Implications

Results indicate a significant positive relationship between entrepreneur-investor intuitive code matching and funding likelihood, such that each unit increase in intuitive code matching is associated with a 25.6% increase in the odds of securing an offer (odds ratio=1.256, $p < 0.001$). By contrast, matching of analytical code did not have a statistically significant effect (odds ratio=1.060, $p=0.121$). This implies that the type of code alignment matters. Code alignment is less influential when that alignment is analytical. Codeswitching frequency is positively associated with offer likelihood (odds ratio=1.134, $p=0.045$). Each instance of codeswitching led to a 13.4% increase in the odds of receiving an offer, indicating that, contrary to our hypothesis, switching between codes resonated with investors. Finally, speed of the first switch did not affect the likelihood of securing an offer (odds ratio=1.025, $p=0.571$). Our study advances the literature by portraying entrepreneurs' narratives as bi-directional rather than one-way communications. Entrepreneurs' frequent codeswitching to align with investors' analytical and intuitive preferences is advantageous, yet when isolated, alignment with intuitive preferences had a significant positive effect on the likelihood of a funding outcome.

CONTACT: Sheila Guo; shengjie.guo-1@ou.edu; University of Oklahoma; 2657 Classen Blvd, Norman, Oklahoma 73071, United States

SUMMARY

TRANSITIONING FROM CURATION TO CHAOS: HOW BACKSTAGE MEDIA DEPICTIONS INFLUENCE NEW VENTURE EXECUTIVE OUTCOMES

Michael D. Hudecheck, University of St. Gallen Institute of Responsible Innovation, Switzerland
Charlotta Sirén, University of St. Gallen, Switzerland

Principal Topic

The emergence of a "post" social media landscape continues to alter the media ecosystems in which new venture founders, CEOs, and other venture executives operate (Seigner, Milanov, Lundmark, & Shepherd, 2023). However, little research has been conducted to explore how new venture executives engage with and manipulate transitional media ecosystems to accrue career-related benefits. Our study, theoretically grounded in the rich organizational literatures on mediatization and dramaturgical practices, explores how and why new venture executives are increasingly portrayed in the media through depictions of "dark" psychological and provocative traits that were historically associated with "backstage" behaviors and representations. More explicitly, we demonstrate a growing shift in new venture executive media coverage from carefully curated "frontstage" portrayals to uncurated "backstage" depictions. We then show that, while this phase transition stems from changes in the media landscape, media-savvy new venture executives are increasingly exploiting the expectations of this post-social media environment for their own ends. In doing so, we contribute to the literatures on reputation, mediatization, and dramaturgical practices.

Method

We explore our theorizing through an in-depth analysis of the media strategies employed by new venture executives, including venture founders, CEOs, and top management team members. First, we collect and curate a dataset of 100K+ full-length online media articles and social media posts from 2015 to 2025. We then employ LLM-assisted content reasoning analysis to establish content scores for Dark Factor of Personality traits (Moshagen, Hilbig, & Zettler, 2018) and a comprehensive list of BIG-5 NEO-PI-R first- and second-order psychological traits, which we aggregate to operationalize Frontstage and Backstage Media Depictions. We incorporate independent variables to operationalize Media Volume, Mediatized Positioning, and Mediatized Responsiveness, as well as dependent variables for Startup Outcomes and Executive Career Outcomes, alongside control variables. We test our hypotheses through mediation analysis and validate our results with robustness and comprehension checks.

Results and Implications

Our results suggest a media coverage shift from frontstage depictions to backstage depictions, which we attribute to the ongoing changes in the media landscape. Our results also suggest a positive relationship between new venture executive media engagement and positive executive career outcomes. In doing so, our results underscore the pivotal role of entrepreneurial venture mediatization practices and provide practical insights into how new venture executives can effectively navigate the transitional media landscape to their advantage.

CONTACT: Michael D. Hudecheck; michael.hudecheck@unisg.ch; University of St. Gallen Institute of Responsible Innovation; Dufourstrasse 40a, St. Gallen, Sankt Gallen 9000, Switzerland

SUMMARY

HOW SIGNALS INFLUENCE ARTIFICIAL INTELLIGENT STARTUPS' FUNDRAISING: A CONFIGURATIONAL APPROACH

Nguyen Thao Khuat, EBS Universität für Wirtschaft und Recht, Germany
Katrin Burmeister-Lamp, EBS Universität für Wirtschaft und Recht, Germany
Matthias Waldkirch, EBS Universität für Wirtschaft und Recht, Germany

Principal Topic

The rapid development of Artificial Intelligence (AI) has reshaped the startup ecosystem but introduced significant challenges in securing investments due to AI's complexity and opacity. Investors often face difficulties in assessing the value and potential of these ventures, leading to heightened information asymmetry. This study explores how AI startups overcome these barriers using signaling theory, focusing on two types of signals: substantive signals – includes AI Complexity, Founding Team Experiences, and Professionalization – and rhetorical signals – consists of Content Complexity, Persuasive Language, and Display of Document. The interplay between these signals is examined to uncover strategies for achieving substantial fundraising amount.

Method

The research employs a configurational approach using Qualitative Comparative Analysis (QCA) to identify pathways that lead to substantial fundraising amount. The study uses data from 215 AI startups that conducted Initial Coin Offerings (ICOs), a novel and information-asymmetry-intensive financing context. Data sources include whitepapers for textual and visual analysis of signals and ICO aggregators for quantitative data on fundraising amounts, team characteristics, and professionalization. QCA was applied to analyze the combined effects of substantive and rhetorical signals on fundraising success.

Results and Implications

The findings reveal three distinct pathways to successful fundraising, each illustrating the diverse strategies startups can adopt. The first pathway, AI-Driven Pioneers, underscores the dominant role of substantive signals. Startups with advanced AI capabilities, highly experienced founding teams, and professionalized processes consistently achieve their fundraising goals. In this configuration, rhetorical signals such as well-structured documents act as enhancers, strengthening the impact of substantive signals. The second pathway, Narrative Leaders, highlights the power of rhetorical signals. Startups with compelling content complexity, persuasive document presentation, and experienced teams can achieve substantial investment even without relying heavily on advanced AI technologies. The third pathway, non-AI Drivers, shows that startups can succeed by focusing on non-technological strengths, such as professionalized operations, effective communication, and team expertise, demonstrating that AI complexity is not always a prerequisite for attracting significant funding. This study contributes to entrepreneurial and signaling theory by demonstrating how different configurations of substantive and rhetorical signals influence investment outcomes, underscoring the principle of equifinality. It challenges the assumption that technological sophistication is the sole driver of investor interest, showing that leadership, communication, and professionalization are equally critical. These insights also offer actionable strategies for AI startups to navigate high-uncertainty fundraising environments, emphasizing the importance of tailoring signaling strategies to align with unique strengths.

CONTACT: Nguyen Thao Khuat; thaonguyen.khuat@ebs.edu; EBS Universität für Wirtschaft und Recht; Münchner Str. 12, Freising, Bayern 85354, Germany

SUMMARY

DON'T BLEED WHEN SHARKS BITE: HOW CONTROLLING NONVERBAL DISPLAY OF EMOTIONS MATTERS IN ENTREPRENEURIAL PITCHING

Jialu Ma, Temple University Fox School of Business, United States

Principal Topic

While conventional wisdom, rooted in communication theory (Mehrabian, 1971), suggests that nonverbal cues (i.e., facial expressions and voice tones) dominate in conveying the emotions that shape audience evaluations, recent entrepreneurship research presents a more nuanced picture on the advantages or disadvantage of nonverbal emotional display for attract investment. Some studies have found positive effects of nonverbal emotional cues on investment outcomes (Jiang et al., 2019; Hu & Ma, 2021), while others have revealed more complex relationships, including inverted-U effects of positive facial expressions (Warnick et al., 2021), negative effects of emotionality of voice tones (Wang et al., 2021), and even non-significant effects of nonverbal enthusiastic cues in conveying passion (Chen et al., 2009). This apparent contradiction in findings has created a significant tension in the literature regarding the comparative advantage of employing nonverbal emotional cues versus verbal emotional cues for funding success. We address this tension by proposing that nonverbal emotional cues matter in shaping investment decisions, but their comparative advantage relative to verbal emotional cues depends on context: presentation versus questioning-and-answering (Q&A).

Method

To empirically test our arguments, we use state-of-the-art deep learning models to measure facial expressions, voice tones, and text sentiment from videos of entrepreneurial pitches featured on the Shark Tank. For each startup, we extract video clips of not only entrepreneurs' presentations but also their responses to angel investors in the Q&A stage. The outcome we predict is the probability of gaining investment offers after the presentation and Q&A as a salient indicator of pitching success.

Results and Implications

We find that nonverbal cues of emotional expression indeed have a comparative disadvantage in presentations. In Q&A, nonverbal cues are advantageous for expressing low-arousal emotions but disadvantageous for high-arousal emotions. These findings contribute to entrepreneurial pitching literature through a holistic and nuanced framework to study emotional display strategy, by directly comparing verbal and nonverbal emotional cues, examining contextual differences in cue effectiveness, and conceptualizing emotions through the lens of arousal beyond valence. For practitioners, our central message is that maintaining composure of emotional expression throughout the investor engagement process is crucial for building trustworthiness and legitimacy in entrepreneurial resource acquisition.

CONTACT: Jialu Ma; jialuma@temple.edu; Temple University Fox School of Business; 1500 N 15th St, Philadelphia, Pennsylvania 19121, United States

SUMMARY

THE MODERATION INFLUENCE OF ACCELERATORS ON STARTUP'S SIGNALING PROCESSES AND RESOURCE ACQUISITION

Jorge-Vinicio Murillo-Rojas, INCAE Business School, Costa Rica
Jan Brinckmann, ESADE Business School, Spain

Principal Topic

External equity investment is crucial for early-stage firm growth and survival (Cassar, 2004), yet few startups secure such funding due to information asymmetry and uncertainty (Sanders & Boivie, 2004; Wiklund et al., 2010). Investors use signals to assess venture quality, as signaling theory suggests these proxies help mitigate information gaps (Spence, 1973). However, signal effectiveness varies (Ahlers et al., 2015). Affiliation with third-party organizations, such as accelerators, might play a crucial role in improving how effectively a startup's quality is communicated to potential investors (Connelly et al., 2011). This study empirically investigates the moderating role of accelerators in improving signal effectiveness for resource acquisition.

Method

This study utilizes data from Emory University's Entrepreneurship Database Program, supported by GALI, to analyze how different signals influence external equity investment acquisition and whether acceleration programs moderate these effects. It examines economic (e.g., prior equity investment), social (e.g., philanthropic funding), innovation (e.g., intellectual property ownership), and founder quality signals (e.g., experience and education). The two-stage Heckman selection method (Heckman, 1979) with a probit model is employed, complemented by average marginal effects estimation and contrast analysis to assess interaction significance (Hoetker, 2007; Norton et al., 2004; Yang et al., 2020).

Results and Implications

Our study suggests that accelerators support new ventures not only by providing resources and services but also by enhancing the effectiveness of signals used to attract investment. Acceleration programs endorse signal quality, train entrepreneurs to communicate signals more effectively, facilitate opportunities for signal transmission, and increase signal exposure. Our findings indicate that accelerators moderate signal effectiveness in startups' resource acquisition processes. Specifically, when firms send economic, social, and innovation signals, the likelihood of acquiring equity investment increases by 7.9% ($p < 0.001$), 2.7% ($p < 0.05$), and 2.5% ($p < 0.05$) more, respectively, if they participated in acceleration programs. Additionally, the effect of management experience and graduate degree signals on equity investment probability is 2.3% ($p < 0.05$) and 3.0% ($p < 0.01$) higher, respectively, for accelerated firms. However, no significant moderation effect was found for entrepreneurial experience as a signal. Robustness tests conducted to check for endogeneity bias provided no evidence of endogeneity. This study contributes to entrepreneurship research by demonstrating that accelerators enhance signal effectiveness in attracting investors. While prior research has examined accelerators' direct impact on startups, this study highlights their moderating role. It also expands on existing literature by proposing mechanisms through which accelerators strengthen signals in diverse geographic and industry contexts.

CONTACT: Jorge-Vinicio Murillo-Rojas; jorge.murillo@incae.edu; INCAE Business School; Campus INCAE Business School, La Garita, Alajuela 00000, Costa Rica

SUMMARY

UNPACKING OWNERSHIP COMPETENCE IN PRIVATE EQUITY: INSIGHTS FROM FAMILY OFFICE INVESTMENTS

Jeroen Verbouw, IE Business School, Spain
Cristina Cruz, IE Business School, Spain
Juan Santaló, IE Business School, Spain

Principal Topic

Family offices (FOs) have become increasingly prominent in private equity markets, managing over $5.9 trillion in assets and playing an active role in direct investments. Despite their growing influence, little is known about how FO ownership affects portfolio firms' strategic decisions. Existing research has extensively examined the effects of private equity (PE) and venture capital (VC) investments, but FOs differ from these investors in key ways. This study examines how FO ownership influences the competitive strategies of portfolio firms, particularly the trade-off between growth and profitability. FOs' unique characteristics—including long investment horizons, idiosyncratic investment behavior, and a cautious risk approach—grant them distinct ownership competences. We argue that these characteristics shape firms' strategic trajectories, prioritizing sustainable profitability over aggressive expansion.

Method

We construct a large-scale matched sample of 1,482 FO investments and 2,772 comparable investments by PE and VC firms in European companies between 1994 and 2023. Using heterogeneous difference-in-difference models, we analyze how FO ownership impacts firm growth and operational efficiency. Our dataset integrates investments from Pitchbook with firm-level financial data from Orbis, allowing us to track pre- and post-investment performance. To mitigate selection bias, we apply multivariate nearest-neighbor matching based on investor type, investment preferences, and firm characteristics.

Results and Implications

First, our findings confirm that FOs exhibit distinct investment characteristics: (1) FO-backed firms remain under FO ownership 35% longer than PE/VC-backed firms, reflecting longer investment horizons; (2) FOs have idiosyncratic investment behavior and deviate more from market investment patterns, as evidenced by significantly higher cosine distances relative to traditional investors; and (3) FO-backed firms have 6% lower financial leverage, indicating a more cautious risk approach. Second, these characteristics translate into a distinct strategic trajectory for portfolio companies. Compared to PE and VC-backed firms, FO-backed firms experience 21% lower growth but a 12% increase in operational efficiency, demonstrating a preference for profitability-first strategies. These effects persist throughout the investment period and are particularly pronounced in FO-led private equity investments. Our study contributes to the strategic governance literature by demonstrating how ownership identity shapes firm decision-making and its outcomes. FOs influence strategic priorities differently than traditional investors and foster long-term, stable value creation over high-risk, high-growth models. In doing so, this research informs entrepreneurial finance by highlighting the heterogeneous effects of different investor types on firm performance. As FOs continue expanding their influence in private markets, understanding their unique governance approach is critical for both scholars and practitioners.

CONTACT: Jeroen Verbouw; Jeroen.Verbouw@ie.edu; IE Business School; Paseo de la Castellana 259, Madrid, Madrid 28046, Spain

SUMMARY

CROSSING THE VALLEY OF DEATH IN SCALE ENTREPRENEURSHIP: A DIGITAL ENTREPRENEURIAL ECOSYSTEM PERSPECTIVE

Qiongrui (Missy) Yao, York College of Pennsylvania, United States

Principal Topic

The purpose of this paper is to shed light on the processes through which scale-economy entrepreneurs cross the valley of death (Wilson et al., 2018). From the perspective of digital entrepreneurial ecosystem (EE) (Sussan & Acs, 2017), we address the research question "How do scale-economy entrepreneurs receive high venture capital (VC) valuations within their national EEs?" Drawing on signaling theory, the adoption of a digital platform is a signal of quality and scalability an entrepreneur would send to VCs to attract funding. We argue that this digital platform adoption-VC valuation relationship is moderated by the venture's operating sector, embedded national EE quality, and home country digitalization.

Method

Scale entrepreneurship is characterized by the rise of unicorns (Audretsch et al., 2021). Accordingly, unicorns represent an appropriate empirical setting for this paper. We collected unicorns from the Crunchbase Private Unicorn Company List including a sample of 1000 unicorns. There were 109 unicorns not funded by VC and thus excluded from the sample. The final sample consisted of 891 unicorns. We ran OLS regressions to test our hypotheses.

Results and Implications

Hypothesis 1, which suggests the positive association between a venture's digital platform adoption and VC valuation, is supported. If a venture adopts a digital platform, its valuation will increase by $2.021 billion on average. Hypothesis 2, predicting that a venture's operation in sectors with higher media attention strengthens the positive relationship between digital platform adoption and VC valuation, is not supported. We find statistically significant support for hypothesis 3 that a venture's embedding in a high-quality national EE will strengthen the positive effect of digital platform adoption on venture valuation. Hypothesis 4, suggesting that a venture's location in a country with a high degree of digitalization strengthens the positive relation between digital platform adoption and venture valuation, is not supported. Interestingly, we find that digital platform ventures located in countries with higher degrees of digitalization receive lower VC valuations. US unicorns account for approximately 55% of the sample. According to the Digital Platform Economy (DPE) Index 2020, the US received the highest DPE score. However, US unicorns received a wide range of VC valuations from $1 billion to $125 billion. A number of US unicorns just had a $1 billion valuation to reach unicorn status (Kotha et al., 2022). The dominance of US unicorns and the practice of making valuation adjustments to enable a handful of ventures to obtain unicorn status might contribute to this finding.

CONTACT: Qiongrui (Missy) Yao; qyao@ycp.edu; York College of Pennsylvania; 441 Country Club Rd, York, Pennsylvania 17403, United States

INTERACTIVE PAPER

AN UPHILL BATTLE: HOW VENTURES USE LENIENCY AND SUCCESSIVE SHIFTS TO MAINTAIN SUPPORT OF EXTERNAL RESOURCE PROVIDERS AMIDST PERSISTENT DISAPPOINTING PERFORMANCE

Mohammed Bendaanane, Technical University Munich, Germany

Principal Topic

Cultural entrepreneurship portrays entrepreneurs as skilled cultural actors (Lounsbury & Glynn, 2001; Lounsbury et al.,2019) who use stories to gain financial resources from external resource providers (Martens et al., 2007; Fisher et al., 2021). However, stories also set future expectations (financial returns) and ventures that fail to meet these expectations must manage the challenges that arise from inability to meet expectations. Existing research has coined several mechanisms to explain how ventures manage shifts and struggles (Hampel et al., 2020; McDonald and Gao, 2019; Garud et al., 2014). However, extant research has often focused on time-bound, single-event obstacles and fall short in explaining how ventures facing successive struggles can still maintain support from external resource providers. I explore this puzzle and ask: how do ventures maintain support from external resource providers in the face of persistent underperformance?

Method

To address this question, I focus on an unusually revelatory of SolarGrid, a social venture founded in 2011 with the goal of profitably bringing electricity to unnelectrified rural areas in Kenya. SolarGrid maintained a constant flow of financial resources with four consecutive VC rounds ($70m). However, the venture struggled financially and was unable to meet the expectations of its funders. Based on a data corpus of 369 single-spaced pages of interviews (16) and extensive archival data, I develop a tentative grounded process model to explain this phenomenon.

Results and Implications

I find that, consistent with prior literature, ventures utilize revised storytelling to maintain support (Garud et al.,2014; Lounsbury et al.,2019). However, the revised stories may not be sufficient (Garud et al.,2024; Garud et al.,2014). My theory posits that ventures can maintain support from external resource providers by pursuing two tactics: lenient organizing and successive shifting. First, lenient organizing entails spreading ventures across geographies and identities. The cross-geographic foothold allow ventures to tap into different resources that would have been harder to access if ventures are based in one context. Similarly, identity leniency allows venture to use identities depending on the type of resource providers targeted and seek resource using multiple identities. Second, operating concomitantly with lenient organizing, successive shifting allows ventures to shift its target resource providers overtime by deploying different identities and geographies. My study contributes theoretically to the literature on cultural entrepreneurship (Lounsbury et al., 2019; Martens et al., 2007; Fisher et al., 2021) and managing shifts (Hampel et al., 2020; McDonald and Gao, 2019; Garud et al., 2014).

CONTACT: Mohammed Bendaanane; mohammed.bendaanane@tum.de; Technical University Munich; Lichtenbergstraße 6, Garching, Bayern 85748, Germany

INTERACTIVE PAPER

VENTURE CAPITAL AND THE FORMATION OF PERFORMANCE ASPIRATIONS

Brecht Depreitere, KU Leuven & Vlerick Business School, Belgium
Veroniek Collewaert, Vlerick Business School, Belgium
Jeroen Neckebrouck, IESE Business School, Spain

Principal Topic

The Behavioral Theory of the Firm (BTOF) suggests organizations set performance aspirations to evaluate current performance and guide strategic actions. A key aspect of BTOF is the speed at which managers update their performance aspirations, a concept applied to both established and entrepreneurial firms. A high updating speed indicates that aspirations are adjusted to reflect recent performance more heavily than more distant past performance. Surprisingly, much of the existing literature assumes a homogeneous updating speed across firms. We challenge this view by proposing that governance structures, particularly Venture Capital (VC) involvement, significantly influence the aspiration updating mechanism. Understanding these differences is crucial, as it provides insights on how VCs impact internal decision-making processes within their portfolio companies.

Method

We constructed a longitudinal dataset of 1,860 European VC-backed ventures. First-round investments were identified using Pitchbook and matched with Orbis Europe to obtain financial data. We then matched these ventures to 44,540 non-VC-backed ventures using coarsened exact matching (matches were determined based on country, industry, firm age, and firm size in the year before VC investment). Using bootstrapping, we estimated the distribution of aspiration updating speeds across four groups: VC- and non-VC-backed firms, both before and after the year of VC entry. We then ran mean comparison tests comparing VC- and non-VC-backed firms, and pre- vs. post-investment periods to assess VC influence on aspiration updating.

Results and Implications

Our findings confirm that aspiration updating speeds are not uniform and are significantly influenced by governance structures, particularly VC involvement. VC-backed firms adjust performance aspirations more rapidly than non-VC-backed firms when performance exceeds aspirations. Conversely, when performance falls below aspirations, VC-backed firms show a slower adjustment of aspirations than non-VC matched counterparts, i.e., maintaining higher performance standards rather than lowering expectations. These findings challenge BTOF's assumption of homogenous aspirations updating speeds across firms and highlight governance as a key factor in the aspiration formation process. Our study reveals how VCs shape strategic decision-making in their portfolio companies. Specifically, we provide insights into how entrepreneurs set targets for acceptable performance, and how this is influenced by external parties like VC investors. Furthermore, our study extends BTOF by introducing VC involvement as a boundary condition affecting the link between performance discrepancies and strategic change. Recognizing heterogeneity in aspiration updating not only refines BTOF's theoretical framework but also provides deeper insights into how VCs actively shape entrepreneurial firms' growth and strategic direction.

CONTACT: Brecht Depreitere; brecht.depreitere@vlerick.com; KU Leuven & Vlerick Business School; 36 Onze-Lievevrouwstraat, Leuven, Vlaams-Brabant 3000, Belgium

ENTREPRENEURS' PERCEPTIONS OF AI-SUPPORTED CREDIT ALLOCATION

Lina Uebbing, Heinrich Heine University, Germany
Lynn Schmodde, Heinrich Heine University, Germany
Marius Claus Wehner, Heinrich Heine University, Germany

ABSTRACT

Access to credit is essential for entrepreneurs, making human bias in lending a serious concern. Artificial Intelligence, as a potential solution, can create a fairer credit evaluation process by reducing human biases. Utilizing signaling theory, we examine (former) entrepreneurs' perceptions (N = 360) of interpersonal justice and innovativeness in different loan evaluation scenarios (human vs. AI-based vs. hybrid). Findings reveal a trade-off: AI-based evaluations were perceived as highly innovative, but reduced interpersonal justice perceptions—especially when combined with male bank specialist. Both perceived interpersonal justice and innovativeness significantly mediated the effect entrepreneurs' intention to pursue the credit allocation process, with interpersonal justice emerging as the stronger driver.

INTRODUCTION

Access to and availability of financing are essential for founding, growth, and the survival of any business on the market (Frimanslund et al., 2023, p. 372; Marlow & Patton, 2005, p. 717). Especially for entrepreneurs, acquiring financial resources is the primary concern (Cumming et al., 2019, p.253), because difficulties in access limit their venture's growth (Stefani et al., 2019, p. 320). For small businesses, bank lending is the key source of external financing (Berger & Udell, 2002, p. 36), underscoring the crucial role that credit institutions and their lending processes play in facilitating entrepreneurs' access to financial resources. However, the lending process, traditionally carried out by humans, is susceptible to biases that may affect fairness and transparency (Lukacik et al., 2022, p. 4).

In the past two decades, artificial intelligence (AI) has been integrated into various industries, including the financial sector (Bahoo et al., 2024, p. 2), and emerged as a potential solution to address the challenges in "critical decision-making contexts" (Narayanan et al., 2024, p. 4), offering the promise of more objective and equitable credit evaluation processes (Jauernig et al., 2022, p. 2; Kumar et al., 2023, p. 480). –AI systems have the potential to reduce discrimination–provided they have balanced training data sets that do not discriminate against certain groups (Köchling & Wehner, 2020, p. 840). For this reason, several banks have recently been considering the use of AI-supported systems for lending, such as ING-DiBa (Koch, 2023). Whereas AI is currently only being used in the retail banking sector, banks are now actively exploring further applications in the corporate banking sector (Koch, 2023; Sinß, 2022).

Despite the growing adoption of AI in banking to enhance predictive accuracy (Sadok et al., 2022, p. 2), it also brings challenges, including potential biases and the necessity for updated regulations to ensure the transparency and fairness of AI algorithms and the data utilized by financial institutions. However, there remains limited understanding of how those who are affected (e.g., entrepreneurs) perceive these innovations (Narayanan et al., 2024, p. 6). Specifically, it is unclear how the integration of AI into the lending process influences their perceptions of fairness (e.g., interpersonal justice) and the credit institution's perceived innovativeness. This gap is particularly relevant given the recent legislative focus on transparency in AI applications, which could affect entrepreneurs' trust in such systems (European Union, 2022).

The interaction between entrepreneurs and credit institutions has traditionally been examined from the perspective of which signals entrepreneurs send to lenders to secure funding (e.g., Zheng et al., 2021, p. 2). However, this study shifts the focus to investigate the opposite direction (Sturm et al., 2025, p. 11), to be precise, which signals lenders, specifically credit institutions, send to their potential entrepreneurial clients.

This perspective is particularly relevant in light of the evolving expectations of entrepreneurs, who increasingly evaluate potential financing partners based not only on financial terms but also on "value-based services" and "personal fit attributes" (Sturm et al., 2025, p. 22). Recent findings indicate that entrepreneurs are even willing to accept lower valuations when investors offer strategic benefits such as network access, industry expertise, and aligned value–underscoring a multidimensional approach to investor selection that goes beyond monetary considerations. These entrepreneurs seek partnerships with fair and trustworthy institutions that embrace innovative approaches, moving beyond traditional roles and practices (Sturm et al., 2025, p. 9).

This study draws on signaling theory (Spence, 1973) to explore the signals sent by credit institutions when integrating AI in credit allocation. We propose that there is a risk that the absence of human interaction in the process could diminish perceptions of interpersonal justice. However, by incorporating AI into their process, banks signal innovativeness to entrepreneurs. Since entrepreneurs are often open to and appreciative of innovativeness, this signaling might provide banks with a competitive advantage. In a competitive financial landscape increasingly influenced by alternative credit lines and fintech innovativeness (Hodula, 2024, p. 5), understanding how entrepreneurs perceive AI in the lending process is critical. Banks striving to maintain profitability and relevance must navigate the delicate balance between leveraging AI for efficiency and ensuring fairness and transparency in their interactions with entrepreneurial clients.

In this study, we aim to compare the preferences of entrepreneurs regarding the use of an AI-driven evaluation method compared to a human evaluation in a bank loan context. More specifically, we aim to examine how (a) entrepreneurs' perceptions of the integration of AI influence interpersonal justice and (b) how the integration of AI affects the perceived innovativeness of a credit institution. This inquiry underscores the interplay between technological innovation, entrepreneurs' perception, and the long-term sustainability of financial institutions.

HYPOTHESIS DEVELOPMENT

Signaling serves as a mechanism to influence audience perceptions through the transmission of informational cues (i.e., signals) from one entity, referred to as the *signaler*, to another, the *receiver*. Signaling, a concept widely explored in management literature (Taj, 2016, p. 339), has recently also gained attention in entrepreneurship research (e.g., Bafera & Kleinert, 2023, p. 2419). Building on prior research, which suggests that signals play a crucial role in shaping perceptions and decision-making (Connelly et al., 2011, p. 39), we apply signaling theory (Spence, 1973) to examine the signals credit institutions convey when integrating AI into their credit allocation processes.

In our study, we examine how different forms of AI integration in the credit loan process are perceived by entrepreneurs. We distinguish between the integration of AI in the pre-selection phase and the use of AI as the sole decision-maker. Additionally, we differentiate between male and female bank specialists. We analyze the entrepreneurs' reactions regarding perceived interpersonal justice and the perceived innovativeness of the bank, as well as how these factors influence their intention to pursue the process. This is particularly important because entrepreneurs may otherwise decide to abandon the process.

Interpersonal justice

"Asynchronous video interviews are a form of one-way, technology-mediated, selection interviewing that continue to grow in popularity" (Lukacik et al., 2022, p. 1). Studies in the human resource management context highlight that while AI-supported selection tools are often perceived as providing consistency across all applicants and enhancing accuracy by potentially mitigating human bias (e.g., Lukacik et al., 2022, p. 4), it remains crucial to examine how applicants subjectively perceive and evaluate the fairness of the selection process when such tools are employed by credit institutions (Narayanan et al., 2024, p. 3). In the context of human resource management, research indicates that AI-supported selection tools are often perceived more skeptically compared to human-based selection processes (Köchling et al., 2022, p. 2110). Given the similarities between human resource management selection processes and credit allocation—both of which involve the evaluation of competencies and characteristics—we anticipate comparable reactions in the entrepreneurial context. A potential explanation lies in the distinction between quantifiable and non-quantifiable factors. While hard facts such as qualifications and experience, assessed during initial stages (e.g., document screening), can be effectively evaluated by algorithms, more nuanced entrepreneurial traits—such as personality, leadership abilities, and attitudes—are challenging to quantify. This can lead applicants to feel that their unique human attributes are overly reduced to numerical data, fostering perceptions of unfairness in AI-supported processes compared to human evaluations.

Additionally, applicants may lack insight into how AI systems assign weights to various factors, particularly in asynchronous video interview scenarios. The opacity of these decision-making processes can exacerbate misunderstandings, further undermining perceptions of fairness. Moreover, a recent literature review by Narayanan et al., (2024) suggests broad agreement in the findings that AI-driven decision-making processes are generally perceived as less interpersonally fair. Thus, we propose that entrepreneurs may view AI-supported credit allocation processes as less fair than those conducted by human evaluators, especially with regard to interpersonal justice, which refers to the extent to which individuals are treated with dignity and respect (Narayanan et al., 2024, p. 6).

Furthermore, Schulte Steinberg & Hohenberger's (2023) study demonstrates that while individuals tend to favor human evaluators over AI evaluators, this preference is observed specifically when the human evaluator is female. Based on this, we hypothesize:

Hypothesis 1: Compared to a male credit specialist evaluation, (a) female credit specialist evaluation, is positively associated, while (b) AI pre-selection combined with a male credit specialist, (c) AI in pre-selection combined with a female credit specialist, and d) AI-based evaluation is negatively associated with interpersonal justice.

Perceived innovativeness

Innovation generally emphasizes the outcomes derived from introducing new elements or combining existing elements in novel ways within a firm's activities (Schumpeter, 1934). In contrast, the term innovativeness pertains to a firm's capacity to embrace and implement new ideas, services, or initiatives. Given that this study examines a credit institution's processes aimed at fostering novelty, the term innovativeness is deemed most appropriate. From the customers' perspective, innovativeness reflects their subjective perception of a business's ability to deliver novel and creative solutions (Kunz et al., 2011, p. 871). The existing body of literature suggests that the adoption of technological innovations has generated favorable outcomes for traditional banks, improving their operational performance, boosting profitability, and contributing to the reduction of credit risk (Lee et al., 2021, p. 470; Zhang et al., 2023, p. 2). Furthermore, the adoption of AI in lending decisions represents a significant innovation in the banking industry, particularly because its application in the retail sector is still in its early stages. The introduction of such technology to corporate banking marks a novel development, positioning banks that adopt AI as forward-thinking and technologically advanced.

Consequently, the use of AI in this context may lead to a higher perceived innovativeness among entrepreneurs. Entrepreneurs are generally characterized by their openness to innovation and their willingness to embrace novel solutions. Research indicates that individuals in entrepreneurial roles tend to value advancements that improve efficiency, accuracy, and transparency, all of which are key promises of AI-supported systems (Riaz et al., 2022, p. 153).

In sum, by utilizing AI, credit institutions may signal innovativeness to entrepreneurs. Given that entrepreneurs are often receptive to innovation themselves (Betta et al., 2010, p. 234), this signaling mechanism may provide credit institutions with a competitive advantage in attracting entrepreneurial clients.

Given their proclivity for welcoming innovative practices, entrepreneurs are likely to view the implementation of AI in lending processes positively and associate it with a higher level of perceived innovativeness. Based on this, we hypothesize:

Hypothesis 2: Compared to a male credit specialist evaluation, (a) female credit specialist evaluation, (b) AI pre-selection combined with a male credit specialist, (c) AI in pre-selection combined with a female credit specialist, and d) AI-based evaluation is positively associated with perceived innovativeness.

When applicants perceive the selection process positively, view the methods employed as fair, and experience favorable emotional responses throughout, they are more likely to develop a positive attitude toward the organization (Bauer et al., 2006, p. 606). Translated to our context, this would imply that a positive perception of the credit institution emerges under these conditions–and therefore, a higher intention to pursue the process. Conversely, dissatisfaction with the selection process can lead candidates to withdraw voluntarily from further consideration. Recent evidence from VC contexts states that entrepreneurs often terminate negotiations when facing "unfair valuations, lacking expertise, and excessive control measures that can derail financing negotiations" (Sturm et al., 2024, p. 27). Building on this, we propose that the interpersonal justice and perceived innovativeness of the process impact the intention to pursue the process, thereby mediating the relationship between the use of AI support in the credit allocation process and the credit institution's overall appeal. Based on this, we hypothesize:

Hypothesis 3: Interpersonal justice is positively related to the intention to pursue the process.

Hypothesis 4: Perceived innovativeness is positively related to the intention to pursue the process.

Hypothesis 5: Interpersonal justice mediates the relationship between evaluation type and intention to pursue the process.

Hypothesis 6: Innovativeness mediates the relationship between evaluation type and intention to pursue the process.

METHOD

To investigate entrepreneurs' perceptions of the use of an AI-driven evaluation method in a bank loan context, we utilized a between-subject design using five hypothetical scenarios as manipulation treatments (Aguinis & Bradley, 2014). Participants were randomly assigned to one of the five scenarios, each representing a unique loan application scenario by the credit institution. A manipulation check was conducted (Shadish et al., 2002, p. 331) following participants' exposure to one of the five vignettes to verify their comprehension of the scenarios.

Sample: n=360; participants who either currently run their own business or have in the past
DV = Intention to Pursue the Process
IVs/Treatments = Female bank specialist, AI pre-selection & male bank specialist, AI pre-selection & female bank specialist, AI-based evaluation; Reference category: male bank specialist
Mediators = Perceived Interpersonal Justice, Perceived Innovativeness
Moderator = Past Discrimination Experience
Controls = Age, Gender
Statistical Analysis = SEM, Moderator Analysis

RESULTS

We estimated a mediation model using lavaan in R. The analysis was conducted as a path analysis, with all constructs specified as manifest variables. In our analysis, we analyzed the effects of different evaluation types intention to pursue the process, mediated by perceived interpersonal justice and innovativeness.

With regard to perceived interpersonal justice, a female bank specialist had no significant impact on perceptions of interpersonal justice ($b = -.06$, $SE = .12$, $p = .338$), compared to the evaluation by male bank specialist, thereby rejecting *Hypothesis 1a.* An AI pre-selection combined with a male bank specialist significantly decreased perceptions of interpersonal justice ($b = -.25$, $SE = .12$, $p = <.001$), thereby supporting

Hypothesis 1b. An AI pre-selection combined with a female bank specialist significantly decreased perceptions of interpersonal justice ($b = -.19$, $SE = .12$, $p = .006$), thereby supporting *Hypothesis 1c*. An AI-based evaluation significantly reduced perceptions of interpersonal justice ($b = -.40$, $SE = .12$, $p < .001$), which supported *Hypothesis 1d*.

Regarding perceived innovativeness, the evaluation by a female bank specialist significantly decreased perceptions of innovativeness ($b = -.14$, $SE = .15$, $p = .013$), therefore, we had to reject *Hypothesis 2a*. An AI pre-selection combined with a male bank specialist significantly increased perceptions of innovativeness ($b = .39$, $SE = .15$, $p < .001$), which was also the case for AI pre-selection combined with a female bank specialist ($b = .43$, $SE = .15$, $p < .001$), which supported both *Hypothesis 2b* and *Hypothesis 2c*. An AI-based evaluation had a positive effect on innovativeness ($b = .47$, $SE = .15$, $p < .001$), thereby supporting *Hypothesis 2d*.

With regard to the effects of the mediators, interpersonal justice positively influenced intention to pursue the process ($b = .53$, $SE = .05$, $p <.001$). Innovativeness also had a significant positive effect ($b = .21$, $SE = .04$, $p = .001$), thereby supporting both *Hypothesis 3* and *Hypothesis 4*.

Finally, the indirect effects of evaluation type on intention to pursue the process through interpersonal justice revealed that AI pre-selection with a male bank specialist ($b = -.13$, $SE = .05$, $p = .001$) and AI pre-selection with a female bank specialist ($b = -.10$, $SE = .05$, $p = .009$) had a significant negative indirect effect , and an AI-based evaluation significantly decreased intention to pursue the process ($b = -.21$, $SE = .06$, $p = <.001$). Therefore, interpersonal justice was only a partial mediator for the relationship between evaluation type and intention to pursue the process, which partially supported *Hypothesis 5*.

With regard to innovativeness, evaluation by a female bank specialist had a significant positive indirect effect ($b = -.03$, $SE = .02$, $p = .050$). AI pre-selection with a male bank specialist had a significant positive indirect effect ($b = .08$, $SE = .04$, $p = .004$), and AI pre-selection with a female bank specialist also had a significant positive indirect effect ($b = .09$, $SE = .04$, $p = .003$). Furthermore, an AI-based evaluation significantly increased intention to pursue the process ($b = .10$, $SE = .05$, $p = .003$). This supported *Hypothesis 6*.

An additional mediation analysis revealed that the negative indirect effect of AI pre-selection with a male bank specialist on the intention to pursue the process via interpersonal justice was weaker among participants with higher discrimination experience ($b = .21$, $p = .024$). The same applied to the fully AI-based evaluation ($b = .21$, $p = .035$), indicating a buffering effect.

DISCUSSION AND IMPLICATIONS

We explore how different evaluation types affect perceived interpersonal justice, perceived innovativeness, and the intention to pursue, especially in AI-supported credit allocation processes.

Perceived Interpersonal Justice

Consistent with prior work on algorithm aversion (Mahmud et al., 2022, p. 2), our findings confirm a general preference for human-based over AI-based evaluations. Contrary to expectations, no significant difference emerged in fairness perceptions between male and female bank specialists, indicating that gender alone did not shape perceptions of interpersonal justice. However, AI pre-selection combined with a female bank specialist significantly reduced fairness perception. One rationale is that the combination of a female and a machine-based evaluation might be perceived as lacking authority and credibility. This perception likely draws on stereotypes that associate authority and decision-making power with men. These gender stereotypes also extend to human-agent and human-robot interactions (Eyssel & Hegel, 2012, p. 2221). Further, Tay et al., (2024) indicate that participants are more likely to accept robots when their perceived gender aligns with traditional expectations. However, the combination of AI and a male bank specialist resulted in the lowest perceived interpersonal justice. A potential explanation could be that men are expected to act independently when making decisions (Chang et al., 2024, p. 111). In this context, reliance on AI might be perceived as inconsistent with stereotypical expectations of male authority. As expected, fully AI-based evaluations were also perceived as the least fair, underscoring concerns around automation in sensitive decision-making contexts.

Perceived innovativeness

Perceived innovativeness was highest for fully AI-based evaluations, followed by AI pre-selection combined with either male or female specialists. In contrast, evaluations by female specialists alone decreased innovativeness perceptions–possibly due to stereotypes associating women with lower technological competence (Chang et al., 2022, p. 111). Overall, our findings suggest that integrating AI enhances perceptions of innovation but may simultaneously undermine interpersonal justice.

Interestingly, the scenario combining AI pre-selection with a female specialist was the most favorable configuration. It boosted innovativeness (second only to full AI) and was associated with fewer fairness concerns than other AI-involving conditions. This suggests that preferences for human evaluators may not derive solely from their human nature but may be linked to traits often stereotypically attributed to women, such as empathy (Schulte Steinberg & Hohenberger, 2023, p. 9). By contrast, AI in pre-selection combined with a male bank specialist was perceived as innovative but less fair, making it a less desirable option overall.

Interpersonal justice and innovativeness exert contrasting effects on the intention to pursue the credit process. Interpersonal justice was the strongest predictor, while innovativeness had a weaker, yet positive, effect. Importantly, high innovativeness could partially—but not fully—compensate for low fairness, highlighting the central role of interpersonal justice in decision-making processes.

Interestingly, the moderated mediation results further reveal that the negative indirect effects of AI pre-selection combined with a male bank specialist and AI-based evaluations on fairness were weaker among participants with prior discrimination experiences. This suggests that individuals with such experiences may view algorithmic decisions as less biased compared to human evaluators (Jago & Laurin, 2022, p. 582).

Finally, the results reveal an ambivalent picture: While fully AI-based evaluations enhance innovativeness, they reduce interpersonal justice—particularly compared to human evaluators. However, individuals with prior discrimination experiences perceive AI-based decisions as slightly less unfair, suggesting a buffering effect. Theoretically, our study contributes to research on algorithmic decision-making by revealing how entrepreneurs perceive interpersonal justice and innovativeness in AI-supported credit allocations, particularly in hybrid human-AI settings. Moreover, our findings expand the understanding of how interpersonal justice and innovativeness function as mediating mechanisms influencing entrepreneurs' intention to pursue credit applications. By shifting the focus to the applicant's perspective, this study contributes to signaling theory by demonstrating how different evaluation types send signals that impact decision-making in financial contexts.

From a practical perspective, our study has implications for banks considering using AI-supported tools. Banks should improve transparency, ensure algorithmic fairness, and provide personalized feedback to mitigate fairness concerns and foster trust in AI-driven processes.

CONTACT: Lina Uebbing; lina.uebbing@hhu.de; Heinrich Heine University; Universitaetsstraße 1, 40225 Duesseldorf, Germany

SUMMARY

WOMEN ENTREPRENEURS' WELL-BEING: A SYSTEMATIC LITERATURE REVIEW

Magdalena Markowska, King's College London, United Kingdom
Ute Stephan, King's College London, United Kingdom

Principal Topic

The purpose of this paper is to take stock of what we know about the well-being of women entrepreneurs. Well-being—broadly defined as the quality of experience and psychological functioning related to navigating an entrepreneurial venture (Wiklund et al., 2019)—is central to effective human functioning (Ryan & Deci, 2001; Ryff, 2017). Yet, current theories of entrepreneurial well-being tend to overlook the gendered nature of entrepreneurial well-being (Stephan, 2018, Caliendo et al., 2022). Women experience more constraints as entrepreneurs and take on more childcare and household responsibilities than men (Adamson, et al., 2023). Yet, despite increased interest in the topic, women entrepreneurs' well-being remains poorly understood.

Method

We performed a systematic literature review (SLR) collating all studies addressing women entrepreneurs and well-being in Scopus and Web of Science from 1983 when first article appeared to December 2024. We defined the search terms and keywords using related SLRs as a reference point (Stephan, 2018, Henry et al., 2016). We used a total of 78 keywords. This search generated 850 articles. We removed 646 articles that were duplicate or did not fulfil our inclusion criteria, and qualitatively coded the remaining 204 publications in detail.

Results and Implications

Our analysis shows that majority of the articles are quantitative with focus on antecedents and consequences of well-being, fewer studies investigated well-being as a process. Women entrepreneurs' well-being is conceptualized broad and encompassing issues of work-family balance, emotional, social and spiritual well-being. This contrasts with much more individual-centric and individual-level conceptualizations of well-being in existing reviews on entrepreneurs' well-being (e.g., Stephan, 2018). Almost by default, research focusing on women entrepreneurs' well-being discusses women's family care responsibilities and emphasizes the importance of family support for their well-being. The reviewed studies suggest that women (vs men) entrepreneurs' well-being is influenced more by non-monetary goals and motivation, especially the motivation of seeking self-fulfillment through entrepreneurship. Introspecting, self-improving, balancing, (re)sourcing and empowering delineate well-being as a process. In terms of consequences, women entrepreneurs' well-being contributes to others' well-being such as their family and local community. This SLR expands our understanding of entrepreneurial well-being as both an achieved and aspired property but also, newly, as a process. Our finding offers a rich and socialized perspective of entrepreneurial well-being whereby well-being is conceived as a multi-domain and multi-level construct which is shaped by context and impacts others. We discuss more general insights for research on entrepreneurial well-being and the value of research in non-WEIRD contexts.

CONTACT: Magdalena Markowska; magdalena.markowska@kcl.ac.uk; King's College London; King's College London, London, England WC2B 4BG, United Kingdom

SUMMARY

EMPOWERING WOMEN ENTREPRENEURS: DISMANTLING ASSUMPTIONS OF DISADVANTAGE AND UNDERPERFORMANCE

Alice Ngo, University of Sheffield, United Kingdom
Phillip Kim, Babson College, United States

Principal Topic

Women entrepreneurs remain persistently under-represented over time. Conventionally, women entrepreneurs have been perceived as disadvantaged compared to men. Two implicit assumptions have dominated this literature: (1) that the conjunction of gender and resource disadvantage necessarily leads to underperformance, and (2) that women entrepreneurs form a homogeneous group, uniformly constrained by resource limitations. These assumptions may be effective for outcome-based comparisons assessing which gender performs better. However, their effectiveness is limited when it comes to generating new insights into both between-gender and within-gender differences in resource configurations for enhancing performance.

Method

This study builds upon insights from the Resource-Based View and adopts a configurational approach to explore (1) how resource (dis)advantages influence gender differences in leveraging external resources for enhancing firm performance (between-gender differences); (2) how resource-poor versus resource-rich women entrepreneurs differ in their approaches to leveraging external resources for enhancing firm performance (within-gender differences). We use Qualitative Comparative Analysis on 243 early-stage for-profit social ventures across 19 accelerator programs from the GALI database. Profitability is used to represent venture performance. This study uses three proxies for internal resources, namely entrepreneurial experience, patents and prior investment. The profitability outcome, gender, and three internal resource conditions are calibrated as crisp sets. Knowledge spillovers - as external resource conditions - are calibrated as fuzzy sets.

Results and Implications

We identify seven distinct pathways to profitability. Based on these findings, we develop three propositions that contribute to the advancement of scholarly discourse on gender differences and women entrepreneurship. While women entrepreneurs are generally disadvantaged compared to men, this disadvantage can paradoxically serve as an advantage for at least two reasons: (1) internal resource constraints heighten their need to seek external resources; (2) their communal and social traits naturally facilitate knowledge sharing through social interactions—one of the primary channels for spillovers within geographically bounded spaces like accelerators. Proposition 1: Women-led ventures benefit more from knowledge spillover from co-located ventures within the same accelerator cohort compared to ventures led by men. Our findings uncover a critical yet underexplored within-gender heterogeneity in resource configuration for enhancing performance. We attempt to theorize and empirically substantiate distinct resource configuration patterns among women-led ventures, differentiating between resource-constrained (weaker) and resource-abundant (stronger) ventures. Proposition 2: Disadvantaged women-led social ventures can achieve profitability only if they can leverage the similar knowledge spillovers from proximate ventures. Proposition 3: Women-led social ventures with more internal resources benefit more from diverse knowledge spillovers.

CONTACT: Alice Ngo; thi.ngo@sheffield.ac.uk; University of Sheffield; Conduit Road, Sheffield, England S10 2TN, United Kingdom

SUMMARY

DOES IT PAY OFF FOR ENTREPRENEURS TO EXPRESS DOMINANCE IN WRITING OR PICTURES? A GENDERED APPROACH

Ana Pérez-Luño, Trinity Business School, Ireland
Fabiola Gerpott, WHU – Otto Beisheim School of Management, Germany
Wladislaw Rivkin, Trinity College Dublin, Ireland
Vera Schweitzer, University of Cologne, Germany
Jakob Stollberger, Vrije Universiteit Amsterdam, Netherlands

Principal Topic
Gender plays a key role in shaping entrepreneurial success, particularly in securing venture capital (VC) funding. Another important determinant of founders' success is the size of their social network. Despite increasing awareness of gender biases in entrepreneurship, our understanding of dominance signaling as a strategic tool for male and female founders remains limited. This study thus focuses on the interplay of gender, social network size, and verbal dominance displays on social media in jointly shaping impact funding success for startup founders. Given societal stereotypes on leadership and gender roles (Eagly & Karau, 2002), we hypothesize that the male and female founders are appreciated differently depending on their social reach and dominance display.

Method
We analyze founder-CEOs of 720 U.S.-based tech startups that raised Series A-D funding rounds in 2021. Funding data is collected from PitchBook, while dominance signals and network size are extracted from X profiles. Founder gender is manually coded from profile pictures. Verbal dominance in tweets is quantified using natural language processing (NLP) (Rychlowska et al., 2017). Network size is operationalized as the number of followers on X (former Twitter). We examine the proposed three-way interaction between gender, social network size and dominance on startup funding through robust hierarchical linear regression.

Results and Implications
Our results suggest that female entrepreneurs with smaller networks receive significantly less funding compared to their male counterparts. However, we do not find a significant difference in funding volume between female and male entrepreneurs who have larger networks. Furthermore, our data highlights a three-way interaction involving gender, network size and dominance on funding volume. High dominance displays benefit female founders with large networks, but penalize those with smaller networks. In contrast, male founders with high dominance are penalized when their network size is larger. When dominance is low, there is no significant funding difference between male and female founders. This research contributes to entrepreneurship research by showing that dominance displays affect male and female founders differently, depending on social network size. It expands emotional display research by highlighting how gender, network reach, and dominance interact to shape funding outcomes. Addressing gender disparities in entrepreneurship, the findings reveal that women benefit from dominance signaling only when supported by strong networks. Using objective behavioral data, the study responds to calls for real-world analyses of leadership behaviors (Fischer & Sitkin, 2023) and offers practical insights for founders navigating gendered expectations in securing VC funding.

CONTACT: Ana Pérez-Luño; perezlua@tcd.ie; Trinity Business School; Trinity Business School, Dublin, Dublin D02 F6N2, Ireland

SUMMARY

BEYOND THE CLOCK: HOW LONG WORKING HOURS RESHAPE ENTREPRENEURS' WELL-BEING UNDER DIFFERENT GENDER AND SOCIOECONOMIC STATUS REALITIES

Ana Pérez-Luño, Trinity Business School, Ireland
Ana Bojica, Universidad de Granada, Spain
Wladislaw Rivkin, Trinity Business School, Ireland

Principal Topic

Entrepreneurs' well-being is increasingly recognized as a key factor influencing business performance and sustainability (Stephan, 2018; Wiklund et al., 2019). While previous research has explored the negative effects of long working hours on well-being (Kivimäki et al., 2015; Virtanen et al., 2012), less is known about how structural factors may nuance this relationship. This study addresses this gap by examining whether gender and SES buffer or exacerbate the stress associated with long working hours and how work demands, in turn, reinforce gender and SES-based disparities among entrepreneurs.

Method

The study uses three complementary datasets: 1. A longitudinal study tracking 153 entrepreneurs across three time periods—before the COVID-19 pandemic (N=420), during (N=218), and in 2023 (N=153). 2. A daily diary study following 140 entrepreneurs over 10 working days. 3. The European Statistics on Income and Living Conditions (EU-SILC) dataset, including 2,445 entrepreneurs. Key variables include working hours, subjective well-being (life satisfaction), gender, and SES indicators (household income and education level). Hierarchical linear regression analyses test the moderating effects of gender and SES, with robustness checks for eudaimonic well-being (Stephan et al., 2023). Control variables include age, marital status, and parental responsibilities.

Results and Implications

Findings confirm that long working hours negatively affect well-being, aligning with prior research (Kivimäki et al., 2015; Virtanen et al., 2012). However, gender and SES significantly moderate this effect. Entrepreneurs with higher SES report greater well-being than lower-SES counterparts despite long hours. A three-way interaction reveals that high-SES entrepreneurs (both male and female) experience lower well-being declines, while low-SES entrepreneurs face greater stress-related deterioration. Contrary to expectations, high-SES women exhibit higher well-being than men as working hours increase (Lan et al., 2024), suggesting that autonomy and financial resources mitigate traditional gendered work-life conflicts. This study challenges traditional gendered assumptions about work-life balance, emphasizing the critical role of SES in shaping well-being outcomes (Brändle & Kuckertz, 2023). Policymakers should tailor mental health and financial support initiatives to assist low-SES entrepreneurs, while future research should explore social class, cultural factors, and eudaimonic well-being models (Piff et al., 2010; Waterman, 2007). The unexpected resilience of high-SES women in long-hour work settings offers a compelling avenue for further study on autonomy, caregiving roles, and entrepreneurial well-being.

CONTACT: Ana Pérez-Luño; perezlua@tcd.ie; Trinity Business School; Trinity Business School, Dublin, Dublin D02 F6N2, Ireland

SUMMARY

SUCCESSFUL MEN AND SUPPORTIVE WOMEN: THE ROLES OF ENTREPRENEUR SUPPORT ORGANIZATION EMPLOYEES

Maija Renko, Aalto University, Finland

Principal Topic

Despite the large body of research on sex differences in business ownership, there is a lack of knowledge about the role of gender in the work of entrepreneur support organizations (ESOs). In this study, I draw from research on gender roles, everyday entrepreneurship practices in diverse communities, and job-crafting research to examine the role and characteristics of those who work at urban entrepreneurship support programs. Such support programs and trainings typically blend ideas from mainstream entrepreneurship education discourse and practices, where gender roles have been well established, with varying elements of community-relevant content. We currenly know little about the role of gender in small-scale neighborhood organizations that deliver such supports: How do these organizations deliver their entrepreneurship trainings and supports, and what kinds of roles do men and women working in these organizations adopt?

Method

This study draws from two separate rounds of qualitative data collection, as well as a survey effort in Chicago. There are over 130 ESOs in the greater Chicagoland area that help entrepreneurs in various ways (The Chicago Community Trust, 2019). I focus on non-profit ESOs that offer cohort-based, entrepreneurship focused training programs in underserved neighborhoods, mostly on the South and West sides of Chicago.

Results and Implications

Data analysis so far points to two major gender-informed themes that emerge from the data. First, ESOs, as institutional agents, act as cultural brokers that enable the entrepreneurs they serve to adapt and follow the norms and conventions of what entrepreneurs are supposed to do. As an example of a role often adopted by a male ESO employee, they train community entrepreneurs on skill areas relevant for traditional business management (finance, accounting, marketing), with focus on practical how-to questions. Women employees often lean more towards one-on-one coaching work with entrepreneurs. These ESO employees often focus on breaking down larger, institutionally relevant tasks to smaller entities that can be completed in a piecemeal fashion. Second, the findings highlight the role of job-crafting by employees as an essential element in how ESOs are able to support entrepreneurs. Combining grounded insights from our qualitative data with a job-crafting perspective (Wrzesniewski & Dutton, 2001) helps explain why and how ESO employees attempt to craft their roles. We contextualize job-crafting theory by identifying situational challenges, gendered norms, and enablers of meaningful work in this context.

CONTACT: Maija Renko; maija.renko@aalto.fi; Aalto University; Ekonominaukio 1, Espoo, Uusimaa 02150, Finland

SUMMARY

FROM INCEPTION TO CELEBRATION: HOW WOMEN IN STEM ARE NAVIGATING THE SWISS ENTREPRENEURIAL ECOSYSTEM

Tatyana Tsukanova, EHL Hospitality Business School / HES-SO, University of Applied Sciences and Arts Western Switzerland, Switzerland
Zoé Giacomi, Faculty of Business and Economics, University of Lausanne, Switzerland
Susana Santos, Jim Moran College of Entrepreneurship, Florida State University, United States

Principal Topic
Science, Technology, Engineering, and Mathematics (STEM) fields drive innovation, contributing to patents, spin-offs, and startups. Despite the recognized importance of gender diversity in innovation, women remain underrepresented in STEM. Entrepreneurial ecosystems (EEs) can shape women's entrepreneurship by fostering inclusive environments. To explore this mechanism, this study examines the challenges and drivers of EEs for women in STEM and proposes a conceptual framework outlining the role of EEs components in supporting women in STEM from pre-launch to venture implementation.

Method
This research employs a multiple exploratory case study approach focusing on women entrepreneurs in STEM fields in Switzerland. Semi-structured interviews with women founders of innovative STEM ventures captured their experiences from pre-launch to implementation. Among 222 Swiss STEM startups identified, women represented 13% of all founders. Thirteen startups with at least one female founder participated in interviews in summer 2024. Data were analyzed using Mayring's qualitative content analysis, combining deductive and inductive coding. Interview data were supplemented with secondary data and publicly available information. A conceptual framework was developed by systematically categorizing themes and patterns, distinguishing stage-specific challenges and support mechanisms within the Swiss EEs.

Results and Implications
The findings identify key actors, startup phases, and components of the EEs. The conceptual framework addresses the unique needs and challenges of women entrepreneurs in STEM, covering pre-launch and implementation phases. The pre-launch phase highlights drivers and obstacles, emphasizing the importance of addressing institutional barriers, developing a positive entrepreneurial culture, and increasing access to resources and support networks. The implementation phase underscores the importance of resource mobilization and partnership building while mitigating challenges. The framework explains the interaction and feedback loops between enabling factors and constraints, demonstrating their synergetic effect. As ventures mature, support mechanisms, including investment, become less accessible, compounded by higher social pressures and work-life balance difficulties. The Swiss EEs provides substantial initial support. However, as ventures progress, women founders encounter more barriers and need stage-specific support. A favorable cultural environment, consistent access to finance, and comprehensive support services throughout the entrepreneurial journey are critical for their success. The proposed framework improves understanding of the ecosystem's impact on women's entrepreneurship and provides insights into adapting EEs to better support underrepresented groups in STEM. This research bridges gaps between STEM, gender studies, and EEs research, offering a holistic view of the challenges and opportunities for fostering diversity in innovative entrepreneurship.

CONTACT: Tatyana Tsukanova; tatyana.tsukanova@gmail.com; EHL Hospitality Business School, HES-SO, University of Applied Sciences and Arts Western Switzerland; Chemin de la Brume 6, Morges, Vaud 1110, Switzerland

SUMMARY

WOMEN AT THE TOP: A CATALYST FOR FORWARD-LOOKING ENTREPRENEURIAL FIRMS

William Wales, University at Albany - SUNY, United States
Jeff Covin, University of Wyoming, United States
Pide Lun, The University of Adelaid, Australia
Anish Purkayastha, The University of Sydney, Australia
Ralf Zurbruegg, The University of Adelaid, Australia

Principal Topic

This study examines how the composition of top management teams (TMTs), particularly the presence of female executives, influences a firm's forward-looking entrepreneurial orientation (EO)—the extent to which an organization communicates and emphasizes a future-oriented strategic posture toward new value creation. Our study is the first to specifically focus on forward-looking EO to capture a strategic orientation focused on 'building for the future' rather than conservatively 'resting on its laurels'. Our study of 20,420 firm-year observations from 2,209 U.S. public firms reveals that female executives catalyze a forward-looking entrepreneurial orientation. Even a single woman within the C-suite moves the needle, with larger effects as female representation grows. We also consider how male CEO power and board entrenchment moderate this relationship. CEO power, stemming from structural and ownership sources, may allow male CEOs to dominate decision-making processes, potentially suppressing female TMT members. Entrenched boards, characterized by long-tenured directors and reduced independence, may be less receptive to novel perspectives and future-oriented strategies proposed by female executives.

Method

This study utilizes textual data from 10-K reports submitted to EDGAR (2000–2020) and relies on Execucomp for TMT characteristics, including the presence of female executives, executive compensation, and CEO traits. Additional controls are sourced from Compustat. Entrepreneurial orientation (EO) is measured using McKenny et al.'s (2018) method, applying their updated lexicon. A firm's EO is computed by tallying the number of EO-focused sentences classified as forward-looking based on accounting practices. Following Corwin et al., (2022) and Post et al., (2022), three measures of women's presence in the TMT are constructed. We employ a two-way fixed effects panel. For robustness, we perform both Impact Threshold for a Confounding Variable (ITCV) and instrumental variable (IV) tests, as well as an entropy balancing technique.

Results and Implications

Female representation in TMTs is a significant driver of forward-looking EO, whereas male presence does not yield the same effect. This relationship remains robust even when accounting for endogenous factors, such as an organization's level of progressiveness. However, our findings indicate that powerful male CEOs and entrenched boards can suppress these positive effects, potentially hindering a firm's future-focused entrepreneurial momentum. This study integrates temporal considerations into the EO literature, offering a measure that distinguishes an organization's emphasis on future opportunities from its past achievements. We propose forward-looking EO as a best practice for refining the conceptualization and measurement of EO in content-analysis-based research.

CONTACT: William Wales; wwales@albany.edu; University at Albany - SUNY; University at Albany, Albany, New York 12047, United States

INTERACTIVE PAPER

THE GENDER GAP IN INNOVATION: HOW ENTREPRENEURS PATENT IN MALE- & FEMALE-DOMINATED INDUSTRIES - EVIDENCE FROM THE UNITED STATES

Kanwal Bokhari, University of Calgary, Canada
Joyce Nabisaalu, University of North Carolina, Greensboro, United States

Principal Topic

Patents serve as a proxy for innovation (Hunt et al., 2013) and represent a quality signal, demonstrating an entrepreneur's ability to develop and protect innovative intellectual property worth protecting (Conti et al., 2013; Hoenig & Henkel, 2015; Vo, 2019). Yet, despite the importance of patents for securing competitive advantage and attracting resources, men consistently secure more patents than women (Hunt et al., 2013). Accordingly, we explore how patenting patterns vary across male- and female-dominated industries, examining how entrepreneurs leverage education, networks, and ecosystem support.

Method

Using the 2022-2023 Entrepreneurship in the Population Survey Project dataset in the United States, we focus on the entrepreneur category to generate a sample of 4,743 entrepreneurs. The dependent variable, patenting, was measured by whether participants indicated they had taken steps to apply for patents (coded as 1 = Yes and 0 = No), with 26.44% selecting Yes. The independent variables are Industry Type: 67.30% of businesses were in male-dominated industries (coded as 1; examples include construction and manufacturing) versus female-dominated industries (like education and healthcare). Gender: 46.07% of participants are female (coded as 1), with 63.89% of women working in male-dominated industries. Education, Entrepreneurial Ecosystem (EE) Support, Networking. We utilized logistic regression as the most suitable technique to evaluate patenting activity.

Results and Implications

Overall, analysis reveals significant gender differences in patenting likelihood, with women demonstrating lower patenting odds than men (OR = 0.823, $p < 0.01$). This pattern persists in female-dominated industries (OR = 0.755, $p < 0.05$), though interestingly, no statistically significant gender differences emerge in male-dominated industries (OR = 0.867, ns). In male-dominated industries, EE support and networks were associated with a higher likelihood of patenting across all educational levels for both genders. We found that in female-dominated industries, EE support increases the patenting likelihood of female entrepreneurs with a graduate degree (Level 4) (OR = 3.400, $p < .01$) compared to men with the same educational attainment. Overall, we advance our understanding of the gender gap in patenting by examining its intersection with industry context and the activity-based masculine-typing of innovation. We argue that this masculine typing creates a double barrier for women entrepreneurs, who face both industry-based and activity-based gender role incongruence. In contrast, men must only navigate industry-based incongruence when underrepresented. Thus, men maintain a higher share of patents across all industry contexts, even where women are the majority, while women consistently obtain a lower share across contexts.

CONTACT: Kanwal Bokhari; kanwal.bokhari@ucalgary.ca; University of Calgary; 2500 University Way, Calgary, Alberta T2N1N4, Canada

INTERACTIVE PAPER

AN INTERSECTIONAL LENS ON BLACK WOMEN ENTREPRENEURS' EXPERIENCES IN CANADA: THE CASE OF IMMIGRANT VS. CANADIAN-BORN ENTREPRENEURS

Wendy Cukier, Diversity Institute, Toronto Metropolitan University, Canada
Tomke Augustin, Diversity Institute, Toronto Metropolitan University, Canada
Guang Mo, Diversity Institute, Toronto Metropolitan University, Canada
Mohamed Elmi, Toronto Metropolitan University, Canada

Principal Topic

This paper applies an intersectional lens on the experiences of immigrant and non-immigrant Black women entrepreneurs in Canada to identify the unique and exacerbated structural barriers they face in the entrepreneurial ecosystem due to their intersecting social categories of race, gender and immigration status (Nkrumah, 2022; Teixeira et al., 2007; Owalla et al., 2020). Significant research suggests that women, immigrant and Black entrepreneurs face serious barriers to start and grow a business (e.g., Brush et al., 2019; Nkrumah, 2022; Okeke-Ihejirika et al., 2023). Most of this research has been conducted in silos, examining either women entrepreneurs, immigrant entrepreneurs, or to a much lesser extent Black entrepreneurs. The consensus between these streams of research is that these entrepreneurs face barriers at all stages and aspects of their businesses, such as facing bias in accessing financing and networks (Kanze et al., 2018; Neumeyer et al., 2019). However, intersectional research (Crenshaw, 1989) shows that barriers and experiences will likely show unique characteristics when combining these streams.

Method

Our sample consists of 700 Black women entrepreneurs in Canada who applied for a pitch competition. Representative of population data, 60.3% were immigrants and 38.7% Canadian-born. We conducted descriptive statistical analysis, ANCOVA to test the differences in motivations, barriers and networks between Canadian-born and immigrant Black women entrepreneurs, and multiple linear regression to add nuances to the results.

Results and Implications

Results show that Black immigrant women entrepreneurs are more likely to be pushed whereas Black Canadian-born women entrepreneurs are more likely to be pulled into entrepreneurship; Black immigrant women entrepreneurs perceive stronger institutional barriers while there was no significant difference between immigrant and Canadian-born Black women entrepreneurs in their perception of organizational-level and individual-level barriers; and Black immigrant women report better community connections. Further, the longer Black immigrant women have lived in Canada, the more likely they are to cite job dissatisfaction as a reason to start their business, suggesting they may have accumulated negative experiences throughout their time in Canada. Our study contributes to the literatures on women, Black and immigrant entrepreneurship by showing that an intersectional quantitative analysis provides crucial nuances to barriers and motivations. We highlight on the specific barriers Black women entrepreneurs face, which are largely in line with barriers for immigrant entrepreneurs, even for Canadian-born women entrepreneurs. Overall, we suggest that additional research can explore the results further, especially with qualitative interpretative research designs.

CONTACT: Wendy Cukier; wcukier@torontomu.ca; Diversity Institute, Toronto Metropolitan University; 55 Dundas St W, Toronto, Ontario M5G 2C3, Canada

INTERACTIVE PAPER

PLEASE SMILE: GENDERED EMOTIONAL LABOR STRATEGIES ENTERING NEW VENTURE BOARDROOMS

Carolin Kaufmann, University of St.Gallen, Switzerland
Dietmar Grichnik, University of St. Gallen, Switzerland
Dean Shepherd, University of Notre-Dame, United States
Joakim Vincent, University of St. Gallen, Switzerland

Principal Topic

Women are drastically underrepresented in new venture boards (Butticè et al., 2023) with existing research focusing on demographic and ideological explanations. Even though emotions shape relationships and predict discriminatory behavior (Talaska et al., 2008) their role as barrier women are facing has not been explored yet. Building on emotional labor theory (Hochschild, 2012) and the influence of social norms (Dewan et al., 2024; Lee et al., 2023) expecting board members to not unconditionally express genuine emotions, we propose a gendered model of emotional reactions when elaborating on the role of women in new venture boardrooms.

Method

We conducted an experimental field study of 140 interviews with board stakeholders, as new ventures are neither affected by gender quota regulations nor embedded within an established company culture. Either a young male or female interview introduced a gender focused questions such as "is it different for women?" and we analyzed emotions expressed facially and verbally afterwards, relying on iMotions and LIWC as established software to quantify emotions expressed. We develop different measures for pre-defined emotional labor strategies, that is modifying emotional expression to meet societal expectations, by calculating distances of genuinely experienced (expressed facially) to self-managed (expressed verbally) emotions.

Results and Implications

Our study provides three main contributions. First, gender discrimination research has predominantly focused on cognitive explanations as mechanisms hindering women's equal participation, such as stereotyping (Brush, Edelman, Manolova, & Welter, 2019) or gender-role incongruent expectations in male-dominated roles (Abra-ham, 2020; Eagly & Karau, 2002). However, the role of emotions remains unexplored, despite their influence on interpersonal relationships and even decision-making in the boardroom (van den Oever & Shropshire, 2024). Our study highlights how men and women engage in different emotional labor strategies, namely suppressing and feigning. These insights deepen the understanding of underlying mechanisms impacting boardroom dynam-ics. Second, we answer recent calls to improve emotional labor measures to not solely being based on self-reports. Based on physio-psychological and linguistic measurements we develop an objective emotional labor measurement. Finally, we acknowledge that emotional labor is not only a phenomenon occurring in the service sector and prove its occurrence in the boardroom as well (Ashtar et al., 2021). We expand emotional labor theo-ry by emphasizing on gendered emotional labor strategies to understand an individual's handling with threaten-ing conditions in societal topics. We extend emotional labor theory from an intrapersonal dimension (negative consequences of emotional labor for individuals) to an interpersonal perspective, highlighting the emergence of interpersonal barriers.

CONTACT: Carolin Kaufmann; carolin.kaufmann@unisg.ch; University of St.Gallen; Dufourstrasse 40a, St. Gallen, Sankt Gallen 9000, Switzerland

INTERACTIVE PAPER

HOW TO ADDRESS THE CHALLENGE OF JOINING A START-UP BOARD? WOMEN BOARD MEMBERS BREAKING ROLE-EXPECTATIONS

Carolin Kaufmann, University of St.Gallen, Switzerland
Manuel Hess, Grenoble Ecole de Management, France
Joakim Vincent, University of St. Gallen, Switzerland

Principal Topic

The formation of new venture boards and their hierarchies is informal (Garg, 2020; Jung et al., 2017) and based on subjective evaluations of competence and expertise (Magee & Galinsky, 2008). As entrepreneurial ecosystems and boardrooms are masculine-stereotyped (Brush et al., 2019; Westphal & Stern, 2007), women face perceptions of lowered competence leading to reduced power in the boardroom (Lahiri et al., 2019; Snellman & Solal, 2023). A fear to confirm these stereotypes is evoked, activating self-doubts and influencing their performance negatively (Steele & Aronson, 1995; Walton et al., 2015). We lack understanding of how women may solve this conflict and still gain attention in the boardroom.

Method

We build on a qualitative gender-intervention analysis based on 58 interviews with founders, active board members, and investors in the new venture ecosystem of the DACH region – the German-speaking area of Europe. This analysis is supplemented with archival data, including newspaper articles and reports from Pitchbook and Gründerszene. Through a multi-round coding process, emerging patterns led to the development of a circular process model that explains gender-specific role expectations, how women gain attention in the boardroom despite stereotype threat, and how these tactics positively influence board dynamics and incrementally transform the ecosystem.

Results and Implications

Role expectations of an ideal new venture board member are mainly influenced by a startup's fear of failure and the prevailing boys-club culture, which is likely to undermine women, because board appointment processes are not transparent and often based on network connections and women that are present are not heard. As a result, the ideal board member is expected to offer expert knowledge, a strong personal fit with the founder(s), and to possess assertive soft skills. Contrary, women board members are expected to possess rather communal soft skills that balance the masculine ideal, empathetic leadership, and complementary strategic skills such as being more pragmatic and task oriented. We define seven distinct tactics women may engage in to cope stereotype threat and to get attention in the boardroom. Dominant stereotype adoption, playing the victim role, undervaluation of self, and overcompensation are unfavorable tactics as they lead to negative outcomes like the activation of stereotypes, resistance to structural change and backlashing board dynamics. Self-reliance, reframing challenges as chances, and rephrasing dominant narratives in the ecosystem, are favorable tactics that challenge individual perceptions of biases, enhance board dynamics, and increase board capital and potentially change the ecosystem from the inside.

CONTACT: Carolin Kaufmann; carolin.kaufmann@unisg.ch; University of St.Gallen; Dufourstrasse 40a, St. Gallen, Sankt Gallen 9000, Switzerland

INTERACTIVE PAPER

GENDERED FINANCE INEQUALITIES: EXPLORING THE EFFECTS OF INSTITUTIONAL AND HUMAN CAPITAL ON FUNDING GAPS ACROSS CULTURES

Shu Yang, Hofstra University, United States
Karen Hughes, University of Alberta, Canada
Kun Liu, Ken State University, United States

Principal Topic

Although previous research on entrepreneurial financing often finds varying gender disparities in funding (e.g., Anglin et al., 2022; Eddleston et al., 2016; Yang et al., 2020), macro-level contextual factors have received less attention. Building on work highlighting structural gender inequalities (e.g., Zhao & Wry, 2016; Hechavarría et al., 2018), we argue that gender funding disparities could not only disadvantage both men and women entrepreneurs but also vary across national contexts, interacting with individual-level moderators. We propose three hypotheses. First, using human capital theory (Becker, 1964), we argue a positive relationship between a lead entrepreneur's prior for-profit experience and equity funding performance (Jiao et al., 2023) as the baseline. Second, drawing on gender role congruity theory (Eagly & Karau, 2002), we hypothesize that being female negatively affects this main effect as experienced female entrepreneurs would raise significantly less equity than their experienced male counterparts. Third, we propose that national gender inequality exacerbates funding gaps for experienced women-led startups (e.g., Thebaud, 2015; Yacus et al., 2019).

Method

We use data from the Global Accelerator Learning Initiative (GALI) and the Gender Inequality Index (GII). GALI tracks startups applying to accelerators globally (2013–2019), with annual follow-ups, providing variables like entrepreneur gender, experience, and equity funding. Our sample includes 5,307 startups from a one-year follow-up, across 281 accelerators in five regions. GII data measures structural gender inequality by country. The dependent variable is equity funding performance (continuous), with prior experience (dummy: 0=no, 1=yes), gender (0=male, 1=female), and GII score (0=low, 1=high) as key variables. Controls account for venture and team factors.

Results and Implications

Using multilevel modeling in Stata 17 to address regional and program nesting, we find marginal support for H1: experience weakly boosts funding. H2 is supported—experienced female entrepreneurs raise less funding than their experienced male counterparts. Surprisingly, H3 is contradicted: in low-inequality countries, experienced women face greater funding penalties, while experienced men benefit more. In other words, in high-inequality contexts, experienced women seem to leverage their human capital better, narrowing gaps. These preliminary findings suggest although gender-equal contexts benefit women in general, they may pose unexpected barriers for experienced women entrepreneurs.

CONTACT: Shu Yang; shu.yang@hofstra.edu; Hofstra University; 20 Newport Parkway, Jersey City, New Jersey 07310, United States

THE INSTITUTE FOR SOCIAL INNOVATION AWARD (BABSON COLLEGE) FOR THE BEST PAPER ON SOCIAL IMPACT

KENYAN PASTORS IN BUSINESS: PATHWAYS FOR SHARING RELIGIOUS FAITH IN ENTREPRENEURSHIP

Sean Dwyer, University of Oklahoma, United States
Mitchell Neubert, Baylor University, United States
Reginald Tucker, University of Oklahoma, United States

ABSTRACT

This study examines how pastor entrepreneurs in Kenya use business ventures as platforms to share religious faith with others. We introduce three mechanisms – *access to community relationships, control over personal time,* and *accumulation of financial resources* – by which the work of owning a small business enables four distinct faith-sharing strategies: *evangelism*, *signaling*, *volunteering*, and *resource supply*. In doing so, we develop a theoretical model demonstrating the pathways by which entrepreneurs fulfill aspirations to help others learn about their religious beliefs. This research provides novel insight into an outcome of entrepreneurial action that has received ample theoretical attention but minimal empirical attention to date.

INTRODUCTION

The recent growth of the religious entrepreneurship subfield (e.g., Smith, McMullen, and Cardon, 2021) has expanded our understanding of the reasons why individuals engage in entrepreneurship. Entrepreneurship literature recognizes that some individuals start businesses "to establish a platform from which they can pursue their political visions and advance particular causes" (Fauchart and Gruber, 2011, p. 944). While there are significant bodies of academic work that examine the ways in which entrepreneurs create social and environmental forms of value (Lumpkin et al., 2013, Vedula et al., 2022), the emerging stream of religion and entrepreneurship literature has opened a new frontier of examining business venture value creation goals aimed at earning a "spiritual return on investment" (Smith et al., 2019, p. 8).

Individuals holding and making use of religious beliefs may seek to fulfill *higher purposes* through their work as entrepreneurs (Dwyer et al., 2025; Pidduck et al., 2024). While the specific venture goals related to such higher purposes encapsulate a wide range of options, empirical research indicates that religious individuals tend to have a proclivity toward creating value that benefits the well-being of others (Khurana et al., 2021; Rietveld, van Burg, 2014). This understanding that religious entrepreneurs are likely to seek to benefit others through their businesses aligns with seminal research in the religion and entrepreneurship literature:

> *Religion likely relates to entrepreneurs setting goals that are not just related to self-serving ideals or profit-maximization, but also to the pursuit of a higher calling or purpose related to one's religion (e.g., do good, help others, show compassion, share wealth, incorporate a day of rest) (Smith et al., 2021, p.4)*

What is especially important to recognize is that while many entrepreneurs, religious or not, aspire to make some sort of positive difference in the world through their businesses (Bosma et al., 2021), that when goals are directly tied to or derived from religious faith, these are likely to "assume a level of primacy within a person's overall goal hierarchy" (Emmons, 1999, p. 96). Accordingly, we can expect faith-based goals to influence entrepreneurial decision-making.

The extent to which religious beliefs drive the pursuit of specific value creation goals in business ventures is contingent upon a variety of factors (e.g., religious affiliation, personal calling), and highlights four potential faith-based goals entrepreneurs are likely to pursue: (1) financial gain for others, (2) service to others, (3) honoring God, and (4) sharing religion (Smith et al., 2019). Of these venture goals, the first two align closely with a general understanding of social value creation from the social entrepreneurship literature (Lumpkin et al., 2013), while the third and fourth are more likely closely intertwined with achieving objectives specific to a particular religion that may fall outside of a general understanding of social value creation (e.g., operating a business in line with religious traditions, sharing religious beliefs to others).

In this study, we seek to examine dynamics related to the pursuit of a distinctly religious value creation goal of sharing religious beliefs with others through a business. Scholars recognize that some religious entrepreneurs use business ventures as a "mechanism to share religion with others" (Smith, Conger, McMullen, and Neubert, 2019, p. 5; Dejardin, Hoogendoorn, Janssen, Rietveld, and Van Burg, 2024). For entrepreneurs who affiliate with many of the largest religions in the world (e.g., Islam, Christianity), the practice of sharing religious beliefs with others represents a common religious practice that is taught and encouraged within these religious traditions (Lindsay and Smith, 2010). However, there is a limited body of research that examines how entrepreneurs achieve these goals in practice.

Accordingly, the purpose of this study is to answer the following research question: *how do religious entrepreneurs use business ventures as platforms to share religious beliefs with others?* To conduct this study, we partnered with a faith-based entrepreneurial training program based in the United States that is training entrepreneurs in Kenya who were associated with a non-profit religious organization focused on equipping individuals to plant churches. The study involved adopting a qualitative, inductive research approach that entailed participatory observation of the training program along with in-depth interviews with leaders of both programs as well as Kenyan pastor entrepreneurs. Supplemental data was also collected related to personal and venture characteristics to provide an in-depth understanding of the pathways entrepreneurs take to share religious beliefs with others.

Through our analysis, we develop a theoretical model explaining how individuals use entrepreneurship as a platform to share religious faith with others. This model includes three primary mechanisms associated with entrepreneurial work that enable individuals to share religious faith with others: *access to community relationships, control over personal time,* and *accumulation of financial resources*. These mechanisms enable entrepreneurs to engage in the following faith-sharing strategies: *evangelism*, *signaling*, *volunteering*, and *resource supply*. Altogether, these findings serve as the foundation for the data-driven model we develop from our research study.

The research contributes to the growing body of research examining the role of religion in entrepreneurship (e.g., Smith et al., 2023). Our primary contribution to this literature concerns building on extant research on business venture value creation goals (Hechavarría et al., 2017; Hoogendoorn et al., 2020) to examine the sharing of religious beliefs with others as a distinct value creation goal and outcome of religious entrepreneurship (Dejardin et al., 2024). While extant empirical research supports the notion that religion is associated with social and environmental value creation goals (Khurana et al., 2021; Rietveld, van Burg, 2014), our research uncovers the micro-foundations of how entrepreneurs create spiritual or religious forms of value through the sharing of religious beliefs with others through a business venture. The study opens up future avenues for research on the role of religion in entrepreneurship.

HYPOTHESIS DEVELOPMENT

Individuals who own and operate business ventures tend to take on an entrepreneurial identity. The fundamental question surrounding entrepreneurial identity concerns "who entrepreneurs are and what they do" in a given context (Mmbaga et al., 2020, p. 3). An entrepreneurial identity can have a profound impact on what drives the ambitions and decision-making process of entrepreneurs. In the religion and entrepreneurship literature, there is growing recognition that entrepreneurs can be influenced by a *religious identity*, which is rooted in connection to "relational identity with God," whereby individuals view "themselves at least partly in a parent-child relationship (as a child of God), working from a received (rather than achieved) identity that is enduring (rather than contingent on performance)" (Smith et al., 2023, p. 10).

This religious identity can have a profound influence on how individuals see themselves as well as their perceived role in society (Dyck, 2014; Emmons, 1999; Gümüsay, 2020) and also provide them with an "ultimate vision of what people should be striving for in their lives" (Pargament and Park, 1995, p. 15). Among entrepreneurs that are strongly influenced by their religious identity, these individuals are likely to pursue value creation goals that align closely with their religious beliefs, which oftentimes entail positively contributing to others (Smith et al., 2019). In recent research, entrepreneurship scholars have referred to such faith-driven goals as "higher purposes" grounded in their religious beliefs (Pidduck et al., 2024, p. 8).

Fulfilling Higher Purposes by Sharing Religious Beliefs

When an individual's entrepreneurial identity is influenced by a religious identity, we might expect these individuals to have a greater propensity to integrate aspects of their religious identity into their business ventures. In many ways, such entrepreneurs are likely to see the everyday activities of running their businesses as "imbued with transcendent meaning and purpose" given their role in relation to God or some sort of higher power (DeAngelis and Ellison, 2017, p. 10).

Recognizing that religious entrepreneurs are likely to place "a focus on value creation for others rather than value capture for themselves" (Smith et al., 2019, p. 5), we can expect such individuals to find distinct ways to fulfill such faith-based goals through their business ventures. Although these goals may, in certain cases, resemble social value creation goals from the social entrepreneurship literature (e.g., Hechavarría et al., 2017), there is evidence that religious entrepreneurs may also seek to achieve value creation goals that are distinctly religious in nature; in particular, the sharing of religious beliefs with others, in pursuit of a "spiritual return on investment" (Smith et al., 2019, p. 8). While there is growing evidence that religious organizations are looking to entrepreneurship as a mechanism to help people share religious beliefs with others (e.g., business-as-mission) (see Smith et al., 2019), there is limited empirical research examining the micro-foundations of how individuals integrate religious faith into their businesses through the sharing of religious faith with others. Accordingly, this paper aims to examine the ways in which religious entrepreneurs use business ventures as platforms to share religious faith with others.

METHOD

Given the limited research examining dynamics related to faith-sharing in the context of entrepreneurship, we determined it appropriate to use an inductive qualitative approach to answer our research question (Gioia et al., 2013; Strauss and Corbin, 1998). This approach allowed us the opportunity to develop data-driven theoretical insights from multiple sources of data.

Research Context

The research context for this study is a faith-based entrepreneurial training program in Kenya, led by *Equipping Org*, which equips individuals with business knowledge, practical skills, and access to financial support to help them operate business ventures. The study focused on Kenyan "pastor entrepreneurs" who were actively launching or operating business ventures while also serving in church leadership roles. These participants were also affiliated with *Pastor Org*, a long-established religious organization dedicated to planting and sustaining churches across Sub-Saharan Africa.

The reason for the partnership between *Equipping Org* and *Pastor Org* was to help address the financial challenges that pastors affiliated with *Pastor Org* had been facing in recent years. While the church planting movement had rapidly accelerated, the growth of the movement was constrained by the lack of financial resources of the pastors. *Equipping Org* mobilized trainers in the United States to travel to Kenya to train the pastor entrepreneurs in key areas of business, such as operations, marketing, and finance, and offered access to fixed-rate microloans. While the training program focused on improving the economic performance of the businesses, the training also emphasized the importance of using the business as a platform for helping them share their religious beliefs with others. Accordingly, the research context was well suited for examining the micro-foundations of faith-sharing in entrepreneurship.

Data Collection

Data collection for this study involved collecting four primary types of data to ensure triangulation (see Jick, 1979). The first data type involved participatory observation by one of the researchers in the study in Kenya. This involved direct participation in training sessions with the participants as well as attending events with leaders from both *Pastor Org* and *Equipping Org*. These observations were detailed in journal entries, representing memos for further analysis. The second form of data collection involved conducting 21 semi-structured interviews (over 17 hours and over 500 transcript pages). This second form of data collection served as the primary source of data.

The third form of data collection entailed analyzing secondary data (over 100 pages) covering the personal and venture backgrounds of the entrepreneurs. These also included insight into the ways in which they planned to share their religious faith with others through their businesses. Finally, the fourth form of data was derived from survey data capturing demographic information from the participants along with insights related to personal calling and motives for engaging in entrepreneurship. Altogether, these data collection efforts served as the basis for data analysis.

Data Analysis

Data analysis involved the use of NVivo 14 for coding purposes. The first author closely reviewed interviews, secondary documents, and survey data to generate initial insights and memos. Following this, the first author engaged in open coding (Strauss & Corbin, 1998), where over one thousand data snippets were labeled to reflect themes grounded in the interviews. These codes were refined into higher-level second-order concepts (Gioia et al., 2013) and involved looking to the literature to help make sense of the data. In the next step, the second-order concepts were organized into aggregate dimensions, producing a data structure that informed the study's theoretical model. This model, presented in the Findings section, illustrates how religious entrepreneurs use business ventures as platforms to share religious beliefs with others.

RESULTS

The findings of our study revealed the pathways by which religious entrepreneurs (i.e., entrepreneurs with a strong religious identity) share religious faith with others through their business ventures. We first present insights related to how the sharing of religious beliefs serves as a means to fulfilling a higher purpose, followed by the mechanisms that enable individuals to pursue one of four pathways to sharing religious faith with others through entrepreneurship.

Sharing Religious Beliefs as a Means to Fulfilling a Higher Purpose

Across the interviews in our study, the pursuit of sharing religious beliefs with others represented a widely accepted value creation goal among the pastor entrepreneurs. Although most of the entrepreneurs operated micro-businesses, with many doing so out of necessity, the entrepreneurs were highly driven to make a positive difference through their businesses through the sharing of their religious beliefs with others.

The entrepreneurs believed that by operating their businesses, they could change the world through sharing religious beliefs with others. The entrepreneurs described their businesses as providing "opportunities" to show love to those in need and help other people learn about their religious faith through

invitations to Bible Studies. For many of these entrepreneurs, the end goal of their businesses was to serve a higher purpose of fulfilling God's plans for their lives.
While the scale of social impact that these entrepreneurs could make was minimal, the ability to create value for a higher power (i.e., serving the interests of a higher power) through sharing religious beliefs imbued these business ventures with significant meaning for the entrepreneurs. In certain cases, the owners of subsistence business ventures (e.g., selling commodities) described their businesses as "kingdom" businesses because of the role they played in serving God's kingdom by helping others learn about the fundamental aspects of their religious beliefs and, therefore, making an impact in this way.

Enabling Mechanisms of Entrepreneurial Work

Entrepreneurship enabled the pastor entrepreneurs in the study to share religious beliefs with others as a result of three key mechanisms associated with business venturing: *access to community relationships, control over personal time,* and *accumulation of financial resources*. In short, by starting and operating a business venture, the pastor entrepreneurs were uniquely equipped to fulfill their higher purpose of sharing religious beliefs with others.

First, the entrepreneurs in the study described their businesses as a means to "access" members of their communities in ways that they would not have been able to otherwise. As contributing members to their communities through the products and services they offered, their businesses served as natural touchpoints for them to engage in *building relationships* with members of the community. Naturally, to share religious beliefs with others requires the presence of "others," and their engagement in entrepreneurship put them in connection with these individuals.

Second, entrepreneurship provided the entrepreneurs with the *control of their personal time* that was necessary to work on their business "when" they wanted to and engage in their personal ministry work "when" they wanted to. While the promise of autonomy represents one of the most common reasons people often engage in entrepreneurship (Gelderen, 2016), the flexibility of entrepreneurship enabled the pastor entrepreneurs to engage in non-work faith-sharing activities as needed. This involved the ability to close down their businesses "when" needed or delegate responsibilities to someone else to run the business "when" needed. Third, entrepreneurship allowed the entrepreneurs a means of *accumulating financial resources* that could be used not only to financially sustain themselves and their families but also non-work initiatives led by themselves or others aimed at the sharing of religious beliefs.

Pathways to Faith-Sharing in Entrepreneurship

Our study revealed four primary pathways to sharing religious beliefs with others, segmented by whether faith-sharing occurred within or beyond the context of the venture, as well as the extent to which the faith-sharing happened directly or indirectly by the entrepreneur. These pathways include *evangelism*, *signaling*, *volunteering*, and *resource supply*.

The *evangelism* pathway involved the entrepreneur directly sharing religious beliefs with others within the context of business. This pathway involved the entrepreneurs leveraging the relationships developed through the business to create opportunities to communicate with business stakeholders about their religious faith. The natural interactions of business activity with members of the community play a primary role in enabling this to occur. The *signaling* pathway entailed entrepreneurs using indirect means of communicating religious beliefs to others through religious symbols (e.g., name of business) or certain behaviours (e.g., treating people kindly) in hopes of eventually invoking conversations with business stakeholders about their religious faith. This pathway is especially advantageous in contexts where community members are less accepting or even hostile toward the particular religion.

The *volunteering* pathway involved entrepreneurs using their businesses as sources of financial support, so that" they had the freedom to engage in sharing activities beyond the scope of the business. In these cases, the control over personal time made available through the nature of owning and operating a small business allowed these individuals to share with people beyond the scope and context of their business. The *resource supply* pathway entailed entrepreneurs giving resources generated from the business to support causes and initiatives aimed at the sharing of religious beliefs with others. In this way, entrepreneurs indirectly engaged in the sharing of religious beliefs through providing resource support to ministries and churches. The businesses not only supported these initiatives with financial resources but also other resources such as land or products belonging to the business.

DISCUSSION AND IMPLICATIONS

The findings of our study provide new insight into an outcome of entrepreneurial action that has received theoretical attention (see Smith et al., 2019) yet limited empirical examination in the entrepreneurship literature. Specifically, we provide insight into the mechanisms of entrepreneurial work as well as insight in the micro foundations of how entrepreneurs, in practice, share religious faith with other people. This study advances our understanding of how individuals use business to create heterogeneous types of value (e.g., Hoogendoorn et al., 2022) by focusing specifically on value creation derived from religious logics and beliefs.

While research shows that religious individuals have a proclivity toward creating more than just economic value through their business ventures (Khurana et al., 2021; Rietveld, van Burg, 2014), this is among the first studies to examine the pathways by which entrepreneurs use businesses to advance the religion-specific mission of sharing religious beliefs with others. In doing so, this study provides evidence of a distinct way religious entrepreneurs fulfill *higher purposes* through within and through their work (Pidduck et al., 2024).

CONTACT: Sean M. Dwyer; seandwyer@ou.edu; University of Oklahoma; 1003 Asp. Ave, Price Hall Room 3050G, Norman, Oklahoma 73019

EVALUATING ENTREPRENEURIAL PERFORMANCE: HOW OBSERVERS' SELF-CATEGORIZATION SHAPES THEIR ASSESSMENTS OF SUCCESSFUL ENTREPRENEURS

Julia Kirsch, Technical University Munich, Germany
Holger Patzelt, Technical University Munich, Germany
Nicola Breugst, Technical University Munich, Germany

ABSTRACT

Entrepreneurs depend on favorable societal evaluations to build legitimacy and succeed. While we know that these evaluations typically depend on observers' perceptions of an entrepreneur's performance, prior research has neglected the possibility that they may also be shaped by how observers perceive themselves. Drawing on self-categorization theory, we develop and test a social evaluation model of entrepreneurial performance, arguing that different performance indicators are valuated differently by observers with high or low entrepreneurial intentions (EI). Using 10,496 assessments from 328 observers in a metric conjoint experiment, our study reveals that entrepreneurs who perform well personally (high income and large societal influence) are evaluated more positively by observers with higher EI, while entrepreneurs whose ventures perform well (high profit and growth) are evaluated more positively by observers with lower EI. These insights highlight the multifaceted nature of entrepreneurial performance from an external observer's perspective and advance the understanding of how observer heterogeneity influences the evaluation of successful entrepreneurs.

INTRODUCTION

As entrepreneurs receive growing attention in the public discourse, societal evaluation of them is becoming increasingly relevant to the emergence of entrepreneurial activity and subsequent entrepreneurial performance outcomes (Anglin et al., 2018; Meek et al., 2010; Nowiński et al., 2020; Schmutzler et al., 2019; Stephan & Pathak, 2016). Entrepreneurs depend on favorable societal evaluations to gain legitimacy and access to essential resources, support structures, and knowledge (Fisher et al., 2017; Fisher & Neubert, 2023; Prochotta et al., 2022; Radu-Lefebvre et al., 2019; Simmons et al., 2014; Stuetzer et al., 2018; Überbacher, 2014). Prior literature indicates that this evaluation is contingent on entrepreneurs' performance such that only those who perform well get the support they need (Cardon et al., 2011; Shepherd & Patzelt, 2015; Simmons et al., 2014). Performance in this regard has thus far mainly been conceptualized as either venture performance, i.e., the venture's profit, survival and growth (Razmus & Laguna, 2018; Shepherd et al., 2019; Shepherd & Wiklund, 2009), or the personal performance of the entrepreneur in terms of their ability to create wealth for themselves as well as for the community in which they are embedded (Lumpkin & Bacq, 2019; Wach et al., 2016).

While extant studies primarily focus on observers' evaluations of entrepreneurs based on objective performance measures, these studies have overlooked the possibility that entrepreneurial evaluations may also be shaped by observers' perceptions of themselves. This is potentially problematic given that evaluators' social identity construction can significantly influence interpersonal evaluation processes (Fisher & Neubert, 2023; Schmitt et al., 2000). For example, prior literature has shown that observers' self-categorization into the entrepreneurial *ingroup* vs. *outgroup* can fundamentally change the way observers perceive and evaluate entrepreneurs (Bottero, 2005; Galak et al., 2011). While for many observers of the general public entrepreneurs may form a distinct outgroup, those observers who are, or strive to become entrepreneurs themselves are likely to consider entrepreneurs their ingroup (Gruber & MacMillan, 2017; Hoang & Gimeno, 2010). What remains unclear is how observers' self-categorization influences their evaluation of entrepreneurs, specifically when they possess information about both the performance of the entrepreneurial venture and that of the entrepreneurs personally. Thus, in the current paper we ask: *To what extent does observers' self-categorization impact their evaluation of entrepreneurs when information on the performance of both, the entrepreneur and their venture is available?*

In the present study, we build on self-categorization theory (Turner, 1999) to explain why observers who tend to self-categorize as (potential) entrepreneurs evaluate entrepreneurs differently compared to those who see themselves in a category different to entrepreneurs. In particular, our theorizing suggests that observers with higher entrepreneurial intentions (EI) evaluate entrepreneurs who are personally performing well (i.e., generating personal high income and societal influence) more favorably, while observers with lower EI evaluate entrepreneurs whose ventures demonstrate stronger performance (i.e., high profit and growth) more favorably. We largely find empirical support for our model based on the analysis of 10,496 evaluations made by 328 observers from the general public in a metric conjoint experiment.

We contribute to the entrepreneurship literature in at least three ways. First, research on observer characteristics and entrepreneurial evaluations has mainly focused on how observers perceive entrepreneurs, rather than their self-perceptions (Fisher & Neubert, 2023). Our findings show that specific observer traits can lead to differing evaluations, emphasizing that assessments are multifaceted. Observers' entrepreneurial intentions are a key factor in understanding these evaluations. Second, existing literature often examines failed entrepreneurs (Cardon et al., 2011). While high performance is viewed as the opposite of failure, our results indicate it consists of various components that affect evaluations differently. This suggests that understanding entrepreneurial success requires a nuanced approach beyond simply contrasting it with failure. Third, our study contributes to the discussion on entrepreneurial role models, indicating that high-performing entrepreneurs inspire others (Abbasianchavari & Moritz, 2021). However, different successful role models are evaluated inconsistently by observers, highlighting the significance of context and audience in effective role modeling.

HYPOTHESIS DEVELOPMENT

Entrepreneurship is inherently a social process where entrepreneurs are continually evaluated by those in their environment. Public image significantly influences societal perceptions of entrepreneurs (Anderson et al., 2009; Lee & Huang, 2018). With increasing media attention, entrepreneurs may for example be viewed as "cultural heroes" (Malach-Pines et al., 2005). Securing public support therefore is essential for obtaining the necessary emotional and financial backing to achieve legitimacy and growth (Anglin et al., 2018). Key evaluation criteria for outside observers are warmth (liking) and competence (Judd et al., 2005). Those perceived favorably tend to be both liked and seen as competent (Fiske et al., 2007). Observable performance is a significant factor in entrepreneurial assessments (Shepherd et al., 2019), with high-performing entrepreneurs generally receiving more positive evaluations (Cardon et al., 2011), as they are seen as vital to economic growth and societal problem-solving (Cohen & Winn, 2007). However, what remains unclear to date is to what extent observers rely on indicators capturing the performance of the entrepreneurial individual versus their venture to form their overall evaluation of entrepreneurs. In the present study, we hypothesize that this evaluation depends on observers' own self-categorization.

The evaluation of entrepreneurs in any given societal context is fundamentally shaped by group dynamics and social affiliations. As many observers of entrepreneurs are not entrepreneurially active themselves, entrepreneurs for them form a distinct societal *outgroup*. The emergence of this perception is addressed in social identity theory (Tajfel, 1978) and further explained by self-categorization theory (Turner, 1999). By self-categorizing as members of particular social groups (*ingroups*), individuals use the comparison to other groups (*outgroups*) to position themselves in the societal landscape. This positioning determines individuals' perceptions of other people in their environment (see Trepte & Loy, 2017). Considering the embeddedness of occupations in social networks and relationships, evaluator's professional identity plays a key role in shaping social evaluations (Fisher & Neubert, 2023; Schmitt et al., 2000). Entrepreneurship can be considered as an occupation connected to a strong self-categorization making the social identity of an aspiring entrepreneur particularly salient (Brändle et al., 2018; Gruber & MacMillan,

2017; Hoang & Gimeno, 2010). One key indicator of observers' occupational self-categorization as entrepreneur are their own entrepreneurial intentions (EI), which are defined as an individual's "self-acknowledged conviction" (Thompson, 2009, p. 687) to set up a new business venture at some future point in time. Hence, observers' EI are indicative of their self-categorization as entrepreneurs. (Gregori et al., 2021; Harmeling, 2011; Murnieks et al., 2014). We thus expect that observers' EI impacts their evaluations of entrepreneurs, including the performance of these entrepreneurs personally as well as that of their ventures. We argue, however, that the self-categorization of observers into the entrepreneurial ingroup or outgroup affects their evaluation of these two facets of entrepreneurial performance in different ways.

An entrepreneur's personal performance encompasses both monetary and non-monetary dimensions. Monetary indicators, such as personal income, reflect wealth created by entrepreneurs (Bublitz & Noseleit, 2014; Florin, 2005), and observers often view high income as a sign of ability, intelligence, and social skills (Ceci & Williams, 1997). Such skills are vital for generating income, enhancing perceptions of entrepreneurs as approachable (Dimitriadis & Koning, 2022). Non-monetary indicators, on the other hand, capture societal impact and civic wealth generated by entrepreneurs (Ali & Cottle, 2021; Bailey & Lumpkin, 2023). Observers appreciate how entrepreneurs improve their communities (Lumpkin & Bacq, 2019).

We argue that observers' entrepreneurial intentions (EI) shape their evaluations of entrepreneurs' personal performance. Those with lower EI are likely to see entrepreneurs as an outgroup, increasing perceived social and cognitive distance (Fiske & Neuberg, 1990), which can lead to harsher evaluations of failed entrepreneurs (Shepherd & Patzelt, 2015). Consequently, they may overlook the personal efforts and challenges that contribute to an entrepreneur's success and may rely on surface-level information instead.

In contrast, observers with higher EI feel more social proximity to entrepreneurs and appreciate their complex journeys (Trepte & Loy, 2017). This proximity facilitates perspective-taking (Ku et al., 2015), allowing them to recognize entrepreneurs' contributions to their own income and societal status, resulting in more positive evaluations. Observers also evaluate members of their in-group based on their performance. High-performing ingroup members enhance positive distinctiveness, leading to favorable evaluations, while low-performing members may face the "black sheep effect" (Marques et al., 1988), where they are judged more harshly by those identifying strongly with the group (Coull et al., 2001). We hypothesize:

H1: Entrepreneurs with more personal income are evaluated more positively by observers with higher EI than by observers with lower EI.

H2: Entrepreneurs with larger societal influence are evaluated more positively by observers with higher EI than by observers with lower EI.

The performance of an entrepreneur's ventures' performance can also be evaluated using monetary and non-monetary indicators. Monetary outcomes typically include sales, profits, or firm valuation (Delmar & Shane, 2006). For observers like the general public, a company's financial performance often serves as a key indicator of value (Razmus & Laguna, 2018), leading to a "financial halo effect" (Brown & Perry, 1994). However, for early-stage ventures, monetary indicators can be hard to interpret or unavailable (Lee & Huang, 2018). Therefore, observers often rely on non-monetary characteristics, such as job creation (Angel et al., 2018; Neumark et al., 2006), which indicate overall regional economic strength.

We hypothesize that observers' entrepreneurial intentions (EI) influence how monetary and non-monetary performance indicators of entrepreneurs' ventures impact evaluations of entrepreneurs. Specifically, we hypothesize that observers with higher EI tend to focus less on venture performance indicators. Their "ego-centric" perspective allows them to appreciate entrepreneurs for their journeys, regardless of specific outcomes, as they recognize the frequent setbacks in new ventures (Funken et al., 2020; Ucbasaran et al., 2013). Prior studies indicate that those with entrepreneurial backgrounds, such as venture capitalists, are less likely to penalize failed ventures and often view contextual factors affecting performance more positively (Cope et al., 2004; Souraki et al., 2022). Thus, high EI observers may see entrepreneurs as likable and competent regardless of performance.

In contrast, observers with lower EI place more importance on venture performance indicators, as these metrics directly impact the community and the entrepreneurial outgroup (Neumark et al., 2006). Successful ventures generate profits that benefit society and create job opportunities (Ács & Szerb, 2007; Bruce et al., 2020). We therefore hypothesize that lower EI observers prioritize venture performance more in their evaluations.

H3: Entrepreneurs whose ventures generate more profit are evaluated more positively by observers with lower EI than observers with higher EI.
H4: Entrepreneurs whose ventures exhibit higher growth are evaluated more positively by observers with lower EI than observers with higher EI.

METHOD

We collected data from 328 participants recruited at diverse public locations across a major German city, aiming for demographic variety. Of 1,938 approached individuals, 381 agreed to participate (20% response rate), and after exclusions (current entrepreneurs, incomplete data), 328 remained. Each participant evaluated 33 hypothetical entrepreneurial profiles, resulting in 10,496 total assessments. The average participant was 39.5 years old, 52% female, 48% held a bachelor's degree or higher, and 92% had no prior entrepreneurial experience.

We used a paper-based instrument consisting of a pre-survey, a conjoint experiment, and a post-survey. The experiment employed a fractional factorial design to create 16 distinct entrepreneur profiles (replicated for reliability, plus 1 practice profile). Profiles varied across five attributes: personal income, societal influence, company profit, employee growth (all high vs. low), and gender. The dependent variable was a combined 7-point Likert scale rating of likability and competence (α = .76). Independent variables included the profile attributes; moderator variables included participants' entrepreneurial intention (EI, α = .91), and control variables included age, gender, education, income, and prior entrepreneurial experience.

To model participants' evaluation policies, we followed Schüler et al., (2023), using an approach that incorporates intraclass correlation coefficients (ICCs) to better capture profile reliability within nested data, improving upon traditional hierarchical linear modeling (HLM).

RESULTS

We tested four hypotheses using 10,496 evaluations from 328 respondents. Hypothesis 1, predicting that observers with higher entrepreneurial intention (EI) evaluate high-income entrepreneurs more positively, was supported. The interaction between income and EI was significant (b = 0.12, $p < .05$, see Appendix A), and the slope for high EI was significantly steeper ($\Delta\beta$ = 0.25, SE = 0.09, z = 2.77, $p < .01$). This pattern is visualized in Appendix A1. Hypothesis 2, suggesting that observers with higher EI rate entrepreneurs with greater societal influence more favorably, was not supported (interaction $p > .05$). Hypothesis 3, proposing that observers with lower EI evaluate high-profit ventures more positively, was supported. The interaction between profit and EI was negative and significant (b = -0.13, $p < .01$), with a significant slope difference ($\Delta\beta$ = -0.26, SE = 0.09, z = 2.88, $p < .01$; see Appendix A2). Hypothesis 4, stating that lower-EI observers evaluate entrepreneurs with higher venture growth more positively, received mixed support. While the interaction was significant (b = -0.05, $p < .05$), the slope difference was only marginally significant ($\Delta\beta$ = -0.18, SE = 0.10, z = 1.76, p = .08; see Appendix A3). Robustness checks excluding and including respondents with past entrepreneurial experience confirmed the direction and significance of interaction effects across models.

DISCUSSION AND IMPLICATIONS

This study makes three key contributions to the literature on the social evaluation of entrepreneurs. First, by introducing a self-categorization perspective (Turner, 1999), we show that observers' own entrepreneurial intentions (EI) shape how they evaluate entrepreneurial performance. Specifically, observers higher in EI evaluate entrepreneurs more favorably when they exhibit personal performance (i.e.,, high income and societal influence), whereas observers lower in EI respond more positively to venture-level performance (i.e., high profit and employee growth). This expands on prior work that focused primarily on attribution theory (Shepherd & Patzelt, 2015) and the similarity hypothesis (Franke et al., 2006; Ding et al., 2023; Küsshauer & Baum, 2023), offering a novel lens through which to understand observer heterogeneity.

Second, our findings reveal new complexity in how observers interpret entrepreneurial success. Whereas earlier work has emphasized the evaluation of failed entrepreneurs (Cardon et al., 2011; Shepherd & Patzelt, 2015), our study distinguishes between personal and venture performance in the context of successful entrepreneurs. This supports the notion that high performance is not a universally valued signal and suggests that evaluations are shaped by more than just the presence or absence of success (Angel et al., 2018; Razmus & Laguna, 2018; Wach et al., 2016).

Third, the study informs research and practice on entrepreneurial role models. Prior studies have shown that high-performing role models influence observers' entrepreneurial intentions (Brunel et al., 2017; Nowiński & Haddoud, 2019), often assuming a one-size-fits-all effect. Our findings suggest a more nuanced reality: the perceived relevance of a role model may depend on the kind of performance they exhibit and the EI of the audience. This insight can inform entrepreneurship education and outreach by tailoring role model narratives to the audience—emphasizing personal achievements in high-EI contexts (e.g., voluntary entrepreneurship programs) and venture outcomes in low-EI settings (e.g., general education courses) (Abbasianchavari & Moritz, 2021).

Despite its strengths, this study has limitations that offer opportunities for future research. While conjoint analysis minimizes bias and offers realistic approximations (Shepherd & Patzelt, 2015; Lohrke et al., 2010), future studies could replicate our findings using real entrepreneurs or richer stimuli (Brown, 1972; Kollmann et al., 2009). Moreover, while we focused on EI as a key form of self-categorization, alternative identity constructs (Fauchart & Gruber, 2011; Cardon et al., 2009) and broader measures of entrepreneurial identity (Farmer et al., 2011; Gregori et al., 2021) may provide further explanatory power. Finally, given that this study was conducted in Germany, future work should explore whether our results hold in more collectivist cultures, where communal outcomes may carry greater weight (Hofstede et al., 2010; Stephan & Uhlaner, 2010).

CONTACT: Julia Kirsch; Julia.Kirsch@tum.de; Technical University of Munich; Lichtenbergstr. 6, 85748 Garching, Germany

Appendix A

Results of Pooled Regressions with Clustered Standard Errors for Overall Evaluation of Entrepreneurs.
Note. N = 10,496 observations nested in 328 observers. * indicates $p < .05$; ** indicates $p < .01$; ***

	Model Results	
	B	**SE**
Intercept	3.06***	0.27
Entrepreneurs' characteristics		
Gender	0.06***	0.01
Income	-0.01	0.04
Influence	0.56***	0.02
Profit	1.60***	0.05
Employee growth	0.86***	0.03
Observers' characteristics		
Entrepreneurial Intentions (*EI*)	0.01	0.04
Observers' Gender	-0.10	0.06
Observers' Age (log)	-0.11	0.08
Tertiary Education	0.06	0.05
High Income	0.14*	0.06
Entrepr. Exp.	0.11	0.08
Self-employment		
EI x Income	0.12*	0.06
EI x Influence	0.07	0.05
EI x Profit	-0.13**	0.04
EI x Growth	-0.05*	0.03
ICC †		
AIC (BIC)	33194.51 (33317.91)	

indicates $p < .01$ † Intraclass correlation coefficient as proposed by Hox (2002).

Appendix B

Moderating effect of observers' EI on the relationships between entrepreneurs' performance indicators and observers' evaluations.

1.

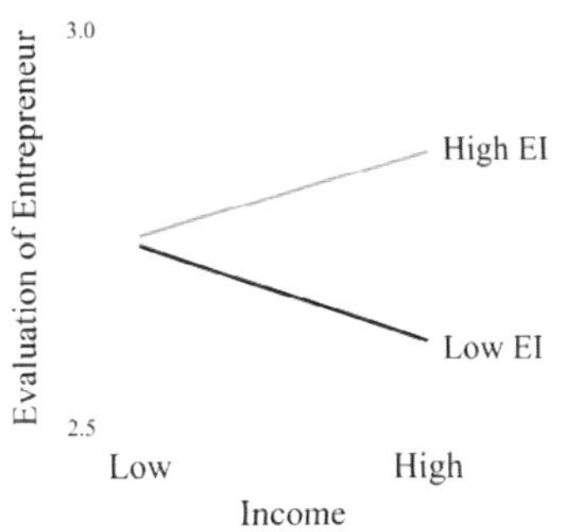

2.

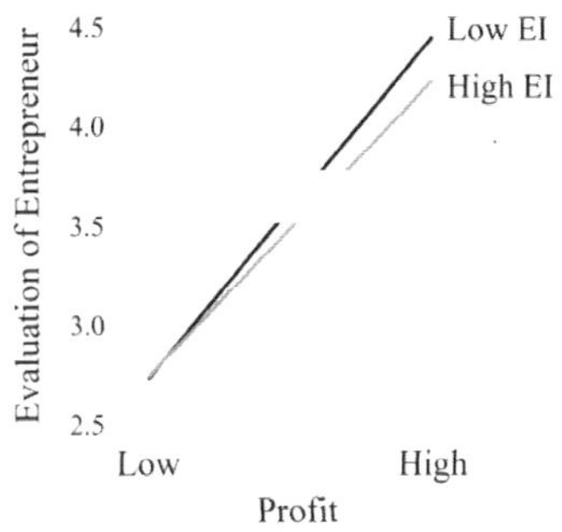

3.

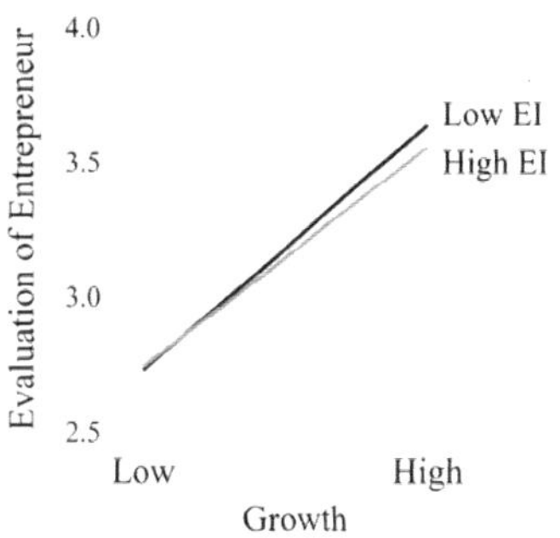

GLIMPSING THE FUTURE: AI-AUGMENTED DECISION-MAKING & THE PROBLEM OF UNCERTAINTY

Judy Rady, Virginia Tech, United States
David Townsend, Virginia Tech, United States

ABSTRACT

Drawing on debates about the boundary conditions of AI in entrepreneurship, this study investigates the role and impact of AI systems as epistemic technologies that help address the problems of Knightian uncertainty in entrepreneurship. We adopt a mixed methods approach: an agent-based simulation study using AI agents to model entrepreneur–AI ensembles to address the problems of Knightian uncertainty and an experimental design to . Our analysis illuminates why some entrepreneur-AI ensembles more effectively tackle Knightian uncertainty and identifies key contingencies that shape outcomes. Overall, this research clarifies the complementary strengths of human–AI partnerships in critical entrepreneurial tasks. Finally, it offers a framework for future exploration and practical guidance for entrepreneurs to take effective action in the age of AI.

INTRODUCTION

A foundational premise in entrepreneurial action research posits that entrepreneurs operate in environments impacted by Knightian uncertainty – a fundamental type of uncertainty that manifests in situations where entrepreneurs do not know what will be possible (or impossible) in the future (Knight, 1921; Townsend, Hunt, & Rady, 2024a). To address the problems of Knightian uncertainty, prior research has identified a variety of tools, strategies, frameworks, and technologies entrepreneurs use to take action in environments they cannot fully understand, know, or predict (Shepherd, Williams, & Patzelt, 2015). In recent studies, a debate is emerging among entrepreneurship scholars regarding the impact of artificial intelligence (AI) tools in shaping how entrepreneurs address the problems of Knightian uncertainty (Ramoglou, Schaefer, Chandra, & McMullen, 2024; Townsend, Hunt, Rady, Manocha, & Jin, 2023, 2024b). While few dispute the transformative impact of AI tools in entrepreneurship (Lévesque, Obschonka, & Nambisan, 2022; Townsend & Hunt, 2019), important questions are emerging regarding the extent to which AI tools are able to tackle the unique epistemic challenges posed by Knightian uncertainty. At the core of this debate are questions regarding the degree to which AI systems, regardless of their relative sophistication and effectiveness, are still subject to important epistemic boundary conditions imposed by Knightian uncertainty (Townsend et al., 2023). The stakes of this debate are practically and economically significant as researchers, policymakers, and practitioners all point to the transformative benefits of both startups and established firms proactively becoming "AI leaders" through early implementation of AI tools to improve, transform, and modernize their decision-making processes (PwC, 2022).

Given the importance of this debate, the purpose of this study is to investigate the role and impact of AI systems as epistemic technologies that enable entrepreneurs to address the problems of Knightian uncertainty. Building on recent debates regarding the boundary conditions of AI in entrepreneurship (e.g., Townsend et al., 2023; Ramoglou et al., 2024; Townsend et al., 2024b), we further advance this debate by focusing on the role of AI-entrepreneur ensembles (Choudhary, Marchetti, Shrestha, & Puranam, 2023; Raisch & Fomina, 2023) in addressing the problems of Knightian uncertainty. We ground the logic of our approach on the premise that the combined, complementary strengths of human entrepreneurs and AI systems working together provide distinct advantages over each individual actor working in isolation on critical entrepreneurial tasks. Towards this end, our study is motivated by two fundamental questions: (1) Why are some entrepreneur-AI ensembles more effective than others in addressing the problems of Knightian uncertainty? (2) What are the critical contingencies that shape the effectiveness of entrepreneur-AI ensembles in addressing the problems of Knightian uncertainty?

We address these questions through a novel mixed methods design, which combines an experimental study to examine the heterogeneous ways entrepreneurs cognitively address the problems of Knightian uncertainty with a complementary study that dynamically simulates with an agent-based model (ABM) how different entrepreneur-AI ensembles address the problems of Knightian uncertainty. In doing so, our proposed study provides several fundamental contributions to research on the complementary roles of human entrepreneurs and AI systems in entrepreneurial action. First, we address a set of critical questions in entrepreneurship regarding the extent to which the complementary strengths of humans and AI systems enable these ensembles to address the problems of Knightian uncertainty more effectively than each actor working in isolation. Second, we identify and explore the important contingencies that shape the effectiveness of different types of entrepreneur-AI ensembles. Lastly, we develop and extend novel computational methods to examine fundamental research questions in the field, providing a useful roadmap for future research to leverage these methods in new studies.

HYPOTHESES DEVELOPMENT

In the field of entrepreneurship, Knightian uncertainty plays a foundational role in theory development (Ramoglou, 2021; Sarasvathy, 2001; Townsend et al., 2018), describing in part how and why the inability to know what is possible or impossible in the future impacts entrepreneurial action (McMullen & Shepherd, 2006). In situations where "decision makers [have an] indefinite number of possibilities to consider" (Packard, Clark, & Klein, 2017: 9), the problems of Knightian uncertainty create both technical and practical constraints on the predictive abilities to anticipate the probably consequences of any actions or decisions they are contemplating. In essence, irrespective of how much information an entrepreneur may gather or regardless of the sophistication of their analytical tool, the problem of "finite" intelligence still limits their capacity to grasp "all the possible combinations of factors that may influence a decision or action" (Townsend et al., 2024a: 460).

Such cognitive limitations hinder the ability of entrepreneurs to resolve key knowledge problems that emanate from continuous and dynamic environmental changes (McKelvie, Haynie, & Gustavsson, 2011), which limits an entrepreneur's ability to "interpret and respond to uncertainty" (McKelvie et al., 2011). To address these cognitive challenges, entrepreneurs use a variety of tools, strategies, frameworks, and technologies to address critical knowledge problems (Grégoire, Corbett, & McMullen, 2011; Grégoire, Cornelissen, Dimov, & Van Burg, 2015; Mitchell et al., 2002, 2004; Townsend et al., 2018). In recent studies, entrepreneurship scholars theorize that AI tools will play an increasingly important role in entrepreneurial task performance across various activities such as customer service, scaling, selling, and even idea production (Chalmers, MacKenzie, & Carter, 2021), content creation (Short & Short, 2023), decision-making (Ramoglou et al., 2024; Townsend et al., 2023, 2024b), enhancing the "quality of doing business" across a growing number of critical tasks (Bogachov, Kwilinski, Miethlich, Bartosova, & Gurnak, 2020: 487).

The advantages of using these tools extend from the growing recognition that intelligent machines are able to "perform tasks and solve complex problems that would normally require (or exceed) human intelligence, reasoning, and prediction power needed to adapt to changing circumstances" (Obschonka & Audretsch, 2020: 4). In this regard, entrepreneurs can utilize AI "as a tool to help them increase their chances of entrepreneurial success" (Shepherd & Majchrzak, 2022: 2). Recent studies in the practitioner literature reflect such optimism as entrepreneurs believe AI enhances their decision-making while reducing their risk of errors (Haan, 2023). Not surprisingly, AI adoption has increased among startups and small businesses by 415% since 2016 (Orion Policy Institute, 2024).

However, despite such optimism, recent studies argue that AI systems are still subject to boundary conditions stemming from key knowledge problems (Obschonka & Levesque, 2024), especially the problems of Knightian uncertainty: actor ignorance, practical indeterminism, agentic novelty, and competitive recursion (Townsend et al., (2023). Collectively, these problems challenge AI systems to "proactively intervene in decision environments without the specification of prior guidelines or rules" (Townsend et al., 2023: 13). Such problems extend from epistemic limitations of AI systems to infer or generalize from previous experiences, lack of control of others' actions within the environment, and the inability to predict the future consequences of their action accurately (Townsend et al., 2023).

Despite these limitations, "this does not mean that these powerful systems are not useful in the field of entrepreneurship" (Townsend et al., 2024b: 1). As recent studies suggest, AI systems can directly enhance entrepreneurial decision-making and new venture performance (Amoako, Omari, Kumi, Agbemabiase, & Asamoah, 2021) by providing powerful and useful tools to enhance entrepreneurial decision-making. In this regard, as entrepreneurs build on a "reciprocal interaction between entrepreneurial people and machine intelligence" (Chalmers et al., 2021: 6) in a manner that combines the complementary inputs and actions of both actors (Choudhary et al., 2023; Raisch & Fomina, 2023) humans and AI systems contribute in complementary ways towards critical entrepreneurial tasks.

Entrepreneur-AI ensembles emerge through a "co-evolutionary process during which humans learn from machines and machines learn from humans" (Raisch & Krakowski, 2021: 10). At the core of this co-evolutionary learning process is a state of conjoined agency where humans and machines work in concert with each other to perform critical organizational tasks (Murray et al., 2021). The emergence of entrepreneur-AI ensembles is contingent on the agentic capabilities – i.e., "a temporally-embedded capacity to *intentionally* constrain, complement, and/or substitute for the work" (Murray et al., 2021: 4) of both humans and AI systems. Both humans and AI systems vary in these capabilities, thus generating a set of fundamental trade-offs in how different entrepreneur-AI systems ensembles are capable (or not) of contending with the problems of Knightian uncertainty. In the following section, we now discuss our proposed study to explore these questions and to shed further light on the emergence and contingent role of entrepreneur-AI ensembles in contending with the problems of Knightian uncertainty.

METHODS

To explore these questions, we use a mixed methods design, combining the strengths of a complementary set of empirical and analytical methods to address our primary research questions. First, we will conduct a survey and experiment study to analyze how different cognitive factors shape how entrepreneurs address to the problems of Knightian uncertainty. This first stage of the analysis illuminates how different configurations of cognitive traits and abilities shape entrepreneurial actions taken in response to the four problems of Knightian uncertainty.

We then complement our initial analysis through the use of emerging analytical approaches that explore novel research questions through the design and testing of agent-based models in entrepreneurship research (Bort, Wiklund, Crawford, Lerner, & Hunt, 2024; Crawford, 2009; Lévesque & Stephan, 2020; Sunny Yang & Chandra, 2013) and in theory development (Davis, Eisenhardt, & Bingham, 2007). In particular, we construct different types of entrepreneur-AI ensembles that we then use to simulate real interactive behaviors in a dynamic environment. While novel, this approach is specifically pertinent to our research inquiry as it combines the strengths of "simulation (which is)…used to create structures that offer a relatively transparent analysis of key causal determinants (McKelvey, 2004a)" (Crawford, 2009: 5), building on recent studies in entrepreneurship using agent-based simulations (Crawford, 2009; Sunny Yang & Chandra, 2013), in an "approach to entrepreneurship (where researchers control the) "design" process and variations of the inner and outer environments can be studied extensively and systematically by its designer (e.g., the researcher)" (Sunny Yang & Chandra, 2013: 211).

However, the process through which we design the ABM model provides several key advantages over existing agent-based methods by enhancing the "observation, planning, and reflection" of these AI agents in a manner that creates realistic patterns of social interaction among human actors (Park et al., 2023). Notably, our proposed mixed methods provide us with the means to address one of the most important limitations of agent-based modeling regarding the imposition of deterministic rules in shaping agent behaviors. In our proposed approach, we can design and analyze more open-ended social behaviors, which we believe will reflect, in significant ways, the actual behaviors of human entrepreneurs.

RESULTS

Consistent with these arguments, we analyzed the impact of AI-augmentation on the ability of different archetypes of entrepreneurs – creators, scientists, profiteers, disruptors, and conservatives – to address the problems of Knightian uncertainty. We first examined forecasting accuracy by computing each archetype's mean prediction error as AI strength increases. Across the board, AI sharply reduces error: "Scientist" and "Profiteer" agents, who initially had the worst forecasts, see their average error plummet by 12–13 percentage points, while "Creator" agents—who began with relatively modest errors—improve by only about 4 points. "Disruptors" and "Conservatives" occupy the middle ground. In effect, AI levels the informational playing field: those who most needed help see the largest gains, compressing performance disparities among archetypes.

Next, we measured profit-outcome dispersion by recording the sample variance of realized payoffs across all simulation rounds. Here, the paradox of AI emerges although agents forecast more accurately, payoffs become more volatile. "Profiteer" agents—the most risk-seeking type—exhibit the highest absolute variance, which more than doubles at maximum AI strength. "Creators" and "Scientists" also experience a 30–40 percent increase in profit variability, whereas "Disruptors" see only a mild uptick. "Conservatives," despite their conservative stance, register the steepest relative rise (from a very low baseline), although they remain the least variable in absolute terms. Thus, AI simultaneously reduces ignorance and amplifies outcome indeterminism, especially for archetypes predisposed to aggressive investment.

To capture agents' capacity to create genuinely new niches, we analyzed radical innovation events per agent. The results highlight a strong selection effect. "Creators," already the most experimental type, deepen their innovative output, increasing from an average of about 16 innovations per run to over 17 at high AI intensity. "Scientists" follow, rising from around 13 to 14.5. By contrast, "Disruptors" remain largely flat, "Profiteers" show only modest growth, and "Conservatives" seldom innovate, even with powerful AI support. This suggests that AI fuels "agentic novelty" primarily in those archetypes whose intrinsic traits predispose them toward exploration; it does not magically transform every agent into an innovator.

Finally, we analyzed how often agents switch between niches as an index of adaptive learning from peers. Here we observe a bifurcated pattern: "Scientists," "Profiteers," and "Conservatives" slightly increase their switch frequency with more AI, indicating a more fluid competitive landscape. In contrast, "Disruptors"—initially the most restless group—actually consolidate more, reducing their switching rate. "Creators" remain essentially unchanged. In other words, AI subtly re-tunes social learning: risk-tolerant types may double down on successful strategies, while others expand their search for under-explored opportunities.

Overall, our four analyses illuminate a critical set of paradoxical challenges facing AI-augmented entrepreneurs as they tackle the problems of Knightian uncertainty: as AI capability grows, agents become better informed (reduced ignorance) yet face greater outcome dispersion (heightened indeterminism), and only a subset of archetypes effectively harness AI for radical experimentation (selective novelty), while peer-learning dynamics shift in contrasting ways across types (competitive recursion). In practical terms, AI empowers more accurate decision-making but also magnifies strategic stakes—and the generative power of AI is fully unlocked only by those agents whose inherent predispositions align with exploration and risk-taking.

DISCUSSION AND IMPLICATIONS

The purpose of this study is to examine how the augmentation of entrepreneurial action with AI affects entrepreneurial agents confronting Knightian uncertainty along four key dimensions: actor ignorance, practical indeterminism, agentic novelty, and competitive recursion. By embedding AI-augmented decision rules within an agent-based simulation, we show that AI fundamentally reshapes both individual forecasting and collective market dynamics, but also surfaces important boundary conditions on its impact on entrepreneurs.

First, we confirm that stronger AI sharply reduces forecasting error for all agent types, but the magnitude of improvement is inversely proportional to initial skill: worst forecasters ("Scientists," "Profiteers") improve the most (~12–13 pp), while already-skilled "Creators" gain only modestly (~4 pp). This leveling effect highlights AI's potential as an epistemic equalizer, buttressing weaker actors' information-processing capacities. From a theoretical standpoint, these results substantiate the promise of AI as a tool to mitigate "actor ignorance" (Townsend et al., 2023), but they also underscore that AI does not equally empower all agents: its marginal benefit is greatest where prior human forecasting skill is lowest.

Paradoxically, our second analysis shows that more accurate forecasts do not translate into more stable outcomes. Instead, profit-outcome variance increases—most dramatically for risk-seeking "Profiteers" (variance more than doubles) and moderately for "Creators" and "Scientists" (+30–40 percent). Even "Conservatives," from a low variance baseline, exhibit the steepest relative rise. These findings speak to a dark side of AI augmentation: by enabling higher-confidence investment and production decisions, AI unleashes amplified payoff dispersion, intensifying "practical indeterminism." In effect, AI compresses ignorance but magnifies outcome swings, which may exacerbate entrepreneurial risk rather than attenuate it.

AI also exhibits a paradoxical impact on radical niche-creation events, revealing that AI fuels "agentic novelty" selectively. "Creators" and "Scientists," endowed with intrinsic exploratory dispositions, substantially increase their innovation counts with stronger AI support, whereas "Disruptors," "Profiteers," and "Conservatives" remain largely inert. These results suggest that AI does not democratize innovation potential. Instead, AI acts as an amplifier of preexisting tendencies toward experimentation. The capacity to harness AI for generating genuinely new market niches thus depends critically on underlying human–machine ensemble characteristics, reinforcing the idea of conjoined agency (Raisch & Krakowski, 2021) rather than AI as a "creative equalizer."

Finally, we show that AI reshapes social learning and adaptive search in nuanced ways. "Disruptors," initially the most active, consolidate more with added AI—reducing their niche-switch frequency—whereas "Scientists," "Profiteers," and "Conservatives" explore more. In contrast, "Creators" are less prone to search. This pattern implies that AI can both dampen and amplify peer-learning loops depending on archetypal traits, fine-tuning the competitive recursion dynamics that underlie market evolution.

Overall, these findings advance theory on the boundary conditions of AI in entrepreneurial action. They corroborate that AI effectively reduces epistemic uncertainty (ignorance) but simultaneously amplifies decision uncertainty (indeterminism), and that its benefits for novelty and adaptive search are moderated by agents' intrinsic dispositions. For practitioners, this suggests that AI adoption should be managed to augment key entrepreneurial strategies: firms seeking to stabilize payoffs may need countervailing risk-management mechanisms, while those aiming for radical innovation should cultivate human-AI teams with high exploratory traits.

In sum, this study illuminates both the empowering and unsettling effects of AI on entrepreneurial uncertainty: while AI narrows knowledge gaps, it also enlarges the variance of outcomes and only selectively unlocks innovation and adaptive search. Understanding these trade-offs is essential for theorizing AI's role in entrepreneurship and for designing human-AI ensembles that truly address the multifaceted challenges of Knightian uncertainty.

CONTACT: David Townsend, dtown@vt.edu; Virginia Tech; 880 West Campus Drive, Pamplin Hall Room 2090, Blacksburg, VA 24061, United States

SUMMARY

ENTREPRENEURIAL TRANSFORMATION IN RESOURCE-CONSTRAINED SETTINGS: INSIGHTS FROM BURUNDI

Sondos G. Abdelgawad, IE Business School, Spain
Mujtaba Ahsan, San Diego State University, United States
Ficawoyi Donou-Adonsou, John Carroll University, United States
Samuel Mathey, FAFEDE, United States

Principal Topic

Research on necessity entrepreneurship in poverty contexts has yielded mixed results. Formal interventions often fail due to poor institutional fit (Claus et al., 2021), while personal initiative (PI) training may not instill lasting self-starting behaviors (Frese, 2024). These approaches focus on top-down institutional changes or preparing entrepreneurs to adapt but do not foster true independence. Our study examines the Independent Venture Program (IVP), a PI intervention that nurtures self-starting behaviors while reducing dependence on external support. Conducted in Burundi, one of the world's poorest nations, we explore how intensive training and long-term peer mentorship shift mindsets and encourage sustainable entrepreneurship. We examine the mentor-entrepreneur relationship, providing insights into resilience-building in resource-scarce environments (Shepherd et al., 2020).

Method

This study draws on qualitative data from field observations and semi-structured interviews with IVP entrepreneurs and mentors in Burundi, where subsistence agriculture dominates and youth underemployment is high. Focusing on a subset of 100 participants across various sectors, we identified three key themes: (1) mindset shift toward self-reliance, (2) practical application of entrepreneurial knowledge, and (3) mentor-entrepreneur dynamics in navigating resource constraints.

Results and Implications

Our findings highlight the role of mindset shifts, practical application, and mentoring in fostering sustainable entrepreneurship in resource-scarce environments. While IVP nurtured self-starting behaviors, challenges remained in translating these shifts into sustained business operations. A mindset shift was central to participants' development of self-reliance. Thierry reflected, "I understood that nobody is going to help me. I have to rely on myself." Similarly, Sylvie and Alexis recognized they could launch businesses without capital, illustrating IVP's effectiveness in fostering entrepreneurial agency. These cognitive transformations enabled participants to navigate financial constraints proactively. However, practical application remained challenging, as entrepreneurs struggled to implement learned strategies. Alain restructured his labor force post-training, while Thierry applied online agricultural techniques to improve productivity. These cases suggest that while agency is crucial, structured guidance enhances skill application. Mentoring relationships provided critical support, but mentors faced difficulties balancing their own ventures with coaching commitments. Some, like Clément's mentor, helped navigate regulatory barriers, yet inconsistent communication sometimes limited impact. Sustained engagement mechanisms could strengthen these relationships. Despite IVP's success, resource constraints remained a barrier to long-term viability. Entrepreneurs struggled with securing steady sales and transitioning to full-time ventures. Sylvie's reliance on sporadic work illustrates these ongoing challenges, underscoring the need for broader structural support. Overall, IVP effectively fostered self-starting behaviors and resilience, yet long-term sustainability requires continued mentorship and ecosystem support.

CONTACT: Sondos G. Abdelgawad; Sondos.Abdelgawad@ie.edu; IE Business School; IE Tower, P. de la Castellana, 259, Madrid, Madrid 28046, Spain

SUMMARY

HARNESSING UNEXPECTED EVENTS TO TACKLE GRAND CHALLENGES: THE EMERGENCE AND CULTIVATION OF SERENDIPITY IN EXTREME POVERTY

Christian Busch, University of Southern California, United States

Principal Topic

Serendipity—the occurrence of unexpected yet valuable discoveries—plays a critical role in entrepreneurial discovery, innovation, and venture development. While prior research has identified serendipity-enhancing factors such as individual and organizational antecedents, prior work has focused on resource-rich contexts such as Silicon Valley or well-funded incubators in emerging markets, and little is known about how entrepreneurs in truly resource-constrained environments leverage serendipity. We explore how entrepreneurial ventures in settings of extreme poverty—marked by pervasive scarcity and systemic shocks—can transform unpredictable events into opportunities to address grand challenges and pursue impact-oriented growth.

Method

To investigate how serendipity emerges in poverty contexts, we conducted a longitudinal, nested case study of a South African entrepreneurial venture addressing SDG 4 (Quality Education). Employing an integrative theory-building approach, we examined two cases of "serendipity gained" and two of "serendipity lost," drawing on interviews, archival materials, and field observations. This design allowed us to analyze the interplay between episodic events and enduring organizational features, offering insight into how entrepreneurial serendipity unfolds under conditions of extreme resource scarcity.

Results and Implications

Our analysis surfaced key mechanisms that enhance organizational serendipity in poverty contexts. These mechanisms explain how ventures can identify and act upon latent, non-obvious resources in highly uncertain environments. We contribute to the entrepreneurship literature by theorizing "necessity-based serendipity," highlighting how serendipitous value can emerge in the absence of traditional entrepreneurial enablers such as financial capital or formal expertise. Second, we extend research on temporality and serendipity, showing that what is interpreted as "serendipitous" may shift over time depending on contextual dynamics. Finally, we contribute to research on grand challenges by illustrating how entrepreneurial ventures in poverty settings locally tackle systemic societal problems through adaptive sensemaking and contextual recombination. Together, we offer a multi-level theory of necessity-based serendipity, providing a framework for understanding how entrepreneurs in extreme settings can turn unexpected shocks into meaningful, context-specific solutions.

CONTACT: Christian Busch; christian.busch@usc.marshall.edu; University of Southern California; 670 N Casas Ave, Pacific Palisades, California 90272, United States

SUMMARY

MASS SHOOTINGS AS AN EXTERNAL DISABLER OF ENTREPRENEURSHIP

Nicolas Gibbens, University of Louisville, United States
Daniel Bennett, University of Louisville, United States
Sheila Guo, University of Oklahoma, United States
Paul Sanchez-Ruiz, Iowa State University, United States

Principal Topic

Mass shootings are sporadic, traumatic events that instill fear and uncertainty in local communities (Greenbaum & Tita, 2004). Affected residents may anticipate recurring violence and adjust their behaviors to mitigate perceived risks (Garofalo, 1979). Although prior research has linked mass shootings to declines in economic output, employment, and human capital (Yousaf, 2020), their impact on entrepreneurial activity remains underexplored. Given that crime generally dampens business activity (Sanchez-Ruiz et al., 2023), understanding how the unexpected shock of a mass shooting influences business exits is crucial. This study therefore asks: How do mass shootings affect local business exit rates?

Method

We integrate fear of crime theory (Garofalo, 1981) with the external enabler framework (Kimjeon & Davidsson, 2021) to develop a conceptual model linking mass shootings to increased business exits. By heightening uncertainty and reducing customer demand (Bowes, 2007), mass shootings prompt more business closures, serving as an external disabler of entrepreneurship (Bennett, 2019). We further examine how institutional factors—local political ideology and state gun control policies—moderate this relationship. To test our model, we assemble a U.S. county-level longitudinal dataset (1978–2022) using data from FBI Supplemental Homicide Report, Business Dynamics Statistics, RAND State Firearm Law Database, MIT Election Lab, and other sources, employing a staggered difference-in-differences design with an event study framework.

Results and Implications

Although data collection and analysis are in progress, we anticipate finding that mass shootings are causally related with increased business exits rates, and that this positive effect is weaker in more political conservative communities and states with less restrictive gun control policies. Our findings will yield important implications for entrepreneurship theory and practice. Practically, these findings will further our understanding of the economic cost of mass shootings by illustrating their potential to disrupt local communities by facilitating business closures. Theoretically, our study contributes to the external enablement literature by introducing mass shootings as a novel form of local environmental change that functions as an external disabler. In doing so, we add a new type of environmental shock to the literature and explore its interaction with multi-level environmental factors, thereby extending the disablement framework advanced by Bennett et al.,(2024). Our study advances theory by illustrating how multi-level institutional interactions can either exacerbate or mitigate the disabling effects of adverse events, enriching our understanding of the dynamic interplay between environmental shocks and entrepreneurial outcomes.

CONTACT: Nicolas Gibbens; NTGIBB01@louisville.edu; University of Louisville; 611 Belamy Circle, Louisville, Kentucky 40208, United States

SUMMARY

THE IMPACT OF STARTUP EMPLOYMENT ON INDIVIDUALS' EARNINGS

Elena Kulchina, North Carolina State University, United States
Florence Honoré, University of Wisconsin-Madison, United States

Principal Topic

Research suggests that employees who have worked at a startup earn less than employees of established firms. We investigate whether the negative effect varies with the heterogeneity of joiners' employment terms and tenure at the startup. Prior literature suggests that former entrepreneurs receive earnings penalties because they are viewed as need-based as opposed to opportunity-based, they are believed to have a low commitment to paid employment, they do not develop skills for established firms, and they face the stigma of failure. Startup employees may receive earning penalties for similar reasons. Thus, we expect that startup employees who move to a startup with an improvement in contract terms (opportunity seekers), signal greater job commitment by longer employment at a startup, and are employed at growing startups should receive a lower income penalty or a reward.

Method

We examine the heterogeneity of long-term earning consequences of startup employment in a sample of Spanish employees observed between 2008 and 2017. We use a difference-in-differences model comparing full-time employees of established firms who joined startups (joiners) with individuals who switched between established firms (switchers) and individuals who stayed with their prior employers (stayers) during the year when joiners and switchers changed jobs. To ensure that our focal and comparison groups are similar to each other, we match individuals on their observed characteristics and earnings in the year prior to the change.

Results and Implications

We find that, on average, startup joiners improve their earnings more than those who stay with their original employers but less than employees who switch established firms, which is consistent with prior work. However, we see significant differences with the heterogeneity of the employee contract terms. Startup joiners who improved their employment terms, such as salary and job stability, when joining a startup (opportunity seekers) increase their long-term earnings as much or even more than switchers of established firms who received the same improvements. Joiners who have a longer tenure at a startup (signal of job commitment) also do better than an average switcher of established firms. Furthermore, we find no negative effect of working at a not-growing versus growing startup, thus suggesting against a failed-startup stigma for employees. Importantly, employees who joined startups with the improvement of their contract terms likely retained their advanced contract terms at their next job. Our results further suggest that employees who joined a startup without improved contract terms likely had few outside options or were future entrepreneurs-in-training.

CONTACT: Elena Kulchina; elena.kulchina@ncsu.edu; North Carolina State University; 620 Halcyon Meadow Dr., Cary, North Carolina 27519, United States

SUMMARY

PROPERTY RIGHTS ENHANCED: LAND USE RIGHTS REFORM IN CHINA THAT STIMULATES ENTREPRENEURSHIP

Kun Liu, Kent State University, United States
Xueru Yang, South China Agricultural University, China

Principal Topic

Despite the importance of property rights (McMullen, et al., 2008; Townsend & Busenitz, 2008), research lacks causal evidence because property rights are often endogenous with entrepreneurship. To examine the causality, we have used a quasi-natural experiment in the setting of land titling reforms in China's agriculture. We develop three hypotheses as follows: Hypothesis 1. The certification of land use rights increases possibilities of farmer entrepreneurship. Hypothesis 2. The effect in H1 would be stronger among those marginalized individuals than those in privileged groups. Hypothesis 3. The effect in H1 would be stronger in areas where more market institutions are available than other areas with less developed market institutions.

Method

The data comes from China Labor-force Dynamic Survey (CLDS) by the Social Science Survey Center of Sun Yat-sen University. The sample includes 46,600 individuals. The dependent variable is a dummy variable for self-employment. The independent variable is the DID. Moderators at the individual level include variables such as social status, membership in the Chinese communist party, migrant experiences, etc. For contextual moderator variables, we included variables that measure whether there are weak or strong business environments.

Results and Implications

Our stagger DID showed that the reform significantly increased the odds of an individual farmer becoming an entrepreneur. This result remained robust when we completed parallel trend testing and dynamic effects analyses. Our results also supported moderation effects: 1) in the post-reform period, individuals with marginalized social status are more likely to become entrepreneurs; and 2) in this period, individuals from more developed areas are more likely to become entrepreneurs. This study makes several contributions. First, using a quasi-natural experiment in China, this study presents clear evidence, albeit indirect, of the causality between strengthened property rights and entrepreneurship. Second, the heterogeneity among farmers matters during the regime change. We focus on social identity and suggest that marginalized individuals benefit more from property rights reforms because they now can play on a leveled field. This finding is interesting because property rights research is mostly at the institutional or macro level (McMullen, et al., 2008) and our study attempts to integrate individual level variances within the property rights framework. Third, to make property rights fully effective requires an integrated system of complementary social institutions. Our results show that in post-reform period farmers are more likely to become entrepreneurs in those areas with more developed market institutions. A holistic approach that considers potential interdependencies among institutions is needed.

CONTACT: Kun Liu; kliu10@kent.edu; Kent State University; 800 E Summit St, Kent, Ohio 44242, United States

SUMMARY

HOW DO MINIMUM WAGES AFFECT NONEMPLOYER ENTREPRENEURSHIP AND TRANSITIONS IN THE UNITED STATES?

Ke Lyu, University of Nevada Reno, United States
Frank Fossen, University of Nevada, Reno, United States

Principal Topic

This study examines how minimum wage policies influence nonemployer businesses—self-employed individuals without employees—in the U.S. labor market. While existing research focuses on traditional employment effects, we address critical gaps in understanding how wage regulations shape entrepreneurial behavior. Specifically, we investigate whether higher minimum wages reduce nonemployer establishments by altering occupational choices, discouraging transitions to employer status, or increasing exits to wage employment/unemployment. Our analysis further explores heterogeneous effects across demographics and industries. This investigation is crucial for policymakers and managers balancing worker welfare with entrepreneurial ecosystem health, particularly amid ongoing debates about federal and state minimum wage increases.

Method

We combine a theoretical occupational choice model with empirical analyses. The model extends prior frameworks by incorporating unemployment decisions, benefits, and capital inputs, predicting that higher minimum wages reduce nonemployer prevalence through improved wage employment attractiveness. Using county-level data from the U.S. Census Bureau's Nonemployer Statistics (2010–2019), we employ two-way fixed effects and boundary pair fixed effects models to isolate causal impacts. The boundary pair approach compares adjacent counties across state borders to control for confounding factors. We implement staggered and stacked difference-in-differences designs to address variation in state-level policy timing. Individual transition outcomes are analyzed using Business Formation Statistics and Current Population Survey data.

Results and Implications

A $1 minimum wage increase reduces nonemployer establishments by 0.5%–0.9%, with spillover effects suppressing self-employment in neighboring counties. Transitions from nonemployer to employer status decline by 12%, while exits to wage employment rise by 8% and unemployment exits increase by 4%. Women show greater resilience, experiencing 30% smaller negative effects on self-employment entry and 15\% higher success rates transitioning from unemployment to self-employment. The transportation/gig economy sector grows by 3.2%—likely due to workers shifting to uncovered ride-sharing roles when traditional employment becomes scarce. These findings suggest minimum wage hikes may inadvertently stifle entrepreneurship by raising hiring barriers while making wage work relatively more attractive. Managers in gig economies should anticipate labor supply fluctuations as policy changes redirect workers between formal and informal sectors. Policymakers must consider gender-specific effects, where female entrepreneurs demonstrate greater adaptability to wage shocks. Methodologically, our boundary pair and staggered DID approaches provide robust templates for evaluating spatially heterogeneous policies. By linking labor costs to entrepreneurial decision-making, this research advances understanding of policy-entrepreneurship interactions and informs equitable wage regulation debates.

CONTACT: Ke Lyu; lvke1220@gmail.com; University of Nevada Reno; 1775 Evans Ave, Apt.C23, Reno, Nevada 89512, United States

SUMMARY

WEATHERING THE STORM: A QUALITATIVE INVESTIGATION OF ENTREPRENEURS COPING WITH ADVERSITY AND ITS UNDERLYING DRIVERS

Erika Ni, University of Mannheim, Germany
Martina Battisti, Grenoble Ecole de Management, France

Principal Topic

Entrepreneurship is inherently challenging (Rauch et al., 2018), posing the question of how some entrepreneurs persist and even flourish despite repeated setbacks and high levels of stress (Stephan, 2018). While coping theoretically links stress to well-being, empirical research is limited to static analyses of problem-vs. emotion-focused coping (e.g., Nikolaev et al., 2023) and inductive classifications of entrepreneurs' coping strategies (e.g., Thompson et al., 2020). In this study, we build on the notion that entrepreneurs can adapt their coping responses to entrepreneurial stressors (Skinner & Zimmer-Gembeck, 2007). Investigating entrepreneurial coping trajectories we seek to understand how entrepreneurs transition toward healthier coping or why they persist with maladaptive strategies.

Method

To this purpose, we conduct a qualitative interpretative study based on 30 semi-structured interviews with experienced founders of growth-oriented ventures, and an additional 8 interviews with key stakeholders entrepreneurs confide in about emotional matters. This sample allowed us to capture the lived experiences of entrepreneurs who repeatedly faced entrepreneurial stress and had the opportunity to adapt their coping, while triangulating these accounts with external observations. For data analysis, we adopt an abductive approach inspired by Gioia et al.,(2013), focusing on participants' lived experiences, while acknowledging that these are situated in a broader context, requiring us to engage in interpretative efforts (Harley & Cornelissen, 2022; Mees-Buss et al., 2022).

Results and Implications

We identify a common maladaptive coping trajectory driven by entrepreneurs' identification with their ventures (Stephan, 2018). Entrepreneurs on this trajectory appraise stressors as threats, experience fear of failure, and employ predominantly maladaptive coping strategies such as activism without reflection, persistent rumination, and unhealthy lifestyle choices. Our analysis reveals three pathways to adaptive coping. First, entrepreneurs who develop identities beyond their ventures—often triggered by major life events—detach their self-worth from business success and perceive the possibility of failure as less threatening, enabling reflected problem-solving and healthier detachment activities. Second, improving emotion management skills through self-reflection or psychological coaching helps entrepreneurs regulate stress responses, fostering both reflected problem-solving and adaptive emotional coping. Third, a perspective shift, where adverse events are interpreted as "normal business," emerges through honest peer exchanges, promoting emotional detachment and a sense of connectedness. By identifying key transition mechanisms – identity shift, emotional self-efficacy, and normalization of stressors – our findings provide a dynamic view of entrepreneurial coping. Practically, this can inform targeted well-being interventions (Williamson et al., 2021) that promote adaptive coping through self-reflection, identity work, or peer support.

CONTACT: Erika Ni; erika.ni@uni-mannheim.de; University of Mannheim; Schloss Ehrenhof Ost, Room 269, Mannheim, Baden-Wurttemberg 68161, Germany

SUMMARY

MEASURING SCIENTIFIC ENTREPRENEURIAL ACTION: SCALE DEVELOPMENT AND INTERNATIONAL VALIDATION

Anna Paulina Schröder, University of St.Gallen | KMU -HSG, Switzerland

Principal Topic

Scientific entrepreneurship (SE), rooted in the Entrepreneurs-as-Scientists perspective (Zellweger & Zenger, 2023), is an approach where entrepreneurs iteratively form, test, and refine business ideas through structured hypothesis-driven experimentation. Despite its increasing relevance in research and practice, the field has lacked a validated measurement instrument to systematically assess scientific entrepreneurial action. This study addresses this gap by developing and validating a 24-item multidimensional scale that captures the three key action categories—belief formation, belief testing, and response definition—along with their associated doubts and heuristics. The study provides a foundational measurement tool for future research, enabling a systematic exploration of scientific entrepreneurship's antecedents, processes, and outcomes.

Method

The SE scale was developed through a rigorous three-stage process to ensure validity, reliability, and global applicability (MacKenzie et al., 2011). First, item generation was guided by the Entrepreneurs-as-Scientists perspective (Zellweger & Zenger, 2023), conceptualizing three core dimensions—belief formation (BF), belief testing (BT), and response definition (BU)—each with three reflective sub-dimensions: action, doubt, and heuristics. A deductive approach led to an initial item pool, refined through expert evaluations (six scholars, four practitioners), producing a 24-item scale. Next, validation was conducted globally. The SE scale was embedded in the GUESSS 2023/2024 survey across 57 countries, with data from 25,205 student entrepreneurs. EFA (72.8% variance explained) and CFA ($\chi^2 = 431.402$, CFI = 0.950, RMSEA = 0.052) confirmed the reflective structure, while PLS-SEM validated the formative relationships. Finally, cross-validation is planned in eight countries (n = 10,436) to ensure robustness.

Results and Implications

The final SE scale is a validated 24-item multidimensional measure, structured across three core dimensions. Specifically, eight items (3 action, 3 doubts, 2 heuristics) were identified for BF, eight (4 action, 2 doubts, 2 heuristics) for BT, and eight (3 action, 2 doubts, 3 heuristics) for BU, providing a detailed assessment of scientific entrepreneurial action. Nomological validation confirmed SE's theoretical relevance. It is positively predicted by entrepreneurial education ($\beta = 0.296$, $p < 0.001$), university environment ($\beta = 0.139$, $p < 0.001$), and resilience ($\beta = 0.268$, $p < 0.001$). SE, in turn, drives firm performance ($\beta = 0.197$, $p < 0.001$), pivoting ($\beta = 0.352$, $p < 0.001$), number of pivots ($\beta = 0.360$, $p < 0.001$), and innovativeness ($\beta = 0.286$, $p < 0.05$). This scale advances measurement in entrepreneurship, operationalizes the Entrepreneurs-as-Scientists perspective, and enables large-scale research. It provides practical applications for educators, policymakers, and entrepreneurs, supporting evidence-based decision-making in venture creation and scientific entrepreneurial behavior.

CONTACT: Anna Paulina Schröder; annapaulina.schroeder@unisg.ch; University of St.Gallen | KMU -HSG; Flurhofstr. 136, St. Gallen, Sankt Gallen 9000, Switzerland

SUMMARY

DANCE WITH WHAT YOU BROUGHT: HOW DO LIFESTYLE ENTREPRENEURS PROVIDE AND CAPTURE VALUE FROM SPECIALIZED RESOURCES?

Benedikt D.C. Seigner, IE University, Spain

Principal Topic

Lifestyle entrepreneurs build ventures around personal interests, blending fulfillment with financial sustainability. Unlike conventional entrepreneurs, they can rely on specialized resources—unique skills or knowledge cultivated through their "special lifestyle" (Ivanycheva et al., 2024). However, these resources are often too narrow for easy scalability or transferability, creating a distinct challenge. Traditional resource-based frameworks, such as effectuation (Sarasvathy, 2001), bricolage (Baker & Nelson, 2005), and co-specialization (Barney, 2018), offer insights into resource recombination but do not fully explain how lifestyle entrepreneurs create and capture value despite their over-specialized resources (Welter et al., 2016). I address this challenge by inductively studying and theorizing mechanisms of how lifestyle entrepreneurs provide and capture value from their highly specialized, non-business resources.

Method

I empirically study this phenomenon in competitive ballroom dancing. Athletes who transition into business are considered lifestyle entrepreneurs (Ivanycheva et al., 2024), and sports contexts offer deep insights into entrepreneurial dynamics (Fonti et al., 2022). I collected data through field observations at three international competitions and training camps, 511 social media posts, 295 pages of official documents, and 18 semi-structured interviews with 21 "elite informants" (Solarino & Aguinis, 2021), including world champions, adjudicators, and event organizers (1,055 minutes total). Drawing on a decade of personal experience as a competitive dancer, I combine autoethnography with inductive grounded theory (Glaser & Strauss, 1967; Strauss & Corbin, 1990).

Results and Implications

Preliminary findings indicate three mechanisms: leveraging, organizing, and supporting. Lifestyle entrepreneurs can leverage their elite status, allowing them to charge premium prices for services (e.g., private lessons in this industry). To address the limited scalability of these services, they create instructor networks or licensing models, distributing their personal vision. Organizing translates industry knowledge into ventures. By focusing on organizing competitions, workshops, and training programs, lifestyle entrepreneurs shift the focus from personal craft to scalable infrastructures. This approach relies on securing alliances with top dancers and adjudicators to gain legitimacy and attract large audiences. Meanwhile, supporting focuses on providing auxiliary services to the lifestyle industry—in dancing, for example, costume design, physiotherapy, or mental coaching tailored to dancers' specialized needs, leading to stable revenue streams. These findings challenge the assumption that lifestyle entrepreneurs forsake scalability (Ivanycheva et al., 2024). Rather, they strategically transform over-specialized resources into assets by creating hierarchical structures, shaping industry norms, and building complementary alliances. Extending resource-based theories (Clough et al., 2019; Sarasvathy, 2001), my findings also hint at more proactive behaviors rather than reactive improvisation under constraint.

CONTACT: Benedikt D.C. Seigner; benedikt.seigner@ie.edu; IE University; Paseo de la Castellana, 259E, Madrid, Madrid 28029, Spain

SUMMARY

THE WAY YOU MAKE ME FEEL: SOLVING COMPLEX PROBLEMS THROUGH EMOTIONS AND SERENDIPITY

Nele M. Terveen, University of Auckland / Stanford University, New Zealand

Principal Topic

Entrepreneurship is an emotionally charged journey characterized by uncertainty, where decisions are influenced by emotions, unexpected events, and limited resources. Effectuation theory (Sarasvathy, 2001) argues that entrepreneurs navigate uncertainty not through prediction but by leveraging available means, forming partnerships, and exploiting contingencies. However, it remains unclear how emotions affect this effectual process, especially in high-stakes sustainability contexts that heighten emotional intensity (Belz & Binder, 2017; Patzelt & Shepherd, 2024).This study explores how emotions shape entrepreneurs' ability to leverage effectual decision-making, particularly in serendipitous moments. Serendipity, or "smart luck" (Busch, 2022), is critical in entrepreneurship, yet its emotional foundations are underexplored. We show how emotional resonance (alignment) and emotional dissonance (misalignment) influence entrepreneurial agency in recognizing and exploiting serendipitous opportunities. Additionally, we examine how AI-assisted sensemaking affects the interplay between emotions, serendipity, and effectuation.

Method

We adopted a multiple-case study design (Eisenhardt, 1989) at a Silicon Valley university, following nine early-stage sustainability entrepreneurs over six months. Data were collected via semi-structured interviews (770 minutes total), AI-assisted reflective journals (372 ChatGPT entries), field observations (65 pages of notes over 12 hours), informal discussions, and contextual messages (63 WhatsApp texts). Using an abductive approach (Glaser & Strauss, 1967), we conducted open coding and created concept-ordered displays (Eisenhardt, 1989) to identify how emotions and serendipity drove entrepreneurial pivots and contingency leveraging.

Results and Implications

I propose an Emotionally Guided Serendipity Model for Effectual Decision-Making. It suggests that positive emotions encourage entrepreneurial agency to seize unexpected opportunities, whereas negative emotions prompt adaptation and strategic pivots. AI-assisted sensemaking expands cognitive processing and externalizes ideas, helping entrepreneurs navigate effectual decisions. I make three theoretical contributions. First, the study expands the role of emotions in effectuation theory by demonstrating how emotions influence pivoting, leveraging contingencies, and co-creating opportunities (Baron & Tang, 2011; Foo et al., 2009; Shepherd & Patzelt, 2024). Second, it integrates effectuation, emotions, and serendipity to advance the understanding of entrepreneurial decision-making under uncertainty (Sarasvathy, 2001; Busch & Barkema, 2020; Ananth et al., 2024). Third, it examines how AI-assisted sensemaking interacts with emotional and effectual processes, shaping entrepreneurial adaptability (Grimes et al., 2023; Raisch & Fomina, 2024; Busch & Grimes, 2023). These findings offer insights for entrepreneurs, educators, and policymakers, emphasizing emotional management, serendipity, and AI integration in decision-making.

CONTACT: Nele M. Terveen; nele.terveen@auckland.ac.nz; University of Auckland / Stanford University; 3/6 Harrison Avenue, Auckland, Auckland 0622, New Zealand

INTERACTIVE PAPER

READY FOR WHAT'S NEXT: WEB 3.0, CULTURE, AND AN ESO'S CREATION OF AN INTERSTITIAL SPACE TO ENHANCE LOCAL ENTREPRENEURSHIP

Brian Bergman, University of Tennessee, Knoxville, United States
Rehab Elkhiyat, University of Tennessee, Knoxville, United States
Scott Kuban, University of Arkansas, United States

Principal Topic

Interstitial spaces – "small-scale settings where individuals from different fields interact occasionally and informally around common activities" (Furnari, 2014: 439) – are garnering interest as important settings for entrepreneurship (Muñoz et al., 2019; Yacoub & Haefliger, 2020), innovation (Cartel et al., 2019), even industry emergence (Aversa et al., 2020). Interstitial spaces are fertile ground for these outcomes because their key features, namely participant diversity, informality, occasionality, allow individuals to break from their daily experience, explore common interests, and experiment with new ways of creating and organizing (Edmondson, 1999; Petriglieri & Petriglieri, 2010). These same features, however, can complicate their formation and sustainment (Smets et al., 2012), as uncertainties regarding the space's purpose, membership, or progress may be difficult to resolve. Separately, cultural entrepreneurship research examines the ways actors leverage cultural resources to legitimize and gain support for their entrepreneurial endeavors (Gehman & Soublière, 2017; Lounsbury & Glynn, 2001, 2019). Considering the challenges associated with creating interstitial spaces, cultural resources may play an underappreciated role in their formation process. Thus, we ask: How do entrepreneurial actors leverage cultural and technological resources to create and sustain an interstitial space?

Method

We explore this question through a qualitative case study of "NextCO" (Yin, 2009), an entrepreneurial support organization located in the US. NextCO is theoretically appropriate because it was founded to "[explore] the future and [launch] new ideas into the world" (web archive) by curating a new, highly diverse "collective" (i.e., an interstitial space). To understand NextCO, we conducted 22 interviews and gathered dozens of internal documents, newsletters, articles, and website archives, which we analyzed using techniques consistent with grounded theory and focused on organizational processes (Charmaz, 2014; Gehman et al., 2018, Gioia et al., 2013; Langley, 1999).

Results and Implications

From our analysis, we present a process model of interstitial space formation and detail the ways through which NextCO fused cultural, technological, and material resources to successfully curate and create an interstitial space around emerging technologies in its community. Ultimately, we build upon the foundational theoretical work on interstitial spaces (Furnari, 2014), shedding light on their formation and broadening the critical role of "catalysts" in these spaces. We also contribute to contemporary discussions on technology entrepreneurship (Guerrero et al., 2024) and non-venture outcomes of entrepreneurship (Thorgren & Williams, 2023) and offer practical insights for communities looking to foster local entrepreneurship and innovation.

CONTACT: Brian Bergman; bbergma1@utk.edu; University of Tennessee, Knoxville; 409A Stokely Management Center, Knoxville, Tennessee 37916, United States

INTERACTIVE PAPER

ORGANIZATIONAL FIELD CONFIGURATION IN THE EVOLUTION OF THE U.S. MAKERSPACE POPULATION

Russell Browder, University of Oklahoma, United States
Howard Aldrich, University of North Carolina, United States
Stella Seyb, Texas Christian University, United States

Principal Topic

When entrepreneurs introduce a new organizational form, their actions can create a new organizational field (Scott, 1995). Research on organizational field configuration has examined the role of events and narratives in shaping a field (Lampel and Meyer, 2008; Wry et al., 2011). Scholars have called for greater understanding of field formation processes in which organizations interact, build a field's structure, develop collective understandings, and translate practices to organizations (Wooten and Hoffman, 2017). This study examines the evolution of independent makerspaces as an organizational field in the U.S. Entrepreneurship research has examined makerspaces as a type of support organization (Bergman and McMullen, 2022) that plays a role in initiating new entrepreneurial ventures (Bao, 2024), promoting consumer and corporate innovation (Browder et al., 2023), and facilitating resourcefulness (Browder et al., 2022).

Method

We constructed longitudinal data on the population of independent U.S. makerspaces from 2005 to 2024. We track makerspace founding rates through stages of field creation and field maintenance, giving special attention to three types of makerspaces and their pursuit of legitimacy (Gorbatai et al., 2021). Data was compiled from three makerspace databases and supplemented with three waves of manual data collection in 2017-2018, 2021, and 2023-2024 by checking websites, social media, and contacting makerspaces to update the database of active organizations.

Results and Implications

The organizational field's evolution resulted in the makerspace population eventually being dominated by 237 nonprofit makerspaces followed by about 50 unincorporated clubs and 30 commercial spaces as of 2023. In the first phase (2005-2011), the initiation of a new organizational field was characterized by pursuit of cognitive legitimacy as the three categories of governance appeared as alternative models. In the second phase (2012-2016), field creation benefitted from growing sociopolitical legitimacy, driven by influential books and federal government recognition. A turbulent third phase (2017-2022) marked a shift to field maintenance amid field-disruptive events. Across the three phases, Kaplan-Meier survival curves revealed a high survival rate for nonprofits, contrasted with about 60% of clubs surviving the first three years, and about 75% of commercial spaces surviving at least five years. This study contributes to entrepreneurship research by providing historical context regarding the limits and potential of makerspaces as entrepreneurial ecosystem actors. The study contributes to the literature on organizational fields by offering a novel perspective on how different paths for a field's development were enabled and constrained over time, extending our understanding of how organizational fields and their central populations evolve.

CONTACT: Russell Browder; russbrowder@ou.edu; University of Oklahoma; 307 W Brooks St, Norman, Oklahoma 73019, United States

INTERACTIVE PAPER

PER ASPERA: EXPLORING KEY CHALLENGES OF SCALE-UPS ACROSS DIFFERENT CONTEXTS

Philipp Bubenzer, HES-SO / ETH Zurich, Switzerland

Principal Topic

This study contributes to our understanding of the unique challenges firms face during scaling. It argues that scaling involves distinct organizational and strategic shifts requiring focused managerial attention (DeSantola & Gulati 2017). It identifies key challenges in process optimization, product development, and managing client and partner relationships. Addressing two key gaps in the literature, this study thus first demonstrates that firms in the scale-up phase face different challenges not present in earlier or later stages of development. Second, the research highlights the critical role that context may play in shaping the scaling process (Coutu, 2014; Palmie et al., 2023). By centering on these questions, the study offers actionable insights for firms navigating the scale-up transition.

Method

The study draws on a large-scale sample of 2,885 firms from various industries and European countries (excluding micro-enterprises}. Multinomial logistic regression was used to assess the relationships between challenges and stage (Pre-scale, Scale-up, or Post-scale). This allowed for a detailed examination of how challenges differ across these stages, while controlling for firm specific variables like age, sector, size and other. Additional interaction effects were tested using binary logistic regression models to explore how firm age influences the relationship between innovation strategies and scaling outcomes.

Results and Implications

Results highlight the role of innovation challenges and innovation partnerships when scaling. Regarding challenges, process innovation challenges emerged as a key characteristic of scaling companies, reinforcing the notion that operational efficiency is a critical factor in successfully navigating the scale-up phase. Conversely, product innovation challenges were positively associated with the pre-scale stage, but negatively with the scale-up stage, suggesting that while product development is vital in early growth, excessive focus on new product creation may divert attention from other crucial scaling activities. Neither distribution nor communication innovation challenges showed significant effects, implying that these factors might not be central in the transition to scaling. In terms of partnerships, firms that relied heavily on clients or public support agencies for innovation were less likely to be in the scale-up stage. This suggests that over-dependence on external actors—whether through client-driven innovation or regulatory complexities—may hinder scaling. While supplier partnerships showed no direct effect, private investment exhibited a marginally negative association, indicating potential tensions between investor expectations and long-term strategic scaling efforts. These findings contribute to a more nuanced understanding of scaling challenges and provide managerial and policy implications, particularly concerning process efficiencies, innovation focus, and partnership strategies.

CONTACT: Philipp Bubenzer; philipp.bubenzer@hefr.ch; HES-SO / ETH Zurich; Chemin du Musée 4, Fribourg, Fribourg 1700, Switzerland

INTERACTIVE PAPER

ENTREPRENEURIAL HUSTLE AND STAKEHOLDER ENROLLMENT: A DUAL-PATH MODEL

Devin Burnell, Texas Christian University, United States
Donald Kuratko, Indiana University, United States
Matthew Marvel, Ball State University, United States
Emily Neubert, Texas Christian University, United States

Principal Topic

While the emerging literature on entrepreneurial hustle—urgent and unorthodox actions intended to address obstacles and opportunities—suggests this behavioral tendency corresponds with success, we know very little about how these actions relate to resource acquisition and why. In this research, we theorize and test a dual-path model that links the dimensions of entrepreneurial hustle—urgency and unorthodoxy—to stakeholders' willingness to invest in an entrepreneur's venture through two key mediators: venture goal progress and product innovativeness. Specifically, we hypothesize venture goal progress mediates the association between entrepreneurs' urgent actions and stakeholders' willingness to invest (Hypothesis 1); and product innovativeness mediates the association between entrepreneurs' unorthodox actions and stakeholders' willingness to invest (Hypothesis 2).

Method

In this study, we targeted entrepreneurial support organizations (ESOs) across the United States that provide mentoring to founders. This approach allowed for the identification of dyads of founders and mentors to provide external sources of assessment of entrepreneurs. Our final sample size was 423 unique dyads of founders and mentors. We measured established measures for our variables and include several controls related to the founder and the venture.

Results and Implications

We first mean-centered all variables before analysis and clustered standard errors by mentors, resulting in 265 clusters. Results of regression analyses without controls suggested the relationship between founder entrepreneurial hustle and stakeholders' willingness to invest was positive and statistically significant. We found that urgency was positively related to venture goal progress, and venture goal progress was positively related to stakeholder's willingness to invest. The bias-corrected 95% CIs for the indirect effects of urgency on willingness to invest through venture goal progress was significant, supporting Hypothesis 1. We also found unorthodoxy was positively related to product innovativeness, and product innovativeness was positively related to stakeholder's willingness to invest. The bias-corrected 95% CIs for the indirect effects of unorthodoxy on willingness to invest through product innovativeness was significant, supporting Hypothesis 2. Results of several robustness tests, including testing with controls and simultaneous inclusion of both mediators (i.e., "net mediation") lend greater support for our hypotheses. These results contribute to research on entrepreneurial hustle by providing evidence that entrepreneurial hustle is related to resource acquisition through goal progress and product innovativeness. More broadly, we extend research that suggests entrepreneurial hustle is a behavioral tendency that enables entrepreneurs to enroll venture stakeholders.

CONTACT: Devin Burnell; d.burnell@tcu.edu; Texas Christian University; 3652 Shelby Drive, Fort Worth, Texas 76109, United States

INTERACTIVE PAPER

HOW ENTREPRENEURS CONCEPTUALIZE THE VALUE OF EDUCATION FOR ENTREPRENEURSHIP: A TEXTUAL ANALYSIS OF REDDIT POSTS ON ENTREPRENEURIAL EDUCATION

Jarrod Holmes Humphrey, Xavier University, United States
Matthew Regele, Xavier University, United States
David Ross, University of Florida, United States

Principal Topic

The value of higher education for active and aspiring entrepreneurs has long been a topic of debate. While higher education is traditionally seen as a pathway to developing human, social, and cultural capital, its relevance to entrepreneurship remains contested. Empirical research on whether higher education improves entrepreneurial outcomes—such as success rates, venture growth, and resilience—has yielded mixed results. Amongst all of these conflicting results and arguments, one perspective has been notably absent: the views of active and aspiring entrepreneurs themselves. How do they perceive the role of higher education in their entrepreneurial journeys? What do they see as its primary value—or limitations?

Method

This study develops a theory of how stakeholders conceptualize the value of higher education for entrepreneurship using data from the social networking website Reddit. Our data covers the subreddit r/Entrepreneur. The r/Entrepreneur subreddit, first launched in 2008, has 2.6 million members and is in the top 1% of subreddits on Reddit in terms of membership. These qualitative data offer unique insights into what entrepreneurs believe about higher education's value—or lack thereof—and the factors shaping these perceptions. By exploring themes such as human and social capital development, signaling and credentialism, and vocational choice, we aim to extend the understanding of how entrepreneurs perceive and navigate educational pathways in shaping self-employment career trajectories.

Results and Implications

Our findings suggest that entrepreneurs conceptualize the value of higher education for entrepreneurial activities along four aggregate dimensions. Specifically, our data structure indicates that (1) traditional entrepreneurial education often fails to meet practicing entrepreneurs' needs. However, for those pursuing higher education, entrepreneurs suggest they (2) leverage university networks for human, social, and financial capital and (3) align their education with the realities of self-employment. Lastly, (4) formal education is seen as beneficial for critical thinking and providing a safety net. Our study makes three theoretical contributions. First, we examine how entrepreneurship is regarded by those it aims to serve, contributing to the broader discussion on its effectiveness. Second, we develop a theory on how differing assumptions about education's value shape the decisions of aspiring and active entrepreneurs regarding higher education and program selection. Finally, we extend theories of education and career choice by detailing how entrepreneurs integrate formal education into their broader career strategies, balancing skill development, risk management, and self-employment realities.

CONTACT: Jarrod Holmes Humphrey; humphreyj5@xavier.edu; Xavier University; 4020 Red Bud Avenue, Cincinnati, Ohio 45229, United States

INTERACTIVE PAPER

THE TRANSITION TO ENTREPRENEURSHIP: HOW BRIDGE EXPERIENCES FACILITATE CONFIDENCE, COMPETENCE, AND CONVICTION IN NASCENT FOUNDERS

Hana Kieger, Indiana University, United States
Devin Burnell, Texas Christian University, United States
Greg Fisher, Indiana University, United States

Principal Topic

Pre-entry experiences refer to active participation in particular activities or events that shape a person's knowledge, skills, and perspective before starting a new venture (e.g., Dencker, Gruber, & Shah, 2009; Wang, 2020). Existing theory argues that pre-entry experiences generate knowledge and motivation relevant to recognizing and evaluating entrepreneurial opportunities (Dencker, Gruber, & Shah, 2009; Wood & Williams, 2014) and can prompt people to enter entrepreneurship under conditions of uncertainty (McMullen & Shepherd, 2006). Prior research has shown that pre-entry experiences can influence post-entry outcomes, such as founders' innovativeness (Klepper & Simmons, 2000), ability to grow in dynamic markets (Chen, Williams, & Agarwal, 2012), and improve the likelihood of venture survival (Dencker, Gruber, & Shah, 2009). This literature has begun to unveil the pre-entry variables that impact entrepreneurial entry, yet we have a limited understanding of what processes and mechanisms enable and best prepare potential entrepreneurs to transition into entrepreneurship.

Method

In Study 1, we performed an inductive study using qualitative data from a podcast called Founders Unfound. We listened to each podcast episode, transcribed the interviews, and coded the interview data starting with an open-coding approach (Corbin & Strauss, 2008). In Study 2, we recruited 98 MBA students who expressed an interest in entrepreneurship and used a study protocol for these participants to intentionally seek out and engage in a bridge experience. This is designed to mimic bridge experiences described by participants in Study 1.

Results and Implications

Study 1 revealed three key mechanisms of bridge experiences: vertical role exposure (exposure to roles with increased leadership responsibility or autonomy), horizontal role exposure (exposure to diverse job tasks), and entrepreneurial role exposure (exposure to a startup environment). These mechanisms led to three key outcomes: realizing competence, confidence, and conviction about entrepreneurship. From Study 2, we uncovered that bridge experiences create entrepreneurial readiness through three distinct phases: intentional design (evaluating skill/confidence gaps and proposing a bridge experience to address them), experiential development (developing entrepreneurial capabilities through cycles of experimentation, external feedback, and adaptation), and transformation (experiencing shifts in entrepreneurial self-efficacy and intentions). The process manifested differently across three groups. Venture Advancers leveraged the experience to overcome specific existing business challenges, Barrier Breakers transformed limiting beliefs into actionable steps, and Career Developers extracted transferable skills for professional growth. This research extends theory by shifting from static pre-entry predictors to a dynamic process, illuminating how bridge experiences facilitate entrepreneurial entry through the development of skills and updated self-perceptions.

CONTACT: Hana Kieger; hkieger@iu.edu; Indiana University; 101 S. Fairfield Dr. #33, Bloomington, Indiana 47404, United States

INTERACTIVE PAPER

FROM MYTH TO MARKET: CO-CONSTRUCTING AND EXPLOITING A CONSPIRACY ENTREPRENEURIAL OPPORTUNITY

Sarah R. Sarjoo, University of Notre Dame, United States
Sally Maitlis, University of Oxford, United Kingdom
Dean Shepherd, University of Notre Dame, United States

Principal Topic

Conspiracy entrepreneurship—the co-construction and exploitation of an opportunity by crafting and disseminating a false narrative for personal commercial gain under the guise of addressing social injustices—is an exemplary form of destructive entrepreneurship. By co-constructing and exploiting a conspiracy entrepreneurial opportunity, a conspiracy entrepreneur causes destruction precisely as Baumol (1990) envisioned for destructive entrepreneurship. Despite significant advances in scholarship on the psychology of conspiracy theories (Bowes et al., 2023; Sunstein and Vermeule, 2009), the origin of such theories remains an open question (Douglas and Sutton, 2023). Thus, while conspiracy theorists can gain substantially from disseminating their theories, how they co-construct and exploit these opportunities remains a mystery.

Method

We take a qualitative, inductive approach to develop a grounded theory of conspiracy entrepreneurship. We focus on a context in which capturing the unfolding co-construction and exploitation of a conspiracy entrepreneurial opportunity was possible: the Lafferty et al., v. Jones (2022) court case in which Alex Jones was sued by relatives of those killed in the Sandy Hook Elementary School shooting. These court case data include more than 1,000 legal documents from court proceedings. We supplement these data with news reports, including articles and documentaries.

Results and Implications

A conspiracy entrepreneur crafts counternarrative elements based on their own agenda and creates a version of what happened that allows a subpopulation to make sense of an event in a particular way. However, these conspiracy narrative elements ultimately add to the overall confusion and uncertainty surrounding the event. We found that event uncertainty can be leveraged as an entrepreneurial opportunity to generate substantial profit for the conspiracy entrepreneur. The conspiracy entrepreneur relies on various tactics to perpetuate people's uncertainty about the focal event to promote the conspiracy entrepreneurial process. These tactics include amplifying discord by generating angertainment, encouraging sensemaking by offering a platform for the audience to voice their opinions and question prevailing accounts, and conspiracy sensegiving whereby the conspiracy entrepreneur employs sequential lies to build the audience's trust and invokes hypothetical scenarios to influence their understanding of the situation. To exploit the conspiracy entrepreneurial opportunity, the entrepreneur builds a loyal community of conspiracy by fueling the anxiety of their audience and effectively leveraging social media to propagate their counternarrative elements. The entrepreneur then monetizes the community for their own benefit by leveraging the conspiracy theory as a marketing tool and stimulating his audience's excitement and product purchases.

CONTACT: Sarah R. Sarjoo; ssarjoo@nd.edu; University of Notre Dame; 133 West Leyte Avenue, Mishawaka, Indiana 46545, United States

INTERACTIVE PAPER

UNICORN POLICIES: HOW DIFFERENT POLICY BUNDLES AFFECT THE EMERGENCE AND GROWTH OF OUTLIER FIRMS

Vitaliy Skorodziyevskiy, University of Louisville, United States
Matthew Farrell, Austin Peay State University, United States

Principal Topic

Scholars have dedicated considerable effort to understanding what drives the emergence of outliers—the largest and fastest-growing ventures such as Facebook or Apple—which is fundamental for theoretical expansion and methodological rigor (e.g., Acs, 2008). Although outlier firms generate huge influences on society and the economy, little is known about contextual impacts on these firms, since empirical research either removes or transforms non-normal observations to achieve significant results (Andriani & McKelvey, 2007). Whereas the literature suggests that endowments, expectations, and engagements are of vital importance to the emergence and growth of outliers (Crawford et al., 2015), less attention has been devoted to understanding the environment in which firms operate and the extent to which alternative policy configurations could promote (predict) the development of these firms. Some scholars argue, though, that initial location, subnational regions, and environmental munificence all influence the highest-performing ventures. We investigate what policy configurations are more conducive to the emergence and growth of outliers.

Method

We collected our sample from the Crunchbase database for all US startups between 2000 and 2020. Our outcome conditions are "unicorn" ventures with either (1) IPO/Acquisition exceeding USD 1 billion or (2) 10 times the VC investment, or (3) configurations leading to a lack of this outcome. Our explanatory conditions: firm size, state-level qualities of infrastructure and education, business tax rate, population density, and political leaning.

Results and Implications

We have preliminarily uncovered findings with respect to the best (and worst) configurations of environments for the emergence and growth of outliers. Based on our paired sample of 82 firms, the results suggest that there are three ideal configurations for outlier performance. The first consistent path for the emergence of outlier firms is for smaller firms in more liberal states with high taxes, good infrastructure, good education, and high population density (i.e., urban areas). The second path is more suited for large firms that operate in conservative states with good infrastructure, poor education, and a low overall tax rate are more likely to become outlier firms. Lastly, large firms could also emerge as outliers in cities with good educational systems, within democratic states with poor infrastructure and low taxes. We discovered two primary environmental configurations that constrain the occurrence of outlier outcomes: (1) when small firms operate in urban environments in liberal states with poor education, poor infrastructure, and lower taxes; (2) when small firms in urban environments are in conservative states with relatively good educational systems.

CONTACT: Vitaliy Skorodziyevskiy; vitaliy.skorodziyevskiy@louisville.edu; University of Louisville; 1949 Poloview Pl, Louisville, Kentucky 40245, United States

SUMMARY

AI ASSISTED CREATIVE ENTREPRENEURIAL PROCESSES IN HOLLYWOOD: HEDGING UNCERTAINTY AND RESOURCE CONSTRAINTS

Vidhi Jain, University of North Texas, United States
Manjula Salimath, University of North Texas, United States

Principal Topic

Creative entrepreneurs such as animators, digital designers and content creators have to overcome a triad of challenges: high cost of technology; high sophistication needed to master cutting-edge technology; and high uncertainty surrounding their adoption. As such, new entrepreneurial firms are at a disadvantage relative to incumbents. Though artificial intelligence (AI) has shown potential to overcome these challenges, a systematic exploration of how it interfaces with entrepreneurial processes is lacking. Conceptually akin to the role played by digital platforms like Etsy in leveling the field for creative entrepreneurs in the craft industry, we posit that another new technology (AI) is having a similar effect, in another creative industry. Technology affordance and constraint theory (TACT) suggests that though technology enables goal fulfillment, some situations can prevent its full realization. By considering AI technology as set of affordance and constraints, we can advance understanding of its interface with creative entrepreneurs and better explore its role within the entrepreneurial process. Hence our research question is: How does AI aid entrepreneurs in the creative industry to deal with resource constraints and uncertainty?

Method

Following a case methodology, we utilize the established alternate template research design method (Langley, 1999), which allows comparison of different theoretical perspective to explain a process. "It provides alternative explanations of the same situation using different theoretical perspectives and assumptions—thereby highlighting elements of each perspective that fit with the data" (Fisher, 2012, p. 1029). We analyzed multi-source secondary qualitative data to answer questions associated with the three entrepreneurial processes, namely, causation, effectuation and bricolage. We identified and selected an AI-driven startup in Hollywood that was notable for its pioneering ability to deliver AI enabled cutting edge visual effects alongside an 80% reduction in time and cost. The context for this startup is interesting given the high uncertainty and technology resistance in Hollywood, which experienced its longest strike in 2023 regarding the use of AI.

Results and Implications

Findings showcase the role of AI in reducing uncertainty and resource constraints for creative entrepreneurs. Theoretically, we contribute towards understanding the effect of AI on entrepreneurial decision processes. Empirically, we extend Lupp's (2023) conceptual layout of the role of machine learning in causation and effectuation by examining how AI helps entrepreneurs navigate highly uncertain environments within the creative industry. Implications for scholars and entrepreneurs in the creative industry are presented.

CONTACT: Vidhi Jain; VidhiHarish.Jain@unt.edu; University of North Texas; 1155 Union Circle #311160, Denton, Texas 76203, United States

SUMMARY

DOCUMENTARIES: A WINDOW INTO THE LIVED EXPERIENCE OF ENTREPRENEURS

Tomoko Yokoi, ETHZ and IMD, Switzerland

Principal Topic

Entrepreneurship is a lived experience shaped by ongoing events, activities, emotions, and thoughts that influence how ventures unfold in real time. However, studying entrepreneurial experiences poses challenges, as traditional methods like retrospective accounts, surveys, or interviews often fail to capture their immediacy and complexity. This paper proposes film and television documentaries as a complementary method for studying entrepreneurs' day-to-day experiences. Documentaries provide a multimodal perspective, combining verbal, visual, auditory, and physical elements to uncover nuanced practices and offer richer insights into the dynamic and multifaceted nature of entrepreneurial life.

Method

Drawing from film studies and research methods using new media, this paper presents a framework that for how researchers can incorporate documentaries into their studies. The framework prompts researchers to i) assess the relevance of documentaries in supporting a particular research inquiry; ii) qualify the source material for its representation of actuality; iii) interpret data to develop concepts, themes or typologies that align with the initial research question.

Results and Implications

Currently there is limited guidance on how to analyze and incorporate documentaries as a medium into entrepreneurship research. This paper address this gap by integrating insights from film studies to 1) distinguish between different documentary types and highlight how each type can be used in different ways by researchers to study different facets of organizational realities and 2) propose different methodological and analytical procedures that are consistent with each documentary type to ensure analytical depth and rigor. By drawing on methodologies from film studies, the incorporation of documentary film analysis into entrepreneurship studies offers new avenues for analysis that go beyond textual or interview-based methods. This analytical lens brings fresh tools for understanding complex organizational phenomena and provides a bridge to explore aesthetic, affective, temporal and symbolic dimensions of entrepreneurship. Documentaries also have the potential to facilitate broader discussions beyond the academic community and support meaningful interactions with society through co-production and dissemination of research insights.

CONTACT: Tomoko Yokoi; tomoko.yokoi@imd.org; ETHZ and IMD; Weinbergstrasse 56/58, Zurich, Zurich 8006, Switzerland

INTERACTIVE PAPER

A NOVEL APPROACH TO MEASURING LATE-STAGE PIVOTS USING LARGE LANGUAGE MODELS (LLM)

Omar Alrubaiaan, Rensselaer Polytechnic Institute, United States
Jason Kuruzovich, Rensselaer Polytechnic Institute, United States

Principal Topic

We introduce a novel method for measuring pivots in startups using LLMs to analyze business description changes over time. While pivoting is a critical entrepreneurial strategy (Blank, 2012; Ries, 2011), prior research has mainly relied on case studies. We leverage CrunchBase (CB) and LLM-based text analysis to automate pivot detection, anchoring our assessments with human raters to validate the approach. Pivoting has been widely discussed in entrepreneurship research and media (Arteaga & Hyland, 2013; Grimes, 2018). While surveys suggest that over 70% of startups pivot (Shepherd & Gruber, 2021), no large-scale automated method exists to track these shifts systematically. Previous work has used firm descriptions in 10-K reports and databases like CB to study company strategies (Shi et al., 2016; Yu, 2020). Advances in NLP, such as BERT and Doc2Vec, allow researchers to quantify textual changes and compare companies using vectorized representations. By applying these techniques, we offer a scalable approach to identifying pivots and exploring their impact on startup outcomes.

Method

Our empirical analysis is based on an extensive dataset of startup profiles from CB. We analyze the startup business descriptions from 2018 and 2023. We vectorized the text by DistilBERT and calculated cosine similarity (CS) scores, eliminating identical descriptions. We sampled 20 companies per 0.05 CS increment (0.5–0.95), excluding startups below 0.5 due to poor text quality. Our final sample consisted of 200 companies manually reviewed for pivot detection. To establish ground truth, five trained undergraduate coders classified pivots using a structured schema, refining their approach through iterative discussions. Their final consensus identified a 25.5% pivot rate. We then designed a structured prompt for LLMs, including definitions, pivot types, and examples, testing it on OpenAI's GPT-3.5-Turbo and GPT-4-1106-Preview. Future tests will include Google's (Gemini-1.5-Flash) and Meta's (LLaMA3-70B-8192).

Results and Implications

Human raters exhibited significant variation in pivot identification, ranging from 15% to 38%, underscoring the inherent challenges of assessing pivots. Through consensus, the final pivot rate was established at 25.5%. LLMs demonstrated near-human accuracy, with GPT-4 achieving 77% accuracy and an AUC of 0.76, compared to the human average of 80%. While higher cosine similarity was generally associated with fewer pivots, exceptions indicate that CS alone is an unreliable pivot detection metric. Our LLM-driven approach offers a scalable, systematic method for studying pivots. Future work will refine prompts, test additional models, and expand the human assessment pool to establish true human accuracy better and determine whether LLMs enhance or diminish assessment reliability.

CONTACT: Omar Alrubaiaan; oalrubiaan@gmail.com; Rensselaer Polytechnic Institute; 27 Esplande St, Selkirk, New York 12158, United States

INTERACTIVE PAPER

SCALABILITY AS AN ORGANIZATIONAL CAPABILITY: DEVELOPING AND TESTING A NEW MEASURE

Anne Domurath, Wilfrid Laurier University, Canada
Nicole Coviello, Wilfrid Laurier University and LUT University, Canada
Holger Patzelt, Technical University Munich, Germany

Principal Topic

Entrepreneurs Reid Hoffman and Chris Yeh (2018, p. 75) observe that "the wisest innovators design operational scalability into their theory." However, what is scalability? Based on a systematic review and critique of the literature, Coviello et al.,(2024) define scalability as an ordinary organizational capability developed by managing and achieving coherence among a firm's technological architecture, organizational architecture, and business model. In particular, they argue that the firm's business model delivers scaling when it is managed together with changes to both the firm's organizational and technological architectures. This means that scalability (not the business model or digitalization per se) delivers scaling. Prior research also suggests that scalability is likely multi-dimensional (Coviello et al., 2024), and it involves the performance of specific functions to accomplish certain tasks. Thus, although some emphasize digital resources when discussing scalability (e.g., Guistiziero et al., 2023), we also know from early research (e.g., DeSantola & Gulati, 2017), that a scaling firm requires internal change to handle greater complexity both externally (in the market) and internally (within the organization). This means establishing routines, standardizing where possible, and sharing best practices. Yet, prior efforts to measure scalability tend to emphasize digital resources and rely on accounting measures or text-based measures from IPO prospectuses and pre-IPO company descriptions (e.g., Somaya & You, 2024). Given the above, Coviello et al., (2024) argue that a new measure is needed to understand scalability as an organizational capability, both theoretically and in practice. Our objective is to develop and test such a measure.

Method

Our project employs a multi-stage approach to developing the new measure of organizational scalability: 1) use of the Delphi Panel method whereby practitioners identify core dimensions of scalability, 2) pattern matching the Delphi results to extant research, 3) initial development and pre-testing of the multi-item measure, 4) testing, refinement, and validation of the measure using primary data collected in three rounds with Prolific.com. The firms in our various samples are young technology ventures, and our informants are entrepreneurs. The participants in the Delphi panel have personal experience with scaling, often more than once.

Results and Implications

Data collection is ongoing and results will be available in May 2025. To the best of our knowledge, this is the first study to conceptualize scalability as an organizational capability. Accordingly, we extend initial efforts to define and assess this concept but do so based on a contemporary definition from Coviello et al.,(2024).

CONTACT: Anne Domurath; adomurath@wlu.ca; Wilfrid Laurier University; 75 University Ave W, Waterloo, Ontario N2T2E6, Canada

THE MEANS CENTER FOR ENTREPRENEURIAL STUDIES AWARD (ILLINOIS STATE) FOR BEST PAPER ON SOCIAL ENTREPRENEURSHIP

THE ROLE OF PROXIMITY AND POLITICAL IDEOLOGY AS EXTERNAL ENABLERS OF SOCIAL ENTREPRENEURSHIP

Mariza Gabriela L. Mesquita, Louisiana State University, United States
Joseph J. Cabral, Louisiana State University, United States

ABSTRACT

Drawing upon external enablers framework, this study examines when and why entrepreneurs initiate ventures in response to different opportunities. Furthermore, by dissecting social dilemmas into root and symptomatic opportunities, we explore the impact of proximity and political ideology on these distinct opportunities. Our findings suggest that entrepreneurs' embeddedness in regions that generate social dilemmas, as well with liberal political ideologies, enhances their ability to enable and trigger venture creation. Interestingly, we do not find evidence of exposure to immediate social dilemmas driving entrepreneurial activity.

INTRODUCTION

Understanding what drives the emergence of social entrepreneurship, which is crucial in addressing social challenges while fostering economic growth, holds both practical and theoretical relevance. To date, the bulk of attention considers how individual-level factors such as skills, passion, knowledge, and experiences drive entrepreneurial endeavors (Bacq et al., 2022; Short et al., 2009; Zahra et al., 2009). However, there remains a limited understanding of the external influences that can either enable or obstruct opportunity recognition and, ultimately, the establishment of social ventures (Austin et al., 2006; Davidsson, 2015). Given the coevolution between entrepreneurs and the broader systems in which they operate, an analysis that considers only internal factors may offer an incomplete understanding of the emergence of new ventures (Davidsson et al., 2020).

In the context of social entrepreneurship, social challenges are a primary driver motivating the establishment of entrepreneurial ventures (Austin et al., 2006; Bacq et al., 2022). However, literature is constrained in that social "problems" are often presented as omnibus challenges, but opportunities exist to alleviate either the root cause of the issue ("root opportunities") or mitigating the impact of those issues ("symptomatic opportunities") (Lucas & Park, 2023). Root opportunities, such as greenhouse gas emissions, are predictable but often indirect, whereas symptomatic opportunities, such as the impacts of natural disasters, are more immediate and tangible but less predictable. Further, a noted complication in the social entrepreneurship and sustainability space is the existence of inequality due to different spatial scopes; in other words, the root sources of negative externalities are often not the same regions, economies, or socioeconomic groups most likely to incur symptomatic outcomes (Althor et al., 2016; Davidsson et al., 2025). These nuances suggest that how a social problem is framed will influence the enabling and triggering of new ventures.

External enablers are crucial in shaping how entrepreneurs recognize and evaluate opportunities. These factors represent changes in the entrepreneurial environment that facilitate the establishment of new ventures by empowering entrepreneurs' actions and stimulating potential customer demand. Examples of external enablers include shifts in technology, politics, and demographics. While they may disrupt the entrepreneurial process, they can also serve as significant triggers for creating new offerings or companies.

Additionally, these changes provide legitimacy to new ventures from both customers and competitors due to the heightened demand they generate (Davidsson & Gruenhagen, 2021; Davidsson et al., 2020; Kimjeon & Davidsson, 2022).

Two external enablers, proximity to opportunity and political ideology, may influence the perception and exploitation of opportunities, promoting venture openness. Proximity refers to the entrepreneur's geographical closeness to opportunities. On the one hand, being nearby provides agents with a deep understanding of the phenomenon and interaction with users and potential customers (Companys & McMullen, 2007). On the other hand, familiarity with the problem may result in accepting existing but inefficient solutions and limiting innovation. As such, more distance can provide new perspectives and a greater inclination to challenge the status quo (Lahiri & Wadhwa, 2021). While locals may overlook an enabler, the same circumstances can be viewed favorably and attract attention from those distant from the problem (Shaw & Gilly, 2000). Specific to the problem formulation process, root opportunities are tied to existing jobs and culture, offering counterforces to innovation. In contrast, symptomatic opportunities are more tangible and widely felt, prompting more higher degrees of opportunity confidence. Thus, there is reason to believe distance to a social dilemma, as well as whether it is a root cause or symptomatic effect of the problem, may trigger different entrepreneurial responses.

The second enabler, political ideology, is crucial in shaping a group's values and perspective. It influences how individuals prioritize individual versus collective outcomes and their views on the government's or free markets' roles in achieving these goals and, therefore, their entrepreneurial action (Barber IV & Blake, 2024; Bennett et al., 2023). This dynamic creates a paradox within social entrepreneurship. Ecological and social challenges such as global warming are often politicized and sometimes dismissed by conservatives, which could limit the recognition of potential opportunities. Yet, conservative ideology values market-based solutions (Maldonado-Bautista et al., 2023), implying that conservative entrepreneurs may more readily identify entrepreneurial options for addressing social dilemmas. On the other end of the political spectrum, liberals generally recognize social issues but tend to favor government intervention, which could hamper entrepreneurs' ability to recognize opportunities and hinder new venture ideas (Graham et al., 2009; Lucas & Park, 2023; Swigart et al., 2020). As a result, although political ideology is expected to influence entrepreneurs' opportunity recognition and entrepreneurial action, the ultimate impact of this factor is unclear.

Thus, this paper aims to understand when and why entrepreneurs launch ventures in response to social dilemmas. Moreover, by breaking social dilemmas into root and symptomatic opportunities and considering how these different types of opportunities are impacted by proximity and political ideology, we hope to better understand the nuances of when and why entrepreneurs launch ventures in response to social challenges.

HYPOTHESIS DEVELOPMENT

Entrepreneurs are at the core of creating new ventures; however, the venture creation process is not possible without opportunities to pursue (Eckhardt & Shane, 2003; Shane & Venkataraman, 2000). Due to their relevance to the entrepreneurship field, scholars have dedicated the last two decades to explaining what constitutes an entrepreneurial opportunity (Davidsson, 2015; Shane & Venkataraman, 2000; Short et al., 2010). Davidsson (2015) argues that three constructs have been advocated as more accurate to the concept of opportunity (Ramoglou & Gartner, 2023). The first is new venture creation, which consists of the entrepreneur's imaginary combination of resources to create a new venture (Shepherd et al., 2021). The second is opportunity confidence, which consists of an entrepreneur's assessment of a stimulus, either internal (e.g., life dream, need for self-realization, family history in business) or external (e.g., governmental incentives, technological trends). Thus, at the same time, entrepreneurs use their subjective perceptions to recognize and assess an opportunity, using their imagination to create possible new products and ventures; external conditions have shown importance to its exploitation and, consequently, the venture creation process, even though still few are explored in literature. In this way, a third construct emerges: external enablers (Davidsson, 2015; Davidsson & Gruenhagen, 2021; Kimjeon & Davidsson, 2022).

External enablers are, as the name suggests, external circumstances that occur at the aggregate level, potentially facilitating entrepreneurs in creating new ventures (Kimjeon & Davidsson, 2022). External enablers include cultural, technological, social, political, and natural-environmental changes (Davidsson, 2015). They should be subjective, selective, uncertain, and interdependent. First, the subjective feature refers to each entrepreneur's perception of an external enabler due to their individual characteristics. Second, it is selective because it depends on the application area, time, and space. Third, the uncertainty of an external enabler is because they are temporary and are subject to the other entrepreneur's action and possible changes in the system. Finally, an external enabler is interdependent with other external circumstances. For example, the potential of new technology is contingent upon infrastructure and regulations. Besides, external enablers vary in two characteristics. The first aspect is scope, which includes sectorial, temporal, spatial, and socio-demographic factors that may directly affect market potential. The second aspect is onset, whose dimensions, predictability, and suddenness imply the readiness of an agent to take entrepreneurial action (Davidsson et al., 2020, 2025). Thus, external enablers are complex when analyzing their effects on new venture creation (Davidsson & Gruenhagen, 2021; Davidsson et al., 2020).

Given the complex type of opportunities, the impact of external enablers on opportunity recognition and new venture creation can vary (Buenstorf, 2007; Davidsson et al., 2020). A problem representing an opportunity can generate a sequence of other opportunities in a cascade effect. For example, industrial activity that enhances greenhouse gas emissions creates opportunities for developing solutions that mitigate carbon emissions and clean energy technologies. It also gives rise to challenges such as rising global temperature and natural disasters, creating entrepreneurial opportunities for resilient infrastructure and disaster alerts. Because external enablers can act as exogenous shifts (Audretsch et al., 2022), we distinguish "root" from "symptomatic" opportunities. Root opportunities stem from primary problems that arise from human activity and their onset is characterized by gradual predictability, such as greenhouse gas emissions, water pollution, and deforestation. On the other hand, symptomatic opportunities arise as secondary effects of root opportunities. Unlike root opportunities, their onset is characterized by lower levels of predictability, sudden enablement, and may lack a direct linkage to human activity, such as natural disasters and pandemics. This distinction allows for a nuanced analysis of how entrepreneurs take advantage of external enablers. Root opportunities give entrepreneurs more time to recognize and plan for them, while symptomatic opportunities require a quick problem-solving response and fast recognition of the opportunity that may trigger new venture creation.

Geographical proximity is one of the external enablers that influence entrepreneurs' opportunity recognition, actively triggering venture openness. When analyzing both root and symptomatic opportunities, the geographic locations of both are often not the same (Althor et al., 2016). Symptomatic opportunities, such as natural disasters, although hard to forecast due to their sudden and unpredictable onset, demand urgent and fast answers to the problems once they happen (Davidsson et al., 2020, 2025). They cause disequilibrium in the social and economic sphere of the location where they occur (Hiatt et al., 2009; Williams & Shepherd, 2016). An extreme situation can lead to a discontinuity of current businesses and discourage plans to launch new ventures due to the higher levels of uncertainty (Grube & Storr, 2018). The situation becomes even more complicated when the region faces deadly events compared to other losses. The trauma caused by deaths can create traumas in communities (Cordero, 2023).

However, even in a challenging circumstance, new endeavors are essential to mitigate the negative impact on economic growth, and the necessity speaks louder. Thus, despite lacking infrastructure and support, direct and indirect investments are made to encourage entrepreneurial activity. Entrepreneurs tend to have a better perception of the opportunities once they are close to the government incentives and the problem itself, even though the circumstances can be challenging from an emotional perspective (Salvato et al., 2020).

From an economic perspective, symptomatic opportunities usually have low barriers to venture launching due to the need for fast response. The need to supply the expectations and market demands is primordial and is predominant in creating solutions (Dean & McMullen, 2007; Dutta, 2017; Nelson & Lima, 2020). Because symptomatic opportunities suddenly create new unmet needs or shifts in the demand curve, the previous ventures may be insufficient to meet the emergent needs, requiring the creation of new solutions to fulfill the need for demand expansion or substitution (Grube & Storr, 2018; Hiatt et al., 2009). Accordingly, because the local community may prioritize support for local businesses within their area to strengthen the local economy and benefit the population, new businesses that emerge in this region often receive rapid support. This backing is crucial to triggering the launching of ventures that are more attractive and viable (Davidsson et al., 2020; Shepherd et al., 2022). Also, in these environments, the entrepreneur's perception and response to an unexpected event are linked to their resilience and work differently than root opportunities (Grube & Storr, 2018; Shepherd & Williams, 2023). In symptomatic opportunity, the disruption of an unexpected event creates a higher awareness of a problem due to the higher spatial, temporal, and socio-demographic scope of the enabler (Davidsson et al., 2020, 2025). In other words, because symptomatic opportunities often involve changes to people's routines, they become more sensitive to the opportunities in their environment. The need to create solutions for themselves and for their community makes it even more difficult for them to become "blind" to a problem.

Thus, the incentives, such as local support, low entry market barriers, ease of access to capital, and the need for solutions in the location, enabled by the agents' proximity to symptomatic opportunities trigger the emergence of innovative solutions such as resilient infrastructure, disaster alerts, and recovery services, for example, while they have a stimulating setting to test and improve their products quickly (Kimjeon & Davidsson, 2022; Shepherd & Williams, 2023). Hence, there are more chances for new ventures with novel solutions to emerge closer to symptomatic opportunities due to a better understanding of the challenges and needs the affected communities face, which helps to create better-tailored products and services. Thus, we propose:

H1: Geographical proximity to symptomatic opportunities is *positively* related to the emergence of new firms.

Although the prediction related to proximity to symptomatic opportunities is straightforward, when considering proximity to root opportunities the relationship is more complex. Due to their onset with gradual predictability, proximity can increase venture launches by enabling businesses to create innovative solutions that disrupt established markets. Overall, proximity to opportunities offers numerous advantages for entrepreneurs, including better access to information, specific knowledge about the problem, access to local resources, and a skilled talent pool (Companys & McMullen, 2007; Pe'er & Keil, 2013). It also facilitates quick and continuous feedback for ventures while developing new products and services (Boschma, 2005; Porter, 1998). While it does not necessarily denote proximity to the customer base, it does lead to a reduction in transaction costs (Troisi & Alfano, 2023). In a broader sense, proximity is vital for utilizing and transferring knowledge, a fundamental resource for fostering innovation (Boschma, 2005; Boschma & Frenken, 2010). In that way, we hypothesize that:

H2a: Geographical proximity to root opportunities is *positively* related to the emergence of new firms.

However, drawbacks to proximity to a problem also exist. For example, the higher the attractiveness of an opportunity and the lower the barriers to accessing it, the higher the concentration of competitors for customers and resources such as human capital and raw materials (Lee, 2009). Consequently, a fight for prices is expected, directly influencing their revenues and profitability (Shaw & Gilly, 2000). At the same time, repeated exposure to a problem can affect how individuals recognize opportunities. Over time, people may become habituated to the issue, resulting in decreased awareness of both the problem and potential solutions. In other words, agents can become "blind" to new opportunities (Lahiri & Wadhwa, 2021). Besides, the persistence of the problem for a long time can also cause the false feeling that there is no solution, causing hopelessness and disinterest in those who want to find other solutions. Finally, entrepreneurs also face resistance due to political, cultural, and economic interests, for example. Systemic agents' ties to jobs and economies deeply embedded in the issue create a disincentive to resolve the problem (Dheer, 2017; Urbano & Alvarez, 2014). In sum, the homogenization of resources and knowledge, added to the systemic encouragement to pursue new solutions hindered by the presence of established competitors, may hinder entrepreneurial action. As such, we propose:

H2b: Geographical proximity to root opportunities is *negatively* related to the emergence of new firms.

Political ideology influences how individuals perceive the environment, particularly regarding the role of government versus the free market (Barber IV & Blake, 2024). The effects of political ideology can impact agents' perception of the potential to create new ventures through regulations, governmental subsidies, ease of access to capital, and access to the international market, for example (Bennett et al., 2024; Tang et al., 2024). Because political ideology reflects internal human beliefs, it plays an essential role in entrepreneurs' perceiving, assessing, and exploiting opportunities (Barber IV & Blake, 2024). As a result, it impacts both formal and informal institutional settings, promoting interactive learning that leads to innovation and the development of new products and services (Bennett et al., 2023; Chung et al., 2024; Etzkowitz & Leydesdorff, 2000). As the perception of the opportunity improves and the environment proves conducive, entrepreneurs seize the moment by launching new ventures to profit from these favorable conditions (Bennett et al., 2024; Urbano et al., 2019).

Two prominent political ideologies are conservatism and liberalism (Jost, 2006). Both can create favorable environments for entrepreneurs but in different ways (Lewis et al., 2021; Tang et al., 2024). Liberalism promotes regulations that prioritize market freedom while also protecting workers (Chin & Semadeni, 2017; Jost, 2006). Additionally, liberal policies are associated with inclusive settings for minorities, increasing the heterogeneity of human capital that can drive innovation (Chin & Semadeni, 2017; Jost, 2006; Swigart et al., 2020). Finally, their encouragement of sustainable solutions fosters the creation of technologies and products that rely on efficient and less aggressive environmental ventures in new markets (Chin et al., 2018). Thus, liberal policies emphasizing innovation, sustainable causes, equity, and community support create a fertile environment for addressing root and symptomatic opportunities. By fostering a regulatory and economic landscape that encourages innovation and supports both long-term and immediate challenges, liberal policies can significantly impact the emergence of new firms that tackle environmental problems and their consequences. Hence, we hypothesize that:

H3a: Political ideology (liberalism) is *positively* related to the emergence of new firms.

However, much like root opportunities, political ideology also exhibits a tension in the literature. Specifically, although liberalism is associated with more attention to social dilemmas, they expect the government to take the lead in the problem-solution through regulations. Conversely, conservative ideology espouses the virtue of market-based solutions where competition drives innovation and new ventures (Swigart et al., 2020). Conservatism believes in minimal government intervention, which can quickly lead to technological advancements, thanks to market competition (Jarrodi et al., 2019). Because conservatism focuses on economic growth and job creation, it can be interesting for companies that do not wish to be strongly affected by the government since they rely on the free and competitive market with politics that encourage tax reduction and stability, which can lower barriers to entry and operational costs, thereby creating a favorable environment for the emergence of new firms. In that way, we propose:

H3b: Political ideology (liberalism) is *negatively* related to the emergence of new firms.

METHOD

The hypotheses were tested in a sample of 31,009 U.S. county-year observations between 2010 and 2022 in the energy industry. This sample is well-suited for analyzing how external enablers affect entrepreneurial recognition and exploitation because the industry is (1) constantly under pressure to develop sustainable solutions due to perceptions of causing a social dilemma, (2) involves localized policies shaped by political ideology, and (3) exhibits both within, and between, variation in regions hosting root and symptomatic opportunities. In the analysis, the treatment is contingent upon whether counties are characterized by root (proxy: greenhouse gas emissions) or symptomatic (proxy: natural disasters) opportunities as well as their local political ideologies, and the outcome of interest is the emergence of new energy ventures within these counties. We account for the population, educational backgrounds, migration rates, unemployment rates, gross domestic product, unionization rates, and income inequality in the counties as control variables.

RESULTS

The findings offer valuable insights into how external enablers influence entrepreneurs' ability to trigger new venture creation. Support for hypotheses 2a and 3a indicates that geographical proximity to the root causes of problems plays a significant role in the emergence of new ventures. Additionally, the political ideology, more precisely liberalism, positively impacts the formation of these ventures in this context. Interestingly, hypothesis 1 is not supported, suggesting that areas with a higher demand for rapid solutions do not necessarily lead to entrepreneurial pursuit and the launch of new ventures. Overall, these results highlight that entrepreneurs interpret social opportunities differently and that external enablers affect how they identify and respond to social dilemmas.

DISCUSSION AND IMPLICATIONS

Findings add to the literature by advancing our understanding of the effects of external enablers on social opportunity recognition. We find that how individuals experience their environment, including the predominant political ideology of the area and their proximity to the problem, is relevant and influences social entrepreneurs' launching of new ventures. Further, this effect is contingent upon the type of social opportunity. Results demonstrate that embeddedness in areas responsible for generating pressures, such as areas with high greenhouse gas emissions, actually foster solutions due to increased entrepreneurs' sensitivity. This suggests that the areas creating social dilemmas are the same ones taking the lead in developing solutions, potentially improving their existing approaches and solutions. Interestingly, being close to symptomatic opportunities seems to have little effect on new venture launches. Speculatively, this may suggest that entrepreneurs perceive disasters as isolated and unpredictable events, which complicates their ability to develop sustainable business plans in urgent situations. Alternatively, the symptomatic experiences may also drive down the regional resource endowments that can be tapped into by entrepreneurs, discouraging venture launches. Finally, our results indicate that liberalism positively influences the recognition of social opportunities. This finding enhances the understanding of the environmental antecedents of social ventures by providing evidence that the dominant political ideology within an individual's environment affects how social entrepreneurs perceive social dilemmas as well as the free market's role in addressing them.

CONTACT: Mariza Gabriela L. Mesquita; mlemos1@lsu.edu; Louisiana State University, Stephenson Department of Entrepreneurship & Information Systems, E. J. Ourso College of Business, LSU, Baton Rouge, LA, USA, 70803

SELF-INTEREST VS. SOCIAL GOOD: DARK TRIAD TRAITS AND DIVERGENT ENTREPRENEURIAL PATHWAYS

Yannick Nicolai Amend, Justus Liebig University Giessen, Germany
Anna-Maria Klein, Julius-Maximilians-University Würzburg, Germany
Monika Christine Schuhmacher, Justus Liebig University Giessen, Germany

ABSTRACT

Despite the growing relevance of personality characteristics in entrepreneurship research, the role of Dark Triad traits in shaping entrepreneurial intentions remains inconsistent. Drawing on social exchange theory, we propose that individuals high in dark traits are drawn to environments that normalize one-sided exchanges (e.g., commercial entrepreneurship), while prosocial and reciprocal contexts (e.g., social entrepreneurship) may be less attractive due to the dark triads' antisocial tendencies. A meta-analysis of 43 independent samples (N = 19,992) reveals positive links between Machiavellianism and psychopathy with commercial, but negative links with social entrepreneurial intention. Narcissism is positively related to commercial and social entrepreneurial intention. Our findings offer novel insights into how dark triad traits with different social exchange expectations shape entrepreneurial intentions.

INTRODUCTION

Entrepreneurship plays an important role in fostering economic growth and advancing social development by promoting innovation, generating employment, and addressing societal challenges through value creation (Cai et al., 2021; Montiel & Clark, 2018). Therefore, a key focus in entrepreneurship research has been understanding the factors that influence entrepreneurial intention, defined as an individual's intent to start a business (Bird, 1988; Cai et al., 2021). While research has traditionally focused on commercial entrepreneurial intentions (CEI), where the emphasis is on self-interested behavior (Brownell et al., 2021), there is growing interest in social entrepreneurial intentions (SEI), where the focus is on societal value (Bacq & Alt, 2018). As SEI differs from CEI especially in its orientation and motivation towards social value creation and in their expectation of reciprocal exchanges (Bacq & Alt, 2018; Hockerts, 2017; Mair & Noboa, 2006), it is necessary to examine these two entrepreneurial intentions as two distinct constructs in the context of personality traits (Voegel, 2018).

Most research to date has focused on the positive traits of entrepreneurs that are consistent with the traditional image of the "good entrepreneur" (Ahmed et al., 2022; Yusof et al., 2024), particularly in the case of social entrepreneurs who are often associated with prosocial motives (Bacq & Alt, 2018; Hockerts, 2017). More recently, however, research has also turned to darker personality traits that may influence CEI and SEI, specifically the dark triad (DT): Machiavellianism, narcissism, and psychopathy (Brownell et al., 2021; Liu et al., 2021; McLarty et al., 2023). Despite their common core of manipulative and egocentric behavior aimed at gaining power and prestige, the three DT traits can be viewed as distinct constructs (Jonason & Webster, 2010; Paulhus & Williams, 2002) that differ int their social interactions with others (O'Boyle et al., 2012). In this context, social exchange theory (SET) provides a theoretical lens to understand how DT traits perceive and engage in social exchanges in different entrepreneurial contexts.

Despite the growing interest in this area, research findings on the relationships between DT traits and entrepreneurial intentions remain mixed. For example, for the relationship between psychopathy and CEI, some studies report predominantly positive correlations (Do & Dadvari, 2017; McLarty et al., 2023) while others find negative ones (Laouiti et al., 2024). Similarly, the relationship between the DT traits and SEI is complex. While Machiavellianism and psychopathy often correlate negatively with SEI, the results on the correlation with narcissism tend to be positive (Wu et al., 2020; Wu et al., 2022). These findings highlight the need to summarize the current state of knowledge and examine how specific DT traits may be differently positioned towards CEI and SEI. Brownell et al., (2021) previously examined the relationship between DT and CEI using meta-analytic tools and found only positive relationships. Their analysis included studies published up to August 2020, comprising 21 articles with 6,842 participants on the relationship between DT traits and CEI. Our study extends this work by distinguishing between CEI and SEI, including a significantly larger sample of 35 articles, and including most recent articles (49% of articles published after August 2020) with 19,922 participants – thus providing a broader and more recent perspective.

HYPOTHESIS DEVELOPMENT

The Dark Triad in Shaping Commercial Entrepreneurial Intentions

Commercial entrepreneurial intention (CEI) refers to the intention to start a business with a focus on commercial benefits (Hockerts, 2017; Mair and Martí, 2006). In the literature, this form of entrepreneurship is often subsumed under the broader concept of entrepreneurial intention (Brownell et al., 2021; Hmieleski & Lerner, 2016; Kramer et al., 2011). To make a clear distinction between commercial entrepreneurship (CE) and social entrepreneurship (SE), this paper explicitly refers to commercially oriented intentions as CEI. These intentions are typically driven by self-interest, with CE providing an environment that tolerates or even rewards self-promoting and one-sided behavior (Bacq & Alt, 2018; Tan et al., 2020).

From the perspective of social exchange theory, social behavior is the result of a social exchange process that is based on reciprocity (Hmieleski & Lerner, 2016; O'Boyle et al., 2012). In contrast, individuals high in the DT tend to prioritize personal goals over reciprocity and mutual benefit (Jonason & Webster, 2010; Paulhus & Williams, 2002). CE often tolerates ethically ambiguous, self-interested behavior (Bacq & Alt, 2018; Tan et al., 2020). As each dark triad trait differs in terms of antisocial tendencies and social interactions, it is important to examine them separately (Jonason & Webster, 2010; Paulhus & Williams, 2002). Therefore, the specific relationships between each of the three DT traits and CEI are discussed below.

Machiavellianism is characterized by manipulative and controlling behavior to achieve personal goals (Dahling et al., 2009; Hanson et al., 2023; Jonason & Webster, 2010). This includes material or instrumental benefits like image, status, power, and wealth (Boddy, 2006; Jones & Paulhus, 2009, 2014; Voegel, 2018). Individuals high in Machiavellianism tend to distrust others (Dahling et al., 2009) and are very adaptable, which makes it easy for them to hide their true intentions (Wu et al., 2019). Based on the Machiavellians' egocentric perspective, literature mostly assumes that CE is a suitable way to pursue their self-centered goals, which is why an overall positive relationship with CEI has been widely confirmed in different cultural environments (Do & Dadvari, 2017; Hoang et al., 2022; Laouiti et al., 2024). Because of their selfish tendencies, Machiavellians see no need to seek reciprocal relationships (Hmieleski & Lerner, 2016). They focus exclusively on achieving their own goals and are prepared to do almost anything to achieve them, while expecting others to do the same (Hmieleski & Lerner, 2016; Taboroši et al., 2023). Therefore, they prefer value appropriation over value creation for others (Hmieleski & Lerner, 2016). The competitive environment of CE (Laouiti et al., 2024; Lien et al., 2022; Wu et al., 2019) tolerates and may even reward their manipulative behavior aimed at self-benefit, providing Machiavellians with avenues for strategic self-advancement (Bacq & Alt, 2018; Brownell et al., 2021) to gain advantages, such as leadership roles or control over employees, while putting their own interests above those of others (Hmieleski & Lerner, 2016; Terek-Stojanović et al., 2023; Wu et al., 2019).

H1a: Higher levels of Machiavellianism are associated with increased Commercial Entrepreneurial Intention.

Psychopathy is characterized by manipulativeness, low empathy, impulsivity, and thrill-seeking behaviors (Decuyper et al., 2009; Jonason & Webster, 2010; Paulhus & Williams, 2002; Tucker, 2016). Those characteristics make it easier for psychopaths to achieve their goals in social interactions (Decuyper et al., 2009; Paulhus & Williams, 2002). Individuals with psychopathic tendencies, like individuals high in Machiavellianism, are motivated by material or instrumental gain, like self-expression, status and dominance (Jones & Paulhus, 2014; Lien et al., 2022). Research suggests that CE appeals to psychopaths, as is provides an environment that allows them to manipulate social interactions for personal and material gain (Cesinger & Kramer, 2011; Kramer et al., 2011; McLarty et al., 2023). Psychopaths are undeterred by risk and uncertainty, which is why they are well suited to and attracted to a competitive environment such as CE, where their exploitative tendencies can thrive (Brownell et al., 2021; Decuyper et al., 2009; Jonason & Webster, 2010; Paulhus & Williams, 2002; Tucker, 2016). While these characteristics of CE may deter others, they make CE appealing to psychopaths, who are motivated by immediate goal achievement (Hmieleski & Lerner, 2016; Paulhus & Williams, 2002; Tucker, 2016). Their lack of empathy and concern for others (Paulhus & Williams, 2002), empower them to engage in exploitative and manipulative behavior, as their focus is solely on the appropriation of value from social exchanges, without the need for value creation or reciprocity (Hmieleski & Lerner, 2016). The tolerance for self-serving behavior and one-side exchange in CE, therefore provides an ideal setting for individuals with high psychopathy scores to exploit social interactions for personal advantage (Bacq & Alt, 2018; Brownell et al., 2021).

H1b: Higher levels of Psychopathy are associated with increased Commercial Entrepreneurial Intention.

Narcissism is characterized by a sense of entitlement and a belief in one's superiority over others (Paulhus & Williams, 2002; Wu et al., 2019). Narcissists seek power, admiration and status (Jonason & Webster, 2010) and display confidence in handling pressure, risk and uncertainty (Al-Ghazali & Afsar, 2021), often paired with a dominant personality (Paulhus & Williams, 2002). Research mostly showed a positive relationship between narcissism and CEI (Cesinger & Kramer, 2011; Hanson et al., 2023; Hmieleski & Lerner, 2016), likely because CE aligns with their pursuit of personal goals (Jonason & Webster, 2010).To achieve these goals, narcissists are willing to manipulate social exchanges (Jones & Paulhus, 2014) and often disregard reciprocity due to their perception of their own superiority (O'Boyle et al., 2012; Paulhus & Williams, 2002). As CE tolerates one-sided, self-serving behavior (Bacq & Alt, 2018; Brownell et al., 2021), it offers a favorable environment for such tactics. Their dominant personality enables them to influence and control social interactions, especially in leadership contexts (Al-Ghazali & Afsar, 2021). Moreover, the entrepreneurial path allows them to bypass hierarchical barriers and fulfill their desire for autonomy and self-determination (Al-Ghazali & Afsar, 2021; Hmieleski & Lerner, 2016). The perception of CE as an admirable career (Wu et al., 2019) further meets their own need for admiration. Confident in managing uncertainty, narcissists are drawn to CEI as a means of gaining power, admiration, and status through socially tolerated, non-reciprocal exchange (Al-Ghazali & Afsar, 2021; Bacq & Alt, 2018; Brownell et al., 2021; Jonason & Webster, 2010; O'Boyle et al., 2012).

H1c: Higher levels of Narcissism are associated with increased Commercial Entrepreneurial Intention.

The Dark Triad in Shaping Social Entrepreneurial Intentions

Motivations for entrepreneurial intentions differ depending on whether the venture has a commercial or social orientation (Hockerts, 2017; Mair & Martí, 2006). Whereas CE emphasizes self-interest, social entrepreneurial intentions (SEI) refer to the intention to start a business with a social mission, associated with a prosocial nature rather than a self-centered nature (Bacq & Alt, 2018; Tan et al., 2020). However, this does not imply that social ventures cannot generate profit (Mair & Martí, 2006). Since the DT traits differ regarding their behavior in social interactions (O'Boyle et al., 2012), the relationships between each of the three DT traits and SEI are discussed below.

Despite their positive association with CEI, both Machiavellianism and psychopathy have been linked to a negative relationship with SEI (Voegel, 2018; Wu et al., 2020). This is primarily due to their self-serving and manipulating strategies, which conflict with the prosocial and reciprocal nature of SEI (Bacq & Alt, 2018; Wu et al., 2020). The lack of prosocial motivation and concern for others' welfare displayed by individuals with tendencies towards Machiavellianism and psychopathy (Azizli et al., 2016; Hanson et al., 2023; Jonason et al., 2012; Lien et al., 2022; McLarty et al., 2023; Wu et al., 2020) make them less inclined to engage in activities that benefit society (Akhtar et al., 2013; Iyayi & Obeki, 2018) or to create social value (Hmieleski & Lerner, 2016; Lien et al., 2022), as expected in SE (Mair & Martí, 2006). The Machiavellian's willingness to do almost everything to achieve their own goals (Taboroši et al., 2023) even if this involves the use of unscrupulous methods or the breaking of social norms (Khawar et al., 2022; McLarty et al., 2023) and the psychopath's antisocial, unemphatic and cold-hearted behavior (Akhtar et al., 2013; Wu et al., 2020), leads both to favor one-sided exchanges over reciprocal ones (Hmieleski & Lerner, 2016), which runs counter to SE's basic intentions (Bacq & Alt, 2018). Therefore, the costs of attending others' needs and acting prosocial (Bacq & Alt, 2018; Hmieleski & Lerner, 2016; Wu et al., 2020), may outweigh the potential personal benefits for individuals high in psychopathy and Machiavellianism (Lien et al., 2022).

H2a: Higher levels of Machiavellianism are associated with decreased Social Entrepreneurial Intention.

H2b: Higher levels of Psychopathy are associated with decreased Social Entrepreneurial Intention.

In contrast to the clearly negative implication of Machiavellianism and psychopathy, narcissism presents a more complex case in the context of SEI. Unlike individuals high in Machiavellianism and psychopathy, the manipulative nature of individuals with high scores in narcissism manifests itself in socially acceptable forms such as charisma and charm, resulting in less destruction of interpersonal relationships to the extent that these are necessary for them to gain admiration (Brownell et al., 2021). To gain social acceptance and admiration, individuals with narcissistic tendencies are willing to temporarily refrain from manipulative behaviors and instead display adaptability and cooperation (Terek-Stojanović et al., 2023), while they are willing to create value for others if it ultimately serves their need fulfillment (Brownell et al., 2021; Brunell et al., 2008; Hmieleski & Lerner, 2016; Terek-Stojanović et al., 2023). Accordingly, individuals with high scores in narcissism endeavor not to be perceived as unscrupulous, even if they are prepared to do anything behind closed doors to achieve their objectives (Hmieleski & Lerner, 2016). This makes SE attractive to them, even if their true motive for becoming a social entrepreneur is ego enhancement rather than the creation of social value (Jones & Paulhus, 2014). By being perceived as helpful and influential, they can fulfill their egocentric goals through this profession (Hmieleski & Lerner, 2016; Kim et al., 2023; Lien et al., 2022; Voegel, 2018), while maintaining the reciprocal nature of SE (Bacq & Alt, 2018; Voegel, 2018). As they are not primarily destructive and tend to subtly exploit their relationships (Brunell et al., 2008; Hmieleski & Lerner, 2016), SE is a suitable option for them to optimize their social exchange, even if it is ultimately reciprocal. However, it is important to note that their motivation is driven by personal fulfillment rather than altruism, which is still a valid basis for the formation of SEI (Mair & Martí, 2006). Thus, individuals with narcissistic tendencies may perceive SE as a way to satisfy their needs by being perceived as helpful and influential (Kim et al., 2023; Lien et al., 2022; Voegel, 2018; Wu et al., 2019) and therefore accept the costs of reciprocal exchange if SE ultimately serves their egoistic goals (Hmieleski & Lerner, 2016).

H2c: Higher levels of Narcissism are associated with increased Social Entrepreneurial Intention.

METHOD

We investigate how DT traits are related to the two forms of entrepreneurial intentions and conducted a systematic review using meta-analytical tools. We first systematically reviewed and synthesized the available literature, then combined statistical results using meta-analytic tools (Eisend, 2020).

To determine the sample for the meta-analysis, we employed a systematic keyword literature search in three large databases, two databases of conference papers and several papers on the dimensions of DT (Dahling et al, 2009; Jonason & Webster, 2010; Jones & Paulhus, 2014; Paulhus & Williams, 2002). All results from the original search were first checked for inclusion in the meta-analysis using the title and abstract, which reduced the initial sample from 4,683 articles to 74 articles. In the next step, the full text was examined, resulting in 36 articles. When effect sizes were missing, the respective authors were contacted. After contacting seven authors and adding nine studies related to the dimensions of DT (Dahling et al., 2009; Jonason & Webster, 2010; Jones & Paulhus, 2014; Paulhus & Williams, 2002), we analyze a sample size of 35 articles with a total of 43 independent samples and N = 19,922 participants. Six of the 35 papers came from non-journal sources, such as dissertations or presentations at conferences.

A two-stage approach was chosen to carry out the meta-analysis. First, a bivariate analysis and then a meta-regression was carried out. To correct the correlations for unreliability, the equations of Hunter and Schmidt (2004) were used. This was conducted at the individual level of the respective sample. Corrections for the construct of entrepreneurial intention were made if the respective study measured it with at least 2 items. If entrepreneurial intention was only measured via a yes/no question, no corrections were made. In these cases, the sample was only corrected for the DT trait. For our calculations, we used the respective correlations of each individual DT trait with entrepreneurial intention as effect sizes. To account for varying sample sizes among the included studies, all individual (corrected) effect sizes were weighted based on their respective inverse variance before being aggregated to the overall effect size (Eisend, 2020). For synthesizing the individual effect sizes, a random-effects model was used. We used the meta-analysis tools in IBM SPSS 29 for the estimations.

RESULTS

We conducted various tests to consider the possibility of publication bias: the trim-and-fill method (Duval & Tweedie, 2000), the Egger's test, the file drawer method (Rosenthal, 1979) and a cumulative meta-analysis. Overall, the corrected effect sizes remained significant in most examinations. Nonetheless, particularly the DT relationships with CEI, suggest a slight overestimation of the original effects, although their overall significance remains. However, the effect for psychopathy is smaller and less consistent in the cumulative meta-analysis, but since all other tests show robustness, the effect could be smaller but still present.

Our results show that regarding the relationship between Machiavellianism, narcissism, psychopathy and CEI, all three characteristics of the DT were significantly positively associated with CEI. The corrected effect sizes of Machiavellianism and narcissism showed medium size correlations (Cohen, 1988) with CEI. For psychopathy we report a smaller corrected effect size. Our results provide support for H1a, H1b, and H1c. The results concerning the relationship between Machiavellianism, psychopathy, narcissism and SEI showed mixed patterns, while all three traits show medium corrected effect sizes (Cohen, 1988). Our examination of the relationships between SEI and Machiavellianism, and between SEI and psychopathy, reveals negative corrected effect sizes for both traits, respectively. In contrast, the relationship between narcissism and SEI was positive and significant. Hence, we find support for H2a, H2b and H2c.

DISCUSSION AND IMPLICATIONS

This paper examined how antisocial traits, specifically the DT (Paulhus & Williams, 2002), influence entrepreneurial intentions in environments with differing expectations of social exchange. Drawing on SET, we propose that the degree of reciprocity required in CE versus SE shapes how attractive these contexts are to individuals with DT traits. While CEI aligns with self-interested, one-sided exchanges, SEI depends on a trait's willingness to engage in reciprocal behavior to achieve personal goals. Addressing previously inconsistent findings (Hanson et al., 2023; Hmieleski & Lerner, 2016; McLarty et al., 2023; Voegel, 2018; Wu et al., 2020), this study provides theoretical and empirical clarity by applying SET to both SEI and CEI. Our study is the first meta-analysis to examine DT traits in relation to SEI and to simultaneously analyze both entrepreneurial intention types within a meta-analysis and through a unified theoretical lens.

We offer three key contributions to entrepreneurship research. Firstly, we extend SET to the context of entrepreneurial intentions, showing how DT traits relate differently to social and commercial entrepreneurial intentions through their engagement with reciprocity and cooperation – core tenets of SET (O'Boyle et al., 2012). Although all three traits share self-serving tendencies, they differ in how they perceive and act within social exchange relationships (Paulhus & Williams, 2002). Clarifying these differences helps explain their distinct links to CEI and SEI. Second, we contribute the conceptual differentiation between CEI and SEI. These forms of intention differ fundamentally in their alignment with social exchange principles: CE permits one-sided, profit-driven behavior (Brownell et al., 2021), while SE demands prosocial, reciprocal engagement (Bacq & Alt, 2018; Mair & Martí, 2006). This distinction helps explain the divergent effects of DT traits on CEI and SEI. Third, our findings reveal how DT traits may vary in their social interaction strategies and how they map onto diverging entrepreneurial contexts. All DT traits are drawn to CEI due to its tolerance for self-serving behavior. However, only narcissists appear willing to engage in reciprocal exchanges – albeit for egoistic reasons and their social dependency for achievement (Hmieleski & Lerner, 2016) – which explains their positive relationship with SEI. In contrast, Machiavellians and psychopaths, who devalue reciprocity (Hmieleski & Lerner, 2016; O'Boyle et al., 2012), are less inclined toward SEI; as SE's reciprocal and prosocial nature limits opportunities for exploitation (Bacq & Alt, 2018; Mair & Martí, 2006).

Future research should treat CE and SE as distinct entrepreneurial environments, especially when considering antisocial or egocentric traits. It is worth exploring whether such individuals can contribute positively to entrepreneurial ecosystems despite their interpersonal tendencies. In particular, understanding their motives and designing guardrails for ethical behavior may help harness their potential while minimizing harmful outcomes.

CONTACT: Yannick N. Amend; yannick.amend@wirtschaft.uni-giessen.de; Licher Straße 74, 35294 Gießen, Germany

EMPOWERING VULNERABLE COMMUNITIES: THE ROLE OF HUMANITARIAN ENTERPRISES IN BRDIGING EMERGENCY RELIEF AND SUSTAINABLE DEVELOPMENT IN CHILDCARE, FAMINE AND EDUCATION

Koteswararao Naik Bhukya, Indian Institute of Management, Mumbai 400087, India
Suraj Panigrahi, Indian Institute of Management, Mumbai 400087, India
Nilofer Gurshith, Indian Institute of Management, Mumbai 400087, India
Simran Malav, Indian Institute of Management, Mumbai 400087, India

ABSTRACT

Despite significant global progress in reducing under-five mortality rates, natural disasters continue to pose a severe threat to child survival by disrupting essential healthcare services, exacerbating the spread of infectious diseases, increasing malnutrition, and exposing children to trauma and unsafe living conditions. This study examines how humanitarian enterprise models integrating microfinance, community-driven interventions, and adaptive service delivery enhance post-disaster recovery and resilience. Using multi-case analyses of disasters, we test three hypotheses. Results confirm significant improvements in livelihood recovery, child nutrition, and resilience. The findings advocate for systemic shifts from short-term aid to sustainable, market-based solutions, emphasizing equity, local empowerment, and cross-sector collaboration to safeguard child health in disaster-prone regions.

INTRODUCTION

Despite a global decline in under-five mortality from 93 deaths per 1,000 live births in 1990 to 38 in 2021 millions of children continue to die from preventable causes. In 2021 alone, an estimated 5 million under-five deaths occurred, highlighting persistent disparities and challenges in child health and survival. In response, governments and global bodies have implemented targeted interventions aligned with sustainable development goal (SDG) number three which talks about good health and well-being, aiming to end preventable child deaths by 2030 through improved healthcare, nutrition, and immunization coverage (Sanhueza et al., 2021). However, these gains remain at risk and fragile in natural disasters, where post-disaster contexts disrupt healthcare delivery, amplify disease exposure, and strain maternal-child services. To sustain SDG progress, there is an urgent need for disaster-resilient health systems and equitable, rapid response recovery mechanisms that protect vulnerable child populations in the aftermath of crises (Melkonyan et al., 2025).

While immediate humanitarian assistance often saves lives during a disaster's acute phase, the long-term recovery of child health services remains critically under-resourced and poorly coordinated (Urbina-Johanson et al., 2025). In vulnerable communities, the withdrawal of aid agencies often leads to the lapse of essential services such as immunization, nutrition programs, and pediatric care, leaving children exposed to prolonged and preventable health risks. Damaged infrastructure depleted local capacity, and declining donor attention further deepen the crisis, leaving children at risk of stunting, disease, and psychosocial trauma (Bettenhausen et al., 2021). This study aims to bridge the existing knowledge gap by examining the post-disaster recovery phase, with a focus on how sustained and equitable childcare systems can be developed to safeguard vulnerable populations once humanitarian interventions begin to taper off.

Traditionally, humanitarian assistance follows a charity-driven model, offering rapid relief in the immediate aftermath of disasters. While vital during emergencies, such models often overlook the long-term developmental needs. The sustainable livelihoods framework (SLF) criticizes this approach, highlighting the limitations of short-term aid in building resilience and sustaining recovery (Natarajan et al., 2022). Recognizing these gaps, the humanitarian sector is witnessing a shift toward humanitarian enterprise models, an evolving approach integrating market-based mechanisms, community engagement, and public-private collaboration to deliver more sustainable outcomes. These models aim to respond to crises and build resilience, strengthen local institutions, and enable affected populations to recover with dignity and autonomy. In this study, we propose a collaborative framework aimed at facilitating the transition from emergency response to long-term recovery by enhancing the continuity and sustainability of child-focused services in disaster-prone settings particularly for children under five, who are among the most at risk.

We conducted case study analyses of global disasters and the services provided by authorities, guided by grounded theory research principles (Corbin & Strauss, 2012). Using a multiple case study approach (Yin, 2009), we explored diverse contexts to enhance generalizability (Eriksson & Kovalainen, 2008). Our research aims to understand how humanitarian enterprises, supported by microfinance institutions (MFIs), contribute to sustainable healthcare, education, and child welfare service delivery. The challenge is not limited to one or two countries; it is a widespread global issue that affects vulnerable communities across different geographies, cultures, and economies. Our research suggests that integrating the humanitarian enterprise model with microfinance and community engagement significantly enhances long-term, sustainable recovery in disaster-prone regions. A core contribution of our work lies in its cross-sectoral approach that links humanitarian assistance with development economics, placing children at the center of resilience strategies. The study offers actionable insights by emphasizing the need for stronger public-private partnerships (PPP), coherent policy alignment, and long-term investment to protect vulnerable populations from recurring humanitarian crises.

HYPOTHESIS DEVELOPMENT

In the aftermath of disasters, affected populations face multidimensional losses, including the destruction of physical infrastructure, loss of income sources, and erosion of social safety nets. While such interventions are essential for immediate survival, they tend to foster dependency and lack the mechanisms to support sustainable recovery. In contrast, microfinance-integrated recovery programs aim to rebuild livelihoods by providing access to financial capital, facilitating entrepreneurial activity, and restoring household income. Grounded in the SLF, this approach emphasizes enhancing financial, human, and social community assets as a pathway to resilience. Furthermore, drawing from social innovation theory (SIT), microfinance enables locally adapted, bottom-up solutions that leverage Indigenous knowledge and networks for recovery (Yunus, 2007; Morduch, 1999). Therefore, this study posits that disaster-affected communities receiving microfinance support will exhibit stronger and more sustainable livelihood outcomes than those relying solely on conventional aid.

Hypothesis H1: Microfinance-integrated recovery programs lead to better livelihood restoration among disaster-affected communities than traditional aid-based approaches.

As communities begin to rebuild livelihoods, the next phase of sustainable recovery requires a stronger focus on localized service delivery mechanisms, particularly those targeting vulnerable sub-populations such as children under five. Health and nutrition are among the first services to be disrupted post-disaster, disproportionately affecting child survival and development. Traditional top-down aid models often struggle to reach these populations effectively due to cultural misalignment, trust deficits, and lack of contextual awareness. In contrast, community-driven humanitarian enterprise models, such as self-managed health cooperatives, women's collectives, and locally governed nutrition hubs, are often more accessible, trusted, and contextually relevant. By empowering local actors and integrating them into service delivery, such models enhance coverage, responsiveness, and cultural acceptability, particularly for the most vulnerable groups.

Hypothesis H2: Community-driven humanitarian enterprise models lead to improved health and nutrition outcomes for children under five.

Post-disaster environments often disrupt essential services such as healthcare and education, particularly affecting children. Adaptive service models like mobile clinics or temporary learning spaces can help ensure service continuity in such fragile contexts. At the same time, unregulated microfinance can deepen vulnerabilities, making regulation critical for linking financial tools with positive social outcomes, such as school attendance or child nutrition. Drawing from Adaptive Systems Theory and Institutional Theory, this study posits that when humanitarian enterprises integrate flexible service delivery with regulated microfinance, they reduce household risk and ensure continuity in vital services.

Hypothesis H3: Humanitarian enterprise models that integrate adaptive service delivery with regulated microfinance improve the continuity of healthcare and education while reducing household risks.

OVERVIEW OF STUDIES

STUDY 1: To empirically test hypothesis H1, this study draws on a comprehensive range of secondary data from post-disaster recovery initiatives implemented in Aceh, Indonesia, following the 2004 tsunami. Aceh is a robust case for comparative analysis, featuring both traditional humanitarian aid interventions and microfinance-integrated recovery programs. The traditional aid included direct relief in food, cash-for-work, and temporary housing. At the same time, microfinance-based interventions were spearheaded by institutions such as Bank Rakyat Indonesia and international NGOs like Mercy Corps, which facilitated microloans, savings groups, and support for small business recovery. The secondary data were sourced from published impact assessment reports, datasets available through UNDP and World Bank evaluations, and peer-reviewed studies on livelihood outcomes in the region. This comprehensive approach includes variables related to income levels, employment status, asset recovery, and household resilience indicators tracked over a 3–5 years recovery period.

Variables. To assess the impact of microfinance-integrated recovery on livelihood restoration, we operationalized livelihood restoration as the dependent variable using a composite index that includes (i) post-disaster income level, (ii) household asset recovery, and (iii) employment stability. The key independent variable is the type of intervention, classified as a binary categorical variable (1 = microfinance-integrated recovery, 0 = traditional aid-only recovery). Control variables include household size, education level of the head of household, and proximity to urban centers (as a proxy for market access).

Method. The analytical strategy involves performing an independent samples t-test to examine whether there is a statistically significant difference in livelihood restoration scores between communities that received microfinance-based recovery assistance and those that relied solely on traditional humanitarian aid. Before proceeding, we conducted a Levene's Test for Equality of Variances to determine if equal variances could be assumed. Since the p-value was >0.05, the assumption of equal variances held, and the standard t-test was used. Additionally, a robustness check using Welch's test yielded consistent results, confirming the reliability of findings.

STUDY 2: To empirically test hypothesis H2, this study utilizes secondary data from post-disaster recovery efforts in the Philippines, particularly following Typhoon Haiyan in 2013. The Philippines provides a pertinent case for examining the impact of community-driven humanitarian enterprise models on health and nutrition outcomes for children under five. In the aftermath of the typhoon, several local organizations and NGOs implemented community-driven health cooperatives, locally led nutrition hubs, and community-based health initiatives to improve child welfare. These initiatives focused on providing essential healthcare, nutrition, and education services through participatory governance and community-led models. The secondary data used in this study were drawn from published evaluations, NGO reports, and datasets available through the Philippines Department of Health, UNICEF, and other relevant organizations, covering the health outcomes and child nutrition indicators over the two years following the disaster. The analysis focuses on regions that received community-driven interventions compared to those that relied solely on conventional, centralized relief efforts.

Variables. To evaluate the impact of community-driven humanitarian enterprise models on health and nutrition outcomes, child health and nutrition outcomes were operationalized as the dependent variables. These outcomes include (i) the percentage of children under five with improved nutritional status (as measured by weight-for-age and height-for-age indicators), (ii) the frequency of child health checkups and vaccinations, and (iii) child mortality rates within the first two years after the disaster. The independent variable, community-driven intervention, is a binary categorical variable (1 = community-driven health/nutrition model, 0 = traditional centralized aid intervention). Control variables include the level of community engagement, availability of local health resources, socioeconomic status of households, and proximity to disaster-affected areas (as a proxy for access to services).

Method. The analysis utilizes propensity score matching (PSM) to account for potential selection bias between communities that received community-driven interventions and those that did not. PSM matches treated and untreated communities based on observed characteristics, ensuring that the comparison groups are as similar as possible, except for the intervention. This method allows us to control for confounding variables such as community size, socioeconomic status, and existing healthcare infrastructure, which could influence the outcomes. After matching, the average treatment effect on the treated (ATT) is calculated to determine the difference in health and nutrition outcomes between the two groups.

STUDY 3: To empirically test the combined hypotheses H3, this study draws on secondary data from the humanitarian response to the 2015 Nepal Earthquake, particularly in the districts of Gorkha and Sindhupalchok. These areas suffered extensive infrastructure collapse, including schools and healthcare facilities. In response, several humanitarian organizations, including Save the Children, CARE Nepal, and UNDP, introduced adaptive service delivery mechanisms such as temporary learning shelters, mobile clinics, and telehealth services. Concurrently, MFIs like Nirdhan and Sana Kisan implemented tightly regulated financial inclusion programs with conditions linked to child welfare indicators such as school attendance and maternal-child health checkups. This dataset includes program evaluation reports from the UN OCHA, Nepal Rastra Bank, and national census follow-ups conducted between 2015 and 2018.

Variables. The dependent variable, "resilience outcome score", was constructed as a composite index including (i) continuity of healthcare and education (measured via post-disaster school attendance and immunization coverage) and (ii) risk reduction (proxied through food security status, household debt default rate, and dropout rate). The key independent variable was an integrated humanitarian model combining adaptive services and regulated microfinance (1 = present, 0 = absent). Control variables included proximity to district headquarters (a proxy for accessibility) and pre-disaster poverty index.

Method. A Multivariate Analysis of Covariance (MANCOVA) was used to assess the effect of integrated interventions on the resilience score while controlling for accessibility and baseline poverty. This approach allows for simultaneous analysis of multiple dependent variables and accounts for covariate effects. The data met assumptions for multivariate normality and homogeneity of variance-covariance matrices.

RESULTS

The results from three empirical studies demonstrate the positive impact of humanitarian enterprise models particularly those integrating microfinance, community participation, and adaptive-regulated interventions on livelihood recovery, child health, and resilience outcomes in post-disaster contexts.

Table 1: Descriptive Statistics for Livelihood Restoration (LRS) Score for Study 1

Group	N	Mean	Std. Deviation	Std. Error Mean
Traditional Aid Only	120	56.40	10.82	0.99
Microfinance-Based Recovery	115	68.75	12.30	1.15

Table 1 presents the descriptive statistics for the livelihood restoration score (LRS) across two groups: those who received only traditional aid and those supported through microfinance-based recovery. The microfinance group had a higher mean LRS (68.75) compared to the traditional aid group (56.40), suggesting better post-disaster livelihood outcomes. Table 2 shows the results of an independent samples t-test, which confirms that this difference is statistically significant ($t = -8.21$, $p < 0.001$). The mean difference of -12.35, with a 95% confidence interval ranging from -15.30 to -9.40, indicates that microfinance-integrated recovery efforts significantly outperform traditional aid in restoring livelihoods. These results strongly support hypothesis H1.

Table 2: Independent Samples t-Test for Study 1

Test	t	df	Sig. (2-tailed)	Mean Diff.	Std. Error Diff.	95% CI Lower	95% CI Upper
Equal Variances Assumed	-8.21	233	0.000	-12.35	1.50	-15.30	-9.40

After performing PSM and estimating the ATT, the results revealed significant improvements in health and nutrition outcomes for children under five in communities receiving community-driven humanitarian enterprise models compared to those relying solely on traditional aid interventions.

Table 3: Matched Sample Characteristics Study 2

Variable	Community-driven (n=50)	Traditional Aid (n=50)	P-value
Community Size	5,500	5,450	0.51
Socioeconomic Status	Medium	Medium	0.94
Healthcare Infrastructure	Moderate	Low	0.03
Household Income (USD)	1,200	1,150	0.48
Proximity to Aid Services	Close	Far	0.02

Table 3 above presents the matched sample characteristics for communities receiving community-driven interventions versus those that received traditional aid. The matching procedure balanced the groups across key variables such as community size, socioeconomic status, and existing healthcare infrastructure.

Table 4: Health and Nutrition Outcomes Study 2

Outcome	Community-driven (n=50)	Traditional Aid (n=50)	Difference (ATT)	P-value
Percentage of Children with Improved Nutrition	72%	60%	+12%	0.02
Health Checkups & Vaccinations	80%	65%	+15%	0.01
Child Mortality Rate	2.3%	4.1%	-1.8%	0.04

Table 4 shows the outcomes of interest: child nutritional status (measured by weight-for-age and height-for-age), health checkup and vaccination rates, and child mortality rates. The results indicate that communities with community-driven models experienced a significant improvement in these outcomes.

Table 5: Test of Adaptive and Regulated Interventions on Resilience Outcomes Study 3

Source	Wilks' Lambda	F	p-value
Adaptive & Regulated Interventions	0.782	4.47	0.004
Accessibility (Covariate)	0.927	1.29	0.256
Baseline Poverty (Covariate)	0.891	1.83	0.167

Table 5 show a significant multivariate effect of adaptive and regulated humanitarian enterprise models on post-earthquake resilience outcomes (Wilks' Lambda = 0.782, $F = 4.47$, $p < 0.01$). Districts where both adaptive services (e.g., mobile clinics, learning shelters) and regulated microfinance were deployed recorded higher continuity in essential services and lower risk of household-level setbacks such as child malnutrition and school dropout. These findings demonstrate that integrating regulated financial support with adaptable service delivery mechanisms contributes to both short-term recovery and long-term resilience.

DISCUSSION AND IMPLICATIONS

This study examined how integrating humanitarian enterprise models anchored in microfinance, community participation, adaptive services, and financial regulation can strengthen recovery outcomes for children under five in disaster-affected communities. The findings from three case-based analyses demonstrate that microfinance-integrated recovery significantly improves livelihoods (H1), community-led models enhance health and nutrition (H2), and a combined approach of adaptive service delivery and regulated financial access bolsters resilience (H3).

Importantly, these results affirm the value of locally driven, context-sensitive, and financially inclusive interventions over traditional top-down aid. However, the analysis also highlights persistent structural issues: lack of sectoral coordination, inconsistent funding for long-term initiatives, and the risk of poorly regulated microfinance systems.

For policymakers and NGOs, the implication is clear: Resilience-building must move beyond relief. It requires designing enterprise-driven interventions that engage local actors, safeguard vulnerable populations, and embed social impact metrics like school retention and child health.

CONTACT: Koteswararao Naik Bhukya, dr.bkrnaik@iimmumbai.ac.in, Indian Institute of Management, Mumbai 400087, India

(PLACE-BASED) ENTREPRENEURIAL RESOURCEFULNESS IN TOTAL INSTITUTION

Sadek Showkat, Loyola University Chicago, United States
Oana Branzei, Western University, Canada

ABSTRACT

Resourcefulness is understood as a social-cognitive behavior of maneuvering "available resources." This study challenges the assumption of "available resources" by examining it among forcibly displaced individuals who have lost all their resources. Drawing on a multi-sited reflexive ethnography in Rohingya refugee camps, our recursive model explains how resourcefulness begins again after all resources have arguably been lost. Our inductive findings bridge the literatures on entrepreneurial resourcefulness and place-based processes to explain how resourcefulness can be (re)produced by *claiming revocable place*. By highlighting the role of place as a resource, we develop a model that introduces a typology of resourcefulness in total institutions: *normative* and *counter-normative*. Methodologically, we contribute by incorporating photo-ethnography to provide a deeper understanding of marginalized settings.

INTRODUCTION

Resourcefulness is defined as a social-cognitive process of maneuvering "available resources" (Williams et al., 2024). Yet, in contexts of forced displacement, such as refugee camps, the core premise of "available resources" becomes untenable, especially when displaced individuals operate within the context of total institutions, which drastically limit what resources individuals can access and how they can use them.

How does resourcefulness resume when individuals had been displaced (Loor et al., 2023)? Post-displacement entrepreneurship heralds the importance of resourcefulness but does not specify how displaced entrepreneurs become resourceful when resources are unavailable. We answer the question of how resourcefulness begins (again) after resources have been lost. We focus on the role of place as a resource. Place is lost through displacement, especially in the context of total institutions. Can it also be found? We do so by extending the resourcefulness literature to include processes that follow forced displacement.

The literature on resourcefulness has already drawn attention to place-based resources (Kimmitt et al., 2023; Moss et al, 2022) and place itself has been conceptualized as a resource. In the context of total institutions, place is not under the control of individuals, but it can be tentatively and temporarily reclaimed for entrepreneurial activities.

Answering this puzzle holds both theoretical and practical importance. From the theoretical standpoint, we show that placelessness can be a starting point for place-based reproduction of resourcefulness suggesting a novel way of studying the source and process of resourcefulness. When so much is lost, and even previously acquired education and work experience may become obsolete (de Lange et al., 2021), place and place-based processes can be turned into resources, especially after forced displacement triggers a sense of placelessness.

To address this question, this study is conducted in the Rohingya refugee camps in Bangladesh, which is currently the largest refugee camp worldwide, providing temporary shelter to one million Rohingya refugees from Myanmar. This research was conducted through a multi-sited ethnography with refugee entrepreneurs in camp settings (Prasad & Shadnam, 2023). Because of primarily dealing with illegally and informally run businesses, the research relies on extensive interviews, photographs, naturalistic observations, and auto-ethnographic field notes. We inductively model highlighting the role of place-based processed in managing the impositions associated with living in a camp.

Based on the multi-sited ethnography in a challenging-to-access context, this study extends, and bridges, the literature on place and entrepreneurial resourcefulness by introducing the concept of *claiming revocable place.* We provide a typology (*normative* and *counter-normative*) and a place-based process model of resourcefulness within a *total institution*—an organizational setting that is relatively removed from society and that encompasses all aspects of its members' daily life (Goffman, 1961). Photo-ethnography contributes by portraying how photographs can be an important, but underutilized vehicle, in entrepreneurship research.

THEORETICAL GROUNDING

The project began by investigating and intersecting the literatures on entrepreneurial resourcefulness (Williams et al., 2021), displacement (Alkhaled & Sasaki, 2022), and refugee camps (De la Chaux et al., 2018). Refugee camps—a perceived form of total institutions—restrict mobility, curtail agency, and impose regulatory oversight. Within such settings, entrepreneurial activities cannot rely on pre-existing resources but must involve crafting relationships with place itself. Although refugee camps are widely acknowledged as initial settlements that both protect refugees and separate them from host communities, there is limited research in the entrepreneurship literature on how entrepreneurs (re)produce resourcefulness within the constraints of these camps. Existing research recognizes different processes of resourcefulness, but there is a lack of clear understanding about how resources are acquired when all resources, including place, have been lost.

RESEARCH CONTEXT

This dissertation-based study draws on a multi-sited reflexive ethnography conducted in Rohingya refugee camps in Bangladesh—the world's largest refugee settlement, housing over one million forcibly displaced individuals. In August to December 2017, nearly 780,000 Rohingya refugees fled an ethnic cleansing campaign in Rakhine State, Myanmar, and sought shelter in Bangladesh, where they now face severe restrictions on movement and limited access to formal and informal employment (Hussam et al., 2022). The research design follows strategies that encourage participant and researcher reflexivity (Alvesson & Kärreman, 2007). The research was carried out over three months and relied on four data sources: 37 semi-structured interviews (with entrepreneurs, camp employees, religious leaders, and community leaders), naturalistic observations (~200 hours), autoethnographic fieldnotes, and photo-ethnography (researcher-led photographs from the field). Fieldwork was conducted across seven camps, and the places presented varied physical, social, and institutional conditions. The first author, a Bangladeshi national who can understand Rohingya language but cannot speak fluently, worked closely with a Rohingya-speaking assistant to gain access. Regulatory restrictions and safety concerns required time-restricted and ethically grounded engagement. The dataset combines rich ethnographic field experiences and visual insights through photo-ethnography—allowing a triangulated understanding of entrepreneurial action under constraint. Data analysis followed a hermeneutic approach across three stages.

DATA ANALYSIS

The data analysis followed a three-stage hermeneutic process grounded in reflexive induction. First, we identified place as a sensitizing construct, and focused on the place-based process shape, limit, or open up particular ways of constructing resourcefulness (Alvesson et al., 2022). This realization led us to reflexively question the assumed "resource availability" and the binary packaging of refugee entrepreneurs as either resilient or agentless (Fathallah et al., 2024). The first author's auto-ethnography and continuous iterative discussion with second author during and after the fieldwork led us to the core construct of claiming revocable place. In the second stage, we constructed two place-based mechanisms inductively (*performing conformity via place* and *safeguarding defiance via place*). In the third stage, we sought to generate plausible explanations for the dynamic patterns by theorizing pathways and feedback loops. In this stage, we repeatedly went back between photographs and interviews. This helped us triangulate the process model across photographs and interview data to make sense of the patterns we observed in the data.

In Stage 1, we identified an empirical anomaly: places in the camps were not "respected," as prior literature might suggest (de la Chaux et al., 2018), but *revocable*—frequently dismantled and reclaimed. In Stage 2, inductive coding across interviews, photos, and fieldnotes revealed two place-based mechanisms: *performing conformity* and *safeguarding defiance*. These mechanisms were not fixed pathways but emerged as contingent responses to institutional threat as individuals repeatedly reclaimed place. In Stage 3, the place-based mechanisms were linked through recursive feedback loops: a *stabilizing loop* (respecting restrictions) and a *destabilizing loop* (infringing restrictions), and a *reversal loop* (repeated displacement). The result is a process model that theorizes *normative* and *counter-normative* resourcefulness as a type of place-based resourcefulness that develops because the persistent constraints of total institution bifurcate how individuals related to place as a potential resource.

FINDINGS

The inductively generated model illustrates the process by which entrepreneurs generate resourcefulness through place. This process is driven by what we call "*claiming revocable place*"—defined as small-scale, easily accessible spatial settings that can be to some extent modified for entrepreneurial activities with the possibility of retaliation by institutional actors who do not sanction, and sometimes punish, such activities. Figure 1 depicts the process of resourcefulness, which we describe below through two core mechanisms.

Figure 1. A Model of Place-based Resourcefulness in a Total Institution

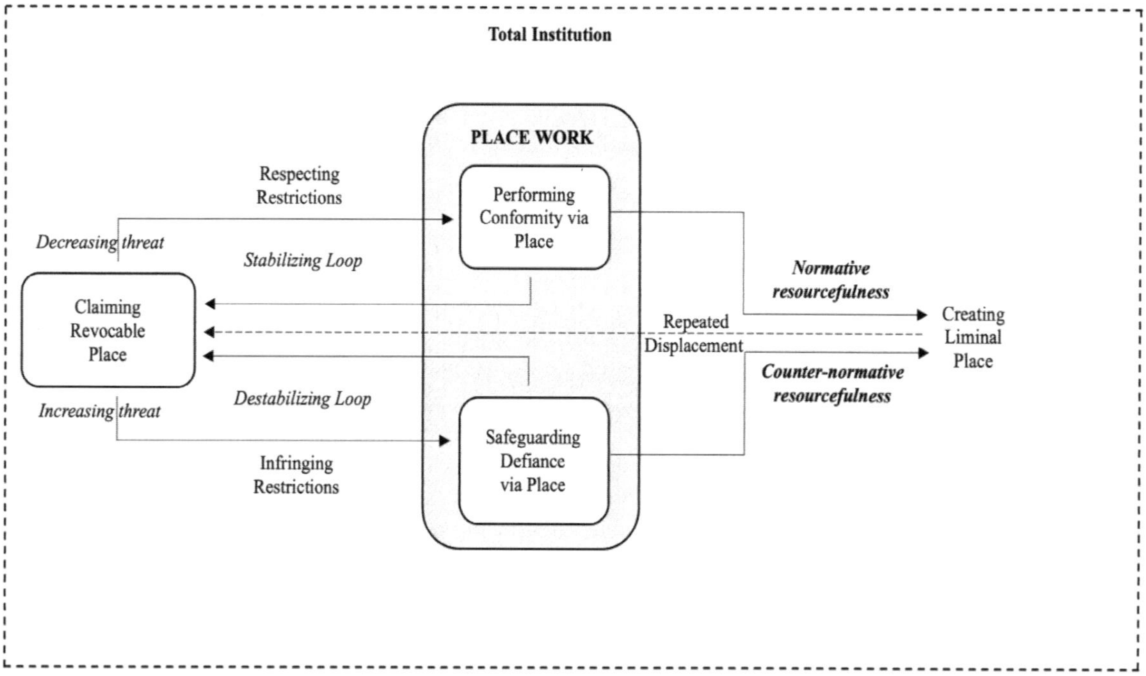

Claiming Revocable Place. Entrepreneurs identified and temporarily occupied small-scale, often risky places—such as roadside corners or household perimeters—for venturing. As one entrepreneur explained how he moved from one place to another during the last five years, continuously identifying new places:
"During that time, a sanitary toilet was being set up under this shop over there. So, I set up my shop there. I used to sell betel leaves, cigarettes, groceries, fish. When my dad set up this shop, he ran it along with a small kid. When this shop was going well, I stopped my shop and started working here. A lot of people used to visit our shop. People from several places used to come to my shop. There were not too many shops around back then. Everyone used to have tea and snacks at our shop."

Performing Conformity via Place. In the stabilizing loop, individuals normalized their ventures through practices compliant with the place. These practices created a sense of security and place-based stability. As explained during an unsolicited conversation with the entrepreneurs:
"Sometimes the police [used to] trouble us... Now, they do not disturb us. That is because we all contribute some money for them. We pay for their tea and snacks. We pay the on-duty police officers. Sometimes they [night patrol team] have day duty too. We have to spend some for them as well. It has turned into a system. There is nothing to do... We have to do it to keep moving on. Or else how will we survive?"

Some of the entrepreneurs kept their place shut to continue their entrepreneurial identity after receiving signage from the camp officials (see Table 1, reference photo 1, for performing conformity via place). Others used the small-scale space to craft themselves as entrepreneurs, shifting from their previous identity as religious leaders (hujur).

Safeguarding Defiance via Place. In the destabilizing loop, entrepreneurs continued business activities despite institutional restrictions, engaging in unauthorized sales of relief goods or working in places (e.g., women's market) going against the societal norms. These actions preserved entrepreneurial identity, even in adverse conditions. For example, they kept selling relief goods (which was not permitted; see Photo 1 under "Safeguarding Defiance via Place" in Table 1), and explained:

"Some products are from relief. I have bought the rest of the products from the market... They (police) understand that we Rohingyas do not have any work to do. Camp 11, 12, and 13 do not get as many products as relief compared to the other blocks. Several people bring their products from other blocks to sell, but if the police find out, they seize the products. They mostly sell these products to other villagers... After I sell my wholesale products, the locals charge some money from the buyers. They charge around 20 to 30 Taka per sack... I never went to the camp to get products."

Each mechanism interacted with transition pathways—either stabilizing or destabilizing—culminating in the creation of liminal places: bounded entrepreneurial settings enabling partial continuity under persistent threat.

The two paths represent ***normative*** and ***counter-normative*** forms of place-based resourcefulness. Place-based *normative resourcefulness* is the social-cognitive process through which individuals conduct accepted, repeated, and recognized engagements with place by aligning place-based practices with tolerated institutional norms, thereby stabilizing entrepreneurial activity. Place-based *counter-normative resourcefulness* is the social-cognitive process through which individuals mobilize unstable, prohibited, or revocable spatial arrangements by navigating or resisting institutional constraints to sustain entrepreneurial activity under threat.

Table 1 presents representative photographs from the field to illustrate how resourcefulness is (re)produced through place-based mechanisms within the refugee camp.

Table 1. Overview of Place-based Mechanisms and Illustrative Data: Quotes and Photographs from the Field

Mechanism	Description	Representative Photo 1	Representative Photo 2
Claiming Revocable Place	Identifying, occupying, and negotiating access to small, precarious places for venture activity within institutional settings where such places can be withdrawn at any time.		
Respecting Restrictions	Stabilizing Pathway		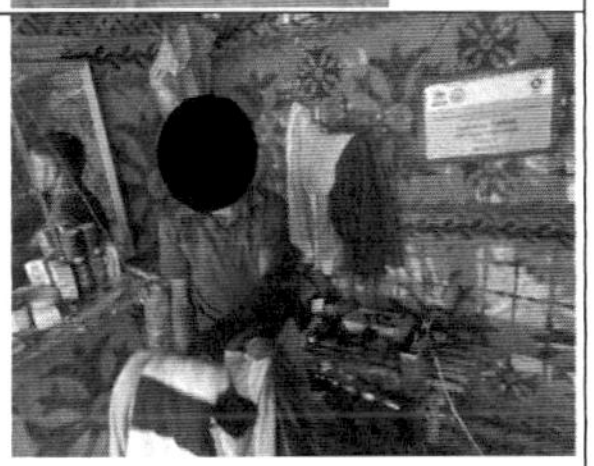

Performing Conformity via Place	Adjusting behaviors, expressions, and possibly identities to align with the prevailing institutional norms of a specific environment		
Infringing Restrictions	De-stabilizing pathway		
Safeguarding Defiance via Place	Using place as an act of resistance. The environment itself can serve as a protective factor, enabling individuals (or groups) to maintain their dissenting behaviors or identities.		
Creating Liminal Place	Deliberately establishing an environment or setting that exists within a transitional or intermediate stage, where traditional norms and structures are temporarily relaxed or redefined.		
Repeated Displacement	Reversal pathway		

DISCUSSIONS AND IMPLICATIONS

Contributions to theory. This study contributes to the literature on entrepreneurial resourcefulness by separating the *process* of resourcefulness from the *availability* of resources. It introduces "revocable place" as a novel place-based construct that explains how forcibly displaced individuals can enact resourcefulness in the absence of stable material resources. Revocable place is particularly relevant to individuals who entrepreneur within total institutions (e.g., slums, indigenous or landless communities under land disputes, and conflict zones). Efforts to claim revocable place are acts of entrepreneurship in and of themselves: few individuals inhabiting refugee camps claim it. Furthermore, revocable place is claimed as a resource in the name of entrepreneurship and is being used for entrepreneurial activities.

Second, the process model reveals how entrepreneurs navigate stabilizing and destabilizing loops, providing a typology for place-based resourcefulness in total institutions. Our typology for place-based resourcefulness in total institution can transferred to other contexts individuals perceive as restrictive and/or punitive. Our findings remain however specific to displacement, which may sensitize individuals to the critical role of place, lost and found.

Third, this study advances methodological contributions by incorporating photographs as part of a visual ethnographic approach to studying entrepreneurship in extremely marginalized settings. Photographs serve not only as visual evidence, but can also reveal the socio-spatial negotiations often missed in text-based data (Munro & Jordan, 2013). While this study does not engage the emotional or participatory dimensions of visual methods (e.g., photovoice), it highlights the potential of photography to reveal symbolic boundaries, informal regulations, and shifting entrepreneurial identities. This opens pathways for future research to integrate visual data in exploring how marginalized entrepreneurs relate to place, structure, and self.

Practical Implications. This study holds significant implications for development practitioners, NGOs, and humanitarian policymakers working in contexts of displacement and marginalization. It shows that even within extreme institutional constraints, *revocable places* can be strategically used by marginalized individuals to foster entrepreneurial activity. These findings suggest that entrepreneurship support efforts should shift focus from standardized training and financial aid models to interventions that enable access to and protection of *micro-places* for economic activity. This includes recognizing and legitimizing household-adjacent stalls, women's-only markets, and informal places created near NGO offices or health clinics. The findings also urge donors and camp managers to reconsider how aid infrastructures often inadvertently criminalize entrepreneurship. Rather than policing informal markets or revoking places arbitrarily, institutions should work to *stabilize place-based opportunities* through dialogic and participatory mechanisms. Finally, for entrepreneurship educators and local enablers, the spatially contingent nature of resourcefulness in camps calls for hyper-localized, spatially attuned entrepreneurial support models that go beyond other types of interventions to include skills and strategies for negotiating place as a resource, even when one does not own or control it.

Limitations and Future Research. This study is based on a first-generation refugee camp, where displacement is recent, institutional controls are strict, and spatial fluidity is high. These features shape the meaning and use of revocable place and may differ substantially in second-generation, long-established, or urban refugee settings where informal markets are better tolerated or even integrated. Future research should examine how revocable places evolve—do they stabilize into semi-formal institutions, remain fragile, or dissolve? Comparing resourcefulness across camp types and resettlement trajectories will help further delineate the boundaries of spatial mechanisms. Additionally, this study adopts an interpretive and researcher-led photo-ethnography. While the ethnographic approach allowed for deep reflexivity and immersion, it also introduces subjectivity. Future research may expand methodological plurality by using participatory visual methods (e.g., photovoice). Finally, this study largely focuses on place-based mechanisms. Future work should integrate intersectional lenses—such as how gender and religion mediate access to revocable place—and explore how other resources may be co-produced through place-based processes.

CONTACT: Sadek Showkat; sshowkat@luc.edu; Quinlan School of Business, Loyola University Chicago, Chicago, IL-60611, United States

[We gratefully acknowledge the financial support from the Responsible Research in Business and Management (RRBM)'s Dare to Care Dissertation Proposal Scholarship and the Center for Free Enterprise at the University of Louisville. We extend thanks to the first author's dissertation committee members, Sophie Bacq, and Fabian Diaz, for their mentorship during his dissertation journey]

THE MULTILEVEL NATURE OF SOCIAL ENTREPRENEURSHIP: MANAGING THE AIDS EPIDEMIC IN BRAZIL

Maria Minniti, Syracuse University, United States
Zachary Rodriguez, Union College, United States
Devin Stein, University of Alabama, United States

ABSTRACT

We investigate how the autonomy and alignment of social entrepreneurs nested within different jurisdictional levels--ranging from local to global--influence the outcomes of public-private collaborations. Exploiting the management of Brazil's AIDS epidemic from 1996 to 2016 as a natural experiment, we analyze health outcomes across municipalities following an entrepreneurial initiative aimed at combating the AIDS epidemic. Our findings suggest that the beneficial impact of increased access to critical HIV/AIDS medication was stronger when local stakeholders had greater agency, and their values were more aligned with the national policy driving the initiative. We contribute to the study of public-private collaborations by demonstrating how the alignment of stakeholders' incentives across levels improves outcomes because it enables social entrepreneurs to contextualize the global policy to local conditions. In other words, by modifying local institutions, local entrepreneurs can better contribute to the implementation of a global initiative. Our findings also show that the mechanisms driving stronger impact include a more efficient use of local resources and complementary local investments. Our study has important implications for advancing our understanding of value creation in hybrid forms of organizing, and of how social entrepreneurs create public value across nested jurisdictional levels.

INTRODUCTION

Entrepreneurship researchers are increasingly studying the complex interactions between diverse stakeholders in public-private collaborations and in other forms of hybrid organizing aimed at achieving social or public objectives (Bacq & Wang, 2024; Patala et al., 2022; Vedula et al., 2022). Existing work has largely focused on the governance arrangements needed to navigate heterogeneous stakeholder interests to achieve collective goals within a given institutional environment, such as formal contracting or relational coordination mechanisms (Bridoux & Stoelhorst, 2014; Dorobantu & Odziemkowska, 2017; Kivleniece & Quélin, 2012). A key insight of this literature is that stakeholder alignment, whether driven by shared values or congruent incentives, plays an important role in enabling collaboration and resource mobilization (Caldwell et al., 2017; George et al., 2016; Grimm & Reinecke, 2024). Taken together, this body of work has provided important insights about managing competing claims to generate economic, social, and public value.

Existing work on public-private collaborations, however, typically looks at alignment across organizations operating primarily at one jurisdictional level (e.g., regional, national), without considering how variations in stakeholder alignment across levels influence the initiative's success. Such a "cross-level alignment" refers to the degree to which social entrepreneurs nested within hierarchical jurisdictional structures support changes at the local, regional, or national levels. This omission is important because national or regional initiatives often rely on some degree of local implementation, and public-private collaborations may face strong frictions if local stakeholders do not share the priorities or values of the initiative being implemented (Kivleniece & Quélin, 2012; Ostrom, 2012). We therefore have yet to understand how stakeholder alignment and support at the local level mediate the impact of public-private collaborations and other forms of hybrid organizing which span not just across sectors (i.e., private, public, non-profit), but also across levels (local, regional, national). Answering this question is important because many large-scale public-private initiatives require cooperation across nested levels of governance, especially to address systemic social and environmental challenges (George et al., 2016; Williams et al., 2017).

Using theories of polycentric governance and co-production (Ostrom, 2010; Voorberg et al., 2015), we contribute to the literature on public-private collaborations (Caldwell et al., 2017; Quélin et al., 2017), and to the social entrepreneurship literature (Bacq & Wang, 2024; Patala et al., 2022; Vedula et al., 2022) by explaining how the effectiveness of large-scale public-private collaborations designed to address complex social issues depends on the alignment of stakeholders across levels, when such initiatives are nested across multilevel governance arrangements. When local actors do not share the values and objectives of a broader institutional change, or do not have the autonomy to support such changes, collaborations face considerable frictions and are less likely to succeed given less stakeholder support and less effective resource mobilization.

The empirical context for our study is Brazil's response to the HIV/AIDS epidemic, where a major policy shift under the 2007 TRIPS exemption expanded local production of antiretrovirals (ARVs) needed to treat HIV/AIDS. We construct an original longitudinal dataset of 116,226 municipality-year observations from 1996-2016, and use a difference-in-difference analysis to analyze how local stakeholder alignment—as proxied by municipal political ideology—influenced the effectiveness of ARV distribution. Our results provide evidence that municipalities ideologically aligned with this policy experienced significantly greater reductions in HIV/AIDS diagnoses and deaths following the TRIPS exemption. The availability of pharmacies and local fiscal capacity further influenced the magnitude of these effects, highlighting the importance of local entrepreneurship for investing in complementary local infrastructure and sustaining local investments as mechanisms driving social value creation. Overall, our work has broad implications for understanding how public-private collaborations and other hybrid forms of organizing can better address grand challenges and other complex social issues.

HYPOTHESES DEVELOPMENT

Stakeholder Empowerment

Public-private collaborations are increasingly being used to address societal challenges (Rufin & Rivera-Santos, 2012; Quélin et al., 2017). Despite a large and growing body of literature investigating the antecedents, processes, and outcomes associated with such collaborations, most of this work has focused on cross-sector tensions and synergies embedded within a single jurisdictional level. In practice, however, public-private collaborations exist within broader contexts where divergences in values across local, regional, and national stakeholders can render implementation inefficient or potentially ineffective (Lucas, 2017). Emerson and Gerlak (2014), for instance, have shown how effective collaboration often hinges on the ability of actors to adapt their institutional environments to local needs and conditions. And yet, adaptation cannot occur if entrepreneurs and other stakeholders do not have the agency to act. Adaptation therefore requires some structural empowerment of all relevant stakeholders within a collaborative system.

Policy changes which empower local entrepreneurs and other stakeholders can enable greater congruence between more global-scale policy objectives and local values. When the impact or activities of public-private collaborations are distributed across nested jurisdictional levels, empowering stakeholders at different levels may help to enhance all actors' joint ability to detect local needs, adapt strategies, and mobilize resources effectively (Ostrom, 2010). In other words, value creation in such contexts is largely contingent on sustained stakeholder involvement across entire organizational ecosystems. Therefore, we hypothesize:

> **H1:** *Policy which empowers local social entrepreneurs increases the impact of public-private collaborations.*

Local Stakeholder Alignment and Purpose

Existing literature has shown that motivation toward entrepreneurial activity arises from various factors, including intrinsic desires, extrinsic rewards, and even a sense of moral responsibility (Shane et al., 2003). In addition, individuals and groups may be motivated to achieve efficiency by the potential for increased productivity, financial gains, or simply the satisfaction of making the most of available resources. In the case of social entrepreneurs, a strong alignment exists between beliefs and action (Zahra et al., 2009). Thus, the alignment between a policy change and their beliefs may motivate increased participation and commitment to collaborations, and to the successful implementation a policy. In turn, this increased identification with the goal of the collaboration leads to the recognition of more and better ways to use extant resources, as well as the willingness to become more vested in the collaboration via complementary local investments. We therefore hypothesize that:

H2: *The impact of public-private collaborations increases when local social entrepreneurs support the policy change that empowers them.*

Mechanism Analysis: Efficiency of Resource Use

Finally, we suggest that both more efficient use of resources and complementary local investments serve as two mechanisms which help explain why the alignment of local social entrepreneurs can help strengthen the impact of public-private collaborations. Entrepreneurs build and mobilize resources which create value in response to opportunities embedded in their external environment (Davidsson, 2015; Kirzner, 1997). When these entrepreneurs are more supportive of policies which guide public-private collaborations, they may be more likely to pursue opportunities associated with those policies. Not only does this shape the resources they invest in, but also helps create strong incentives to be efficient with the use of those resources (Cohen & Winn, 2007; McMullen & Warnick, 2016). Therefore, we hypothesize:

H3: *When local social entrepreneurs support a policy change, the more efficient use of local resources increases the impact of public-private collaborations.*

H4: *When local social entrepreneurs support a policy change, the emergence of local complementary investments increases the impact of public-private collaborations.*

METHOD

The empirical context for our study is Brazil's response to the HIV/AIDS epidemic, where a major policy shift under the 2007 TRIPS exemption expanded local production of antiretrovirals (ARVs) needed to treat HIV/AIDS. This provides a natural experiment which allows us to examine how much municipal ideological alignment with a policy--enhancing universal access to healthcare--shaped the success of public-private collaborations aimed at reducing HIV/AIDs cases and deaths. We construct an original longitudinal dataset of 116,226 municipality-year observations from 1996-2016, and use a difference-in-difference analysis to analyze how local stakeholder alignment—as proxied by municipal political ideology—influenced the effectiveness of ARV distribution. We also test the mechanisms underlying this relationship by assessing how the presence of local pharmacies and municipal tax revenues contributed to these effects.

Sample: n=116,226 municipality-year observations across 5,389 municipalities from 1996-2016
DV = Annual AIDS diagnoses and deaths
IVs = Municipal ideological score, municipal tax revenues, number of pharmacies
Controls = Municipal population, average municipal age, distance to capital
Statistical Analysis = Difference-in-differences

RESULTS

Table 1 shows our empirical findings for all hypotheses. Given the complexity of interpreting the results of these models with multiple interaction terms, Figures 1 and 2 provide marginal effects plots to more clearly convey our findings. Note also that our use of the inverse hyperbolic sine transformation on our dependent variables (as an alternative to a log transformation which retains values of zero) prevents us from being able to easily and directly interpret each individual coefficient.

Our results broadly support Hypothesis 1, indicating that the TRIPS exemption policy improved public health outcomes on average. Consistent with Hypothesis 2, however, there is also considerable heterogeneity in this effect which varies by political context. Figures 1 and 2 display marginal effects of the policy across the spectrum of municipal political ideology. In both cases, left-leaning municipalities experience substantially greater reductions in diagnoses and deaths following the reform. In contrast, the marginal effect of the reform is negligible—or even slightly positive—in right-leaning municipalities. Note that, while several variables and interaction terms in the regression tables (e.g., Post CL × Left in Model 2) are positive and may appear to suggest worse outcomes, these should not be interpreted in isolation. Instead, the full marginal effect—combining baseline, interaction, and post-policy terms—shows that left-leaning municipalities saw the largest declines in AIDS outcomes, consistent with stronger implementation and local support.

For our mechanism analysis in Hypotheses 3 and 4, we introduce measures of local pharmacies (Models 3–4) and municipal tax revenue (Models 5–6) as proxies for existing capacity and potential for complementary investment. Pharmacy availability significantly amplifies the reform's effect in municipalities with both right- and left-leaning governments, but the effect is notably stronger in left-leaning areas. We thus find support for Hypothesis 3, where local capacity improves efficiency of implementation. Similarly, Model 5–6 provide support for Hypothesis 4, suggesting that complementary local investments, facilitated by greater fiscal resources, increase the impact of public-private collaborations. In sum, the reform was most effective in left-leaning municipalities, and we find evidence that at least part of this effect can be explained by more complementary local investments and more efficient use of local tax revenues. The results underscore the importance of stakeholder alignment and local capacity in translating national policy into desirable health outcomes.

DISCUSSION AND IMPLICATIONS

Our results provide evidence that municipalities ideologically aligned with this policy experienced significantly greater reductions in HIV/AIDS diagnoses and deaths following the TRIPS exemption. The availability of pharmacies and local fiscal capacity further influenced the magnitude of these effects, highlighting the importance of complementary local infrastructure and sustained local investments as mechanisms driving social value creation.

We make several contributions to the literatures on social entrepreneurship and public-private collaborations. First, we extend social entrepreneurship theory by demonstrating that increases in local autonomy, coupled with cross-level alignment between local, regional, and national stakeholders is a critical yet overlooked factor shaping public-private collaboration success. Second, we contribute to the literature on public-private collaborations by explaining how cross-level stakeholder alignment functions as a coordination mechanism that enables better policy implementation and resource mobilization, particularly when initiatives rely on local implementation. Third, we address calls to integrate resource-based and stakeholder perspectives by showing that sustained stakeholder commitment across levels of implementation is essential for effective resource mobilization (Barney, 2018; McGahan, 2021). Overall, our work has broad implications for understanding how large-scale public-private collaborations and other hybrid forms of organizing can better address grand challenges and other complex social issues.

CONTACT: Devin Stein; dstein2@ua.edu; University of Alabama; Tuscaloosa, AL 35487, UNITED STATES

FIGURE 1 Effect of TRIPS Compulsory Licenses on New AIDS Diagnoses

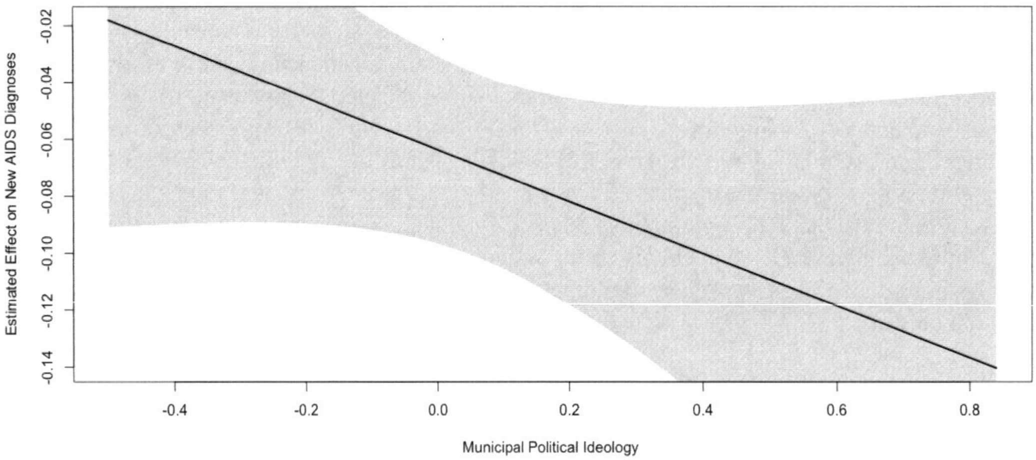

FIGURE 2 Effect of TRIPS Compulsory Licenses on AIDS Deaths

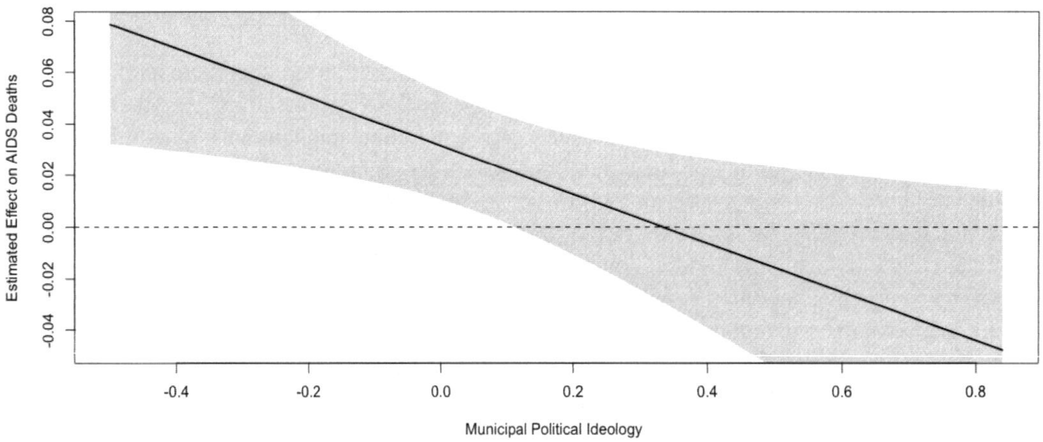

TABLE 1 - Empirical Results

Dependent Variable:	$\sinh^{-1}$(Diagnoses)	$\sinh^{-1}$(Deaths)	$\sinh^{-1}$(Diagnoses)	$\sinh^{-1}$(Deaths)	$\sinh^{-1}$(Diagnoses)	$\sinh^{-1}$(Deaths)
Hypothesis Tested:	H1 & H2		H3		H4	
Model:	1	2	3	4	5	6
Post CL	0.2556***	0.0723***	0.0549*	0.0603***	0.0588*	0.0627***
	-0.0152	-0.0099	-0.0302	-0.0192	-0.0301	-0.0192
Right	-0.0484***	-0.0085	-0.0911**	-0.013	-0.0907**	-0.0128
	-0.0184	-0.0121	-0.0404	-0.0257	-0.0404	-0.0258
Left	0.2681***	0.1080***	0.2817***	0.0959***	0.2690***	0.0862***
	-0.0119	-0.0078	-0.0238	-0.0152	-0.0238	-0.0152
Post CL × Right	0.00003	-0.0015***	0.0042***	0.0006*	0.0094***	0.0043***
	-0.0002	-0.0001	-0.0005	-0.0003	-0.0005	-0.0003
Post CL × Left	-0.0035***	0.0010***	0.0086***	-0.0055***	0.0093***	-0.0052***
	-0.0002	-0.0001	-0.0005	-0.0003	-0.0005	-0.0003
Pharmacies			0.0041***	-0.0009	0.0041***	-0.0009
			-0.0009	-0.0006	-0.0009	-0.0006
Municipal Tax Revenue			-0.3326***	-0.2041***	-0.1220**	-0.0774**
			-0.0628	-0.0399	-0.0589	-0.0376
Post CL x Right x Pharmacies			0.0672***	0.0408***		
			-0.005	-0.0032		
Post CL x Left x Pharmacies			0.0078***	0.0060***		
			-0.0003	-0.0002		
Post CL x Right x Municipal Tax Revenue					0.000000***	0.000000***
					0	0
Post CL x Left x Municipal Tax Revenue					0.0000***	0.0000***
					0	0
State FEs	Yes	Yes	Yes	Yes	Yes	Yes
1YR Lag AIDS Dx Trend	Yes	Yes	Yes	Yes	Yes	Yes
1YR Lag AIDS Death Trend	Yes	Yes	Yes	Yes	Yes	Yes
Population Controls	Yes	Yes	Yes	Yes	Yes	Yes
Fit Statistics						
Observations	108,321	108,321	28,005	28,005	28,005	28,005
R^2	0.243	0.2919	0.375	0.4746	0.3764	0.4731
Adjusted R^2	0.2428	0.2917	0.3742	0.4739	0.3756	0.4724

Notes: Clustered (Municipality) standard-errors in parentheses.
*Significance Levels: ***: 0.01, **: 0.05, *: 0.1*

SUMMARY

SEEING IS BELIEVING? EXPLORING UNCERTAIN SOCIAL IMPACT THROUGH VISUAL DATA FROM MINI-GRIDS

Alessia Argiolas, Technical University Munich, Germany
Hans Rawhouser, UNLV, United States

Principal Topic

Researchers and practitioners are increasingly interested in measuring social impact (Rawhouser et al., 2019), yet existing studies largely overlook uncertainty—a key factor in entrepreneurial decision-making—despite recognized behavioral challenges in assessing whether and how prosocial activities lead to meaningful outcomes (Nason et al., 2018). Combining visual and verbal data, this study examines how agents navigate the uncertainty of identifying unlimited impacts from specific prosocial actions.

Method

We conducted a qualitative, multiple-case study of seven social enterprises providing electricity to rural communities in India and Kenya. Between November 2023 and September 2024, we visited 19 solar mini-grid sites. To understand how different actors identify social impacts, we asked participants to produce photos and drawings (Ray & Smith, 2012) and gathered verbal accounts through 125 semi-structured interviews. Using grounded theory methodology (Gioia et al., 2013), we uncovered key heuristics that explain how agents evaluate social impact under conditions of uncertainty.

Results and Implications

We propose a theoretical framework to explain how agents involved in social impact initiatives use heuristics to bracket what counts as social impact under social impact uncertainty. We identify four distinct heuristics: two apply to open-ended impact frames — "attraction by visual gravity" (understanding impact through novel, progressive visual cues) and "reassurance by impact momentum markers" (recognizing impact through visual indicators of duration and self-proliferation). Two heuristics apply to closed-ended frames — "screen for signal strength" (focusing on obvious impacts in underserved areas) and "match impact model and measures" (connecting to established frameworks like the SDGs). Our findings reveal patterns in how social entrepreneurs and beneficiaries approach social impact uncertainty. Beneficiaries predominantly perceive social impact outcomes as open-ended, reflecting their broad perspective on the diverse changes they experience. Entrepreneurs tend to perceive social impact outcomes as closed-ended, aligning activities with specific accountability frameworks and agendas. However, we also observe instances where these roles shift. Overall we theorize a "social impact bracketing process," where heuristics help connect observed changes to uncertain social impact outcomes. At the same time, "impact model complexities" filter responses, defining boundaries for excluding what is not "social" and determining what to include in open-ended sets and exclude from closed-ended sets. This study provides a theoretical foundation for understanding social impact as decision-making under uncertainty, providing explanations about how entrepreneurs and beneficiaries navigate and define impact. By incorporating typically excluded beneficiaries and leveraging novel visual methods, it offers new insights into the ways social impact is recognized, measured, and interpreted.

CONTACT: Alessia Argiolas; alessia.argiolas@tum.de; Technical University Munich; Lindenstraße 29, Freising, Bayern 83534, Germany

SUMMARY

SOCIALLY RESPONSIBLE FOR WHOM? ENTREPRENEURSHIP AND THE DARK SIDE OF SOCIAL NETWORK SUPPORT

Amir Emami, Faculty of Management, Kharazmi University, Iran
Richard Hunt, Virginia Polytechnic Institute & State University, United States
Yasaman Shamohammadi, Kharazmi University, Iran

Principal Topic

Entrepreneurial actors serve as key contributors to societal prosperity by channeling productive resources into value-generating activities, yet research inadequately addresses how social networks influence the balance between individual and collective benefits in socially responsible entrepreneurship. While productive entrepreneurship (PE) and social network support (SNS) are linked to economic growth and well-being (e.g., Baumol, 1990; Davis et al., 2021), their interplay with personal social responsibility (PSR)—defined as morally driven ethical actions (Crane et al., 2019)—remains understudied. This is a critical omission insofar as SNS-driven self-worth and life satisfaction (Escartí et al., 2010; Klyver et al., 2020) both play a key role in how entrepreneurs accept and maintain social responsibility. Accordingly, we develop and test new theory related to the complex relationship between PE, SNS, and PSR by posing the question: What drives PSR among entrepreneurs, and how do social networks amplify or undermine it? While existing research has generally taken a "more is always better" approach to the role and impact of network support, we envision critical boundary conditions that constrain the favorable effects of SNS and PE on PSR. Specifically, we propose a "dark side" wherein SNS actually works to the detriment of PSR in contexts with relatively weak rule-of-law.

Method

We tested three hypotheses using structural equation modeling on data from 402 early-stage entrepreneurs in Iran, a context marked by weak legal frameworks. PSR was operationalized as a multidimensional construct (ethical, environmental, philanthropic, and social contributions) following Davis et al., (2021).

Results and Implications

Findings reveal that social network support (SNS) and productive entrepreneurship (PE) enhance personal social responsibility (PSR), with PE exerting stronger effects (H1/H2 supported). However, in weak rule-of-law contexts (e.g., Iran), SNS undermines the PE-PSR link (H3 supported), as networks normalize unethical practices, prioritizing group norms over societal welfare. This "dark side" reveals how SNS can decouple PE from moral responsibility. Theoretically, we (1) advance a dual view of SNS—both supportive and ethically corrosive—via moral disengagement theory (Bandura, 1999); (2) contextualize PSR, showing institutional voids transform networks into conduits for destructive behaviors; and (3) challenge universal assumptions about SNS's benefits. Practically, policymakers must balance network support with ethical accountability (e.g., embedding ethics training in mentorship programs). For scholars, the findings emphasize individual agency and contextual contingencies, particularly in post-colonial settings where social ties override collective goals (Javidan & Dastmalchian, 2003). This redefines socially responsible entrepreneurship as a negotiation between individual intent and network influence.

CONTACT: Amir Emami; a.emami@khu.ac.ir; Faculty of Management, Kharazmi University; Somayyeh St. No. 242,, Tehran, Tehran 1599964511, Iran

SUMMARY

DUAL IMPACT PATIENCE: UNDERSTANDING INVESTOR SHORT-TERMISM FOR FINANCIAL RETURNS AND SOCIAL OUTCOMES

Cédric Gutierrez, Bocconi University, Italy
Dónal Crilly, London Business School, United Kingdom
Keun Woo Jeong, London Business School, United Kingdom

Principal Topic

Social entrepreneurs face the challenge of communicating the long-term monetary and social prospects of their venture to investors who may have more immediate concerns. Existing research largely treats investor impatience as uniform and focuses mostly on financial outcomes. In contrast, we introduce the concept of dual impact patience, arguing that investor short-termism is multidimensional, varying across financial and social domains as well as through their interaction. Drawing on three studies, we test how the time horizon of financial and social goals (and their interaction) affect investment in social ventures and further explore possible mechanisms.

Method

We conducted three interconnected studies. Study 1 used an incentivized choice-based experiment with experienced investors to assess individual discounting of future financial versus social benefits. Study 2 analyzed more than 1,400 social enterprise pitches from a crowdfunding platform to analyze how the temporal framing of financial and social goals jointly affects fundraising outcomes. Study 3 is a policy-capturing experiment using a 2×2 factorial design, in which we manipulated the time horizon of both social and financial outcomes.

Results and Implications

Our findings reveal that investor short-termism is multidimensional rather than uniform, with different temporal preferences for financial and social outcomes. Study 1 shows that investors exhibit greater patience for social returns compared to financial ones, particularly when social benefits are framed in non-monetary terms. Analysis of crowdfunding data in Study 2 confirmed this pattern, revealing that emphasizing long-term financial goals significantly reduced investment, while the negative effect of distant social goals was less pronounced and partially contingent on financial timelines. The policy-capturing experiment established causality, demonstrating that the negative impact of time horizon on investment was three to four times stronger for financial goals than for social ones. Importantly, the negative effect of distant social goals was amplified when financial returns were expected in the short term. We identified two key perceptual mechanisms mediating the relationship between time horizons and investment decisions: perceived impact and abstractness. Distant social objectives were viewed as more abstract and less impactful, explaining approximately 13-22% of the investment reduction. These findings highlight the critical role of temporal framing in social entrepreneurship and offer insights for social entrepreneurs on how to effectively communicate the long-term financial and societal benefits of their ventures to investors.

CONTACT: Cédric Gutierrez; cedric.gutierrez@unibocconi.it; Bocconi University; Via Sarfatti 35, Milan, Lombardia 20136, Italy

SUMMARY

CAN ILLEGAL ENTREPRENEURSHIP CONTRIBUTE TO FAIRNESS? THE CASE OF THE SHADOW LIBRARY SCI-HUB ON THE DISTRIBUTION OF SCIENTIFIC KNOWLEDGE

Andrew J. Isaak, Vienna University of Economics and Business (WU Wien), Austria
Hibat-Allah Ezzahid, HEM Business School, Morocco
Sonja Sperber, Vienna University of Economics and Business (WU Wien), Austria

Principal Topic

Shadow libraries are a form of illegal entrepreneurship which aims to foster a fair and equal distribution of scientific knowledge. While the practice of implementing shadow libraries like Sci-Hub is considered to be illegal as it undermines publishers' established (fee-based) data-bases in academia that rely on subscriptions by educational institutions, it sets out to tackle the global inequality in terms of access to knowledge-based resources by enabling their free distribution to all. Also from authors' perspective, reaching a wider audience with their work being broadly distributed, read and frequently cited is beneficial, hence they may well profit from the distribution of their scientific work via shadow libraries. Yet, the illegal practice of shadow libraries clearly uncovers a dilemma between publishers' interest in protecting their business model as the main status quo distributor of academic publications, and the attempt to create fairness for all researchers independent of their financial status. So far, publishers have responded to this development with a shift towards an open access model, which comprises high fees as source of revenue for publishers, but these fees are now mainly passed on to institutions which cover them via institutional open access agreements.

Method

Through the lens of resource scarcity theory, our study mainly focuses on the legitimacy paradox of such mission-driven illegal entrepreneurship: while shadow libraries are affected by surface illegality (regulative illegitimacy), they nonetheless experience societal acceptance (normative legitimacy). Hereby, we follow a qualitative inductive methodology with a case study approach. Our purposeful sampling method entails the collection of over 30 interviews with stakeholders (e.g., associate editors of scientific journals, senior representatives of university libraries, prolific authors) surrounding Sci-Hub as the worldwide leading shadow library by access and usage statistics, providing free full access to published scientific articles by bypassing publishers' paywalls. We triangulate our interview transcripts with archival data, particularly video transcripts with Sci-Hub's founder, Russian activist Alexandra Elbakyan, that provide unique insights into the motivation behind the founding of the platform and its impact on the traditional scientific publishing landscape.

Results and Implications

Among others, we demonstrate that shadow libraries play a significant role in making research more accessible, particularly for those with limited resources. Overall, we contribute to the resource-based view by providing a better understanding of how and under which institutional conditions resource scarcity can motivate and justify illegal entrepreneurship. Our study also has practical implications for universities, publishers and policy-makers.

CONTACT: Andrew J. Isaak; andrew.isaak@wu.ac.at; Vienna University of Economics and Business (WU Wien); Eva-Maria-Mazzucco Platz 4/22, Vienna, Wien 1220, Austria

SUMMARY

ORGANIZING FOR 'DISTANCED HELPING' IN SOCIAL VENTURES: THE CASE OF A FINNISH LOGISTICS HUB MOBILIZING AID FOR WAR-TORN UKRAINE

Anastasia Koptsyukh, Aalto University, Finland

Principal Topic

In extreme environments, some ventures do not merely function: they endure, adapt, and push forward, no matter the obstacles. Disaster-born ventures like DispatchHope (pseudonym), a bottom-up Finnish socio-humanitarian logistics initiative launched in response to the 2022 Russian invasion of Ukraine, point to autonomous organizing in entrepreneurship, trust-based operations, and relentless execution. The processes that sustain such ventures are not simply logistical; they are deeply rooted in preparedness, history, and a collective ethos of bravery. This case also reflects a broader historical kinship between Finland and Ukraine: two nations that have faced Russian aggression throughout history, including during the Soviet era. The response to Russia's war in Ukraine is not merely a humanitarian effort; it carries significant historical and geopolitical weight, reinforcing a shared understanding of what it means to resist, endure, and take action in the face of existential threats. Finland's support for Ukraine extends beyond governmental policies. Ventures like DispatchHope, engaging both local residents and the Ukrainian diaspora, exemplify how compassionate response (Shepherd & Williams, 2014) is operationalized and how aid and resources are mobilized (Williams et al., 2021) beyond geographical borders, transforming solidarity into practical, structured action.

Method

This qualitative study draws on 41 interviews, 153 longitudinal organizational memos, and secondary data (Langley, 1999), to examine how DispatchHope developed and refined its operations amid uncertainty. From securing critical supplies to innovating in stakeholder engagement and operational processes, the venture designed effective distribution channels to sustain operations across volatile borders. Real-time decision-making, adaptive partnerships, and decentralized execution enabled DispatchHope to maintain unmatched momentum despite resource constraints and external threats.

Results and Implications

This case aligns with broader cultural and historical patterns of bravery and adversity-driven entrepreneurship. Akin to Ukrainians crowdfunding a satellite for defense intelligence (Reuters, 2022), an act that transcends strategy and becomes a statement of defiance and self-reliance, DispatchHope's processes are similarly driven by a refusal to wait, a refusal to break, and an unshakable belief in moving forward, regardless of conditions. Through the lens of organizing for 'distanced helping' in social ventures, this study shows how adaptive self-governance, trust-based operational models, and historically forged strength enable ventures to sustain large-scale humanitarian efforts across borders, transforming solidarity into structured, enduring action in times of crisis. By studying this persistence, we uncover historical bravery, rather than innovation alone, as a defining feature of entrepreneurial survival and impact in times of crisis.

CONTACT: Anastasia Koptsyukh; anastasia.koptsyukh@aalto.fi; Aalto University; Selkämerenkatu 12, Helsinki, Uusimaa 00180, Finland

SUMMARY

HOW DO ENTREPRENEURS FORM SOCIAL BUSINESS OPPORTUNITY BELIEFS? AN OPPORTUNITY ACTUALIZATION PERSPECTIVE

Alexander Meister, ESCP Business School, Germany
René Mauer, ESCP Business School, Germany
Sönke Mestwerdt, University of Manchester, United Kingdom
Matthias Mrożewski, ESCP Business School, Germany
Christoph Seckler, ESCP Business School, Germany

Principal Topic

In an attempt to resolve the lack of conceptual clarity and debate on entrepreneurial opportunity creation and discovery processes, the opportunity actualization perspective has gained considerable resonance in recent years (McMullen et al., 2024; Ramoglou & McMullen, 2024). This perspective introduces a novel view of entrepreneurial opportunities that are driven by demand and emerge in a pre-existing environment that creates the conditions for what is possible (Gras et al., 2020; Ramoglou & Tsang, 2016). However, so far, the actualization perspective has exclusively theorized "about 'opportunity' when discussing it within the context of [...] for-profit entrepreneurship" (Ramoglou & McMullen, 2024, p. 290). Thus, we seek to extend the actualization perspective by developing a broader understanding of entrepreneurial opportunities, specifically examining how opportunity beliefs are formed in the context of social business opportunities that encompass both economic and social dimensions.

Method

To gain insight into how entrepreneurs form entrepreneurial opportunity beliefs that merge social and economic motivations, we adopted a qualitative research approach that draws on two data sources: semi-structured in-depth interviews with 24 impact entrepreneurs and think-aloud verbal protocols from an imaginary decision-making task conducted after the interview. Based on our findings, we are able to develop a holistic framework of social business opportunity belief formation that integrates non-market and market considerations.

Results and Implications

We contribute to the opportunity actualization perspective (e.g., McMullen et al., 2024; Ramoglou et al., 2023) and the debate on social business opportunities (e.g., Hu et al., 2020) in different ways. First, we expand the market-oriented focus to include non-market and non-economic considerations. This makes the actualization perspective even more broadly applicable, as the understanding encompasses not only profit-oriented opportunities but also opportunities that are economically viable and socially responsible (Hägg et al., 2024; Romoglou et al., 2023). Second, we contribute to the literature on opportunity belief formation (e.g., McMullen et al., 2024; Wood et al., 2014). Here, we emphasize the importance of perceived breadth and depth of the support from the stakeholder environment relevant to the integration of social and commercial opportunity beliefs (e.g., Bacq & Wang, 2024; Bosse et al., 2023). Third, we are the first to introduce the opportunity actualization perspective to the discourse of social opportunities (e.g., Hu et al., 2020; Yitshaki et al., 2022) and show how this theoretic angle can resolve the conceptual challenges that the creation and discovery perspective bring to the discourse (González et al., 2017; Sassmannshausen & Volkmann, 2018).

CONTACT: Alexander Meister; ameister@escp.eu; ESCP Business School; Heubnerweg 8-10, Berlin, Berlin 14059, Germany

SUMMARY

SPONTANEOUS EMERGENCE OF NON-MARKET INSTITUTIONS AND VENTURING

Sadek Showkat, Loyola University Chicago, United States
Trent Williams, Brigham Young University, United States

Principal Topic

Literature on institutional emergence typically assumes that market institutions emerge from a higher-level institutional order (Bylund & McCaffrey, 2017). Only limited research in management and entrepreneurship has examined how micro-level actors contribute to the emergence of institutions after a collapse (see Haxhi et al., 2024). Even within this limited (and conceptual) research, scholarship has been driven primarily by two assumptions. First, scholars have explored how actors at the micro-level interact for the emergence of "market institutions"—defined as a set of actors or structures that judge, govern, or organize other actors in the field (Lawrence & Phillips, 2019). As a consequence, despite having a myriad of non-market institutions, such as family institutions, community institutions, and religious institutions (Nason & Bothello, 2022), in the Global South and informal economies, we have not explored how these non-market institutions emerge (or even change or flow). Second, due to the focus on the emergence of market institutions and assuming those as controlling institutions for the venturing process, we do not know how the non-market institutions emerge to influence venturing—the creation of new ventures (Shepherd & Williams, 2014). Thus, this paper asks: How do individuals shape the norms, authority structures, and 'boundaries' of emergent non-market institutions that influence venturing?

Method

We conducted a multi-sited ethnography in the world's largest refugee camp in Cox's Bazar, Bangladesh–a first-generation refugee camp consisting of 900,000 refugees. Our emergent model is based on ethnographic fieldwork in the refugee camps with 51 interviews (37 entrepreneurs, two religious leaders, three community leaders, and nine camp employees, including NGO executives and field-level officers), more than 200 photos, and autoethnographic field notes. The interview duration was 15 minutes to 1 hour. We conceptualize the co-emergence of non-market institutions and ventures, focusing on the role of individuals. We took a multi-modal approach to our data analysis by combining visual data (through photographs) with semi-structured interviews and auto-ethnographic notes.

Results and Implications

We adopted a hermeneutically oriented approach to data analysis (Mees-Buss et al.,2022). In the first stage, we focused on the literature on post-disaster venturing (Thorgren & Williams, 2023; Williams & Shepherd, 2016), non-market institutions (Nason & Bothello, 2022), and institutional emergence after collapse (Haxhi et al., 2024). Through the dual-phase model, we show that some non-market institutions (especially family and community) can change after a collapse. We further explain that newly emerged non-market institutions (religious and spatial) play a significant role in the typology of new ventures.

CONTACT: Sadek Showkat; sshowkat@luc.edu; Loyola University Chicago; 16 E Pearson St., Chicago, Illinois 60607, United States

INTERACTIVE PAPER

CRITICAL CONFIGURATIONS: A POWER LAW PERSPECTIVE ON THE EMERGENCE OF OUTLIERS AMONG SUSTAINABILITY-ORIENTED VENTURES

Matthew Farrell, Austin Peay State University, United States
Christopher Crawford, University of Tampa, United States
Xuanye Li, New Mexico Tech, United States
Vitaliy Skorodziyevskiy, University of Louisville, United States
Chris Willis, University of Mary Washington, United States

Principal Topic

There is a growing interest in the field to understand the behavior and outcomes of sustainability-oriented ventures. However, very little empirical research has been done on ventures that also focus on non-economic goals. Yet, outliers– i.e., those ventures that achieve significant growth and profitability–are arguably best positioned to make a major social contribution by fulfilling their non-economic goals. Therefore, the purpose of our study is to answer the research question: Under what conditions do the highest performing sustainability-oriented ventures—the outliers—emerge?

Method

We leverage data from four reputable databases: Crunchbase, US Census, Tax Foundation, and MIT Election Data and Science Lab. We utilized crisp set qualitative comparative analysis (csQCA) on the data aggregated from 360 ventures, all of which were active in an industry that dealt directly with sustainability concerns (e.g., wind energy, recycling).

Results and Implications

Our results produced two configurations with acceptable fit values. In the first, the corporate tax rate and the number of employees are present, while the rest of the causal conditions are absent. The second configuration is similar to the first, with the exception that political centrism and population density are present. The literature suggests that VC funding is a geographically driven phenomenon; interestingly, so is American politics, with conservative political leanings being significantly stronger in rural areas and neighborhoods tending to cluster along these political lines. Therefore, we posit that in more centrist states, sustainability-oriented ventures will originate in urban areas. In contrast, ventures from states that have more left-wing policy orientations can develop in more remote areas. Consequently, our results reveal a "Goldilocks zone" of political centrism wherein sustainability-oriented outlier ventures from either remote areas in left-wing states or urban areas in centrist states can thrive. As a practical implication, we argue that sustainable enterprises would be best served by entering population-dense areas in centrist states. This can allow these ventures to achieve the greatest economic success, permitting their largest possible sustainable contributions. However, our results shows that consistently successful ventures in left-wing states can flourish in more remote areas, suggesting possible downsides to such collocation with VCs. The contribution of our study is two-fold. First, we suggest that not all environments are conducive for sustainability-oriented ventures, especially for those ventures that intend to go public (i.e., outliers). Second, we add to the growing literature on outliers by studying sustainability-oriented outlier ventures.

CONTACT: Matt Farrell; farrellm@apsu.edu; Austin Peay State University; 158 Matheson Drive, Clarksville, Tennessee 37043, United States

INTERACTIVE PAPER

ENTREPRENEURIAL EVENT SYSTEMS AND EXTERNAL ENABLEMENT: UNPACKING EVENT-VENTURE CONFIGURATIONS FOR SOCIALLY BENEFICIAL INNOVATION

Madeleine Meurer, Erasmus University, Netherlands
Per Davidsson, Jönköping International Business School, Sweden
Mohamed Genedy, Jönköping International Business School, Sweden
Lucia Naldi, Jönköping International Business School, Sweden

Principal Topic

Event System Theory (EST) offers an understanding of how events shape organizational behaviours and outcomes. EST emphasizes the importance of event characteristics - such as magnitude, novelty, disruption, and criticality - which collectively determine an event's overall strength and its capacity to elicit responses from organizations. EST further posits that these characteristics interact additively, making the aggregate strength of an event a determinant of the urgency and scale of organizational responses. If we apply these assumptions to entrepreneurial ventures, we would, thus, assume that stronger events drive more significant socially beneficial innovations as ventures respond to the heightened urgency and scale of grand challenges. Yet, research on external enablement (EE) emphasizes that the relationship between event characteristics and entrepreneurial outcomes is contingent on the match between an event and a venture. In this paper, we, thus, argue that instead of viewing event characteristics as additive, we should understand them within a framework of complex interactions, where the configuration of event characteristics and venture attributes can synergize or conflict, profoundly influencing socially beneficial innovation in entrepreneurial ventures. We, therefore, ask, "how do different event-venture configurations foster socially beneficial innovation?"

Method

The results of our study reveal which and how event-venture configurations foster socially beneficial innovation. Therefore, our prospective contributions are three-fold. First, we aim to extend EST by moving beyond the traditional focus on additive event strength. Second, we contribute to the literature on EE by theorizing how EE functions as systems rather than isolated disruptions. Third, we suggest a novel, longitudinal, approach to QCA.

Results and Implications

The results of our study reveal which and how event-venture configurations foster socially beneficial innovation, guiding our contributions in two important ways. First, we extend EST by emphasizing the complex interplay between event characteristics and venture attributes. This extension provides a broader theoretical lens for interpreting how events influence innovation. Second, we contribute to the literature on EE by theorizing how EE function as interconnected systems. This perspective shifts the discussion from single EE impacts to a systemic view.

CONTACT: Madeleine Meurer; meurer@rsm.nl; Erasmus University; Burgemeester Oudlaan 50, Rotterdam, Zuid-Holland 3062PA, Netherlands

INTERACTIVE PAPER

NAVIGATING INCLUSION: THE ROLE OF SOCIAL PERCEPTIONS AND ATTITUDES IN SHAPING ENTREPRENEURIAL ECOSYSTEMS FOR ENTREPRENEURS WITH DISABILITIES

Mirza Tihic, Syracuse University, United States
Alexander McKelvie, Syracuse University, United States

Principal Topic

Entrepreneurial ecosystems (EEs) show regional variations in impact (Isenberg, 2016) but often neglect diversity, particularly for entrepreneurs with disabilities (EWD). Moreover, much of the focus has been on high-growth, technology-based ventures within advanced economies, often at the expense of understanding how these ecosystems can foster or hinder participation among diverse entrepreneurs (Spiegel, 2017; Wurth et al., 2022). Using Social Identity Theory, we examine how social perceptions influence EWD inclusion in EEs. EWD face societal barriers and stigma about their capabilities (Ahl & Marlow, 2021; Tihic et al., 2021; Wiklund et al., 2018), which can constrain their entrepreneurial identity. Our study contributes to ecosystem research (Abootorabi et al., 2023; Ahmed et al., 2022; Stephan et al., 2023; Volkmann et al., 2021) by analyzing EE inclusiveness for EWD. Through survey data on attitudes toward entrepreneurship and disability, we investigate whether EEs are accessible to EWD and how support systems impact participation. The main research question is: How do social perceptions and attitudes toward EWD shape their inclusion or exclusion in EEs? This exploration enriches the current debate on diversity within EE and provides a critical lens through which to assess their effectiveness in fostering a more inclusive entrepreneurial environment.

Method

To investigate how social perceptions influence EE inclusiveness for entrepreneurs with disabilities (EWD), we survey more than 400 ecosystem participants who impact EWD outcomes. Our methodology employs previously validated multi-response items measuring attitudes toward disability and entrepreneurship, informed by 51 semi-structured interviews. The analysis uses OLS and probit regressions to assess relationships between social perceptions and ecosystem inclusiveness.

Results and Implications

This study offers critical insights at the intersection of social perceptions toward entrepreneurship and entrepreneurial ecosystem (EE) inclusiveness. The findings expand Social Identity Theory by providing empirical evidence on how EEs affect entrepreneurs with disabilities (EWD) and their perceived roles within broader social structures, enhancing our understanding of how marginalized groups (Caldwell et al., 2020; Darcy et al., 2023; Jammaers & Zanoni, 2020) construct entrepreneurial identities despite social barriers. Practically, our results inform ecosystem models by demonstrating how perceptions impact inclusiveness. These insights enable ecosystem actors to develop targeted interventions supporting EWD while simultaneously serving other disadvantaged groups. For policymakers, our findings facilitate addressing systemic biases hindering EWD inclusion. This research contributes to diversity and inclusion theorizing within entrepreneurial ecosystems (Cavallo et al., 2019; Darcy et al., 2020; Krueger, 2021; Tula et al., 2024).

CONTACT: Mirza Tihic; mtihic@syr.edu; Syracuse University; 101 Waverly Ave, Syracuse, New York 13210, United States

INTERACTIVE PAPER

INNOVATIVE PATHWAYS FOR IMPACT: BUSINESS MODELS FOR DIGITAL SOCIAL VENTURES

Mmekidmfon Umanah, King's College London, United Kingdom

Principal Topic

Digital social ventures (DSVs) have emerged at the intersection of technological innovation and social entrepreneurship, leveraging digital innovations to address critical societal and environmental challenges. However, despite the increasing relevance of DSVs, our understanding remains limited regarding how these ventures organize their business models and how different approaches to business model organization influence their ability to create and scale social impact. The purpose of this paper is to address this gap by exploring two key research questions: What are the distinct business model archetypes for DSVs? and How do these archetypes influence the impact paths of DSVs?

Method

This study combines deductive and inductive approaches, building on business model design literature and insights from empirical case studies. Specifically, we use conceptual theory development—theory building through logical reasoning —to systematically derive propositions about business models archetypes and their relationship to social impact paths. We anchor our analysis in the broader literatures on business models, digital technologies and social entrepreneurship. This conceptual foundation is then validated and expanded with empirical evidence from case studies of 30 DSVs operating in Nigeria. Thus, rather than building theory inductively from our data, we use it to describe and exemplify both the business model typologies of DSVs (see e.g., De Cuyper et al., 2024) and their implications for social impact.

Results and Implications

Our research presents a conceptual framework for understanding DSVs through two fundamental dimensions; engagement approach (participatory vs. receptive) and technological integration (native vs. blended). By juxtaposing these two dimensions, we propose that DSVs can be categorized across four broad archetypes: Digital Drivers, Digital Architects, Digital Integrators, and Digital Enablers. We further exemplify each archetype using empirical data. Drawing on positive social change theory, we propose how each archetype influences ventures' impact pathways through breadth and depth of reach. Our study offers three key theoretical contributions. We advance the literature on digital social entrepreneurship by introducing a novel typology that captures how DSVs structure their business models. Second, we extend recent work on social venture business models by providing a framework for understanding how social ventures can organize themselves based on their technological capabilities and engagement approaches. Lastly, we extend the positive social change framework by demonstrating how digital technologies enable distinct pathways for achieving surface-level and deep-level change in emerging markets. In practice, we believe we have offered insights that can inform the design of effective interventions, in resource-constrained environments like Nigeria.

CONTACT: Mmekidmfon Umanah; mmekidmfonabasi.umanah@kcl.ac.uk; King's College London; 102A Larden Road, London, England W37SX, United Kingdom

INVESTORS AND PIVOTING: HOW INVESTORS ARE PUSHING AND SHAKING ENTREPRENEURS INTO UNCERTAINTY

Janine Crivelli, University of St.Gallen, Switzerland
Manuel Hess, Grenoble Ecole de Management, Switzerland
Joakim Wincent, Hanken School of Economics and University of St.Gallen, Switzerland

ABSTRACT

Entrepreneurial pivots are often framed as founder-driven responses to market uncertainty. This study challenges that view by theorizing pivoting as a relational process shaped by founder–investor dynamics. Drawing on an inductive theory-building cross-case analysis of pivots, we explore how investor relationships and funding timing influence pivot initiation, type, and execution. We find that founders seeking investment engage in more radical, product-focused pivots, while existing investors drive parallel pivots aimed at scalability—often through customer segment shifts. Crucially, pivot implementation hinges on founders' perceptions of investor expertise and trust. Our findings advance a relational theory of pivoting, recentering investors in the strategic adaptation of early-stage ventures.

INTRODUCTION

Entrepreneurial ventures often face rapidly evolving markets that demand strategic flexibility and frequent recalibration of initial plans. One particularly consequential form of strategic adaptation is the entrepreneurial pivot—a deliberate shift in a venture's product, business model, or target market in response to customer feedback, performance signals, or environmental shifts (Burnell, Stevenson, & Fisher, 2023). Pivoting is a common practice among startups, often associated with improved product-market fit, accelerated learning, and enhanced scalability (McDonald & Gao, 2019; Zellweger & Zenger, 2023). Conceptually, pivoting has been linked to a scientific process of hypothesis testing (Zellweger & Zenger, 2023), involving iterative experimentation, feedback integration, and realignment with market needs (Murray & Tripsas, 2004; Bingham & Eisenhardt, 2011; Zott & Amit, 2008).

Although existing research has highlighted the benefits of pivots during the early stages of venture development, it tends to focus primarily on founder-driven decisions, emphasizing cognitive mechanisms, learning from failure, and psychological ownership (Grimes, 2018; Kirtley & O'Mahony, 2020). Within this founder-centric framing, the pivot is typically treated as a relatively autonomous decision, responsive to customer feedback and internal performance assessments. However, this framing overlooks a crucial and increasingly relevant aspect of venture life: the involvement of external stakeholders, particularly investors, who often wield significant influence over strategic decisions once they have committed capital and secured governance rights.

The omission of investor influence from much of the pivoting literature presents a notable gap. While prior studies have explored how investors support early-stage experimentation and provide legitimacy to ventures (Ko & McKelvie, 2018), they offer limited insight into how investors affect a founder's ability—or willingness—to pivot after funding rounds. Given that investors acquire equity and often board representation, they are not just resource providers but active co-creators of venture strategy (Wasserman, 2017). Once an investment narrative is in place, founders may face implicit or explicit constraints on pivoting, particularly if a strategic shift appears to deviate from the investor's original expectations (Zuckerman, 1999). Some investors may encourage pivots as a sign of adaptability, while others may resist them due to concerns about legitimacy loss, reputational risk, or sunk cost biases (Fisher, Kotha & Lahiri, 2016; Garud, Schildt & Lant, 2014).

This study seeks to fill this important gap by examining the following research question: How do investor relations and the financial support investors provide influence a founder's decision to pivot? This question is particularly timely given the growing prevalence of investor-backed ventures and the increasing complexity of founder-investor dynamics during venture growth. In addressing this question, we aim to move beyond a binary view of investors as either enablers or constraints, and instead provide a more granular understanding of how investor relations shape the timing, type, and trajectory of pivot decisions.

To explore this phenomenon, we adopt a theory-building, inductive research design using a multiple case study approach (Eisenhardt, 1989a). Our data consists of 49 ventures that faced the decision whether to pivot, drawing from semi-structured interviews with founders, investors, and venture coaches, supplemented by archival and digital records.

Our findings contribute to the entrepreneurial pivot literature by illuminating how relational dynamics with investors shape both the decision to pivot and its implementation. We identify varying pivot patterns—including founder-initiated product pivots and investor-initiated customer pivots—depending on investor engagement and relational trust. We further develop a process model showing how emotional states like desperation, founders' perceptions of investor value, and resource constraints influence whether and how pivots are pursued. These findings also offer new insights into legitimacy theory, resource dependence theory, and psychological ownership in the context of venture governance. Our study reconceptualizes pivoting as a collaborative, context-dependent process rather than an isolated act of founder adaptation.

THEORETICAL BACKGROUND

Research on pivoting remains predominantly founder-centric, emphasizing individual-level cognitive and behavioral explanations. Founders are portrayed as the primary agents of strategic change, with pivoting decisions attributed to their learning from market feedback (Blank, 2013), identity alignment (Snihur & Clarysse, 2022), or emotional attachments to venture ideas (Cardon et al., 2009; Pierce et al., 2001). Legitimacy theory further suggests that founders may resist pivoting if it threatens their external credibility, particularly in the eyes of investors and business partners (Garud et al., 2013; Zott & Huy, 2007). This theoretical framing implies that pivots occur only when founders can maintain consistency with a venture's overarching mission, thus avoiding potential legitimacy penalties.

Yet, ventures rarely operate in isolation. External stakeholders—especially investors—play a central role in shaping venture trajectories. Resource dependence theory posits that ventures rely heavily on external capital to access markets, talent, and legitimacy (Pfeffer & Salancik, 1978). This dependence introduces asymmetries in decision-making power, where investors can influence, delay, or even override strategic choices such as pivots (Katila, Rosenberger, & Eisenhardt, 2008). Indeed, recent scholarship shows that investors not only contribute capital but also provide strategic guidance, industry networks, and symbolic endorsements (Huang & Knight, 2017; Ko & McKelvie, 2018). These contributions, while enabling, may also constrain entrepreneurial autonomy, particularly when investors hold fixed expectations about the venture's original direction (Felin et al., 2024).

Investor influence can also interact with founders' emotional and cognitive dynamics. Psychological ownership theory explains that founders often develop strong personal identification with their ventures, treating them as extensions of self (Pierce et al., 2001). This emotional attachment can make pivots feel like a loss of identity, particularly if prompted or imposed by external stakeholders. However, the presence of supportive investors—those perceived as experienced, trustworthy, and value-adding—can mitigate founder defensiveness and facilitate constructive reframing of the venture's strategic narrative (Clough et al., 2019; Shepherd, 2003). Conversely, a mismatch between founder and investor expectations can result in conflict, disengagement, or strategic paralysis (Wasserman, 2012). Importantly, psychological ownership is not limited to founders. Investors—especially those involved early and closely—may also develop strong attachments to the venture, viewing it as "theirs" due to their financial, emotional, and reputational investments (Avey et al., 2009; Bernhard & O'Driscoll, 2011). These investors are often enrolled through a compelling founding narrative that shapes their expectations about the venture's purpose, market, and trajectory (Navis & Glynn, 2011). When this narrative shifts due to a strategic pivot—such as a change in industry, product category, or customer segment—investors may perceive a breach of identity continuity, resulting in resistance or even active opposition to the pivot (Park et al., 2020; Knight, Grimes, & Gehman, 2024). Thus, investor pushback may not stem solely from concerns about feasibility or returns but also from a sense of violated psychological ownership, further complicating the relational dynamics of strategic change.

Despite these important insights, current theories tend to treat investor involvement as static (either present or absent) and unidimensional (either supportive or constraining). This perspective underestimates the heterogeneity of investor roles and the fluid, relational nature of founder-investor interactions. Recent work by Knight, Grimes, and Gehman (2024) emphasizes the importance of cultural alignment and narrative resonance in sustaining investor commitment across strategic transitions. Their findings highlight how repeated pivots—if not credibly framed—can erode investor confidence and reduce resource flows. These dynamics point to a broader need for theorizing pivots not just as founder-driven acts of agility, but as negotiated outcomes shaped by interpersonal trust, power asymmetries, and temporal pressures.

Together, legitimacy, psychological ownership, and resource dependence theories offer valuable yet partial explanations of pivoting decisions. To fully understand how and why ventures pivot, particularly in investor-backed contexts, we must consider the relational and emotional interdependencies between founders and investors. This study addresses this gap by exploring how investor relations shape both the decision to pivot and the nature of its implementation.

METHOD

Due to the dearth of prior research on the flexibility of ventures, an inductive, multi-case study was conducted to generate new theory (Edmondson & McManus, 2007). The analysis of multiple cases enabled the development of a more generalizable theoretical framework than would be possible from a single case (Eisenhardt & Graebner, 2007).

For this inductive study, we rely on four data sources: (1) semi-structured interviews with the founders, (2) semi-structured interviews with different types of investors, (3) semi-structured interviews with coaches of ventures that did not raise funds from investors, and (4) secondary sources. In total, we interviewed 31 founders and 15 investors who were involved with the 31 founders interviewed earlier. In addition, we interviewed four venture coaches and also examined venture information available online (data banks like Pitchbook, Crunchbase, etc.), newspaper articles, and internal documents, such as minutes of investor meetings, as available. In line with established guidelines for multiple-case theory building (Eisenhardt, 1989a; Eisenhardt & Graebner, 2007), we employed within-case and cross-case analyses, avoiding a priori hypotheses.

FINDINGS

Much of the existing literature frames pivots as entrepreneurially driven, autonomous decisions made by founders responding to internal experimentation or external feedback (e.g., Ries, 2011; Grimes, 2018). However, this understanding overlooks the role of investors, who often influence strategic decisions, particularly when ventures are in resource-dependent positions. Our inductive analysis offers an alternative, relational perspective: pivots are frequently shaped—and at times initiated—by investor involvement. This reorientation challenges the notion of pivots as founder-exclusive strategic actions, highlighting instead a co-constructed process where investor dynamics can amplify, constrain, or redirect pivot trajectories. The data reveals that founders' perceptions of venture performance and emotional responses to adversity play a critical role in shaping their strategic behavior, particularly how they engage with investors. These pivots, however, vary significantly in type and underlying drivers depending on the nature of founder-investor interactions. Our findings are organized into four interrelated themes, each building toward a more relational theory of pivoting in investor-backed ventures.

First, we observed two dominant pivot types: product pivots, typically founder-initiated, and customer pivots, often investor-initiated. Product pivots emerge from founders' close interactions with users and iterative learning processes. These involve reengineering the core offering to enhance its value proposition. In contrast, customer pivots are driven by investors' strategic inputs—based on their market experience or industry networks—and involve shifting the venture's target segment to unlock new scaling potential. These distinctions illustrate differing orientations: founders focus on improving what they have built, while investors emphasize aligning with broader market opportunities. This initial insight sets the foundation for a more complex understanding of how investor involvement shapes strategic change.
Proposition 1: Investor-initiated customer pivots are more likely to yield scalability and market expansion, whereas founder-initiated product pivots are likelier to enhance the value proposition through product change.

Second, related to these two types of pivots, the data indicates that founders' responses to adversity are closely linked to their engagement with investors regarding potential pivots. Founders who recognized negative or declining performance in their ventures and experienced a sense of desperation actively engaged with existing investors or, in some cases, sought to establish new investor relationships. While desperation seemed an initial lens through which to view founder engagement with investors, a more nuanced pattern emerged. Prior work by Zuzul and Tripsas (2020) suggests that revolutionary founders, motivated by industry-changing visions and innovations, resist pivots that would dilute their mission and introduce incremental change. In contrast, discoverers are framed as more flexible and responsive to change. However, our data complicate this binary. We find that also non-revolutionary founders—those without highly novel business models or disruptive ambitions and innovations—also resisted pivots, even under significant financial pressure and investor encouragement.

These founders did not display the symbolic identity rigidity of revolutionaries, but rather, a strong—at times overconfident—belief in their personal expertise and vision. They perceived investor suggestions as misinformed and retained faith in their own judgment, often preferring incremental improvements over structural change. This behavior was particularly prevalent among founders with domain-specific knowledge, who saw investor feedback lacking technical depth or contextual understanding. This insight challenges identity-based assumptions in the pivot literature and highlights how confidence—rather than ideology—can anchor ventures in suboptimal strategic paths.
Proposition 2: Even among non-revolutionary founders, resistance to investor-driven pivots may arise from high confidence in one's expertise and a strong attachment to the initial business model. This form of pivot inflexibility is not driven by mission-based identity, but by epistemic self-assurance and founder control orientation.

Third, while existing theories highlight the constraining effects of legitimacy-seeking behavior on entrepreneurial flexibility (Garud et al., 2014; Wood & McKinley, 2017), our data reveal a more nuanced reality. Our data show that the timing of investor involvement relative to the pivot significantly influences who initiates the change—and why. Founders often initiated pivots in ventures without prior investor backing to signal legitimacy to potential investors. In these cases, pivoting became a symbolic act of responsiveness and strategic flexibility aimed at demonstrating a willingness to overcome stagnation, financial distress, or increase scalability. Founders used pivots to reframe the venture's trajectory and appeal to external capital providers as adaptive and investable.

Conversely, in ventures with established investor relationships, pivot initiatives frequently originated from investors. However, these pivots only materialized when founder-investor trust was high. In cases where founders perceived investors as lacking relevant experience or being overly rigid, pivot discussions stalled or failed to result in action.
Proposition 3: The initiation and execution of pivots are shaped by legitimacy dynamics and relational trust. Founders without investors may initiate pivots to signal adaptability and attract investment, whereas in investor-backed ventures, pivots require trust-based alignment to proceed and succeed.

Our analysis suggests that a complex interplay of founder characteristics, investor involvement, and resource constraints shapes the implementation of a pivot. Incremental changes are more common among ventures without investor relations, reflecting a conservative approach to strategic adjustments. In contrast, ventures seeking new investors often pursue radical or parallel pivots driven by customer feedback and market opportunities. Parallel pivots mean that ventures innovate and experiment with change while maintaining continuity in their currently existing operations. This aligns with investor preferences for scalable and resilient business models. This leads to several aspects affecting pivot implementation. This final insight offers a novel framework (see Table 1, see Appendix) for understanding how ventures calibrate the scope and risk of strategic change based on market needs and the architecture of their relational and financial dependencies.

Proposition 4: The degree of founder-investor relationship and resource availability affect risk-taking behavior in pivoting decisions: no investor relationship and organic resource availability associate with incremental pivots, weak investor relationships and resource scarcity associate with radical pivots, strong investor relationships and resource support associate with parallel, lower risk pivots.

DISCUSSION AND IMPLICATIONS

This study offers a relational view of entrepreneurial pivoting by foregrounding the co-constitutive roles of founders and investors in shaping whether, when, and how pivots occur. Rather than seeing pivots as founder-driven strategic inflection points based solely on market feedback, our findings suggest that pivots are socially embedded processes shaped by emotional states, perceived performance, and resource dependencies. We identify four interrelated dynamics—desperation, investor role, investor quality, and pivot type—that together inform a process theory of investor-influenced pivoting.

Theoretically, this work contributes to the entrepreneurial pivot literature by extending the focus from founder cognition (e.g., Grimes, 2018; Kirtley & O'Mahony, 2020) to relational processes. It builds on Knight et al., (2024) by empirically showing how investor narratives and founder perceptions co-shape strategic change. We challenge assumptions from legitimacy theory that investor involvement primarily constrains pivoting. On the contrary, our findings demonstrate that investor presence can facilitate more ambitious pivots—particularly when founders feel desperate and perceive investors as value-adding. In line with resource dependence theory (Pfeffer & Salancik, 1978), our data suggest that desperation reframes the founder's risk calculus, lowering resistance to external influence and increasing receptivity to investor-driven change.

Even founders with a strong sense of psychological ownership—reflected in their desire to retain control over the venture's business model, product, and strategic direction—may become willing to engage with investors and initiate a strategic pivot when confronted with existential pressure. While psychological ownership is typically associated with efforts to protect and preserve the status quo (Pierce et al., 2001; Avey et al., 2009), our findings reveal that this protective tendency can become malleable under conditions of threat, such as financial distress or repeated failure. We contribute to psychological ownership theory by demonstrating its conditional relevance in the entrepreneurial context: founders who closely identify with their ventures and are inclined to retain unilateral control appear willing to relinquish parts of that control when survival and future growth are at stake—particularly when trust in investors is high (Bernard et al., 2021; Clarysse et al., 2013). This reveals an emotional paradox: although psychological ownership typically motivates founders to resist external influence, acute desperation may erode its boundary-preserving function (Pierce & Jussila, 2011), thereby increasing founders' openness to radical strategic change—especially when supported by credible and trusted investors.

Practically, these findings have implications for both founders and investors. For founders, understanding the emotional triggers of strategic openness, especially desperation, is vital for timing investor engagement. Founders should not view desperation solely as a liability but as a window for transformative collaboration. Conversely, investors should recognize that their influence depends on financial control, relational trust, and perceived expertise. Merely injecting capital is insufficient to drive meaningful pivots; legitimacy as a strategic co-thinker is critical.

Our findings also speak to venture accelerators, incubators, and board advisors. When founders resist pivoting despite negative signals, interventions should focus on recalibrating emotional framing and assessing the quality of investor engagement, rather than solely emphasizing market logic. Furthermore, policymakers and ecosystem architects should consider supporting relational capital, not just financial capital, as a driver of innovation in early-stage ventures.

Several directions emerge for future research. First, the emotional antecedents of strategic change remain underexplored. While we highlight desperation as a catalyst, other affective states (e.g., hope, anxiety, pride) may shape pivoting behavior in distinctive ways. Second, the concept of "pivoting styles" warrants attention. Our data suggest that founder-initiated product pivots differ substantially from investor-initiated customer pivots in content and underlying motivations and risks, which should be tested quantitatively in a next step.

In conclusion, entrepreneurial pivots are not isolated acts of visionary leadership but socially negotiated processes embedded in webs of trust, control, and perceived necessity. By theorizing pivoting as a relational adaptation, our study advances entrepreneurship research toward a richer, more processual understanding of how ventures navigate uncertainty, not alone, but together with those who hold a stake in their survival and success.

CONTACT: Janine Crivelli; janine.crivelli@unisg.ch; University of St.Gallen; University of St.Gallen, Global Center for Entrepreneurship and Innovation, Dufourstrasse 40A, 9000 St.Gallen, Switzerland

APPENDIX

Founder-Investor Relationship	Resource Availability	Risk-Taking Behavior	Pivot Type	Strategic Implication
No investor relationship	Organic (self-funded)	Low	Incremental pivot	Preserves control; small-scope changes to refine the offering
Weak investor relationship	Scarce (high need for capital)	High	Radical pivot	Driven by desperation, large-scale reorientation to secure survival and funding
Strong investor relationship	Abundant (strong investor support)	Moderate	Parallel pivot	Enables dual-track experimentation while maintaining continuity and control

Table 1: A framework for understanding how founders calibrate the scope and risk of pivots in interaction with investors

MANAGING TECHNICAL DEBT IN B2B DIGITAL STARTUPS

Ting Li, Ivey Business School, Western University, Canada
Darren Meister, Ivey Business School, Western University, Canada

ABSTRACT

A significant challenge for business-to-business (B2B) digital startups is transitioning from serving their initial clients to accommodating a growing client base. Once these startups secure a few key clients, they often become infrastructurally constrained, making it difficult to meet the demands of new clients. Drawing upon the technical debt (TD) and digital capabilities literature, we examine why B2B digital startups incur TD and how it can be strategically managed through matching digital capabilities to accelerate growth. Through a longitudinal case study of three Canadian B2B digital startups, we find that TD is crucial for rapid growth but must be balanced with digital deployment and adaptation capabilities. Our findings emphasize the importance of maintaining a debt-capability equilibrium for digital startups and introduce three distinct company personas to illustrate different approaches to managing TD.

INTRODUCTION

Every entrepreneur wants to launch a business with a flawless infrastructure using cutting-edge tools and industry-best practices. Yet, reality often tells a different story. Tight deadlines and the pressure to deliver viable proofs of concept for funding and resources often lead to shortcuts. While essential for survival, these shortcuts accumulate as TD (Ernst et al., 2015; Klinger et al., 2011). Like financial debt, TD involves choosing quicker, simpler solutions now at the expense of additional cost, work, and complications later (Alves et al., 2016; Tom et al., 2013).

For digital startups, TD stands for the cumulative buildup of technological and informational obligations they face during the maintenance and further development of their digital platforms and infrastructure (Rolland et al., 2018, p. 420). Originally introduced as a software engineering metaphor, TD describes the future costs of modifying or maintaining an information system due to development shortcuts (e.g., reducing testing time) (Cunningham, 1992; Kruchten et al., 2012). It can stem from either current actions (mistakes and errors) or future constraints (lack of flexibility or missed opportunity) (Z. Li et al., 2015). While extensively studied in software engineering, TD remains underexplored in entrepreneurship research. Given their interdisciplinary backgrounds, the authors see an opportunity to position TD as a strategic tool for informing digital infrastructure design and shaping digital capability development in startups.

Many digital startups become stuck at the business transition stage when their initial digital infrastructure, once sufficient for business validation, struggles to support sustained growth (Klotins et al., 2018; Picken, 2017). TD frequently accumulates as a result of rapid development and scaling, where the urgency to enter the market or deliver new features leads to compromises in code quality and system architecture (Klotins et al., 2018; Nguyen-Duc et al., 2021). Understanding how to effectively manage TD is therefore critical for ensuring startup survival and sustained growth. In this study, we aim to address the following research question: *How can B2B digital startups strategically manage TD to achieve transition success?*

This study focuses on B2B startups, which serve fewer customers than B2C companies, making the impact of TD from onboarding new clients more noticeable. These startups also face long and complex sales cycles and frequently co-develop products with key clients and partners. As a result, they are particularly vulnerable to infrastructure shortcomings when adapting to emerging and less compatible needs during business transitions.

BACKGROUND LITERATURE

Startup Lifecycle

The entrepreneurship literature identifies three major stages: validation, transition, and scaling (Picken, 2017; Sanasi et al., 2023). This study focuses on the transition stage, which remains underexplored despite its significance in managing TD as startups evolve from minimum viable products to full offerings (Nair et al., 2020; Picken, 2017). As well, scaling challenges continue to be a central theme in entrepreneurship studies (Bortolini et al., 2021; Zajko, 2017).

Transition-stage startups typically have fewer than 50 employees, operate with a lean, decentralized structure, and prioritize speed and flexibility over rigid processes (Dos Santos et al., 2011; Haase & Eberl, 2019). Their digital infrastructure is often built by founders, early employees, or contractors, with limited documentation and transparency. However, this less-structured approach frequently results in a misalignment between evolving business needs and the capabilities of the infrastructure (Street & Meister, 2004), resulting in the accumulation of TD.

Technical Debt (TD)

Ward Cunningham (1992, p. 30) introduced the concept of TD to describe how immature code, while functional, can lead to inflexibility and costly fixes over time. He likened it to financial debt–providing short-term gains but it must eventually be repaid. Over time, the term has evolved to represent the trade-off between expediency and long-term consequences in software development.

Common types of TD include code debt (poorly written code that is hard to maintain), design debt (suboptimal architectural choices), documentation debt (incomplete or outdated records), and testing debt (insufficient testing leading to undetected issues) (Alves et al., 2016; Tom et al., 2013). While excessive TD can be costly, a manageable amount can accelerate feature development and early product launches (Rolland et al., 2018). However, effective TD management requires collaboration between technical and non-technical stakeholders (Wiese & Borowa, 2023). TD often becomes apparent in digital startups during the business transition (Klotins et al., 2018), when early compromises made under time and resource constraints start to hinder growth. For example, an initially low-cost data hosting solution may later require complex migrations as the customer base expands. Understanding and managing TD is therefore essential for startups to scale successfully.

Digital Capabilities

We employ digital capabilities as the guiding framework to explore effective strategies for managing TD. The concept encompasses a complex bundle of digital resources and digitally enabled processes that collectively drive superior performance (Pavlou & El Sawy, 2010; Wade & Hulland, 2004). *Digital capabilities* are defined as the ability to effectively use digital technologies to deliver and uphold a functional digital infrastructure while aligning with company-specific priorities (T. Li & Chan, 2019; Pavlou & El Sawy, 2006).

In this study, we examine two key digital capabilities relevant to startups: digital deployment and digital adaptation. *Digital deployment capability* refers to a startup's proficiency in adopting, integrating, and retiring components of digital infrastructure (Balocco et al., 2019; T. Li & Chan, 2019). *Digital adaptation capability* represents a startup's ability to utilize existing and emerging technologies for trial and experimentation of new solutions that address evolving business requirements (Bohn & Kundisch, 2020; Ghezzi & Cavallo, 2020).

During the transition stage, digital startups often shift their focus from acquiring proof of concept and building initial infrastructure to enhancing their key offerings and solidifying the supporting digital infrastructure to meet scaling requirements (Ries, 2011; Teece, 2014). Digital deployment capability ensures smooth implementation and integration of technologies to support growth (Devaraj & Kohli, 2003), while digital adaptation capability allows startups to create digitally enabled intangibles, often in the forms of new infrastructural components, processes, and routines that are more effective and/or efficient than the previous versions (Gregory et al., 2015; Hanelt et al., 2021). Together, these two capabilities help startups strategically manage TD while ensuring infrastructural relevance in a dynamic business environment.

METHOD

This study stems from a large research project that examines the digital capabilities of digital startups. Over a span of five years, the first author tracked over 40 digital startups, identifying a recuring challenge: many B2B digital startups struggle after acquiring their initial clients, often constrained by infrastructural limitations rooted in earlier decisions. For instance, some tailor their digital solutions too narrowly to a specific client set, restricting broader market appeal, while others adopt seemingly cost-effective infrastructure choices that later hinder scalability and customization.

To understand these challenges, we conducted an in-depth case study of three B2B digital startups, that met the following criteria: (1) classified as a digital startup, (2) self-identified as being in the business transition stage, (3) had 10–49 employees, and (4) visible TD. We analyzed five years of digital capability evolution (2018–2023) and supplemented this with 1–2 semi-structured interviews with founding team members, totaling 19 hours. These interviews explored the TD creation and resolution, and its connection to digital capability development.

Startup A is a fintech company offering small business banking solutions. Startup B is an edtech providing entrepreneurial training to individuals from underrepresented regions. Startup C is a med-tech company producing portable, analytics-driven products for rapid drug and chemical detection.

We adopted the coding technique of the grounded theory approach for data analysis (Myers, 2019; Strauss & Corbin, 1998). Grounded theory has proved to be extremely useful in developing context-based, process-oriented descriptions and explanations (Urquhart et al., 2010). Our analysis focuses on how TD is managed and its interaction with digital capabilities. Specifically, we examine three key dialectic relationships: debt-infrastructure, debt-capability, and capability-infrastructure, uncovering nuances in how startups navigate these interdependencies.

RESULTS

Our analysis reveals that the development and use of digital capabilities are essential mechanisms for keeping TD under control. We found rich interaction patterns between TD, digital capabilities, and digital infrastructure design. Specifically, maintaining an equilibrium between TD and digital capabilities (hereafter referred to as debt-capability equilibrium) enables effective management of TD for successful business transitions. Figure 1 presents the key findings as spiral interaction cycles linking TD, digital capabilities, and digital infrastructure. The spiral form, rather than a closed loop, reflects the ongoing evolution and reinforcement of digital infrastructure and capabilities in startups.

At a high level, the design of the digital infrastructure defines the scope of TD. Decisions related to IS outsourcing, vendor choices, product co-development, and other related issues may result in the buildup of TD. Infrastructure design can also entail proactive control of TD through mechanisms such as strategic avoidance or rapid iterations (i.e., quick-out, quick-discover, and quick-fix). For example, a startup may delay customizing a non-essential feature until it has a larger client base, ensuring the change aligns with broader market needs rather than a niche segment.

TD can both hinder and enable digital infrastructure development and reconfiguration. Across all three startups, we observed increasing infrastructural complexity, making it more challenging to integrate new functions and features. Accumulated TD may slow down the pace of innovation, increase the maintenance burden, and compromise system quality and productivity. Nonetheless, all startups embraced some degree of TD to accelerate their speed to market, reduce infrastructure development costs, and take advantage of powerful yet affordable digital solutions from third-party managed service providers (MSPs). Given its inevitability during business transitions, the key challenge lies in maintaining a manageable level of TD with clear, feasible paths to resolution.

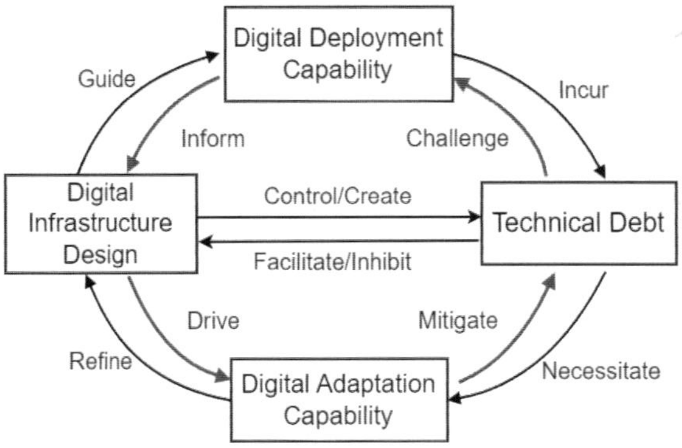

Figure 1. The Debt-Capability Equilibrium

The debt-capability equilibrium can serve as an effective tool for understanding and determining whether TD is controlled at an optimal level. Figure 1 shows two distinct but interconnected loops that describe the dynamic interactions between digital infrastructure design, TD, digital deployment capability, and digital adaptation capability. These bidirectional relationships emphasize both forward and reverse impacts.

In the forward impact loop, startups' infrastructural choices guide the deployment of digital solutions. Whether relying on MSPs or focusing on developing proprietary ones, the deployment of digital infrastructure components is likely to incur TD (e.g., inconsistent coding practices leading to high maintenance costs). It is worth noting that startups with strong digital deployment capabilities are less likely to incur unanticipated TD (e.g., system bugs from integration). However, they still encounter expected TD. When it starts to affect infrastructure quality and performance, TD necessitates modifications. Startups will then leverage their digital adaptation capability to refine or pivot as necessary, thereby sustaining functional and competent digital infrastructure.

In the reverse impact loop, TD challenges the established digital deployment processes and can lead to implementation or integration failures. This is particularly common for digital startups during business transitions as they experience early growth (Klotins et al., 2018). For instance, a startup may adopt multiple software solutions for different operations (e.g., *QuickBooks* for accounting, *Knit* for payroll, and *Dext* for expenses) and initially transfer data manually between these systems. This workflow can easily become unmanageable as the startup scales from a few clients to dozens. The difficulties in deployment will inform future design or necessary updates of digital infrastructure. This, in turn, motivates startups to explore and experiment with accessible digital technologies to address deployment challenges identified, thereby building and utilizing their digital adaptation capability. When done successfully, these adaptation initiatives will mitigate TD.

TD MANAGEMENT PERSONAS

To illustrate the value and practical implications of achieving a debt-capability equilibrium, we present three company personas representing distinct debt management styles observed in our sample: balanced innovator, debt-heavy struggler, and cautious developer, as depicted in Figure 2.

Startup A exemplifies a *balanced innovator*, maintaining a balance between TD and digital capabilities. Founded in 2015 by two software engineers, it provides seamless and affordable global payment solutions, particularly catering to immigrant-heavy markets. From its inception, startup A demonstrated strong digital capabilities, launching its first product quickly, expanding from 2 to 18 employees, and reaching the business transition stage within three years.

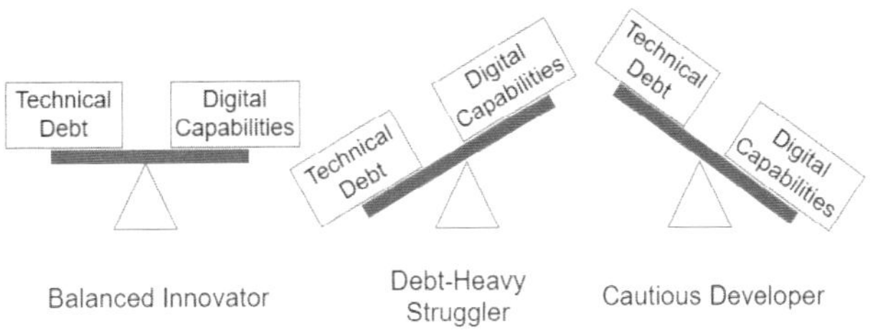

Figure 2. TD Management Personas

An important challenge occurred when it pivoted from individual remittance services to B2B payments. This shift necessitated handling larger data volumes, complex transactions, regulatory compliance (e.g., HIPAA, PIPEDA), and enhanced security. The engineering team devoted significant time and resources to resolving related challenges, thereby driving the development and utilization of their digital adaptation capability, filling gaps in their individual-oriented infrastructure and proactively fixed bugs (i.e., digital adaptation capability helps mitigate TD).

However, when expanding into Europe, the banking data exchange APIs were not compliant with local regulations such as GDPR (i.e., TD challenges the digital deployment capability). Lacking time for full integration, startup A took a strategic shortcut and developed a standalone infrastructure for European clients. This allowed them to secure a key client while incurring anticipated TD in the form of higher maintenance costs and infrastructure inconsistencies. As revenue from Europe grew, the company hired two engineers to integrate the parallel systems and resolve this debt. This balanced approach enables ongoing improvements in the company's digital infrastructure. At the time of this study, startup A is on the verge of being considered a scaling business.

Startup B embodies the characteristics of a *debt-heavy struggler*. Startups in this category are burdened by high TD but lack the digital capabilities to manage it effectively. This imbalance creates significant challenges, increasing failure risk. As an education platform provider, startup B mobilizes university resources to train aspiring entrepreneurs in countries such as Kenya. For example, faculty members from different universities contribute to a session of a 12-session online course titled "*Business Validation in Emerging African Markets*." Startup B then compiles these sessions and partners with local universities to ensure course accessibility on its platform.

Lacking a technical background and unable to afford in-house software engineers, the founder initially outsourced platform development to a freelance programmer. The early product was good enough to secure funding and onboard a few university partners. As the company expanded, a new CIO further outsourced infrastructure components and system integration tasks to multiple contractors (i.e., digital infrastructural design guides deployment). As the founder stated, "[The CIO] put together an infrastructure that allows us to outsource even more than the initial freelancer. Now we are outsourcing in Eastern Europe and India, and we can outsource as much as we need."

While outsourcing reduced costs and accelerated growth, it introduced challenges at the business transition stage. Fragmented development practices increased system complexity, making the platform difficult to maintain (i.e., TD necessitates adaptation). Frequent bugs and system failures disrupted operations, at one point leaving the platform completely non-functional for an entire month (i.e., failed adaptation unable to resolve infrastructure malfunctions). As of this study, Startup B had stalled for over two years, struggling to overcome its accumulated TD.

Startup C is classified as a *cautious developer*, possessing strong digital capabilities while maintaining minimal TD. While this debt-averse approach mitigates some risks, it can also slow growth by limiting the strategic use of TD. Startup C specializes in proprietary drug-detecting algorithms, for frontline use. The company's tech-savvy founding team avoids incurring TD. For example, they use Slack to document all software development activities (i.e., digital infrastructure design can help keep TD under control). They also manufacture their own spectroscopy devices for drug identification. However, according to the co-founding CTO, their portable spectrometer hardware is similar to other tools on the market. The company's differentiation comes from their software that more accurately determines the presence of drugs.

During the early transition stage, Startup C experienced vendor lock-in. While developing its minimum viable product, the company relied on a small MSP for data hosting and result visualization. This dependency enabled a faster go-to-market strategy by allowing the team to focus on its core intellectual property, particularly the algorithms used for drug detection. However, as the company quickly acquired new clients, the original MSP struggled with infrastructure updates and customization demands (e.g., integration with client data portals). The accumulated TD became so problematic that the company had to switch vendors. The migration process involved multiple phases, setbacks, and hiring two sets of contractors. Fortunately, the migration was completed with minimal disruption to clients. Since then, the founders have been extremely cautious about decisions that might incur TD. For example, they chose to develop many digital solutions in-house rather than using off-the-shelf products or available market solutions, fearing infrastructural constraints again (i.e., digital adaptation capability mitigates TD).

This conservative approach has trade-offs. For instance, a group of potential clients requested some tabulation features for the drug-checking output dashboard. Instead of using ready-made options from Azure, the company built an in-house solution to support future feature development and integration. By the time the update was completed, the clients had chosen a competitor. As of this writing, Startup C is progressing slowly but steadily. The founder recognizes that their commitment to minimizing TD may have contributed to the company's measured pace of growth.

DISCUSSION

This paper explores the value and implications of TD in B2B digital startups, drawing three key lessons from our findings. First, digital startups must strategically incur TD during early stages. The experiences of the three sample startups highlight that neither excessive nor minimal TD is ideal–maintaining the debt-capability equilibrium is essential for sustainable growth. Startup A, as a balanced innovator, leveraged digital capabilities and controlled TD to fuel continual digital infrastructure improvements. In contrast, startup B, a debt-heavy struggler, faced growth hindrances due to excessive TD from outsourcing, while Startup C, a cautious developer, missed market opportunities due to their debt-averse strategy.

Second, first clients can jeopardize a startup if the company becomes too infrastructurally locked in by their specific requirements. This situation represents a form of TD that is particularly challenging for startups during the transition stage. Like vendor lock-in, where a startup becomes overly dependent on a single provider (as seen with Startup C's MSP switch), startups tailoring their offerings too closely to early clients may restrict market potential. B2B startups are especially vulnerable, as early-stage co-development with key clients may lead to excessive customization.

Third, the strategic timing of TD incurrence and resolution is crucial. Maintaining a balanced debt-capability equilibrium hinges on the timing of debt increase and reduction. During business transitions, startups should make deliberate, needs-based decisions regarding TD, strategically addressing its consequences. Developing and applying digital adaptation capabilities can be both costly and challenging, underscoring the importance of planning.

This paper identifies future research directions including validating the debt-capability equilibrium and startup personas presented, examining if balanced innovators consistently outperform cautious developers across contexts, and exploring the distinction between known and inadvertent technical debt. While our findings highlight digital adaptation capability's role in managing unanticipated TD, more research is needed to understand these dynamics and advance TD management in startups.

CONTACT: Ting Li; tili@ivey.ca; Ivey Business School, Western University; 1255 Western Rd, London, ON N6G 0N1, Canada.

LEGITIMACY SPILLOVERS AFTER BLOCKBUSTER SUCCESS: HOW ATYPICALITY AND CATEGORIZATION SHAPE INVESTOR DECISIONS

Naomi Moonen, Tilburg University (JADS), The Netherlands
Donato Cutolo, IE University, Spain
Alexander Vossen, Tilburg University, The Netherlands
Ksenia Podoynitsyna, Vlerick Business School, Belgium

ABSTRACT

Blockbuster successes, such as the release of ChatGPT, generate legitimacy spillovers, altering how investors evaluate other ventures. But do all ventures benefit equally and are these benefits limited to those within the same category? We draw on theories of categorization, organizational atypicality, and entrepreneurial finance to examine how a venture's category affiliation and degree of atypicality shape its ability to capitalize on legitimacy spillovers. Using Crunchbase data on over 200,000 U.S.-based ventures, we study how ChatGPT's release affected funding outcomes for nearly 10,000 AI ventures and a matched sample of non-AI ventures. Our difference-in-differences analysis shows that typical ventures within the AI category benefited the most from legitimacy spillovers in the immediate aftermath of ChatGPT's release. In contrast, atypical non-AI ventures experienced a surge in funding, suggesting that investor attention spilled over across category boundaries, particularly toward ventures perceived as distinct and innovative. These findings nuance our understanding of legitimacy spillovers, showing how they distribute unequally depending on a venture's categorical position and the broader dynamics of investor perception.

Keywords: Legitimacy Spillovers, Categories, Atypicality, Investor Funding, ChatGPT, AI

INTRODUCTION

The extraordinary success of a new venture can reshape its broader entrepreneurial ecosystem. Such blockbuster successes, which are highly visible, widely celebrated achievements that redefine market expectations and attract outsized stakeholder attention (Wang et al., 2023), can trigger broad shifts in perceptions and resource flows. One mechanism through which this occurs is legitimacy spillovers: when the prominence of one venture enhances the perceived legitimacy of others (Soublière & Gehman, 2020; Aldrich, 1994; Suchman, 1995). Although prior research suggests that ventures in the same category benefit from such spillovers, we ask whether these benefits extend beyond category boundaries and whether all category members benefit equally.

Understanding how legitimacy spillovers operate has become increasingly important as entrepreneurial ecosystems grow more complex and interdependent. In emerging technologies such as AI, blockchain, and biotech, category boundaries are fluid and rapidly evolving (Kennedy & Fiss, 2013). Blockbuster events not only legitimize certain ventures, but also reshape broader market perceptions, influencing investment patterns, entrepreneurial strategies, and the evolution of entire sectors. Yet, little is known about which ventures are best positioned to capitalize on these moments of market redefinition, particularly when novelty and distinctiveness come into play.

To address this, we investigate how ChatGPT's release in November 2022 reshaped funding dynamics for new ventures, especially those in and around the AI category. We argue that legitimacy spillovers depend on both a venture's category affiliation and its organizational atypicality, that is, the extent to which it deviates from categorical norms. We propose that while typical ventures benefit from increased cognitive legitimacy, atypical ventures may also benefit when investors, driven by hype and fear of missing out, broaden their search for novel investment opportunities. These arguments offer a more nuanced view of how legitimacy flows and how investors evaluate entrepreneurial opportunities. In doing so, we challenge the common assumption that legitimacy spillovers benefit all category members equally and extend theory by highlighting atypicality as a critical moderator of these effects across and beyond categorical boundaries.

HYPOTHESIS DEVELOPMENT

Legitimacy spillovers are rooted in the cognitive processes through which audiences evaluate organizations. When a new venture achieves exceptional success, such as OpenAI with its release of ChatGPT, it can enhance the legitimacy of other ventures within its category by increasing the visibility and comprehensibility of the category as a whole. Categories serve as cognitive frames that help investors simplify evaluation (Zuckerman, 1999; Vergne & Wry, 2014). As a category gains cognitive legitimacy through a salient success, investors become more familiar with its features and expectations, leading to positive resource flows toward other ventures that fit the category.

Following a salient success, audiences experience a cognitive anchoring effect, reinforcing the salience and desirability of category-conforming ventures (Durand & Paolella, 2013; Hsu et al., 2009). Ventures that are easily recognizable as prototypical category members benefit from reduced evaluation ambiguity (Hsu, 2006; Zuckerman, 1999). As a result, we expect that category-affiliated ventures will be especially well-positioned to leverage legitimacy spillovers in the aftermath of a blockbuster success.

Hypothesis 1. *New ventures within the same category as an extraordinarily successful new venture are better able to harness legitimacy spillovers, resulting in increased funding performance relative to comparable ventures that are not members of this category.*

Yet, not all category members are equally well-positioned to benefit from these spillovers. Organizational atypicality, which is the degree to which a venture deviates from prototypical representations of its category, influences how readily it can be evaluated and understood. Typical ventures, which align closely with established categorical norms, are more cognitively accessible to investors and thus more likely to be perceived as legitimate, especially in the wake of increased category salience. In contrast, atypical ventures face higher evaluation ambiguity and are less likely to benefit from increased category-level legitimacy (Zuckerman, 1999; Kovacs & Hannan, 2015). Following salient success, typical ventures are cognitively easier for investors to categorize and associate with the successful prototype, while atypical ventures require greater cognitive effort and are perceived as riskier bets.

Hypothesis 2. *Within the same category as the extraordinarily successful new venture, typical ventures are better able to harness legitimacy spillovers than atypical ventures.*

While category membership provides a direct link to legitimacy spillovers, high-profile events can also influence how investors evaluate ventures outside the focal category. As investor attention is heightened following salient success, evaluation criteria may shift. Categories provide essential cognitive shortcuts, but salient successes often destabilize established evaluative schemas (Navis & Glynn, 2011; Hsu & Grodal, 2021). Investors, motivated by hype and the fear of missing out (Friederich et al., 2024; Logue & Grimes, 2022), may expand their search beyond category boundaries to find the next breakthrough. When seeking the next transformative innovation, they may begin to favor atypical ventures precisely because these ventures span or defy existing categories (Uzzi et al., 2013). In these contexts, atypical ventures, which are often positioned at the intersection of multiple, less familiar categories—signal novelty and disruptive potential (Cattani et al., 2017; Taeuscher et al., 2021). Their deviation from norms can be interpreted as a strategic asset under conditions of uncertainty and heightened attention (Pontikes, 2012). Atypicality, often a liability under stable conditions, becomes a strategic asset when investors reinterpret deviation as a signal of novelty, flexibility, and potential market disruption. Thus, we expect that atypical ventures outside the blockbuster category are especially well-positioned to capture legitimacy spillovers.

Hypothesis 3. *Among ventures that do not share the successful venture's category affiliation but are otherwise comparable to those within that category, atypical ventures experience a stronger positive effect on their subsequent funding performance than do typical ventures.*

METHOD

To test these hypotheses, we leverage the release of ChatGPT as a natural experiment. We compile a panel dataset of 209,469 U.S.-based new ventures from Crunchbase, covering the period from November 2021 to December 2023. We focus on ventures affiliated with AI categories as treated units and employ coarsened exact matching (CEM) to identify a control group of 9,931 non-AI ventures with similar pre-treatment characteristics.

Our dependent variable is new funding received per six-month period (in thousands of USD). Atypicality is measured based on the average distance between the subordinate categories a venture spans, using Jaccard similarity and Shepard's law (Shepard, 1987; Kovacs & Hannan, 2015). Ventures in the top 10% of this measure are coded as atypical. We estimate a difference-in-differences model with venture and time fixed effects, and use both fixed-effects OLS and Poisson regressions to accommodate zero-inflated funding data.

RESULTS

Regression results are reported in Table 1. Our results support all three hypotheses. First, we find that AI ventures received significantly more funding than matched non-AI ventures after ChatGPT's release (Poisson coefficient = 0.64, $p < 0.05$), corresponding to an 89.2% increase. This supports Hypothesis 1.

Second, we find a negative and significant interaction between ChatGPT release and atypicality within AI ventures ($p < 0.01$), indicating that typical AI ventures benefited more from the event. Atypical AI ventures saw a relative decline of 34.2% in expected funding compared to their typical counterparts, supporting Hypothesis 2.

Third, we observe a positive effect of atypicality for non-AI ventures after ChatGPT's release. The three-way interaction (Atypicality x ChatGPT Release x AI) is significant and negative ($p < 0.01$), indicating that atypicality increases funding for non-AI ventures but not for AI ventures. This supports Hypothesis 3. The three-way interaction is illustrated in Figure 1.

DISCUSSION AND IMPLICATIONS

Our findings reveal that the impact of a salient entrepreneurial success extends beyond its category and is shaped by the interplay of category affiliation and atypicality. Within-category spillovers are strongest for typical ventures that align with emerging category norms. In contrast, atypical ventures outside the focal category benefit as investor attention broadens.

This study contributes to research on categorization, legitimacy, and entrepreneurial finance by showing that legitimacy spillovers are not evenly distributed. Rather, they are contingent on a venture's position relative to the focal event. We extend theories of organizational atypicality by demonstrating that it can be a source of advantage in the right contextual conditions, especially during moments of hype, when distinctiveness becomes a signal of innovation potential.

Our results also contribute to research on entrepreneurial identity by showing that distinctiveness can serve as a strategic lever during periods of heightened attention (Navis & Glynn, 2010). By crafting identities that emphasize innovative category-spanning features, ventures can position themselves to capitalize on the broader reevaluation of market opportunities that follows a blockbuster event.

For entrepreneurs, our findings highlight the strategic value of category positioning. Ventures can benefit from salient events even if they are not in the focal category, particularly when they emphasize distinctiveness and innovation. For investors, the findings suggest that the halo of blockbuster successes may cloud judgment, underscoring the need to look beyond category boundaries and reassess the value of atypical ventures in dynamic markets.

Overall, we show that legitimacy spillovers are both selective and cross-cutting. They reward typicality within categories, but atypicality across them. In doing so, we provide a novel lens on how attention reshapes resource flows in entrepreneurial ecosystems.

CONTACT: Naomi Moonen; n.e.l.moonen@jads.nl; Tilburg University, Jheronimus Academy of Data Science (JADS); Sint Janssingel 92, 5211 DK 's-Hertogenbosch, The Netherlands.

TABLE 1. RESULTS OF FIXED EFFECTS REGRESSION MODELS

Dependent Variable:	Expected New Funding					
Model:	(1) OLS	(2) OLS	(3) OLS	(4) Poisson	(5) Poisson	(6) Poisson
Variables						
ChatGPT Release × AI	734.6**		1,079.8**	0.64**		0.91***
	(359.2)		(537.3)	(0.28)		(0.34)
Atypicality		611.7	-816.3		0.68	-1.4
		(1,697.6)	(1,193.0)		(1.3)	(1.6)
Atypicality × ChatGPT Release		-1,344.2**	-280.5		-1.1***	-0.23
		(623.8)	(260.4)		(0.34)	(0.24)
Atypicality × AI			1,645.6			2.2
			(2,144.1)			(2.1)
Atypicality × ChatGPT Release × AI			-1,041.4*			-0.94**
			(621.7)			(0.43)
Funding acquired until t-1	-250.8*	-223.0	-257.7**	-0.07***	-0.07**	-0.07***
	(128.3)	(139.6)	(129.9)	(0.02)	(0.03)	(0.02)
Months since last funding	-66.4	-112.0	-68.8	-0.04***	-0.05***	-0.04***
	(56.1)	(91.0)	(57.0)	(0.01)	(0.02)	(0.01)
Fixed-effects						
New Venture	Yes	Yes	Yes	Yes	Yes	Yes
Time	Yes	Yes	Yes	Yes	Yes	Yes
Fit statistics						
Observations	78,936	39,468	78,936	78,936	39,468	78,936

Clustered (New Venture) standard-errors in parentheses
*Signif. Codes: ***: 0.01, **: 0.05, *: 0.1*

Notes: Models 1 to 3 present the results of the Fixed Effects OLS regression models, while Models 4 to 6 show the results of the Fixed Effects Poisson regression models. The main effects of ChatGPT Release and AI Membership are absorbed by the respective time/unit Fixed Effects. Many control variables used in the CEM matching process were omitted from the final models. Models 2 and 5 focus solely on the AI sample.

FIGURE 1. THREE-WAY INTERACTION EFFECT OF ATYPICALITY, AI AFFILIATION, AND CHATGPT RELEASE ON NEW VENTURE FUNDING

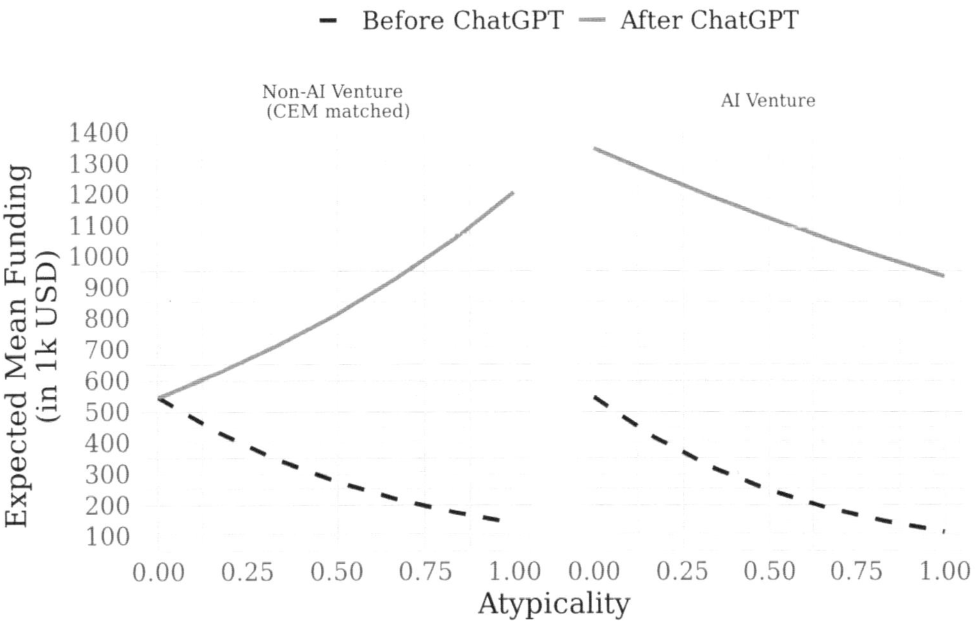

Notes: Visualization of a Poisson FE regression examining the three-way interaction between atypicality, AI-category membership, and the release of ChatGPT on funding performance. The plot is faceted by AI-category membership, with non-AI ventures displayed on the left and AI ventures on the right.

PUFFING UP A PRELAUNCH PROJECT: GENERATING AUDIENCE ENGAGEMENT VIA ENTREPRENEURIAL FRAMING

Greg Fisher, Indiana University, United States
Shannon Younger, University of Arkansas, United States
Alex Kier, Washington State University, United States

ABSTRACT

To bring something new into being, entrepreneurs often need to generate interest in and engagement with what they are doing across diverse audiences, even before delivering the venture's offering for the first time. To achieve this, they may rely on entrepreneurial framing. But what types of frames can entrepreneurs use to appeal to diverse audiences, and how might different framing elements combine with one another to generate audience engagement? To better understand these questions, we examine the framing tactics and configurations used by the founders of the Premier Lacrosse League when launching their new sports league for the first time. We link these features to varying levels of audience engagement, thereby uncovering how entrepreneurial framing can generate interest in and engagement with a new venture.

INTRODUCTION

To bring something new into being, entrepreneurs often need to foster early stakeholder support by generating excitement for and engagement with their entrepreneurial endeavor, even before delivering their core offering for the first time (Fisher, Josefy & Neubert, 2024). To do so, they may engage in entrepreneurial framing (Snihur, Thomas, Garud & Phillips, 2022), which entails using "rhetorical devices in communication to mobilize support and minimize resistance" (Cornelissen & Werner, 2014: 185). Entrepreneurial framing is, therefore, a key mechanism for getting diverse audiences excited about and engaged with something that does not yet exist (Fisher, Kuratko, Bloodgood & Hornsby, 2017).

Although prior research highlights the importance of entrepreneurial framing and its potential for generating excitement about and engagement with an entrepreneurial offering across diverse audiences (e.g., Fisher et al., 2024; Snihur et al., 2022), scholars have yet to uncover which entrepreneurial framing elements are useful for this purpose and how such framing tactics may combine to generate desired engagement with key audiences. Three factors make it difficult for entrepreneurs to generate early stakeholder engagement. First, entrepreneurs must often generate engagement before having a specific, concrete product or service to point to in support of their framing claims (e.g., Murray & Fisher, 2023). Second, various framing elements and multiple communication modes can be jointly employed to generate engagement (Snihur et al., 2022; Logue & Grimes, 2022), and these framing elements and communication modes may interact with one another in unexpected ways to generate audience responses. Yet work to this point has largely treated frames singly (e.g., Lee & Huang, 2018; Pan, Li, Chen & Chen, 2020) or in contrast to one another (e.g., Taeuscher & Rothe, 2024) rather than as multilayered, interactional actions, claims, and discourses that can be employed simultaneously. Third, generating engagement with a promised offering necessitates appealing to a broad, diverse stakeholder audience (Fisher et al., 2017). This audience may include potential customers, employees, investors, suppliers, partners, and journalists, among others. Yet "framing complications can arise when there is a multiplicity of framing audiences with divergent interests" (Snihur et al., 2022: 593), confounding the development of effective entrepreneurial frames.

These challenges prompt us to address the following broad questions in this research: What entrepreneurial frames do entrepreneurs use in their attempts to generate engagement with a venture with a yet-to-be-delivered offering, and what combinations of entrepreneurial framing elements are associated with higher levels of engagement with an entrepreneurial venture that is yet to deliver its offering for the first time?

To address these questions, we examine the frames used to generate engagement with diverse audiences in the founding of the Premier Lacrosse League (PLL). In 2018, brothers Paul and Mike Rabil set out to create a new professional lacrosse league that would be a direct competitor to the existing league in the United States. To do this, they needed to generate interest and support from diverse stakeholders, including players, investors, sponsors, media partners, and fans. We examine these efforts by first assessing the entrepreneurial framing tactics used to appeal to diverse audiences then analyzing how various entrepreneurial framing configurations are associated with audience engagement. In so doing, we identify various equifinal entrepreneurial framing configurations for generating engagement with a venture with a new offering that has yet to be delivered. Our study contributes to the literature on entrepreneurial framing by identifying how entrepreneurs frame their efforts to appeal to diverse audiences and how entrepreneurial framing elements uniquely combine to generate varying levels of audience engagement.

BACKGROUND THEORY

Entrepreneurial framing is "the use of rhetorical devices in communication to mobilize support and minimize resistance" (Cornelissen & Werner, 2014, p. 185). Entrepreneurs use framing to construct meaning around their venture and influence audience emotion and engagement in their entrepreneurial endeavors (Seigner, Milanov, Lundmark & Shepherd, 2023). Not only are frames used to meet short-term, venture-specific needs (e.g., fundraising, sales pitches, partnership proposals, or project kickoffs) (Snihur et al., 2022), but they also position a venture in the minds of audiences as part of a longer-term, ongoing entrepreneurial journey (Fisher et al., 2016; McMullen & Dimov, 2013). Indeed, as Snihur and colleagues note, "by establishing distinctiveness, coherence, and resonance through framing, entrepreneurs enact entrepreneurial opportunities" (2022: 583).

Framing directs audiences' attention to selected salient features of an entrepreneurial venture and signals its relatedness to the market (Navis & Glynn, 2011). To do so, founders utilize different framing modes, forms of framing language, and areas of framing emphasis (Snihur et al., 2022). Frames may connect a venture with its context (Hargadon & Douglas, 2001), serving as the means through which a venture (or market) is made meaningful and relevant within a broader cultural context (Gehman & Soublière, 2017; Lounsbury & Glynn, 2019).

Entrepreneurs can use frames to make elements of a venture or market category seem familiar and understandable by invoking images and meanings that audiences recognize (Fisher et al., 2017; Hargadon & Douglas, 2001; Lounsbury & Glynn, 2001). In this way, entrepreneurs frame for resonance, aligning with audiences' beliefs, values, and aspirations to mitigate audience uncertainties (Snihur et al., 2022) and attract the resources that they need to build a venture or market category (Zimmerman & Zeitz, 2002; Navis & Glynn, 2010). Additionally, frames can highlight distinctive elements of a new venture to make it stand out from competitors (Navis & Glynn, 2011; Snihur et al., 2022). Research has shown that entrepreneurs can use frames to draw attention to a venture's distinctive offerings (McKnight & Zietsma, 2018), market leadership position (Snihur et al., 2018), or environmentally friendly practices (Weber et al., 2008).

Snihur and colleagues (2022) developed a conceptual framework linking entrepreneurial framing content and processes to venture and field legitimation based on an extensive review of the entrepreneurial framing literature. Their model reflects that entrepreneurial framing aimed at distinctiveness, coherence, and resonance fosters the legitimation of new ventures. While this review clarified what we already know about entrepreneurial framing, it also highlighted some critical challenges. Specifically, these scholars identified "gaps in our understanding" around how framing applies across time, across audiences, and across geography (Snihur et al, 2022: 593). These are important issues to address because doing so allows for more complexity and dynamism to be injected into framing research, shifting us toward a more realistic and practical understanding of entrepreneurial framing. These challenges necessitate how we account for multi-layered framing elements and examine entrepreneurial framing across broad audiences to understand how framing elements are used and how they coalesce in interesting ways to generate audience excitement and resonance. Hence, we examine the entrepreneurial frames entrepreneurs use in their attempts to generate engagement with a venture with a yet-to-be-delivered offering and consider what combinations of entrepreneurial framing elements are associated with higher levels of audience engagement.

METHOD

We conducted a study with two interrelated parts, focusing on the creation of the Premier Lacrosse League (PLL) by Paul and Mike Rabil in 2018. At the time of founding, Paul Rabil was a high-profile lacrosse player who had played professionally for ten years following a successful collegiate career. In his ten years in the existing professional lacrosse league, Major League Lacrosse (MLL), Paul became dissatisfied with the experience of players (e.g., low pay, poor facilities) and the lack of media exposure generated by the league to represent and grow the game. So, he and his older brother Mike, a serial entrepreneur, set out to create a new professional lacrosse league.

The Rabil brothers established the PLL in a different format from most other sports leagues. Instead of having independent team owners, the PLL initially owned all the teams centrally. This ownership structure made the PLL responsible for employing all players and coaches, attracting sponsorship, coordinating games and schedules, and negotiating media rights. Another distinctive facet of this new league was that teams were not assigned to specific cities or locations. Instead, the Rabil brothers created a geographically independent brand for each team and established what they referred to as a "tour-based model."

Our period of interest is from when they first publicly announced the creation of the PLL in October 2018 to the first set of games in the new league in June 2019. This timeframe represents the period when the product (a new league) has been promised but not yet delivered. During this time, the Rabil brothers were directly competing with the MLL for players, sponsors, and media exposure, so they needed to generate "a collective vision and promise of a possible future" (Logue & Grimes, 2022: 1055) to engage diverse stakeholders to get excited about an offering that was not yet in existence (Fisher et al., 2024).

In the first part of the study, we employed an abductive approach to uncover entrepreneurial framing tactics and approaches for fostering engagement and interest across diverse audiences. We collected all the news articles (45), podcasts (11), and press releases (13) published about the PLL between October 2018 and June 2019. Our data analysis evolved in three stages. The first stage entailed organizing the data into a timeline of events. The second stage entailed analyzing the framing reflected within each data item (press release, article, podcast), which we then used for the third stage to analyze framing across different data items over time.

In the second part of the study, we employed qualitative comparative analysis (QCA) to assess framing configurations in YouTube videos generated by the PLL about their new offering in relation to audience engagement with those videos in our focal timeframe. QCA is a set-theoretic method that uses Boolean algebra and algorithms to account for combinations of factors that are associated with key outcomes (Fiss, 2011). QCA is well suited to analyzing multiple pathways to a particular outcome (Fiss, 2007; Kier & McMullen, 2020) because it captures the three elements of causal complexity—conjunction, equifinality, and asymmetry (Ragin, 2008b).

For this analysis, we examined all 61 YouTube videos produced and published by the PLL between October 2018 and June 2019. We measured audience engagement based on normalized scores of views, likes, and comments for each video relative to others released at a similar time. We also coded each video for framing emphasis (sport, business, or fun), T-shaped messaging, visual elements, message length, and clustered timing (whether videos were released close to other videos).

RESULTS

Our data analysis revealed that unique entrepreneurial framing techniques were used to engage with diverse audiences to generate engagement throughout the development of the PLL. The PLL founders engaged with multiple stakeholders through a strategy we dubbed *T-shaped messaging*. Additionally, they utilized *customized storytelling around a central character* in their messages to connect with various audiences. Furthermore, the PLL appeared to coordinate *strategic information releases*, timing their messages in accordance with other information they had to share to generate broader audience attention to their work.

The first main finding is that the PLL founding team used what we call *T*-shaped messaging. In this approach, the horizontal bar of the *T* represents broad communication to multiple stakeholders in a single message. When sharing messages, PLL founders typically included language appealing to all their external stakeholders—players, media partners, investors, sponsors, and fans. Additionally, they would communicate in depth with one specific stakeholder group in each message, using detailed, nuanced examples that represent the vertical bar of the *T*. While the horizontal bar shows breadth across diverse stakeholders, the vertical bar shows depth in appealing to one focal group. The stakeholder targeted with in-depth messaging usually aligned with the outlet where the message appeared. For instance, when the league was first announced by Bloomberg, a "financial news organization," Paul dedicated significant time on Bloomberg's podcast connecting with investors by discussing entrepreneurial opportunities and their innovative business model.

The second main finding that emerged from the analysis was the extent to which a key person was referenced in media items and how that key person was incorporated into a broader narrative. In many instances, Paul was that central and prominent character for the PLL, and he fulfilled this role in several useful ways—as a character to rally behind, as the embodiment of multiple roles that resonate with the PLL's various audiences, and as a storyteller himself.

Paul's face and name were strongly associated with the new league, and this seemed to facilitate reporting centered around a story with a *central character*. Paul was frequently referred to as a multi-hyphenate, with his various roles and experiences consistently highlighted. His experience as an athlete was often combined with that of his role as an entrepreneur, linking them into one cohesive storyline. Paul himself utilized his personal experiences extensively in his messages, often tailoring his selection of stories to reach a particular audience.

From our data analysis, we also found information releases to be clustered, occurring in sporadic bursts. Upon deeper examination, we found these to be primarily grouped around precipitating information—information significant enough to the PLL to be shared externally. Through this pattern it became clear that the founding team strategically held back certain information about the league so they could regularly and consistently have novel material for their own content, which would also give the media impetus to publish about the league.

In the second part of our study, we analyzed the framing configurations in PLL YouTube videos and their relationship to audience engagement using qualitative comparative analysis. This analysis revealed several equifinal configurations associated with high levels of audience engagement, demonstrating that different combinations of framing elements can lead to similar outcomes. The QCA analysis identified five configurations associated with high audience engagement and six associated with low engagement. From this analysis, we identified four primary framing configurations that generate high levels of audience engagement.

The first configuration, which we call the *business documentary framing* configuration, reveals novel and interesting information about the often-undisclosed business aspects of building a sports league. Moreover, this information is presented in a visually appealing way. The videos may vary in length, but they tend to err on the longer side, and they may or may not include non-business-focused information. Additionally, when released, they are not clustered with other videos but are instead released as isolated, standalone videos.

The second configuration, *business transactional framing*, accounts for business-focused videos with a strong and explicit incentive for the audience to engage. While this type of video shows strong engagement, it appears to be more of a forced and transactional audience engagement rather than natural or organic audience engagement. These videos bait audience engagement to access promised giveaways and, hence, result in inflated engagement that is not necessarily reflective of genuine interest in what the venture is doing.

The third configuration, *high-action sports sequence framing*, accounts for videos that put the sport of lacrosse on display in a visually appealing and engaging way. These videos are released when there are sport-related events and are clustered with other (mostly sports) videos. These videos appear to fuel the fire of the core offering, in this case, the sport of lacrosse, and they do so with a strong visual representation of the sport and with some length and substance to each video.

The fourth configuration, *fun presenter chemistry framing*, reflects fun, upbeat videos that have a different look and feel from the other configurations. They are not as visually appealing as the high-engagement sport and business videos, but they reflect the personalities of the people on screen and, importantly, the chemistry between the individuals in the video. They also include humorous elements.

DISCUSSION AND IMPLICATIONS

From an entrepreneurial framing perspective, our empirical and theoretical insights highlight how framing emphasis and mode (Snihur et al., 2022) can combine and configure to generate higher (versus lower) levels of audience engagement (Seigner et al., 2023). Our analysis highlights that four configurations appear to generate high levels of audience engagement. Each configuration can be contrasted with similar configurations that generate low engagement to isolate the mechanisms driving engagement.

The main takeaway from examining these different framing configurations is that the entrepreneurial framing recipe for generating audience engagement when producing videos about an early-stage entrepreneurial endeavor is varied. There are nuanced differences between the entrepreneurial framing configurations for creating engaging videos, and overly simplifying what works and does not work is likely to result in unintended and possibly undesirable outcomes. Entrepreneurs need to be aware that a "one size fits all" approach to entrepreneurial framing is ineffective, as different framing elements need to be combined in unique ways to generate desirable levels of audience engagement (Seigner et al., 2023).

Most research on entrepreneurial framing has linked it to legitimation, either at the venture or field level. Snihur and colleagues (2022) point out that entrepreneurial framing helps legitimate firms and the broader institutional milieu (e.g., a market category) in which firms are embedded, noting that this "legitimation process is aimed at fostering a belief among resource-holding audiences that the new venture, its offering and business model, and its field and the related practices and regulations, are desirable and congruent with social expectations, values, and norms (Ashforth & Gibbs, 1990; Etzioni, 1987; Überbacher, 2014)" (p. 582-583). We extend and elaborate on this entrepreneurial framing theory by showing that entrepreneurial framing serves a purpose *beyond* entrepreneurial legitimation. We demonstrate how entrepreneurial framing can influence the extent to which audiences engage with a new venture.

Furthermore, although we adopt the broad entrepreneurial framing framework proposed by Snihur and colleagues (2022), distinguishing framing elements such as framing mode and emphasis, we also adapt and evolve this framework by identifying other key distinguishing framing elements. In doing so, we connect an important theoretical framework to a specific research context and use this new research setting to extend and elaborate theory (Fisher & Aguinis, 2017). For example, within framing mode, we identify that the length and the visual elements of entrepreneurial messages are key distinguishing features that impact audience engagement. This connects to work in marketing around video features and digital consumer engagement (e.g., Munaro et al., 2021). Entrepreneurs must decide whether to frame their communication as short versus longer-form messages and whether to include a visual element. These decisions have important implications for other aspects of entrepreneurial framing.

Moreover, we identified that framing emphasis varies based on the context of entrepreneurial framing.

We assert that it can be valuable to derive various framing categories from the data inductively. For our study, this entailed distinguishing between an emphasis on sport, business, or fun, which aligns with the core offering of the PLL—developing a league that furthers the sport of lacrosse with the express purpose of engaging and amusing fans. Snihur and colleagues (2022: 583) identified "novelty, familiarity, public interest" as examples of emphasis frames, and recent research identified "social impact framing" (emphasizing the benefits of the venture to society) and "innovativeness framing" (emphasizing its innovativeness) as other examples of emphasis frames (Taeuscher & Rothe, 2024). The broad insight here is that emphasis frames are very context-dependent, and scholars will need to weigh the advantages and disadvantages of deductively versus inductively deriving emphasis frames as the field continues operationalizing the entrepreneurial framing framework (Snihur et al., 2022).

Additionally, we explicitly used our research setting to directly address two complications within the existing entrepreneurial framing literature and a third indirectly. Snihur et al., (2022) noted that difficulties arise in employing entrepreneurial framing across audiences, across time, and geography. Through our concept of *T*-shaped messaging, and its inclusion in various message configurations, we address and provide an illustrative framework for how entrepreneurs can effectively use framing to appeal to diverse audiences. Additionally, the express consideration of strategically timed releases of information addresses framing complications across time, suggesting that the timing of information can be manipulated to benefit new ventures and potentially extend interest and engagement over time. While we do not explicitly speak to the entrepreneurial framing complication across geography, our context indirectly enabled us to measure messaging delivered to a broadly dispersed audience. This suggests that there are universal framing principles available to generally drive engagement.

Our research also contains some limitations that provide opportunities for future research. First, our research context involves an entrepreneurial team that needed to develop their venture in a very public setting. However, not all ventures are developed in such a public setting. Some ventures operate in stealth mode or out of the public eye until their first product or service is ready to launch (Shi & Weber, 2022). The theoretical and empirical insights generated here and the broader need for generating attention around and engagement with an entrepreneurial endeavor may be irrelevant for entrepreneurial ventures that wish to remain out of the public eye and do not need broad stakeholder support before launch. It would be useful to examine how their entrepreneurial framing compares to a venture like the PLL.

Second, entrepreneurial framing has mostly been examined as a mechanism for establishing legitimacy for a new venture or market (Snihur et al., 2022). Legitimacy is defined as a "generalized perception or assumption that the actions of an entity are desirable, proper, or appropriate within some socially constructed system of norms, values, beliefs, and definitions" (Suchman, 1995: 574). Here, we consider how framing is used for a different purpose—to generate audience engagement. Entrepreneurial legitimacy and engagement are different constructs, yet there may be some overlap. Future research could provide valuable insights by examining the similarities and differences in entrepreneurial framing for legitimacy versus engagement and exploring how framing approaches for each are interconnected.

In conclusion, this study enhances our understanding of how entrepreneurs can effectively use framing to generate audience engagement before delivering their initial offering. By identifying specific framing tactics and configurations associated with high engagement, it provides valuable insights for both entrepreneurship theory and practice.

CONTACT: Shannon Younger; syounger@walton.uark.edu; University of Arkansas; 220 N. McIlroy Ave. Ste. 301, University of Arkansas, Fayetteville, AR 72701-1201

SUMMARY

THE LIABILITY OF HYBRIDITY: EXPLORING THE IMPACT OF HYBRID MISSIONS ON M&A IN ACADEMIC ENTREPRENEURIAL VENTURES

Hooman Abootorabi, Towson University, United States
Sergey Anokhin, University of Louisville, United States

Principal Topic

Mergers and acquisitions (M&A) are key strategies for business expansion, restructuring, and exit, especially among new ventures pursuing growth. While firm-specific imperatives drive M&A likelihood, the success of these transactions depends on cultural and strategic integration. Recent research highlights that many ventures adopt hybrid missions—simultaneously pursuing social and economic goals—which can influence their attractiveness in M&A deals. However, the impact of hybridity on M&A outcomes is underexplored. Academic spin-offs (ASOs), which commercialize university-developed technologies, frequently use M&A as a strategy for growth or exit. Due to their inherent dual focus on societal and commercial objectives, ASOs represent an ideal setting for examining whether pursuing hybrid missions affect M&A likelihood. Drawing on the institutional logic framework, this study posits that hybrid missions lower the probability of M&A events due to the integration complexities and goal misalignments—a phenomenon described as the "liability of hybridity." Further, this study investigates whether factors such as venture capital (VC) support, public grants, and patents can offset these liabilities by enhancing new ventures' legitimacy and resource base.

Method

We utilize a longitudinal dataset of 275 Norwegian ASOs tracked from 2000 to 2015, encompassing 1,658 firm-year observations. Business plans were scrutinized to categorize firms according to their social, economic, or hybrid orientations. Data regarding ownership structures and exit events were extracted from financial records and national business registries, while various control variables were obtained from public and institutional sources. To test our hypotheses, we employ Cox regression to evaluate M&A likelihood, supplemented by logit regression and Kaplan-Meier survival analyses. Moderators, including VC equity percentage, public grant amounts, and patent counts, are used as proxies for firm legitimacy, resource availability, and growth potential.

Results and Implications

Our findings show that ASOs with hybrid missions are less likely to be involved in M&A, indicating that the misalignment of dual goals can complicate integration. In contrast, ASOs with primarily economic missions have a higher probability of being acquired or merged. However, the 'liability of hybridity' is mitigated when ASOs secure venture capital investments, receive public grants, and hold a patent portfolio. These results offer three distinct contributions: to the M&A literature by demonstrating that the specific institutional logics pursued by firms shape their acquisition probabilities, to the hybridity and institutional logic literature by understanding the strategic implications of dual logics, and to the academic entrepreneurship literature by offering practical insights into factors that influence the attractiveness and acquisition likelihood of ASOs.

CONTACT: Hooman Abootorabi; habootorabi@towson.edu; Towson University; 212 WASHINGTON AVE, Towson, Maryland 21204, United States

SUMMARY

WAYFINDING AROUND INSTITUTIONS: HOW ENTREPRENEURS IN BALTIMORE'S BLACK BUTTERFLY NAVIGATE INSTITUTIONAL BARRIERS

Yolanda Christophe, Syracuse University, United States
Suntae Kim, Johns Hopkins University, United States

Principal Topic

In economically challenged places, minority entrepreneurs often face significant barriers to resources set up by gatekeeping institutions, such as banks, corporations, and local governments. This study examines how the minority entrepreneurs navigate and overcome these barriers to secure essential resources in the context of Baltimore, a city marked by deep-seated racial segregation and discrimination.

Method

We conducted an interview-based study to explore how Baltimore's minority entrepreneurs navigate the entrepreneurial resource space in the deeply segregated city. Our ongoing data collection includes 45 semi-structured interviews with minority entrepreneurs and representatives of support organizations, as well as attendance at four community convenings with local leaders. Entrepreneurs (n=30) represented sectors such as food services, manufacturing, arts, cannabis, consumer packaged goods, professional services, tailoring, among others. Entrepreneurship support organization representatives included bank representatives (n=3), program/support managers (n=7), and community leaders (n=5). We are currently collecting and analyzing data following the principles of grounded theory approach.

Results and Implications

Findings reveal that the journeys of entrepreneurs in Baltimore—especially those originating from distressed "Black Butterfly" areas—are characterized by facing numerous barriers and trying to get around them. They face significant barriers to resources, such as financial capital, customer access, regulatory approvals, trained workforce, and physical locations, which are heavily controlled by gatekeeping institutions, such as banks, large corporations, local governments, and property owners. Facing barriers to resources, many of Baltimore's entrepreneurs engage in a wayfinding process, following meandering paths on which they leverage existing assets and connections to resourcefully get around the barriers. For example, when a founder of a consumer-packaged goods startup faced barriers to access grocery store shelves due to a large distributor's gatekeeping, they followed a chain of emergent connections (a community grocery store → an entrepreneurship talk show host → a food services provider for schools→ a local midsize distributor) to ultimately gain access to an alternative customer base (university cafeterias). Our interviews reveal many stories of this kind, from which we glean various unconventional paths entrepreneurs charted to get around institutional barriers to resources. Our research offers two key contributions: First, while prior studies predominantly focus on financial barriers, we document a broader array of institutional obstacles faced by minority entrepreneurs. Second, we shift focus from barriers to the processes entrepreneurs use to navigate them. While prior studies highlight institutional obstacles, we reveal how entrepreneurs engage in a multi-step wayfinding process, leveraging connections and assets creatively, to access resources despite structural impediments imposed by institutions.

CONTACT: Yolanda Christophe; ychristo@syr.edu; Syracuse University; 1 Websters Landing, Syracuse, New York 13202, United States

SUMMARY

FROM COSTLY SIGNALS TO RELIABLE VOICES: HOW ONLINE INFLUENCERS RESHAPE NEW VENTURE LEGITIMACY SIGNALS

Kevin D. Heupel, Oklahoma State University, United States
Duygu Phillips, University of Delaware, United States
Matthew Rutherford, Oklahoma State University, United States

Principal Topic

Signaling theory traditionally posits that separating equilibria arise when signals are costly to produce, ensuring that only high-quality signalers can afford them. However, this perspective overlooks signal reliability—signal fit and signaler honesty—as an alternative mechanism. While costly signals deter low-quality signalers, reliable signals achieve separation by ensuring accurate and truthful information transmission. Digital environments challenge cost-based assumptions, as low-cost yet highly reliable signals, such as influencer endorsements, effectively establish legitimacy. We theorize that ventures can create separating equilibria by amplifying signal reliability rather than cost, enabling spokespersons to establish legitimacy through perceived fit and honesty, independent of cost structures.

Method

Two between-subjects experiments using Prolific participants. Experiment 1 examined how different spokesperson types (celebrity, employee, founder, online influencer) impacted cognitive, moral, and pragmatic legitimacy for a fictional new venture. Experiment 2 manipulated signal fit and signaler honesty to assess their respective impacts. Legitimacy was measured using Alexiou and Wiggins' (2019) validated scales. Venture reputation and social status served as controls. ANOVA with post-hoc comparisons assessed mean differences across conditions.

Results and Implications

Experiment 1 revealed that online influencers ($M = 5.53$, $SD = 0.55$) significantly outperformed other signalers in establishing pragmatic legitimacy ($F = 4.02$, $p = 0.01$): founders ($M = 4.89$, $SD = 0.84$), celebrities ($M = 4.83$, $SD = 0.84$), and employees ($M = 5.03$, $SD = 0.66$). However, no significant differences emerged across spokespersons for cognitive or moral legitimacy. Experiment 2 demonstrated that amplifying signal fit nullified the advantage of online influencers, resulting in comparable pragmatic legitimacy levels across spokespersons ($F = 1.58$, $p = 0.19$). However, significant differences emerged when amplifying signaler honesty ($F = 3.35$, $p = 0.01$). Employees ($M = 4.95$, $SD = 0.85$) and founders ($M = 5.15$, $SD = 0.55$) exhibited similar levels of signaler honesty to that of online influencers, but celebrities scored significantly lower ($M = 4.80$, $SD = 0.88$). These findings suggest that ventures can strategically enhance spokesperson effectiveness by increasing perceived fit and honesty, thereby shifting the signaling environment from separating to pooling equilibria. Our study advances signaling theory beyond its traditional focus on cost by theorizing how signal reliability can create separating equilibria. We demonstrate how new ventures can enhance legitimacy by strategically amplifying signal fit and honesty. Finally, we contribute to applied legitimacy strategies by identifying mechanisms through which ventures can optimize spokesperson selection and signal amplification to navigate digital environments.

CONTACT: Kevin D. Heupel; kevin.heupel@okstate.edu; Oklahoma State University; 1131 Chandler Street, Stillwater, Oklahoma 74074, United States

SUMMARY

HUMBLE ENTREPRENEURS, HAPPY STAKEHOLDERS: HOW HUMILITY INFLUENCES STAKEHOLDERS IN RESOURCE ACQUISITION

Huiqing Ju, Iowa State University, United States
Ileana Maldonado-Bautista, Iowa State University, United States
Annaleena Parhankangas, Iowa State University, United States
Paul Sanchez-Ruiz, Iowa State University, United States

Principal Topic

Resource acquisition success extends beyond signaling a compelling idea or presenting a well-crafted business plan. It is a complex social process that depends on an entrepreneur's ability to establish credibility, build relationships, and engage early-stage investors through communication (Clough et al., 2017; Huang & Knight, 2017). While entrepreneurial storytelling and narratives shape investor perceptions, less is known about how subtle linguistic alignment between entrepreneurs and investors influences resource acquisition success. We introduce humility rhetoric congruence—the alignment of humility cues in language between entrepreneurs and investors—as a key tool driving resource acquisition. Building on behavioral and social sciences research on language style matching (Ireland et al., 2011), we argue that linguistic mirroring of humility rhetoric fosters interpersonal connection, enhances perceived authenticity, and builds trust. When entrepreneurs express humility in ways that resonate with investors, it strengthens investor engagement and improves resource acquisition success. We further propose that investor sentiment—specifically authenticity and clout—mediates the relationship between humility rhetoric congruence and resource acquisition success.

Method

We tested our hypotheses on a sample of 361 equity-based crowdfunding campaigns listed on Wefunder from 2018 to 2022. The dependent variable, crowdfunding success, was measured using the natural log of the raised funding (Ahlers et al., 2015). The independent variable, humility rhetoric congruence, was computed based on a formula adapted from the language style matching (Ireland & Pennebaker, 2010) to capture the alignment in humility rhetoric (Sanchez-Ruiz et al., 2024). The mediators, authenticity and clout of investors, were operationalized by the corresponding aggregated variables in LIWC (Kacewicz et al., 2014; Newman et al., 2003).

Results and Implications

We used Hayes' (2013) PROCESS macro to test our hypotheses. Results show that humility rhetoric congruence is positively associated with crowdfunding success ($b = 1.39$, $p < 0.001$). This relationship is mediated by investor sentiment via two distinct pathways (index = 0.42, 95% CI [0.18, 0.70]). Entrepreneurs' effortful display of humility enhances investors' authenticity ($b = 21.33$, $p < 0.001$), which positively affects the funding outcome ($b = 0.02$, $p < 0.001$), while simultaneously reducing investors' clout ($b = -10.59$, $p < 0.05$), which negatively impacts the funding outcome ($b = -0.01$, $p < 0.01$). These results support our theorizing regarding the resource acquisition process being shaped by the communicative efforts and relational dynamics between entrepreneurs and stakeholders (Kalvapalle et al., 2024).

CONTACT: Huiqing Ju; hju@iastate.edu; Iowa State University; Gerdin Business Bldg, 3219, 2167 Union Dr., Ames, Iowa 50011, United States

SUMMARY

THE MAKING OF A "SKILLED CULTURAL OPERATOR": ETHNICITY INFERENCES AND BUSINESS NAMING DECISIONS

David S. Lucas, Syracuse University, United States
Joel Gehman, George Washington University, United States

Principal Topic

The cultural entrepreneurship lens suggests that entrepreneurs need to become "skilled cultural operators," drawing on cultural resources that resonate with key audiences as they seek to acquire resources to launch and grow their ventures. This is particularly important for individuals who may be outsiders or from marginalized groups. Immigrants, for instance, are known to face a liability of newness in the host country context. As such, they are likely to face additional hurdles in achieving cultural resonance and can even face concerns of discrimination. In other words, entrepreneurs from marginalized identities are not only looking to achieve cultural resonance but also to mitigate cultural dissonance. The purpose of this paper is to study how entrepreneurs seek to overcome cultural dissonance in their ventures. Our central premise is that entrepreneurs will strategically withhold cultural resources that would generate dissonance with key audiences as they seek to craft a legitimate venture identity. We theorize on this premise in the context of business naming decisions—specifically, the practice of eponymy, which is the choice to name a business after its founder. We suggest that founders whose surnames are perceived as more "foreign" will be less likely to eponymize their firms—and that these patterns will be stronger in areas with greater nationalism, for more familiar surnames, and for industries where eponymy is prevalent.

Method

We test these ideas in a sample of roughly two million founders of new firms in the United Kingdom from 2016 to 2019. We use linear probability models to predict the probability of choosing eponymy. Our core independent variable, surname foreignness—the probability that a surname is belongs to a non-domestic origin founder—is constructed using surname-nationality pairings from all business owners in the United Kingdom. Our key moderators are nationalism (based on the local geography's Brexit vote shares), surname prevalence (the frequency of a surname), and the eponymy rate in the firm's primary industry. We use a host of controls at the individual, geography, temporal, and sector levels. We also employ an instrumental variable analysis and a host of robustness checks with alternative measures and specifications.

Results and Implications

We find support for our theory. Founders with more 'foreign' surnames are considerably less likely to eponymize their businesses, and these patterns are stronger in more nationalist cultural contexts, for more prevalent surnames, and for industries where eponymy is common. This work has implications for cultural entrepreneurship theory, signaling theory, and for research on discrimination in entrepreneurship.

CONTACT: David S. Lucas; dlucas01@syr.edu; Syracuse University; 287 Oneida River Rd, Pennellville, New York 13132, United States

SUMMARY

CELEBRITY ENTREPRENEUR TURNAROUND STRATEGY: THE CASE OF WREXHAM ASSOCIATION FOOTBALL CLUB

Scott L. Newbert, Baruch College, United States
Tanurima Dutta, Baruch College, United States
Clau Sganzerla, Texas Tech University, United States
Theodore Waldron, Louisiana State University, United States

Principal Topic

An increasing number of celebrities are successfully turning around failing organizations. Understanding this form of "celebrity entrepreneurship" (Hunter and Davidsson, 2007) is important as turnaround is one of the most challenging, complex crises that all organizations will inevitably experience (Whetten, 1980). Unfortunately, the economically rational approach to turnaround espoused in extant strategy research (e.g., Pearce and Robbins, 1993; Trahms et al., 2013) does not apply to celebrity entrepreneurship given that celebrities "behave in nonconforming ways, and audience excitement about the change their behaviors represent, rather than rational evaluations of their actual performance or social standing, dominates decision-making" (Pollock et al., 2019: 451). In light of the unique focus of celebrities, we view the turnarounds they lead through a social evaluation lens (Rindova et al., 2006) in order to develop new theory that explains celebrity entrepreneur turnaround.

Method

We ground our exploration and subsequent theorizing in an inductive qualitative analysis of the turnaround of Wrexham Association Football Club (AFC) by actors Rob McElhenney and Ryan Reynolds. Wrexham AFC is a Welsh football club that boasted a rich history of success that fell on hard times starting in the early 1980s and lasting for nearly 40 years. Amid this decades-long decline that saw the club relegated to the lowest tier of English professional football, McElhenney and Reynolds bought the club in 2020 and began moving it back up the professional ranks. To understand the approach used by these celebrity entrepreneurs, we collected 444 pages of transcripts from the television docuseries, "Welcome to Wrexham," 26,753 tweets from X/Twitter, 4,264 Instagram posts/comments, 5,353 TikTok posts/comments, 4,330 Facebook posts/comments, 81 IMDb reviews, and 645 transcriptions of YouTube videos. We analyze our data using well-accepted qualitative methods (e.g., Gioia et al., 2013).

Results and Implications

Our theory suggests that celebrity entrepreneurs engaged in turnaround will seek positive emotional evaluations from their stakeholders in three stages: generating hype among the public-at-large, inspiring help from critical resource gatekeepers, and instilling hope within the local community. We contribute to the literature by identifying both a new theoretical mechanism (social evaluation) to explain turnaround, thereby offering a novel perspective regarding the motivations driving turnaround behaviors and the breadth of stakeholders to which those behaviors are directed, and a new context (turnaround) in which celebrity has the potential to create value, suggesting that positive stakeholder assessments may serve as essential strategic tools for helping organizations recover from extended periods of distress.

CONTACT: Scott L. Newbert; scott.newbert@baruch.cuny.edu; Baruch College; 1 Bernard Baruch Way, New York, New York 10010, United States

SUMMARY

CASTING SHADOWS: ACQUIRED STIGMA AS A WEAPON AGAINST NEW VENTURES

Shinwon Noh, University of St. Thomas, United States
David Deeds, University of St. Thomas, United States

Principal Topic

While research shows stigma can both constrain and motivate entrepreneurial opportunities, we introduce the concept of "acquired stigma" - negative associations new ventures inadvertently inherit when creating industries similar to stigmatized sectors, despite their own practices. This is evident in e-cigarette startups struggling to distinguish themselves from traditional tobacco products. Unlike the "Halo Effect" where association with high-status incumbents increases legitimacy, we examine how industry stigmatization creates barriers protecting incumbents by decreasing new entrants' status and resource access. Our study explores: how stigmatized incumbents' actions affect categorization of innovative products; strategies incumbents use to extend stigma to newcomers; and how acquired stigma impacts entrepreneurs' ability to access resources and bring disruptive innovations to market.

Method

Our qualitative study examines dynamics between tobacco companies and e-cigarette ventures through multiple data sources. We conducted: a) Historical analysis of tobacco industry stigmatization b) Examination of financial reports, marketing materials, and public statements from both sectors c) Analysis of regulatory documents and public health reports d) Investigation of e-cigarette startups' differentiation strategies We analyzed thousands of media reports through Nexis-Uni, Business Source Premier, MedLine, and PubMed, focusing on entrepreneurial activities in the e-cigarette industry. Data from the E-Cigarette Forum's annual survey of 10,000 ex-smokers provided insights into consumer perceptions of innovative products introduced by entrepreneurs.

Results and Implications

The study identifies three incumbent strategies for extending stigma to e-cigarette startups: Imitation and Acquisition: Tobacco companies entered the e-cigarette market through internal development or acquisition, blurring distinctions between traditional cigarettes and innovative products. Image Spillover: Incumbents replicated traditional cigarette marketing for e-cigarettes, hindering entrepreneurs' differentiation efforts. Quality Degradation: Big Tobacco released low-performance e-cigarettes, potentially undermining consumer confidence in the entire category. These actions increased regulatory scrutiny and compliance costs, created public misconceptions about e-cigarettes' relative harm, and reduced venture capital investment in startups. The research reveals an ongoing identity struggle affecting entrepreneurs' legitimacy and differentiation efforts. Acquired stigma effectively raises industry entry barriers, protecting incumbents by deterring entrepreneurial growth and limiting capital access. Future research will develop a comprehensive stigma typology to benefit research on stigma in entrepreneurship with applications beyond tobacco to other stigmatized industries.

CONTACT: Shinwon Noh; snoh@stthomas.edu; University of St. Thomas; 2115 Summit ave, St Paul, Minnesota 55105, United States

SUMMARY

THE IMPACT OF CORPORATE VENTURE CAPITAL INVESTMENT TEAM CHARACTERISTICS ON PORTFOLIO OUTCOMES AND SYNDICATION BEHAVIOR

Michael Salge, RWTH Aachen University, Germany
Carolin Krieweth, RWTH Aachen University, Germany

Principal Topic

In practice, corporate venture capital (CVC) units exhibit significant variation in the makeup of their investment teams. In some industries, teams are composed of corporate industry veterans and in others, professionals with extensive VC experience are more common (Souitaris & Zerbinati, 2014). Our paper seeks to investigate the impact of CVC investment team career backgrounds on portfolio outcomes and syndication behaviors. We explore how experience at the corporate parent, as a founder, or in VC investing, shapes their investment decisions. We examine how career backgrounds influence two critical aspects: (i) portfolio outcomes, including key metrics such as time to exit, risk profile, and overall exit success; and (ii) syndication behavior, such as the size of investment syndicates, the reputation of co-investors, and the likelihood of the CVC acting as the lead investor in these deals.

Method

To empirically test our hypotheses, we leverage data from over 3,000 funding rounds led by CVC units, sourced from the Refinitiv Eikon database. This is combined with a unique dataset where we hand-collected the career backgrounds from professional network websites, of more than 800 CVC professionals from over 50 CVC units of S&P 500 firms. We also utilize the Compustat database to incorporate additional control variables and employ regression analysis to investigate the impact of team characteristics on CVC investment behavior.

Results and Implications

Our analyses reveal that CVC units with teams dominated by professionals with VC and startup experience, achieve faster exits, invest more exploratory and tend to partner with larger, more reputable syndicates. Teams with deeper corporate sector knowledge, invest more closely aligned with the parent's strategic agenda. These results give several implications. For practice, our study helps corporations make informed hiring decisions for their CVC teams and align their recruiting choices with their desired innovation outcomes. Corporates that seek to establish CVC units with a more financially-focused strategy should prioritize hiring professionals with previous VC or founding experience, as they seem to better leverage their broader networks in the venture space for forging powerful syndicates and be more exploratory in their investment decisions. For academic research, drawing on the attention-based-view, we provide insights into how novel team-level antecedents influence portfolio and syndication outcomes within the context of CVC. Our work extends existing research on e.g., the CVC unit head (Bendig et al., 2024) and the top management team (Anhokin et al., 2016), providing insights into how the entire CVC team shapes portfolio outcomes.

CONTACT: Michael Salge; michaelsalge@time.rwth-aachen.de; RWTH Aachen University; Bettinastraße 31, Frankfurt am Main, Hessen 60325, Germany

SUMMARY

VENTURE CAPITAL AND TALENT STRATEGY: UNPACKING THE SHIFT FROM GENERALISTS TO SPECIALISTS IN STARTUP HIRING PRACTICES

Simon (Seongbin) Yoon, University of California Irvine, United States

Principal Topic

Startups face a critical tension in hiring generalists versus specialists as they scale. Specialists provide deep expertise and a clear market identity (Zuckerman et al., 2003), while generalists offer adaptability and cross-functional coordination (Karim & Williams, 2012; Nagle & Teodoridis, 2020). Prior research presents conflicting views on when startups should transition from generalists to specialists. Path-dependent theory suggests startups initially attract specialists, adding generalists later (Burton et al., 2007; Ferguson et al., 2016), while life-cycle theory argues that early-stage generalists are replaced by specialists as firms mature (Hellman & Puri, 2002; Boeker & Karachalil, 2002). However, prior research has largely focused on founders and top management teams (TMTs), overlooking hiring dynamics at lower levels. Additionally, studies emphasize functional divisions without fully considering hierarchical structures. A more comprehensive approach is needed to understand how startups balance generalists and specialists across different levels. Startups navigate multiple developmental stages, each requiring distinct skill sets. In early stages, firms rely on specialists to address technical and operational challenges critical for product development and market entry (Shane & Stuart, 2002; Beckman et al., 2007). However, as firms scale, coordination and decision-making become more complex, increasing the need for employees who integrate across functions (Penrose, 1959; Greiner, 1972). As firms move through more stages, they shift from specialized expertise to broader coordination. Early-stage firms hire specialists to solve technical challenges, but later-stage firms favor generalists who can manage interdependencies and cross-functional tasks. Moreover, in early stages, generalists dominate leadership roles due to resource constraints, while specialists fill technical positions. As firms mature, investors push for experienced specialists in senior roles, whereas generalists become more common in junior positions to support cross-functional collaboration. Thus, startups with more developmental stages tend to hire more specialists for higher-level positions while favoring generalists at lower levels.

Method

For this research, I use three unique datasets—LinkedIn online resume (380M employee profiles), Crunchbase (1.2M startups), and Pitchbook (600K startups)—and employ an OLS fixed-effects model at the firm and year-month levels, clustered at the firm level.

Results and Implications

We test Hypothesis 1 and find startups with a greater number of developmental stages are likely to hire fewer specialists. For Hypothesis 2, startups with a greater number of developmental stages are likely to hire more specialists for higher-level positions but fewer specialists for lower-level positions.

CONTACT: Simon (Seongbin) Yoon; seongbiy@uci.edu; University of California Irvine; 2633, Verano Place, Irvine, California 92617, United States

INTERACTIVE PAPER

ORGANIZING FOR SERENDIPITY: HOW AND WHY COMPANIES USE DISCOVERY AND VALIDATION HEURISTICS TO LEVERAGE THE VALUE IN THE UNEXPECTED

Christian Busch, University of Southern California, United States
Jill Juergensen, University of Surrey, United Kingdom

Principal Topic

Serendipity—the notion of making surprising and unexpected discoveries—often plays a key role in organizational success, and many innovations such as microwaves or Post-it notes were discovered unexpectedly rather than through strategic planning. Recent research highlights that while unexpected discoveries by definition cannot be predicted, companies can make it more likely via means such as a supportive culture. However, companies often struggle to enable serendipitous innovation at scale. Common barriers include lack of legitimacy for unexpectedly emerging ideas, resistance to change, and fear of distraction. As a result, ventures have rarely succeeded in cultivating serendipity at scale. Yet, in our research with a fast-growing emerging market enterprise, we observed surprisingly frequent serendipitous innovation. This led us to ask: How and why can companies organize for unexpected positive discoveries and overcome typical barriers to serendipitous innovation?

Method

To explore this question, we conducted a longitudinal case study of a Chinese company. Data included semi-structured interviews, extensive archival materials, and participant observations. Our analysis involved iterative coding, moving from open to axial coding and finally aggregating themes into conceptual dimensions. This approach allowed us to capture dynamic, transferable insights on how serendipity can be fostered in an organizational context, at scale.

Results and Implications

Our findings show that the company developed simple but effective heuristics—rules of thumb—that helped identify and harness unexpected ideas while overcoming the typical barriers to serendipitous innovation. First, *discovery heuristics* promoted organizational fragmentation and heightened awareness, encouraging employees to notice unexpectedly emerging needs and reduce functional fixedness. Second, *emergence validation heuristics* reduced perceived risk by legitimizing unexpected ideas and balancing demand- and supply-side concerns. Third, *survival validation heuristics* aligned incentives and staffing to reduce risk aversion and maintain focus. These heuristics made it possible to cultivate serendipity, at scale. Our study contributes to the literature by offering a model for scaling serendipitous innovation and extending prior work on corporate entrepreneurship. We show that creative heuristics can help overcome organizational inertia and support entrepreneurial behavior across teams that help foster serendipity. Overall, we develop a *theory of scaling serendipity*—demonstrating how ventures can consistently harness the value in the unexpected.

CONTACT: Christian Busch; christian.busch@usc.marshall.edu; University of Southern California; 670 N Casas Ave, Pacific Palisades, California 90272, United States

INTERACTIVE PAPER

SURPRISE AS A NEXUS OF PRACTICES: DEVELOPING A PRACTICE THEORY PERSPECTIVE ON THE "LEMONADE PRINCIPLE" OF EFFECTUATION

Betsy Campbell, Penn State University, United States

Principal Topic

Surprise, like uncertainty, has long been associated with entrepreneurship. Scholars initially framed external surprises as something to avoid. Then, with the "Lemonade Principle" (Sarasvathy, 2011), surprises became conceptualized as hidden opportunities. What is missing is how surprise is recognized and made relevant by founders. This paper views entrepreneurial surprise through the lenses of practice theories, a set of frameworks that conceptualize and investigate social phenomena (Sandberg & Dall'Alba, 2009). It offers an entrepreneurship-as-practice perspective (Champenois et al., 2020; Teague et al., 2021; Campbell, 2024)) of entrepreneurial surprise as a nexus of observable inter-related practices.

Method

Researchers of practice attend to the dynamic, complex, and idiosyncratic aspects of entrepreneurial work (Anderson & Starnawska, 2008) and favor methodological approaches that grapple with the details of entrepreneurial work as it is done by real people in real contexts (Johannisson, 2011). Ethnomethodology, the study of the situated and ordinary processes that enable people to participate competently in an activity (Garfinkel, 1967; Goffman, 1967; Sacks, 1984) is such an approach. Using the ethnomethodological traditions of Conversation Analysis (CA) and interaction order (Garfinkel, 1967; Goffman, 1967; Sacks, 1992), this paper considers entrepreneurial surprise as a matter that is revealed to entrepreneurs (and analysts) through the structure and order of multi-modal interaction. The paper uses CA to reveal the practical means that allow entrepreneurs to recognize, negotiate, and address matters of topical and functional surprise through the sequence of situated interactions (Schegloff, 2006). The data includes over 100 hours of self-recorded private conversations between co-founders who have had previous entrepreneurial success. It captures the naturally-occurring intra-team meetings as the teams progress through the early stages of new venture creation, including unexpected circumstances and developments.

Results and Implications

The paper offers a fresh means to advance theorizing on entrepreneurial surprise by considering surprise as a members' phenomenon (Llewellyn & Pence, 2009; Campbell, 2021). By using ethnomethodological traditions within a practice theories framework, it neither glosses over surprise as an external phenomenon nor as an intrapersonal (cognitive) process. Instead it reveals the interactional means by which surprise is recognized and negotiated by founders for themselves. The paper proposes that entrepreneurial surprise is: (a) endogenously connected with the enactment of entrepreneurship practices; (b) distributed among members of entrepreneurship practices, and (c) emergent, dynamic, and situated. It advances the entrepreneurship-as-practice literature and has implications for practitioners and educators.

CONTACT: Betsy Campbell; bjc28@psu.edu; Penn State University; 700 East Hudson Ave, Altoona, Pennsylvania 16602, United States

INTERACTIVE PAPER

FRAMING THE UNKNOWABLE: THE ENTREPRENEURIAL RHETORIC OF NEW VENTURE GROWTH

Ann-Sophie Kowalewski, IESE Business School, Spain

Principal Topic

We still know little about the role of language in framing a venture's most uncertain features. While prior research has focused on legitimizing novel or atypical ideas, this paper asks whether and how founders' linguistic framing of future growth enhances the persuasive power of investor presentations. Theories of framing and language expectancy (LET) suggest the importance of both audiences' expectations and entrepreneurs' communication (Burgoon & Miller, 1985; Garud et al., 2023; Goffman, 1974). Accordingly, I develop and test three hypotheses about entrepreneurs' use of rhetoric devices to guide audiences' attention to garner support (Cornelissen & Werner, 2014; Snihur et al., 2022) and investors' interpretive and evaluative processes.

Method

This study taps into proprietary data from 595 nascent ventures between 2014 and 2023, including transcripts equivalent to 151.2 hours of presentations, in-situ investor evaluations, and supplementary information at the firm and founder levels. The construction of Growth Emphasis—the proportion a pitch dedicated to future growth—leverages attention mechanisms and sentence transformers to capture the texts' semantic and contextual meaning in high-dimensional vector representations (Vaswani et al., 2017). Three moderators capture entrepreneurial framing, measuring concrete and authentic language, as well as novelty frames through keyword search in LIWC-22. The analysis includes hierarchical OLS regressions alongside robustness checks with alternative model specifications, subsample analyses, and alternative measures.

Results and Implications

Findings support to all hypotheses, underscoring the important theoretical distinction between persuading stakeholders to engage with something unknown to them and something that is so uncertain that related messages qualify as unknowable (Huang & Pearce, 2015): First, framing future growth in early-stage pitch presentations concretely enhances resonance and positively relates to investor evaluations of the venture. Second, frame ambiguity weakens this positive association and emphasizing frames like novelty accentuate the related evaluative loss. In contrast, authentic language conveys genuineness and strengthens investors' confidence in emerging impressions, reinforcing positive evaluations. Overall, this paper advances the burgeoning stream on the role of language in entrepreneurship and impression management by shedding novel light on framing ambiguity on persuasiveness and revealing previously underexplored mechanisms. Second, it responds to increasing calls to integrate the perspectives of entrepreneurs and their audiences, facilitating a detailed account of the nexus between strategic communication and evaluative processes (Kalvapalle, Phillips, & Cornelissen, 2024). Third, it extends LET by identifying frame ambiguity as a relevant boundary condition to the potential upside of expectancy violations. Additionally, the findings offer actionable guidance for entrepreneurs, investors, and third parties.

CONTACT: Ann-Sophie Kowalewski; askowalewski@iese.edu; IESE Business School; IESE Business School (Building H), Barcelona, Catalonia 08017, Spain

INTERACTIVE PAPER

THE ROLE OF EXTERNAL FEEDBACK WITHIN NEW VENTURE DEVELOPMENT

Vanessa Lind, Ruhr University Bochum, Germany
Petra Andries, Ghent University, Belgium
Syeda Nimra Batool, Royal Holloway University of London, United Kingdom
Stefanie Bröring, Ruhr-University Bochum, Germany

Principal Topic

External actors, such as potential customers, investors, and industry experts, play a crucial role in identifying promising business models for entrepreneurs (Andries et al., 2021). Thus, feedback provided by these actors may positively shape the development process of the venture (Snihur et al., 2017). Importantly, identifying which stakeholders are most relevant for the venture's development is complicated (Ramoglu et al., 2023). What remains underexplored is how and why entrepreneurs differ in whom they ask for feedback and how that affects the feedback process. We therefore address the research question: "How do entrepreneurs seek for feedback and integrate external actors in the business model development process?" Thereby we advance theorizing on feedback within venture creation, helping to predict effects of feedback and strategies of feedback seeking (Combs et al., 2024).

Method

We employ an exploratory case study design. We selected 12 new ventures that have a high likelihood of needing, seeking, and receiving feedback on their business models. We focus on technology-based university spin-offs, which face unique challenges in finding a technology-market fit (Vohora et al., 2004). By identifying keywords and phrases related to feedback providers and their evaluations (Corbin & Strauss, 1990) and comparing patterns within and across ventures, we observed significant differences in how teams interacted with their stakeholders.

Results and Implications

Our findings indicate that entrepreneurs vary in their motivations for seeking feedback, which influences their choice of feedback providers. Entrepreneurs who acknowledge insufficient knowledge for venture development seek feedback from perceived in-group members. This aligned feedback-seeking approach increases the likelihood of valuing and incorporating the feedback. On the other side, transactional feedback-seeking entrepreneurs that had a narrow approach of feedback-seeking chose stakeholders that they perceived as out-group members. This led to entrepreneurs incorporating feedback only in specific situations and after thorough evaluation of feedback feasibility. Based on our findings, we aim to contribute to the literature on entrepreneurial feedback processes by explaining how the motivation for feedback seeking influences the selection of feedback providers. Previous research has substantially focused on modes of interaction between entrepreneurial teams and feedback providers and how entrepreneurs make sense of feedback contents (Andries et al., 2021; Shepherd et al., 2022). Building on this, we aim to illuminate how entrepreneurs seek and choose external feedback providers for business model development. Our findings also provide insights into why entrepreneurs differ in terms of stakeholder integration and feedback acceptance (Ramoglou et al., 2023).

CONTACT: Vanessa Lind; vanessa.lind@rub.de; Ruhr University Bochum; Suttner-Nobel-Allee 4, Bochum, Nordrhein-Westfalen 44803, Germany

INTERACTIVE PAPER

INTERNALIZING TECHNICAL AI EXPERTISE: HOW MACHINE LEARNING ENGINEERS SHAPE AI USE IN VC-FUNDED STARTUPS

Matthias Qian, ESMT Berlin, Germany

Principal Topic

Adopting a new technology represents an important strategic choice for firms, but translating technological investments into sustained productivity gains critically hinges on overcoming substantial internal integration and coordination costs (Brynjolfsson & Hitt, 2000). Firms adopting artificial intelligence (AI) can deploy it narrowly in isolated tasks without necessarily incurring significant coordination challenges, but unleashing AI's full potential typically requires embedding it deeply into interdependent processes spanning multiple organizational functions (Agrawal, Goldfarb, & Gans, 2022; Bresnahan; 2021). The latter approach introduces substantial internal challenges related to differentiated knowledge structures, specialized expertise, and complex interdependencies that must be coordinated effectively across organizational boundaries (Henderson and Clark, 1990). Although prior research emphasizes the strategic importance of internal capabilities for technology adoption, exactly how firms can successfully manage and overcome these integration challenges remains unclear. This paper addresses this central tension by examining how internalizing technical AI expertise—particularly through hiring dedicated machine learning engineers—enables firms to reduce integration costs, bridge internal coordination barriers, and more effectively embed AI within organizational practices.

Method

We construct a unique dataset from the near universe of job postings of US startups. We operationalize the internalization of technical AI capabilities through the hiring of dedicated machine learning engineers. As such, we expect the organizational response to the hiring of machine learning engineers to be more direct and immediate in the context of startups, due to the lack of existing routines, structures, or capabilities that resist strategic change. We apply natural language processing of the vacancy text to develop a nuanced understanding of AI use within firms. We construct new-to-the-world measures, which include measures of the number of AI users, the organizational functions with AI users, and job characteristics of the AI users.

Results and Implications

We estimate a difference-in-difference model, assessing how the patterns of AI shift following the hire of a machine learning engineer, relative to a matched sample of AI using startups without machine learning engineers, with similar human capital, founded in the same year and location. We establish that the hiring of machine learning engineers precipitates an immediate and substantial increase in the number of AI users in the firm. In addition, following the internalization of technical AI capabilities, the AI use spreads through organizational functions, with additional organizational functions adopting AI. Third, we establish that internalizing technical AI expertise changes the coordination patterns of AI users, who prioritize interactions with other AI using functions.

CONTACT: Matthias Qian; matthias.qian@esmt.org; ESMT Berlin; Schlossplatz 1, Berlin, Berlin 10178, Germany

INTERACTIVE PAPER

KICKING UP DUST: HOW DEVIANCE FUELS ENTREPRENEURIAL SUCCESS

Matthew D. Regele, Xavier University - Williams College of Business, United States
Stephen Lippmann, Miami University, United States
Lindsay Schakenbach Regele, Miami University, United States

Principal Topic

Entrepreneurs are often assumed to require legitimacy to secure resources and survive. Yet, some entrepreneurs strategically violate norms to gain influence and control within their ecosystems. This study examines strategic deviance, the intentional and sustained use of norm violation as an entrepreneurial strategy. Outlaw Motorcycle Clubs (OMCs), such as the Hells Angels under Sonny Barger's leadership, provide an ideal setting for studying how deviance can be leveraged for organizational longevity and dominance. OMCs openly flout legal and social norms, yet they persist. Their strategic use of deviance parallels the tactics of disruptive entrepreneurs like Elon Musk (Tesla, SpaceX) and Travis Kalanick (Uber), who challenge institutional constraints to reshape industries.

Method

The study employs an inductive qualitative approach, using archival research and netnographic analysis. Archival data include autobiographies, legal cases, media interviews, and documentaries related to Outlaw Motorcycle Clubs. Netnographic data were collected from online forums and social media groups discussing OMCs. The data were analyzed using grounded theory and AI-assisted thematic coding to identify patterns related to deviance, legitimacy, and power.

Results and Implications

Our findings reveal that OMCs do not simply reject legitimacy but strategically toggle between norm violation and norm compliance, a dynamic conceptualized as legitimacy ambidexterity. Internally, deviance serves as a mechanism for cohesion and control. OMCs institutionalize rule-breaking through initiation rituals, strict internal codes, and displays of defiance, strengthening group loyalty and reinforcing a shared identity. Rather than undermining stability, these practices foster deep commitment among members and create a resilient organizational culture. Externally, OMC leaders strategically use deviance as a branding tool to enhance their public image, deter competition, and establish dominance. These tactics mirror the behaviors of regulatory entrepreneurs, who challenge industry norms and exploit legal gray areas to accelerate growth and disrupt markets. Theoretically, this study advances research on entrepreneurial strategy, legitimacy, and deviance by demonstrating that norm violation can be an enduring entrepreneurial tool rather than a liability. Prior research assumes that legitimacy-seeking is a primary driver of organizational survival, but OMCs illustrate how sustained deviance can also be a source of advantage. This study also introduces legitimacy ambidexterity as a conceptual framework to explain how organizations operating in contested spaces can navigate institutional constraints by oscillating between legitimacy-seeking and norm-breaking behaviors. These insights extend beyond OMCs, offering implications for disruptive firms, activist organizations, and other enterprises that operate at the edge of social and legal acceptability.

CONTACT: Matthew D. Regele; regelem@xavier.edu; Xavier University - Williams College of Business; Xavier University, Cincinnati, Ohio 45207, United States

INTERACTIVE PAPER

HOW FRAMING INFLUENCES ENTREPRENEURIAL ACTION: GEN AI LAUNCHES BY OPEN AI AND GOOGLE

Raj K. Shankar, Great Lakes Institute of Management Chennai, India
Sanchita Kuchi, Great Lakes Institute of Management Gurgaon, India
Simarjeet Singh, Great Lakes Institute of Management Gurgaon, India

Principal Topic

Organizations and new ventures engage in entrepreneurial action to bring new innovations to market (McMullen, Ingram & Adams, 2020). Upon launch, such innovations diffuse among users in predictable ways (Rogers, 1995). But the diffusion/adoption of recent GenAI launches by OpenAI and Google defies prior theoretical expectations, prompting need for advancement of innovation theory (Cornelisson et al., 2024). While Google's launch followed the standard diffusion of innovation process, OpenAI took a different approach: launched themselves as an experimental tool, became available directly to end users (not innovators and/or early adopters), and used a narrative which appealed to the emotional side of the audience. OpenAI's ChatGPT is clearly the market leader (Similiarweb, 2024). Prior research in strategy and entrepreneurship shows that framing choices can influence entrepreneurial outcomes (Clarke, Cornelisson & Healey, 2019; Falchetti, Cattani & Ferriani, 2020). Yet, empirical explorations on the role of framing in innovation diffusion remains limited (Snihur et al., 2022).

Method

The empirical setting for our study is the launch of two competing Generative AI offerings—ChatGPT and Bard/Gemini. Our data comprises: (1) launch announcements sourced from Factiva, company websites, and social media platforms; (2) user adoption trends post-launch drawn from web traffic repositories (SimilarWeb, SEMrush); (3) user reactions and engagement on social media platforms like Twitter, Reddit, Youtube, etc. We employ a combination of web traffic analytics and content analysis through established methodologies in Computer-Aided Text Analysis (McKenny et al., 2018) complemented with manual review. We analyze company communications using keywords related to social impact and innovativeness, and user-generated responses through cognitive and emotional keyword lists drawing on the linguistic dictionaries (McKenny & Short, 2018; Moss et al., 2016; and Henry, 2008).

Results and Implications

We find that social impact frames enacted through identification mechanisms result in rapid mass adoption of a radical innovation vis-à-vis innovativeness frames enacted through a familiarity mechanism. Our results have at least three key implications. First, our multiple pathways model promises to offer a more nuanced explanation of how and why framing matters/expands to/the diffusion of radical innovations (Kratochvil, 2025). Second, we contribute to the corporate entrepreneurship literature by pointing to two factors that corporations can choose (frames and framing mechanisms) to act entrepreneurially (Covin & Miles, 2007; McMullen et al., 2020). Third, we contribute to the entrepreneurial framing literature, by empirically illuminating how frames and framing performativity mechanisms influence innovation adoption in the context of radical innovations (Snihur et al., 2022; Taeuscher & Rothe, 2023).

CONTACT: Raj K. Shankar; rajkrishnan.s@greatlakes.edu.in; Great Lakes Institute of Management Chennai; Great Lakes Institute of India, Chennai, Tamil Nadu 603102, India

INTERACTIVE PAPER

VOLUNTARY RECALLS, SEVERITY, AND TIMING: NAVIGATING DELISTING RISK FOR NEWLY PUBLIC FIRMS

Mohammad S K Shazzad Hossain Sikdar, Colorado Mesa University, United States
Arvin Sahaym, Washington State University, United States

Principal Topic

Product recalls pose a significant threat to firms, particularly newly public firms (NPFs). Recalls disrupt operations, damage reputations, and cause financial losses (Thirumalai & Sinha, 2011), potentially leading to delisting, a critical risk for NPFs that may lead to permanent market exit (Carpenter et al., 2003). This research explores whether NPFs face higher delisting risks compared to established firms after product recalls. Drawing on Expectancy Violation Theory (Burgoon,1993) and integrating insights of Legitimacy Theory (Deephouse & Carter, 2005), we hypothesize that NPFs are more likely to be delisted following a product recall compared to established firms. Furthermore, voluntary recalls, recall severity, and timing amplify this risk. The research also examines a three-way interaction between being NPFs, recall severity, and voluntary recalls on the likelihood of delisting.

Method

The final sample consists of 159 unique U.S. biotech firms with 2,624 product recall events between 2001 and 2016. To test our hypotheses, we used logistic regression as the dependent variable, Delisting, is dichotomous (1 = delisted). The independent variable is whether a firm is a newly public firm (Atinc et al., 2017; Carpenter et al., 2003). The moderating variables include voluntary recall, recall severity, and timing of recalls. We used FDA recall database, FDA recall classifications, the WRDS database, and Nexis Uni to operationalize and measure the key variables for the analysis.

Results and Implications

The results indicate that NPFs are 2.87 times more likely to be delisted compared to established firms following a product recall, underscoring their heightened vulnerability. Notably, NPFs issuing voluntary recalls are 8.8 times more likely to be delisted than established firms, revealing a potential paradox: transparency may inadvertently signal organizational weakness. Severe recalls amplify delisting risk by 2.2 times, emphasizing the necessity of robust quality control and risk mitigation. Contrary to expectations, recall timing does not significantly influence delisting outcomes. Finally, NPFs facing severe voluntary recalls are 3.56 times more likely to be delisted compared to established firms under the same conditions, highlighting the amplified scrutiny NPFs endure under complex circumstances. These findings have significant implications. NPFs must prioritize rigorous quality assurance protocols and develop comprehensive crisis management strategies, particularly concerning voluntary recall decisions. Investors should acknowledge the elevated delisting risk associated with NPFs, especially those encountering severe or voluntary recalls. Regulatory bodies might consider implementing tailored oversight mechanisms for NPFs navigating recall situations.

CONTACT: Mohammad S. K. Shazzad Hossain Sikdar; mhossainsikda@coloradomesa.edu; Colorado Mesa University; 2454 Window Rock Lane, Grand Junction, Colorado 81505, United States

INTERACTIVE PAPER

JUDGING A BUSINESS BY ITS COVER: STAKEHOLDER APPROACH AND SUPPLY CHAIN IMPACT ON EARLY-STAGE VENTURE FUNDING

Jaime Suarez, Baylor University, United States
Zulay Buchs, University of Technology Sydney, United States
Joao Rossi Mazzoni, Baylor University, United States

Principal Topic

The purpose of this paper is to examine how early-stage ventures (ET ventures) frame their supply chain strategies in business plans and how this influences investor interest. Drawing on stakeholder theory (Freeman, 1984), supply chain integration (Christopher, 2011), legitimacy research (Zimmerman & Zeitz, 2002), and investment decision-making (Wöhler & Haase, 2022), the study explores whether prioritizing suppliers as strategic stakeholders enhances venture success. The central research question explores how early-stage ventures prioritize supply chains as key stakeholders and how this prioritization impacts investor decisions.

Method

We employ a mixed methods approach, combining thematic analysis (Braun & Clarke, 2006) and grounded theory (Charmaz, 2006) to analyze over 700 business plans from a venture competition. The qualitative phase identifies recurring themes in supply chain articulation using text coding, focusing on four key dimensions: (1) Stakeholder Identification, assessing how often and in what capacity supply chain partners appear in business plans; (2) Stakeholder Prioritization, analysing whether supply chain partners are positioned as critical to venture success; (3) Supply Chain Framing in Investor Communication, examining how entrepreneurs strategically present supply chain relationships to mitigate risk and signal competitive advantage; and (4) Legitimacy and Trust Building, investigating how entrepreneurs leverage supply chain relationships to enhance credibility with investors. The quantitative phase examines how ventures structure their stakeholder strategies and how these influence judges' venture viability and scalability assessments. Rather than applying cluster analysis rigidly, we categorize ventures based on the extent to which their business plans integrate supply chain considerations across the four key dimensions. We then use regression models to test the impact of supply chain articulation on judges' evaluations.

Results and Implications

Initial results show that judges prefer supply chain resilience, supplier partnerships, and risk mitigation. Industry-specific patterns emerge, with ventures in manufacturing, biotech, and regulated sectors benefiting most from strategic supply chain positioning. Startups that frame suppliers as innovation partners or scalability enablers and perceived execution risk reducers are also favoured. This study extends stakeholder theory by demonstrating that supply chain prioritization influences investment decisions. It provides entrepreneurs with a framework to strategically present supply chain integration in business plans, improving judges' preferences. Investors, in turn, gain a more comprehensive criterion for evaluating venture viability beyond product and team. By bridging supply chain management and venture competition outcomes, this research advances the understanding of how stakeholder integration influences startup viability.

CONTACT: Jaime Suarez; jaime_suarez1@baylor.edu; Baylor University; 1311 S 5th St, Waco, Texas 76706, United States

INTERACTIVE PAPER

SYNDICATION STRATEGY OF CORPORATE VENTURE CAPITAL FIRMS: THE IMPACTS OF INDUSTRIAL AND GEOGRAPHIC DISTANCE

Timiry R. Tian, Old Dominion University, United States
Chris Willis, University of Mary Washington, United States
Jing Zhang, Old Dominion University, United States

Principal Topic

Corporations often take minority ownership positions in high-tech startups as a means of obtaining their cutting-edge technologies, commonly referred to as Corporate Venture Capital (CVC) investment (Dushnitsky & Lenox, 2005). While the CVC literature has studied how CVC investment facilitates corporate innovation and new venture growth, few studies have examined CVC firms' syndication strategy and its impacts. To address this gap, this paper applies the knowledge-based alliance literature to explore how industrial and geographic distances between CVCs and startups influence CVC firms' choice between syndicating with other CVC firms or independent venture capital (IVC) investor partners.

Method

We built a dataset of 12,612 CVC investment deals into 7,526 startups by 117 CVC firms that are part of S&P 500 corporations during 2000-2022. Data was sourced from PitchBook, a leading private database for venture capital data. CVC firms' financial data were collected from Standard & Poor's Compustat. Other economic indicators and development indices were collected from the World Bank. We limited VC rounds to "Seed Round," "Early-Stage VC," "Later Stage VC," and "PE Growth/Expansion" deals, excluding real estate and financial services sectors. Using logistic regression, we analyzed the impact of industrial and geographic distances on syndication choices. Control variables included startup characteristics (e.g., age, industry), CVC firms (e.g., prior investment experience, R&D intensity), startup geographic locations (e.g., GDP per capita growth rate), and deal characteristics (e.g., investment round, deal size). Robustness checks with alternative specifications confirmed consistency.

Results and Implications

We predict that when CVC firms need to acquire knowledge, they are more likely to syndicate with other CVC rather than IVC partners. Our findings confirmed that a larger knowledge distance between CVC and startups leads to syndication with more CVC rather than IVC partners. Furthermore, we predict that the syndication choices of CVC firms differ significantly between technology-intensive and market-intensive industries. Our findings indicate that syndication with CVC partners occurs more frequently in technology-intensive industries, while syndication with IVC partners occurs more often in market-intensive industries. Additionally, we find that CVC syndication strategies vary between domestic and foreign markets. CVC firms tend to syndicate more with IVC partners when investing in foreign markets. Prior literature suggests that syndication among VC investors adds complementary technological and market knowledge and extends VCs' networks to acquire critical knowledge and resources. Our findings contribute to an understanding of the CVC and IVC partners' distinct roles in helping the focal CVC firm acquire or access technological or market knowledge.

CONTACT: Timiry R. Tian; rtian001@odu.edu; Old Dominion University; 4273 Llewellyn Ave, Apt 408, Norfolk, Virginia 23504-1136, United States

INTERACTIVE PAPER

TOWARD MULTIDIMENSIONAL DISTINCTIVENESS: BALANCING AESTHETICS AND INTERACTIVITY IN TABLETOP BOARD GAMES

Jim Whitbeck, Oregon State University, United States
Violetta Bacon-Gerasymenko, Oregon State University, United States

Principal Topic

The tension between competitive benefits of differentiation and institutional pressures toward conformity is well-established in organizational theory and strategic management (Deephouse, 1999; Zhao et al., 2017; Haans, 2019). In the context of board games, core mechanics represent essential features that define how players interact with the game system and with each other. These mechanics are subject to the same countervailing forces identified in prior work: differentiation enables unique market positions and reduced competition (Porter, 1991; Baum & Mezias, 1992), while conformity pressures reward adherence to established norms and penalize deviation (DiMaggio & Powell, 1983; Zuckerman, 1999). Following Haans (2019), these mechanisms operate multiplicatively rather than additively. At low levels of core distinctiveness, the benefits of differentiation outweigh legitimacy concerns, leading to improved performance. However, beyond a certain threshold, the mounting legitimacy challenges begin to dominate the diminishing returns to differentiation. As Zhao et al.,(2018, p. 590) note in their study of video games, "exemplars provide tangible information cues regarding market demand, focalize attention of producers and consumers, and shape the trajectories of category evolution."

Method

This study examines the relationship between game distinctiveness and user ratings over time. The dependent variable is the average user rating, calculated at 1, 2, 3, 4, and 5 years after publication using individual user ratings. The independent variable and moderator are core and peripheral distinctiveness, measured through principal component analysis (PCA) on each distinctiveness dimension to consolidate categories. Distinctiveness is operationalized as the Euclidean distance between each focal game and exemplar games, with values Winsorized at the 5th percentile to account for outliers. Exemplars are defined as category-relevant (family or strategy) award-winning games from the year prior to the focal game's publication. The analysis controls for various game characteristics including year published, price, player count , complexity rating, playing time, minimum age , number of artists, and number of publishers. We additionally control for design team fixed effects.

Results and Implications

Our analysis reveals an inverted U-shaped relationship between mechanical distinctiveness and game performance at the 2-year mark, suggesting moderate mechanical innovation optimizes ratings. However, this relationship is significantly moderated by genre distinctiveness, with the curve flipping to a U-shape at high genre distinctiveness levels. This indicates that within novel genre contexts, either highly conventional or highly innovative mechanical designs outperform moderately distinctive ones, demonstrating that optimal distinctiveness strategies depend critically on a game's genre positioning.

CONTACT: Jim Whitbeck; whitbeja@oregonstate.edu; Oregon State University; 902 NW OAK AVE, Corvallis, Oregon 97330, United States

INTERACTIVE PAPER

RECOGNIZABILITY VS. RELIABILITY: EVALUATION DIFFERENCES FOR GENERALISTS ACROSS STARTUPS' HIRING STAGES

Simon (Seongbin) Yoon, University of California Irvine (UCI), United States
Cristiano Bellavitis, Syracuse University, United States
Ming Leung, University of California, Irvine (UCI), United States
David Park, Syracuse University, United States

Principal Topic

Startups face unique challenges in hiring as they balance the need for flexible generalists and specialized experts. This tension is especially relevant in today's labor market, where online platforms and outbound recruiting have transformed hiring practices (Elfenbein & Sterling, 2018). While generalists bring broad skill sets, research suggests they often face disadvantages, with two key explanations for this penalty: at the first stage (selection), where they may be excluded from the shortlist (Zuckerman et al., 2003), and at the second stage (evaluation), where they are less likely to receive a hiring offer (Leung, 2014). However, the specific stage at which generalists are penalized remains unclear. Building on a two-stage hiring model—Selection and Evaluation (Payne, 1976)—we examine how startups navigate this trade-off. Generalists may attract more initial interest due to their versatility, yet specialists tend to be favored in the final hiring decision when technical expertise becomes crucial (Leung, 2014). Outbound recruiting, a key startup hiring strategy, allows firms to proactively engage potential candidates rather than relying on traditional job postings (Elfenbein & Sterling, 2018). This approach helps startups identify a broad talent pool but also introduces complexities in choosing between generalists and specialists. While generalists may be more visible early in the process, specialists often gain an edge when deeper skills are required. This study explores how startups manage this hiring paradox. We analyze how outbound recruiting influences candidate selection and evaluation, shedding light on why generalists may receive more initial attention but specialists are ultimately preferred. Understanding these dynamics offers valuable insights into startup hiring strategies and the evolving role of generalists and specialists in entrepreneurial growth.

Method

Using data from a major online networking platform (2016–2020), we analyze 1,319 high-tech startups and 8,899 candidates. This proprietary dataset tracks the full hiring process, from employer outreach to candidate responses and offers. We assess specialization's impact using logistic regression with startup and year-month fixed effects and clustered standard errors for startups. Robustness checks using a Linear Probability Model yielded consistent results.

Results and Implications

The analysis shows that greater specialization breadth increases the likelihood of securing an interview, supporting H1. However, it decreases the chances of receiving a job offer, supporting H2. This finding suggests that while candidates with diverse skills attract initial interest, they may be perceived as less focused, ultimately reducing their hiring prospects in startups.

CONTACT: Simon (Seongbin) Yoon; seongbiy@uci.edu; University of California Irvine (UCI); 2633, Verano Place, Irvine, California 92617, United States

THE NATIONAL FEDERATION OF INDEPENDENT BUSINESS (NFIB) AWARD FOR THE BEST PAPER ON THE GENERAL TOPIC OF ENTREPRENEURSHIP

BALANCING EMOTION AND OWNERSHIP: A STUDY OF ENTREPRENEURIAL PASSION IN EQUITY DIVISION

Yannick Nicolai Amend, Justus Liebig University Giessen, Germany
Petrit Ademi, University of St. Gallen, Switzerland
Monika Christine Schuhmacher, Justus Liebig University Giessen, Germany

ABSTRACT

How do founders split their equity? We address this question using a metric conjoint experiment involving 198 U.S. entrepreneurs who made 4,752 equity allocation decisions. Drawing on resource acquisition and similarity attraction theories, we show that equity offers reflect candidates' resources, interpersonal similarity, and equity demands. We further explore how entrepreneurs' characteristics, particularly entrepreneurial passion (EP), shape these decisions. Notably, EP differentially moderates several effects: entrepreneurs passionate about inventing offer more equity to similar co-founders, while those passionate about founding offer less. Our findings extend research on new venture team formation and entrepreneurial decision-making by highlighting the strategic and emotional complexity behind equity splits.

INTRODUCTION

In nascent entrepreneurial ventures, lead entrepreneurs hold the strategic position of evaluating potential co-founders (Lazar et al., 2020) and offering equity to them (Ruef, 2009). The inclusion of co-founders represents a critical entrepreneurial decision as it defines team composition (Preller et al., 2020), alters operational processes and hierarchical structures (Klotz et al., 2014), and influences new venture performance (Lazar et al., 2022). Referred to as the ventures' "first deal" (Hellmann and Wasserman, 2017, p. 1), the equity split among co-founders has been shown to potentially strengthen team identification (Weissenböck et al., 2024) and heighten entrepreneurial efforts (Bitler et al., 2005). However, equity allocations may also lead to conflicts, negative interaction spirals, and even exits (Breugst, et al., 2015; Gregori & Parastuty, 2021).

Despite the critical importance of equity allocation decisions in founding teams (Hellmann and Thiele, 2015; Mueller & Hennicke, 2024; Wasserman, 2017), research has primarily focused on the consequences of equity allocation. Surprisingly, there is limited exploration of the antecedents, particularly how co-founders' resources, interpersonal attributes, or demands shape entrepreneurs' allocations, and how entrepreneurs' characteristics factor into these strategic decisions. Although previous research has emphasized that entrepreneurs with greater entrepreneurial experience (Yan & Zhang, 2024) and specific industry experience (Kotha & George, 2012) tend to retain more equity, the underlying decision-making process remains poorly understood. We are missing insights on how co-founders shape the equity offered to them as well as how the characteristics of lead entrepreneurs, aside from their experience, contribute to equity allocation decisions. The aim of this study is to examine the factors influencing entrepreneurs' equity offers to co-founders. Specifically, our work integrates recent advances in entrepreneurial team formation literature with insights into entrepreneurial passion (EP) research, which suggests that EP may significantly shape early-stage decision-making (Dalborg et al., 2015; Stroe et al., 2018). While prior research has demonstrated that co-founders prefer joining lead entrepreneurs that display similar EPs (Fu et al., 2022), it is compelling to explore how entrepreneurs' EP affects their equity offer to potential co-founders.

HYPOTHESIS DEVELOPMENT

Although recent data confirms that dyadic teams remain the most prevalent configuration, equity is typically not distributed equally between two co-founders. Nevertheless, equal equity splits have become increasingly common over the past decade (Walker & Dowd, 2025). Despite these trends in practice, we still lack a robust theoretical and empirical understanding of how entrepreneurs arrive at equity allocation decisions.

Therefore, we draw on the most established formation strategies: interpersonal similarity attraction and resource seeking (Lazar et al., 2020; 2022). Thereby, we reflect recent suggestions that lead entrepreneurs employ selection strategies simultaneously (Lazar et al., 2022; Shah et al., 2019). Notably, research underscores that equity serves as the cornerstone of founder compensation (Åstebro, 2012; Hall & Woodward, 2010; Wasserman, 2006), providing a powerful incentive for venture development (Kagan et al., 2020). Also, equity allocations' impact on financial rewards (Åstebro, 2012) and co-founder attraction (Chandler et al., 2005; Wasserman, 2012) highlight its central role in nascent decision-making. Considering the importance of equity and its function as remuneration for co-founders' potential contributions, we integrate candidates' equity demand as important commitment and negotiation factor informing entrepreneurs' decision-making. Particularly, because equity shapes power dynamics and prospective direction of a newly founded venture (Breugst et al., 2015; Hellmann & Wasserman, 2017), we aim to explore how personal aspects influence individual equity offerings.

Entrepreneurial Passion in Equity Offerings

With growing attention to the psychological underpinnings of entrepreneurial decision-making, research has begun to uncover how affective states and identity considerations shape entrepreneurs' decisions. One particularly salient construct is entrepreneurial passion (EP), defined as intense positive feelings toward entrepreneurial activities that are central to an entrepreneurs' identity (Cardon et al., 2009). EP is increasingly recognized not only as a motivational force but also as a signal to external stakeholders (Chen et al., 2009). On one hand, passionate entrepreneurs may enhance their appeal to investors and potential co-founders through enthusiasm and perceived commitment, thus facilitating access to financial or human capital (Fu et al., 2022; Warnick et al., 2018). On the other hand, this same passion may bias entrepreneurs' assessments of firm valuation or their readiness to cede control, potentially resulting in suboptimal equity structuring (Breugst et al., 2012). This duality points to a deeper paradox: while EP can serve as a social signal aligned with impression management, it can also impair cognitive calibration in high-stakes financial decisions. The tension between these signaling and cognitive functions of passion calls for a deeper investigation into the mechanisms through which passionate entrepreneurs navigate equity offerings. Particularly, decision-making scenarios may offer promising pathways for illuminating how entrepreneurs reconcile affect-driven conviction with strategic trade-offs when distributing equity. Accordingly, we examine how EP, particularly the dimensions of inventing and founding which are the most relevant in nascent ventures, influences the equity offers made by lead entrepreneurs to potential co-founders.

Entrepreneurial Passion for Inventing in Equity Offerings

Entrepreneurial equity allocation is not solely determined by economic criteria (Yan & Zhang, 2024); it is deeply intertwined with cognitive and affective processes that shape how founders interpret and evaluate potential co-founders (Brattström, 2023; Breugst et al., 2015). One powerful affective construct influencing this evaluation is *entrepreneurial passion for inventing*, defined as intense positive affect experienced when engaging in activities related to recognizing and exploring new product or service ideas (Cardon et al., 2009). Passion for inventing reflects a founder's deep identity connection with the creative aspects of entrepreneurship and often shapes their perceptions, preferences, and evaluative criteria (Murnieks et al., 2020). When lead entrepreneurs exhibit higher levels of passion for inventing, they may be particularly attuned to cues that support their own venture vision and reinforce a shared sense of purpose, driving more generous equity allocations to candidates who appear aligned with these goals. On one hand, interpersonal similarity in venture vision and personal values has been consistently linked to greater perceived cohesion, predictability, and trust within founding teams (Lazar et al., 2020; Ensley et al., 2002). These cues signal congruent mental models and smoother collaboration (Forbes et al., 2006), often leading to favorable assessments in the context of team formation. For individual with high EP for inventing, these cues may take on amplified meaning, as they are associated with lower relationship conflict (Boone et al., 2020). In addition, because their identity is tightly bound to the act of inventing, they may be more likely to interpret such similarity attributes as indicative of a strong co-creative fit—someone who not only understands the founder's vision but also fuels it. As a result, we propose that the equity granted to co-founder candidates increases with interpersonal similarity, and that this relationship is strengthened when the lead entrepreneur experiences higher levels of EP for inventing. Thus, we hypothesize:

H1a: The positive relationship between co-founder candidates' similarity of venture vision and the equity offered by lead entrepreneurs is strengthened when lead entrepreneurs exhibit higher levels of entrepreneurial passion for inventing.

H1b: The positive relationship between co-founder candidates' similarity of personal values and the equity offered by lead entrepreneurs is strengthened when lead entrepreneurs exhibit higher levels of entrepreneurial passion for inventing.

Beyond interpersonal alignment, resource-based attributes (i.e., domain expertise and network size) also play a crucial role in co-founder decisions (Shah et al., 2019; Lazar et al., 2020). Candidates with domain expertise bring technical and market-specific knowledge that can support the realization of novel ideas, while those with larger networks offer greater access to external resources, early adopters, or strategic partners. Both attributes are instrumental for venture creation, but their salience may differ depending on the lead entrepreneur's motivational orientation. Entrepreneurs with high EP for inventing are particularly driven by advancing their ideas and creating something novel. In this context, candidates who possess resources that enable innovation or amplify the reach of a novel product are perceived as more valuable and deserving of higher equity stakes. Passion for inventing enhances cognitive focus and commitment toward the goal of innovation (Cardon et al., 2009), potentially leading passionate founders to prioritize candidates who offer strategic resources (i.e., domain expertise or network size) to the inventing process. Accordingly, we argue that EP for inventing not only intensifies the desire to surround oneself with similar individuals but also increases the perceived value of resource-bearing candidates whose attributes directly facilitate the inventive process. Thus, we hypothesize:

H2a: The positive relationship between co-founder candidates' domain expertise and the equity offered by lead entrepreneurs is strengthened when lead entrepreneurs exhibit higher levels of entrepreneurial passion for inventing.

H2b: The positive relationship between co-founder candidates' network size and the equity offered by lead entrepreneurs is strengthened when lead entrepreneurs exhibit higher levels of entrepreneurial passion for inventing.

Entrepreneurial Passion for Founding in Equity Offerings

EP for founding refers to a deep emotional and identity-based engagement with the process of establishing new ventures, including resource mobilization, team formation, and formalizing business infrastructure (Cardon et al., 2009). Entrepreneurs high in this form of EP are often driven by the challenge and excitement of launching and shaping new ventures rather than maintaining stable interpersonal dynamics (Obschonka et al., 2017).

While similarity in values and vision tends to enhance team cohesion, trust, and alignment (O'Reilly et al., 1989; Forbes et al., 2006), high founding passion may render these factors less influential in equity decisions. This is because founding-passionate entrepreneurs may focus more on the dynamic execution of venture creation than on long-term interpersonal harmony. The drive to build something novel from scratch can foster a preference for individuals who are differentiated rather than aligned—especially when sameness is perceived as introducing redundancy, cognitive lock-in, or groupthink (Beckman et al., 2003). Moreover, entrepreneurs who strongly identify as "founders" may perceive themselves as the ultimate drivers of vision and culture, reducing the need to seek interpersonal alignment from others. As a result, they may discount interpersonal similarity attributes when offering equity, instead optimizing for speed and flexibility. Thus, we hypothesize:

H3a: The positive relationship between co-founder candidates' similarity of venture vision and the equity offered by lead entrepreneurs is weakened when lead entrepreneurs exhibit higher levels of entrepreneurial passion for founding.

H3b: The positive relationship between co-founder candidates' similarity of personal values and the equity offered by lead entrepreneurs is weakened when lead entrepreneurs exhibit higher levels of entrepreneurial passion for founding.

In parallel, EP for founding may also alter how resource-based attributes—such as domain expertise and network size—are weighted in equity decisions. While these attributes typically enhance a candidate's attractiveness by exhibiting high functional contribution (Klotz et al., 2014), lead entrepreneurs with high EP for founding may emphasize founder centrality and rapid execution over collaborative resource sharing.

Specifically, high domain expertise may be viewed as a double-edged sword: while beneficial for task performance, it may introduce normative rigidity or power asymmetries that challenge the founder's control over venture direction (Audretsch & Fiedler, 2023; Marvel & Lumpkin, 2007). Similarly, candidates with large networks may be perceived as bringing in outside influence, increasing visibility, or adding coordination complexity that slows down execution during the critical founding stage. Founding-passionate entrepreneurs often thrive in contexts of autonomy and rapid iteration (Cardon et al., 2009) and may therefore prefer to preserve control rather than dilute it through resource-based dependencies. Consequently, the usually positive impact of resourceful co-founders on equity allocation may be attenuated when lead entrepreneurs are highly passionate about founding, reflecting a strategic preference for founder-centric team configurations. Thus, we propose:

H4a: The positive relationship between co-founder candidates' domain expertise and the equity offered by lead entrepreneurs is weakened when lead entrepreneurs exhibit higher levels of entrepreneurial passion for founding.

H4b: The positive relationship between co-founder candidates' network size and the equity offered by lead entrepreneurs is weakened when lead entrepreneurs exhibit higher levels of entrepreneurial passion for founding.

METHOD

To study participants' willingness to offer equity to potential co-founders, we employed a metric conjoint experiment as seen in recent entrepreneurship studies (e.g., Ademi et al., 2022). Following established conjoint designs, our study contains three parts: the scenario and task description, the equity offering tasks, and a post-experimental survey. Our experiment assumes that participants have a promising entrepreneurial idea for which they decided to look for a co-founder as part of the venture creation process. We outline that co-founder candidate profiles differ solely in the following five attributes: (1) domain expertise, (2) network size, (3) similarity of venture vision, (4) similarity of personal values, and (5) equity demand. We include key co-founder selection attributes from the established selection strategies, namely resource acquisition (1,2), and similarity attraction (3,4) (Gray et al., 2024; Lazar et al., 2020). As we theoretically and empirically explore equity divisions, we included (5) equity demand of the co-founder as negotiation factor drawing upon research showing that demands influence perceptions in decision-making (Kahneman, 1992; Maaravi & Segal, 2022). Participants were instructed to evaluate co-founder candidates with respect to the equity (in %) they are willing to offer to them.

RESULTS

We tested our hypotheses utilizing hierarchical linear modeling (HLM) as it has the capability to account for effects across different levels (Aguinis et al., 2013). Each participant assessed sixteen co-founder profiles and eight replicated profiles, resulting in a total of 3,168 decisions for the main test and 1,584 for the retest. We underscore the confidence in our results with a test-retest reliability that is in line with extant studies (e.g., Kier et al., 2022; Souakri et al., 2022). In the following, we present the results of hypothesis testing aimed at disentangling the intricacies of lead entrepreneurs' decision-making processes in offering equity to potential co-founders.

Our analysis reveals that lead entrepreneurs offer significantly less equity when they possess prior entrepreneurial experience or have experienced venture failure before. As expected, we establish that the similarity attraction strategy, characterized by similarity of venture vision and personal values, and the resource acquisition strategy, operationalized through domain expertise and network size positively influence initial equity offerings of lead entrepreneurs. We further show that entrepreneurs offer more equity to potential co-founders that exhibit a higher equity demand.

We proposed that the positive relationship between interpersonal similarity (in the form of (a) similarity of venture vision and (b) similarity of personal values) and the equity offered by lead entrepreneurs would be strengthened when entrepreneurs exhibit higher levels of EP for inventing. Both interaction terms were statistically significant and positive, indicating that entrepreneurs passionate about inventing were more likely to allocate equity to candidates who shared interpersonal similarities. Thus, H1a and H1b are supported. H2a and H2b posited that the positive relationship between co-founder candidates' resource attributes (a) domain expertise vision and (b) network size and equity offered would be amplified under higher levels of EP for inventing. Results show that EP for inventing only positively moderates the relationship between domain expertise and equity offered (H2a), while the moderation effect concerning network size was not statistically significant (H2b). Thus, offering only partial support for H2.

Regarding EP for founding, we proposed that the positive relationship between interpersonal similarity and equity offered would be weakened when entrepreneurs exhibit higher levels of EP for founding. The moderation effects were both significant and negative, indicating that EP for founding attenuates the influence of interpersonal similarity attributes on equity offerings. Thus, H3a and H3b are supported. Finally, H4a and H4b predicted that the positive relationship between candidates' resource attributes and equity offered would be weakened by higher EP for founding. However, neither moderation effect is significant, providing no support for H4a and H4b.

Overall, our data indicates that EP for inventing increases equity offers based on both interpersonal similarity and resource attributes, while passion for founding attenuated the effects of interpersonal similarity. This evidence points to a nuanced role of EP, whereby its distinct dimensions differentially guide cognitive attention and decision weight in equity decisions by entrepreneurs.

DISCUSSION AND IMPLICATIONS

Taken together, our results suggest that equity allocation decisions are not only strategic but also deeply shaped by the motivational and cognitive profiles of lead entrepreneurs, particularly their EP. We find that the extent to which candidates' contributions—expressed as interpersonal similarity and resource attributes—translate into equity offerings is significantly contingent on the specific dimension of EP held by the lead entrepreneur. This highlights the individualized nature of equity allocation decisions, wherein passion functions as a lens through which candidates are interpreted and valued. By disentangling the roles of EP for inventing and founding, our study enriches emerging perspectives on entrepreneurial decision-making and offers a differentiated view on how emotional and cognitive orientations guide strategic decisions during team formation.

We offer three notable contributions to the current theory and literature on entrepreneurship. First, we advance the literature on new venture team formation by unpacking how equity allocation decisions emerge in the co-founder selection process. While existing work emphasizes post-hoc equity outcomes or collective team dynamics (Hellmann & Thiele, 2015; Wasserman, 2017), we shift focus to the dyadic, decision-making level of equity offers. By using a metric conjoint experiment grounded in resource acquisition and similarity attraction (Lazar et al., 2020; Gray et al., 2024), we respond to recent calls to analyse how these selection strategies operate in tandem (Patzelt et al., 2021; Shah et al., 2019). Our findings suggest that entrepreneurs consider not only what co-founders offer (resources or alignment) but also what they demand (i.e., equity), introducing a novel behavioural dimension to the decision-making around team formation. Second and relatedly, we contribute to research on entrepreneurial decision-making by incorporating equity demand as largely overlooked but important determinant in equity decisions. Prior research has treated equity split as an outcome rather than a dynamic process (Breugst et al., 2015; Chandler et al., 2005). We conceptualize equity demand as a decision factor and empirically show that entrepreneurs respond to it with increased equity offers, supporting a more exchange-oriented view of early team formation. Third, we extend entrepreneurial passion research by illustrating its moderating role in equity decisions, thus shifting attention from the external stakeholder lens (e.g., Fu et al., 2022) to internal venture processes. Specifically, we demonstrate that different passion profiles drive systematic differences in equity offered, particularly in response to interpersonal similarity. Passion for inventing increases generosity toward similar others, while passion for founding induces a more strategic, perhaps protective, equity stance. These findings position EP as an affective-cognitive filter shaping early strategic decisions and invite future work to integrate emotional dispositions more centrally in models of venture formation.

Our findings also carry practical implications for entrepreneurs and aspiring co-founders navigating early equity negotiations. Firstly, aspiring co-founders should recognize that equity demands are not neutral requests but meaningful signals in the formation process. Our findings suggest that lead entrepreneurs tend to reward equity demands. Secondly, lead entrepreneurs should be aware of how their own psychological dispositions (i.e., failure experiences, EP) impacting their offers. For example, entrepreneurs high in EP for founding may risk undervaluing relational alignment. Thirdly, accelerators, investors, and startup advisors should consider incorporating structured decision-support tools (e.g., conjoint exercises, offer simulations) into venture development programs. Such tools can help founders clarify their own selection logics, weigh trade-offs more explicitly, and make more consistent equity offers, reducing post-formation conflict and improving long-term team cohesion and performance.

In conclusion, we hope this study serves as a catalyst for deeper theoretical and empirical engagement with equity allocation as a decision-making process, rather than a post-hoc distributional outcome, in the examination of entrepreneurial team formation.

CONTACT: Yannick N. Amend; yannick.amend@wirtschaft.uni-giessen.de; Licher Straße 74, 35294 Gießen, Germany

LIFE COURSE SYNCHRONIZATION IN ENTREPRENEURIAL TEAMS AND NEW VENTURE SURVIVAL

Mateja Andric, University of St. Gallen, Switzerland and Stockholm School of Economics, Sweden

ABSTRACT

The amount of time and energy entrepreneurs are able and willing to contribute toward the business can change whenever they transition into a new life stage. This creates unique challenges for entrepreneurial teams, given that team members' priorities and contributions to the new venture may shift asynchronously over time if the timing of major life transitions differs between team members. Building on life course theory, we introduce novel theorizing on how the degree of synchronization between entrepreneurial team members' life courses affects new venture survival, thereby jointly theorizing the effects of life course synchronization on both team-level and firm-level survival. We hypothesize an inverted U-shaped relationship between life course synchronization in entrepreneurial teams and the likelihood of both team dissolution and new venture closure, thus arguing for a pronounced likelihood of new venture survival at very low and very high levels of life course synchronization. Drawing on population and firm register data from Statistics Sweden, we find empirical support for our conjectures.

INTRODUCTION

New ventures are often started and run by entrepreneurial teams (Lazar et al., 2020). Accordingly, there is a rich stream of research investigating how entrepreneurial teams can improve their team productivity (cf. Patzelt, Preller, & Breugst, 2021; Yang, Bao, & Aldrich, 2020) and how they should be demographically composed to support the success of the new venture (Jin et al., 2017; Klotz, Hmieleski, Bradley, & Busenitz, 2014). Surprisingly, prior literature has paid very limited attention to the fact that entrepreneurial teams are temporally embedded in individual team members' private life courses, which may or may not evolve synchronously over time. In particular, startups are nowadays often started by teams of young individuals, such as university students or recent graduates (cf. Lazar et al., 2020), who are at an early stage in their lives at the time of new venture foundation and who may subsequently undergo major life transitions, such as starting a relationship, marrying, and having children. Given that life transitions come with new social roles, responsibilities, and life priorities (Elder & Shanahan, 2006), the amount of time and energy entrepreneurs are able and willing to contribute toward the business may change as they transition to a new life stage (Jayawarna, Marlow, & Swail, 2021). In the case of entrepreneurial teams, this leads to the unique challenge that individual team members' capacity and willingness to invest time and effort into the new venture may change asymmetrically over time if the timing of major life transitions differs between team members. However, the simultaneous transition into a new life stage may come with challenges too, such as when the simultaneous entry into parenthood by all team members cumulatively reduces the team's capacity to look after the business. The *relative timing* of each team member's life transitions may thus carry important yet ambiguous consequences for the functioning of the team and thus also for the success and survival of the new venture.

We, therefore, investigate how the degree of temporal overlap between entrepreneurial team members' life courses affects new venture survival, which constitutes an important outcome for new ventures given their high failure rates (Soto-Simeone, Sirén, & Antretter, 2020). Specifically, we build on life course theory (Elder, Johnson, & Crosnoe, 2003) to introduce the concept of life course synchronization, which we define as the degree of similarity in the timing of major life transitions and life stages between members of a team. Team members' life courses may range from being highly synchronous (e.g., when team members transition into parenthood simultaneously), to being highly asynchronous (e.g., when one team member starts a family early whereas the other team members remain single and childless). We jointly examine two survival outcomes that are particularly relevant in the new venture context: The likelihood of team dissolution (i.e., the exit of a team member) as a reflection of team-level survival, and the likelihood of new venture closure as a reflection of firm-level survival. We propose that the degree of life course synchronization among entrepreneurial team members has a curvilinear, inverted U-shaped effect on both the likelihood of new venture team dissolution and new venture closure, thus arguing for an U-shaped relationship between life course synchronization and both team-level and firm-level dimensions of new venture survival. We probe entrepreneur-firm matched register data from Statistics Sweden using sequence analysis techniques and multinomial logit regressions, and find support for our conjectures.

Our study contributes to the literature on entrepreneurial teams (Boss, Dahlander, Ihl, & Jayaraman, 2021; Lazar et al., 2022; Patzelt et al., 2021; Yang et al., 2020) by introducing life course synchronization as an important but to date neglected predictor of new venture survival. We also add to research on entrepreneurial life courses and life events (Andric, Hsueh, Zellweger, & Hatak, 2024; Jayawarna et al., 2021; Yu, Stephan, & Bao, 2023) and literature on the impact of parenthood on entrepreneurship (Delecourt & Fitzpatrick, 2021; Yang, Kacperczyk, & Naldi, 2024) by advancing our understanding of how the interrelation between temporal structures of entrepreneurial team members' lives shapes both team- and firm-level outcomes in new ventures.

HYPOTHESIS DEVELOPMENT

Prior research recognizes that new ventures are embedded in the private lives of entrepreneurs. The family embeddedness perspective, for instance, suggests that the entrepreneur's family exerts a strong influence on the new venture creation process (Aldrich, Brumana, Campopiano, & Minola, 2021; Aldrich & Cliff, 2003). Given that entrepreneurs' family situation varies across different stages of their lives, research suggests that entrepreneurs experience varying levels of work-family conflict and changing work-family priorities over time across their life course, which ultimately also affects how (successfully) they run their business (Davis & Shaver, 2012; Jayawarna et al., 2021; Jennings & McDougald, 2007). Particularly strong life changes are triggered by the transition into parenthood, which tends to impose new family responsibilities and time demands on entrepreneurs that can reduce the amount of energy and time they can devote to the business (Delecourt & Fitzpatrick, 2021; Jennings & McDougald, 2007). Literature has also found that marriage negatively affects women's self-employment earnings (Marshall & Flaig, 2014), suggesting that entering a relationship constitutes another major change in entrepreneurs' lives that can affect their businesses.

While prior research recognizes the importance of entrepreneurs' life events and life stages for new ventures, we to date have a very limited understanding of the implications of entrepreneurs' life courses for new ventures led by entrepreneurial teams, where multiple entrepreneurs with potentially diverging life courses work together. Life course theory (Elder, 1998; Elder et al., 2003) provides a useful basis to understand how the degree of overlap between entrepreneurs' life courses can affect new venture survival. Life course theory conceptualizes human life courses as an interweaving of different age-graded sequences of roles and states that are embedded in social relationships and contexts (Elder, 1994). By providing insights both on social and temporal processes explaining how human lives develop (Elder, 1998; Elder & Shanahan, 2006), life course theory constitutes a holistic theoretical basis for understanding the temporal structure and evolution of entrepreneurial life courses as well as the implications for entrepreneurs' firms. A core principle in life course theory is thereby the notion of "linked lives", which suggests that individuals' lives are linked through social relations. The principle of life linkages is of high relevance in new venture teams, considering that the strength of linkages between entrepreneurs through their co-membership of an entrepreneurial team may subside once they form a family and thus start other, new life linkages with novel responsibilities and priorities. Life course theory also describes the concept of synchronization, suggesting that in order "to cope with simultaneous, linked trajectories, the scheduling of events and obligations becomes a basic task in managing resources and pressures." (Elder & Shanahan, 2006, p. 685). For entrepreneurial teams, the extent to which major life transitions such as entering a relationship and becoming a partner are synchronized between members of the entrepreneurial team may thus determine how well entrepreneurs manage their business, but also how they navigate the dual pressures from the business and the family.

In our theorizing, we focus on the entry into a romantic relationship and the transition into parenthood, given that each of these events involve the addition of a linked life through the emergence of a new relationship (to the partner or the child, respectively). Possible life stages thus range from being a childless single, being in a relationship without children, being a parent in a relationship, over to being a single parent—any change between these life stages thereby constitutes a major life transition in the lives of entrepreneurs. We suggest that the extent to which these life stages and transitions are synchronized between entrepreneurial has a curvilinear effect on both the likelihood of new venture team dissolution and of new venture closure. Specifically, we propose that high levels of synchronization carry benefits in the form of temporally aligned shifts in life priorities between entrepreneurial team members that facilitate ongoing goal alignment as well as mutual understanding and support within the team. The aligned interests are not only beneficial for team cohesion and thus reducing the likelihood of team dissolution, but also reduce the likelihood of new venture closure due to new venture team members' aligned investment of resources and effort towards joint strategic objectives. Benefits also exist at low levels of life course synchronization—or asynchronization—in the form of a temporal diversification of challenging life stages such as early parenthood. Taking turns in moving through different life stages (rather than experiencing the same life stage simultaneously) enables continuity within the team and also at the level of the venture, as team members can step in to support the business when others are undergoing personal challenges. However, at moderate levels of life course synchronization, the benefits of both synchronization and asynchronization diminish. In this intermediate state, misalignments in life priorities and inconsistent mutual support reduce the team's ability to work effectively, leading to lower new venture survival in the form of a higher propensity for both new venture team dissolution and new venture closure.

Hypothesis 1: Life course synchronization has an inverse U-shaped relationship with new venture team dissolution.

Hypothesis 2: Life course synchronization has an inverse U-shaped relationship with new venture closure.

METHOD

We use longitudinal register data from Statistics Sweden covering the time span from 2004 to 2020 and match individual-level population data with firm-level data to identify new ventures that are led by entrepreneurial teams. Specifically, we use data from the individual-level registry LISA (Longitudinell integrationsdatabas för Sjukförsäkrings-och Arbetsmarknadsstudier) to identify individuals who are owner-managers in a firm and subsequently identify firms that are run by two owner-managers at least once during the available observation period (2004-2020). Within this set of firms that are led by an entrepreneurial team, we focus on those that were initially run by two entrepreneurs, as this simple yet also most prevalent two-person setting allows for a straightforward dyadic comparison of two life courses and thus a coherent testing of our hypotheses that are of a dyadic nature. Furthermore, we restrict our sample to firms in which both entrepreneurs were in their early adulthood (at a maximum age of 40) when they first worked together as entrepreneurial team members in the respective firm. We do not impose any restrictions with respect to whether the firm was run by an entrepreneurial team from the beginning on, instead also including firms if they were initially run by only one entrepreneur and subsequently joined by a second entrepreneur. Also, we keep firms in the sample also when further entrepreneurs joined the team at a later point in time. However, in our computation of measures of life course synchronization, we exclusively focus on the life courses of the initial two team members (subsequently referred to as the core dyad) and capture the presence/addition of further team members over time by a corresponding control variable. Importantly, we exclude firms in which the core dyad consisted of couple entrepreneurs, as their life courses are fully synchronized given their romantic relationship.

Using individual-level demographic data from the LISA database for the entrepreneurs in the core dyad, we derive a range of team-level variables, including variables on the demographic composition of the core dyad and variables measuring the level of synchronization between the two entrepreneurs' life courses (see subsequent variable descriptions for details). We match this team-level data with firm-level data from the database RAMS (Registerbaserad arbetsmarknadsstatistik), which allows us to derive measures of firm survival. Our sample comprises almost 40,000 firm-year observations.

Our independent variable is *life course asynchronization* (i.e., the inverse of life course synchronization), which measures the degree of dissimilarity between the life courses of the entrepreneurs in the core dyad. We used the inverse of synchronization rather than synchronization itself because analyses of sequences (including life course sequences) are usually performed based on measures of distance (rather than similarity) between pairs of sequences. We derived this measure using sequence analysis techniques and specifically by creating a distance measure using the optimal matching algorithm (Abbott & Tsay, 2000; Mikolai & Lyons-Amos, 2017). To derive this measure, we characterized entrepreneurs' life courses through four states representing different life stages: (1) single and childless, (2) in cohabiting partnership and childless, (3) in cohabiting partnership and parent, and (4) single parent. We thus created sequences for each dyad member indicating at each year what state the entrepreneur is in, the sequence thereby starting in the year when both entrepreneurs started working as team members in the respective firm for the first time, and ending in the year when both entrepreneurs were working together in the firm for the last time. The life course sequences thus end either due to the exit of one of the members of the core dyad, due to firm closure, or due to discontinuation as a result of merger/sale or division. We then identified for each year in which the firm operated how dissimilar the life courses of the dyad have been up until that year, the degree of dissimilarity thereby corresponding to the optimal matching distance, which we derived using a simple substitution matrix with distances between any two states being set to 1. For instance, a core dyad that has been working together for five years and whose entrepreneurs were in a different life stage from each other in three out of the five years will exhibit a distance value of 3 (this would, for instance, be the case when both were childless singles in the first two years, and one of the two entrepreneurs shifted to having a partner for the last three years, whereas the other entrepreneur remained a childless single). We derived the sequence distance for each year in which the firm operated and only focused on teams that have existed for at least three years to have a sufficient length of sequences. Finally, we normalized the sequence distance by sequence length by dividing the distance (3 in the previous example) by the length of the sequence as of the year in question (5 in the previous example). The final measure of life course asynchronization thus take values ranging from 0, reflecting full overlap in life courses and thus complete life course synchronization, to 1, reflecting zero overlap in life courses in the core dyad and thus complete life course asynchronization.

For our dependent variable, we used a categorical variable measuring whether the sequence continues next year (i.e., whether the firm continues to operate with both entrepreneurs from the core dyad) or whether the sequence ends next year due to team member exit, firm closure, or other reasons of discontinuation. Specifically, this variable takes the value of 0 for firms that continue to operate and where both entrepreneurs from the core dyad remain in the team next year; 1 for firms that continue to operate but where the team dissolves due to one of the two members of the core dyad exiting the firm; 2 for firms that close down; and 3 for firms that cease to exist due to a merger or division. The latter category is not of interest with respect to our hypotheses but nevertheless separately measured and included to establish a clear comparison between the two outcomes of interest (team member exit and firm closure) relative to the counterfactual of continuation.

We included the following control variables: (1) a categorical variable measuring whether 0, 1, or 2 out of the two entrepreneurs have had a child as per sequence start to capture variation *within* groups of teams that had similar initial conditions in terms of presence or absence of children; (2) a dummy variable indicating whether any of the entrepreneurs has had a newborn child recently (in this year or the last year), to avoid confounding by heightened work-family conflict arising from having to care for a newborn baby; (3) the difference in years of education between the members of the core dyad; (4) the maximum age in the core dyad as per sequence start year to account for the fact that variation in life course asynchronization may be lower for older-aged entrepreneurs who already have a partner and child at the start of the sequence; (5) a categorical variable measuring the gender constellation (i.e., whether the core dyad consists of 0, 1, or 2 female entrepreneurs); (6) number of entrepreneurial team members, to account for whether any further entrepreneurs joined the team; (7) sequence length, to account for systematic differences in sequence distance across sequences of different length; (8) a dummy variable indicating whether the firm existed before the sequence started, to account for whether the firm was founded by the core dyad; (9) a dummy variable indicating whether the firm was initially run by only one entrepreneur before the other entrepreneur joined; (10) a dummy variable indicating whether the firm is incorporated; (11) firm size (log of number of employees), and (12) firm age.

We ran multinomial logit models with cluster-robust standard errors (at the firm level) to test the effect of life course asynchronization on next year's likelihood of team dissolution and firm closure. Given that we theorize a curvilinear effect, we include both life course asynchronization and its quadratic in the model in order to estimate a quadratic polynomial function.

RESULTS

Our preliminary analyses suggest an inverted U-shaped relationship between life course asynchronization and new venture team dissolution, in support of Hypothesis 1. Furthermore, we also find evidence for an inverted U-shaped relationship between life course asynchronization and new venture closure, in support of Hypothesis 2.

Our results stay robust when defining partnership more strictly as being in a registered partnership or being married (as opposed to the looser definition of being a cohabiting couple). Also, our results remain robust when using alternative delineations of life stages, such as when distinguishing states based on (a) the presence versus absence of children; (b) the number of children the entrepreneur has; and (c) the presence versus absence of partner. Our results also remain robust when running the analysis only on the subset of firms where team size never exceeded two members; when performing the analyses only among firms whose team members initially had zero children (i.e., who started in a similarly early life stage before settling and starting a family); and when performing our analysis only among teams in which at least one member has already had a child (suggesting that our results are not confounded by the mere presence versus absence of any entrepreneurs that are parents). Finally, if our theorizing holds, the theorized curvilinear effects should be weaker in the case of female-only teams, since women entrepreneurs may be socialized to be more understanding of each other's life circumstances and the work-family integration challenges associated with important life transitions such as parenthood. Indeed, we find that our results are primarily driven by core dyads that are male-only ore of mixed gender, as opposed to female-only teams.

DISCUSSION AND IMPLICATIONS

By introducing novel theory and evidence on the relation between life course synchronization and new venture survival, we offer several contributions to literature.

First, our study contributes to research on entrepreneurial teams (Patzelt et al., 2021; Preller, Patzelt, & Breugst, 2020) by introducing life course synchronization in new venture teams as a novel predictor of new venture survival. We shift the attention away from static demographic team member characteristics toward the dynamics inherent in entrepreneurial team members' lives, and point to the importance of the relative timing of team members' private life transitions for the longevity of new venture teams and their businesses. While many entrepreneurial teams are formed based on friendship, homophily, or match in functional experience (cf. Ruef, Aldrich, & Carter, 2003), our result suggest that an initial demographic and functional fit between team members may not be sufficient for building a successful and surviving business, as life courses of team members may develop in diverging ways that can be harmful to team and firm survival if teams get "stuck in the middle" at intermediate levels of life course synchronization. In this vein, our findings suggest that evaluating the mutual fit based on future anticipated life courses at the team formation phase may be a valuable criterion for entrepreneurial teams to assess whether their life courses are mutually compatible in terms of either high or low (but not intermediate) life course synchronization.

Second, we contribute to the literature on entrepreneurial life courses and life events (Andric et al., 2024; Jayawarna et al., 2021; Yu et al., 2023) as well as to research on the effects of parenthood on entrepreneurship (Delecourt & Fitzpatrick, 2021; Yang et al., 2024). Whereas prior studies have focused on life events and life stages of individual entrepreneurs, our study suggests that the implications of entrepreneurial life courses for new ventures are more complex and nuanced in team-led ventures. Specifically, our findings show that not only the life situation of the individual entrepreneur, but also the timing of life transitions relative to their team members' life transitions matters for new venture outcomes. This insight also suggests that starting a business in a team instead of as a sole entrepreneur may be a viable alternative for individuals to ensure continuity in their business in the face of challenging phases in their private lives, provided entrepreneurial team members mutually coordinate the timing of important life transitions.

Finally, our study also contributes to practice by informing entrepreneurial team members to what extent they should synchronize their life transitions to increase the likelihood of survival of their new venture team and firm. As such, we introduce life course synchronization as a tool through which entrepreneurial teams can better navigate their work-life interface and more effectively manage their businesses in the face of changing life circumstances.

CONTACT: Mateja Andric; mateja.andric@unisg.ch; University of St.Gallen; Dufourstrasse 40a, 9000 St. Gallen, Switzerland.

TYING A KNOT: A PROCESS MODEL OF ORGANIZATIONAL CULTURE CREATION IN NEW VENTURE TEAMS

David Nawrath, Technical University of Munich, Germany

ABSTRACT

Organizational culture is critical for long-term success, yet its creation in new ventures remains poorly understood. Addressing this gap, we conducted a longitudinal multiple-case study of ten team-founded ventures and propose the *Knot Model of Organizational Culture Creation*. The model shows how teams navigate misalignments through *looping* (discussion), *wrapping* (refinement), and *cinching* (formalization), weaving together diverse perspectives, industry norms, and external pressures. When founders prioritize personal imprinting, a process we call *self-mirroring entrepreneurship*, the team enters a *pairing* stage that disrupts cohesion and fosters sub-cultures. Our study underscores the importance of team cognition for cohesive culture creation and contributes to research on organizational culture, emergent organizing, and team dynamics in early ventures.

INTRODUCTION

When a new venture forms, founding teams engage in emergent organizing activities that shape its long-term trajectory (Lichtenstein, Dooley, & Lumpkin, 2006). One of the most critical is culture creation, defined as shared values and norms that guide behavior (O'Reilly & Chatman, 1996). Culture forms an organization's enduring core (Pardo-del-Val & Martinez-Fuentes, 2003), shaping beliefs, behaviors (Schein, 1983), innovation (Russell, 1989), knowledge sharing (Azeem et al., 2021), and even investor appeal (Hallen & Eisenhardt, 2012).

Although culture's importance is well established, most studies focus on established firms (Kim, Toh, & Baik, 2022) or solo founders embedding personal values (Pettigrew, 1979), overlooking team-founded ventures. Yet such teams, where two or more individuals co-found, co-manage, and co-own (Lazar et al., 2020), are increasingly common (Clarysse & Moray, 2004). They bring cognitive diversity (Shepherd, Breugst, & Patzelt, 2023), but we know little about how they collectively create culture.

To address this, we conducted a longitudinal, inductive study of ten team-founded ventures, collecting 306 interviews and supplementary data over 9-13 months per venture. We find that culture creation unfolds in two phases: individual reflection, followed by a team-level process triggered by external events. Teams iterate through *looping* (discussion), *wrapping* (refinement), and *cinching* (formalization) to align diverse views. When founders lead with personal imprinting, a dynamic we term *pairing*, cohesion suffers, and sub-cultures emerge.

Our *Knot Model of Culture Creation* shows how team dynamics, founder values, and industry norms interweave to shape culture. The outcome depends on emerging team cognition (Grand et al., 2016; Kozlowski & Chao, 2012): high cognition fosters alignment, low cognition leads to fragmentation.

We contribute to (1) culture creation (e.g., Kim et al., 2022; Schein, 1983), (2) emergent organizing (e.g., Shepherd, Souitaris, & Gruber, 2020), and (3) team cognition, by showing how collective culture creation unfolds through high-stakes interpersonal integration.

THEORETICAL FOUNDATION

Organizational Culture Creation in New Ventures. Organizational culture guides behavior and shapes internal and external dynamics (Schein & Schein, 2016). While well-studied in established firms, where leaders embed values into stable structures (Pettigrew, 1979; Kim et al., 2022), culture in new ventures must be built from scratch amid uncertainty (Shepherd, Breugst, & Patzelt, 2023). Solo founder studies emphasize personal imprinting (Schein, 1983; O'Reilly et al., 2014), but team-founded ventures require shared cultural construction, shaped by social interaction and grounding mechanisms (Echterhoff, Higgins, & John, 2009; Kashima, 2016). Although recent theory highlights team-based culture creation (Shepherd et al., 2023), empirical insights into these dynamics remain scarce.

Emergent Organizing in Founding Teams. Founding teams are increasingly common and often outperform solo founders (Lechler, 2001). Their success depends on integrating diverse personalities and skills (Howell, Bingham, & Hendricks, 2022). Early-stage collaboration involves emergent organizing: configuring relationships and routines in real time (Shepherd, Souitaris, & Gruber, 2020). These involve task-related processes like strategy (West, 2007) and interpersonal processes that influence cohesion and longevity (Marks et al., 2001). Since founding teams lack formal structure, members must self-manage relationships (Weick, 1988; Lanzara, 2007). Yet, empirical work on interpersonal dynamics in founding teams is limited (Breugst & Preller, 2021).

Team Cognition Emergence. Effective teamwork requires shared cognition, how teams organize, integrate, and apply knowledge (Toader & Martin, 2023; Cannon-Bowers & Salas, 2001). Although widely acknowledged, little research explores how team cognition develops over time (Fiore et al., 2010; Kozlowski & Klein, 2000). It emerges through dynamic interactions, where individual perspectives merge (Morgeson & Hofmann, 1999; Cooke, 2015). Grand et al., (2016) show that iterative knowledge sharing builds task-related cognition but neglect interpersonal processes. Culture creation offers a lens to understand how shared cognition forms through team interaction (Ashforth, Rogers, & Corley, 2011; Hong et al., 2000), transitioning from "I think" to "we think" (Kashima, 2016).

METHOD

To explore how founding teams create organizational culture, we used an inductive, longitudinal multiple-case study design (Denzin & Lincoln, 2011; Langley, 1999). This approach enabled us to capture cultural dynamics over time and compare across different team contexts (Langley et al., 2013).
Sampling. Our sampling followed four steps. First, we targeted emerging ventures, defined as having completed at least one of two gestation activities in the past 12 months (Davidsson & Gruenhagen, 2020), assuming they had not yet developed an organizational culture. We screened five incubators in a leading European ecosystem, as these environments often support venture gestation (Liao, Welsch, & Tan, 2005). Second, we focused on team-founded ventures, excluding solo founders to study collective culture creation (Lazar et al., 2020). Third, we contacted 68 eligible teams and received initial agreement from 19. To reduce impression management bias (Gordon & DiTomaso, 1992), we included only those where all founding team members agreed to participate. This led to 13 ventures with 36 founders in total. The ventures were, on average, 2.4 months old, had 2-4 founders (mean = 2.8), and 1–3 employees (mean = 2.0). All operated in IT-related industries and were in early development stages (Kazanjian, 1988).

Data Collection. Our main data source was monthly semi-structured interviews with all founders over 9–13 months (avg. 10.5), continuing until teams consistently articulated shared cultural elements. We adjusted our interview guide iteratively (Gioia et al., 2013), initially exploring individual cultural beliefs, later focusing on team-level codification and alignment. We conducted 306 interviews (avg. 34 minutes), yielding over 170 hours of audio and 3,200 single-spaced transcript pages. Interviews were mostly conducted in German and translated by one author and three assistants, two of whom were native English speakers.

In addition, we used non-participant observation. We visited each venture one to two times per month, attending formal and informal team interactions, such as meetings, workshops, and lunch breaks. These visits ranged from 30 minutes to multiple days and helped contextualize team dynamics and validate interview insights. Observations were documented in field notes.

We also collected secondary data (Yin, 2009), including internal documents (e.g., business plans, pitch decks), meeting notes, and public materials (e.g., social media). This helped triangulate data, validate accounts, and construct timelines for each venture's culture creation.

Data Analysis. We analyzed data using an inductive process study approach (Langley, 1999), starting with identifying key activities in each venture's cultural development. We created timelines that traced external triggers, team responses, and the evolution of norms and values. We used open coding at the sentence and paragraph level to identify early cultural actions and misalignment discussions. First-order codes included actions like "aligning with industry norms". Next, we analyzed team dynamics to understand mechanisms driving cohesion or fragmentation. Teams with high cognitive alignment emphasized collaboration ("constructive problem-solving," "openness to disagreement"), while fragmented teams revealed conflicting narratives. This stage highlighted the central role of emergent team cognition.

Across cases, we identified 81 recurring first-order codes, which we refined iteratively. To ensure rigor, we triangulated data types: interviews gave access to individual perspectives, observations revealed team interactions, and documents confirmed event timing. We also bracketed our assumptions and actively looked for disconfirming evidence. One unexpected insight was that misalignment sometimes manifested through subtle behavior, such as selectively promoting certain values in public communication. We then grouped the 81 codes into 25 second-order themes using organizing strategies (Langley, 1999; Strauss & Corbin, 1998). For example, "alignment with stakeholders' expectations" and "strategic commitment to values" formed the theme *cinching*. We used narrative summaries and visual timelines to compare across ventures and identify shared patterns and divergent paths. Finally, we developed six aggregate dimensions, capturing how individual and environmental factors translated into collective cultural dynamics. Temporal bracketing helped us visualize sequences of action (Langley, 1999), while model iteration refined our theoretical insights (York, Hargrave, & Pacheco, 2015). Illustration 1 provides an overview of our data structure. To ensure reliability, two independent coders reviewed the data structure. We achieved 81.2% interrater agreement, resolving discrepancies through discussion (Fauchart & Gruber, 2011). Theoretical saturation was reached after the ten primary cases (see Illustration 2); revisiting the three excluded cases yielded no new themes (Suddaby, 2006).

RESULTS

Creating a new venture's culture unfolds in two stages: individual and team-driven culture creation. Initially, founders individually form cultural aspirations, shaped by personality and industry norms. These remain unspoken until an external trigger prompts shared discussion, marking the beginning of the team-driven phase. The *Knot Model of Culture Creation* captures this shift, consisting of the stages of *pairing*, *looping*, *wrapping*, and *cinching* (see Illustration 3).

Phase I - Individual Culture Creation. Culture creation begins subtly, with each founder imagining what the venture should stand for. These aspirations often reflect personal values, David emphasizes "quality is non-negotiable" (Int. II), while Jin imagines the culture to "be who I am" (Int. III). Yet not all founders seek a perfect match between personality and culture. Batmid, for instance, excludes personal risk-taking: "I want this venture to be curious about innovation… but not focus on taking too much risk" (Int. I). Founders also assess how their values fit with industry norms. Diedrich recognizes the need to adapt: "We need to be agile in everything we do… I hate to adopt but… I do not have a choice" (Int. II). While some adjust values accordingly, others, like Marisa, struggle with inner conflict: "I do not know if this… will lead to a conflict in me" (Int. III). These aspirations remain uncoordinated until an external trigger, such as investor critique or customer confusion, initiates discussion. Arushi recalls, "We realized that we should speak about the values we are portraying" (Int. II). With that, the team culture creation phase begins.

Phase II - Team Culture Creation: Once individual aspirations are voiced, misalignments between founders become visible. Ilias notes, "Ines and I have fundamentally different views about how the culture at NAG should look like" (Int. IV). Teams that navigate these differences constructively rely on emerging team cognition, the ability to align diverse perspectives through trust and collaboration. Darius captures this dynamic: "We know our differences… we can do it" (Int. IV). Trust and shared purpose support this alignment. "We genuinely enjoy spending time together," Bob shares (Int. III), while Marisa recalls, "We felt we could rely on one another… not just for our skills, but for the way we work and communicate" (Int. II). In contrast, teams built primarily on complementary skills often struggle. As Niklas explains, "We were brought together… not necessarily because we shared similar values" (Int. IV).

In teams with low team cognition, cultural alignment often begins with *pairing*, as founders with similar priorities informally ally to push their vision. "I needed backup to push my vision through," Nizwa recalls (Int. IV). These subgroups quietly shape cultural elements within specific domains, often without team-wide consensus.

As culture becomes a shared agenda, all teams engage in *looping*, an iterative exchange aimed at aligning values. For some, this marks the beginning of culture creation; for others, it follows earlier subgroup dynamics. Founders begin to shift focus from individual ideals to external demands. "Personal wishes… did not play any role," Batmid reflects (Int. III). Imke adds, "We also have (future) investors to satisfy" (Int. VI).

Next, teams move into *wrapping*, where cultural elements are tested and refined through day-to-day work. This phase reveals tensions between agreed values and actual behavior. "We had problems following [the values]," Benjamin admits (Int. VII). Darius expresses discomfort: "I felt that I need to act like I am a different person" (Int. V). These frictions prompt teams to clarify or adapt norms, balancing idealism with practical fit.

Finally, in *cinching*, teams consolidate and operationalize the culture, aligning it with strategic goals and stakeholder expectations. Elias reflects, "The team is pulling together… compared to the early days" (Int. VII). Norms are revised to improve external resonance, such as Elke's team modifying a value to appeal to applicants (Int. VIII). Once finalized, the culture is made visible and rarely changed. Dennis captures this sense of closure: "You need to finish up this whole thing… and decide what you want to stand for" (Int. IX).

The outcomes of the culture creation process varied depending on the level of emerging team cognition. Teams with high alignment developed a collective culture that reflected shared values and was openly communicated across internal and external audiences. Founders codified these values through visible artifacts, such as screensavers or pitch decks, and expressed a strong sense of ownership and pride. As Dennis put it, "At some point, you need to finish up this whole thing… and decide what you want to stand for" (Int. IX). In contrast, teams with unresolved tensions created a fragmented culture, in which a dominant cultural narrative coexisted with a more discreet hidden subculture maintained by a subgroup of founders. These founders subtly advanced their own priorities in areas like hiring or performance evaluation, often without the broader team's awareness. "We do not celebrate that we developed other principles… we implement them in the dark," Clarisse admitted (Int. IX). While the primary culture served as the public face of the venture, the hidden subculture allowed some founders to continue pursuing their personal vision beneath the surface.

DISCUSSION AND IMPLICATIONS

This study examined how founding teams create organizational culture in new ventures. We found that culture creation is typically triggered by external events and shaped more by industry norms than by individual personalities. While some founders seek to imprint personal values, this often leads to fragmentation, particularly in skill-driven teams with low emerging team cognition.

Organizational Culture Creation in New Ventures. We contribute to the culture creation literature (e.g., Schein, 1983; Kim et al., 2022; Shepherd et al., 2023) by offering a process model that shows how culture is collectively constructed in early-stage teams. Unlike research on established firms emphasizing leader-driven values (Pettigrew, 1979), we show that early culture reflects stakeholder demands and is negotiated across stages, *looping*, *wrapping*, and *cinching*. In low cognition teams, *pairing* leads to subgroups and hidden subcultures, challenging the dominant culture (Jermier et al., 1991). Our findings show subcultures can emerge early, driven by individual values rather than structural divides.

Emergent Organizing. We extend emergent organizing theory (Lichtenstein et al., 2006; Shepherd et al., 2020) by showing how interpersonal processes arise in response to external triggers. Culture emerges as a lasting social mechanism, not from formal structures (Courtright et al., 2017), but through dynamic team interactions. We introduce *self-mirroring entrepreneurship* to describe how some founders assert identity through culture, often at odds with team alignment.

Team Cognition Emergence. Building on Grand et al., (2016) and Kozlowski & Chao (2012), we show that team cognition develops through culture creation. It may be shared or configural, depending on team alignment. In some cases, fabricated cognition masks disagreement, creating cognitive faultlines (Lau & Murnighan, 1998). Our findings highlight that interaction quality, not frequency (Pearsall et al., 2010), drives cognitive integration.

Practical Implications. Founding teams should surface cultural aspirations early to prevent misalignment and subcultures. Early conversations support collective cognition and cultural cohesion. Stakeholders should influence culture during formative phases, when it is most open to shaping (Rauch, 2025).

CONTACT: David Nawrath; david.nawrath@tum.de; Technical University of Munich; Arcisstr. 21, 80333 Munich, Germany

ILLUSTRATIONS

Illustration 1: Data Structure

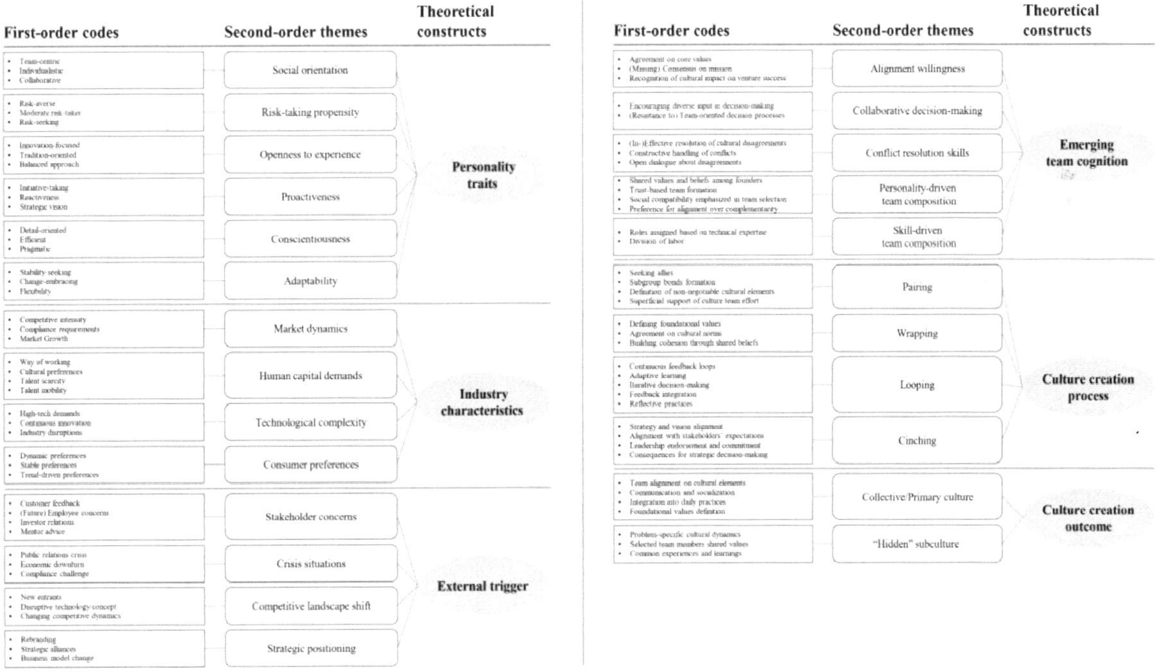

Illustration 2: Sampling Overview

Venture	Name	Age	Gender	Educational Background	Founding Year	Industry	# of Employees	Culture Creation Pathway
TOK	Anna	23	Female	Business	2023	Software	1	Collective culture creation
	Arushi	26	Female	Engineering				
OSK	Batmid	29	Male	Engineering	2022	Finance	2	Collective culture creation
	Benjamin	31	Male	Engineering				
	Bob	26	Male	Humanities				
KYO	Claus	28	Male	Business	2023	Software	1	Fragmented culture creation
	Candice	28	Female	Computer Science				
	Clarisse	25	Female	Business				
	Cosimo	27	Male	Computer Science				
BEP	Darius	34	Male	Business	2023	Software	3	Collective culture creation
	David	29	Male	Engineering				
	Dennis	31	Male	Law				
	Diedrich	29	Male	Engineering				
HIR	Elke	26	Female	Engineering	2022	Biotech	5	Collective culture creation
	Elias	25	Male	Business				
FUK	Friedrich	29	Male	Computer Science	2023	Education	1	Fragmented culture creation
	Fritz	32	Male	Engineering				
NAG	Ilias	31	Female	Business	2023	Energy	3	Collective culture creation
	Imke	35	Female	Economics				
	Ines	29	Female	Business				
HAK	Jakob	29	Male	Engineering	2023	Energy	0	Fragmented culture creation
	Jin	25	Female	Architecture				
	Jonas	24	Male	Engineering				
ASA	Mahmoud	30	Male	Economics	2022	Life Science	1	Collective culture creation
	Maria	27	Female	Computer Science				
NAH	Nils	34	Male	Computer Science	2023	Legal	3	Fragmented culture creation
	Niklas	29	Male	Computer Science				
	Nizwa	33	Female	Business				

Illustration 3: The Knot Model of Organizational Culture Creation

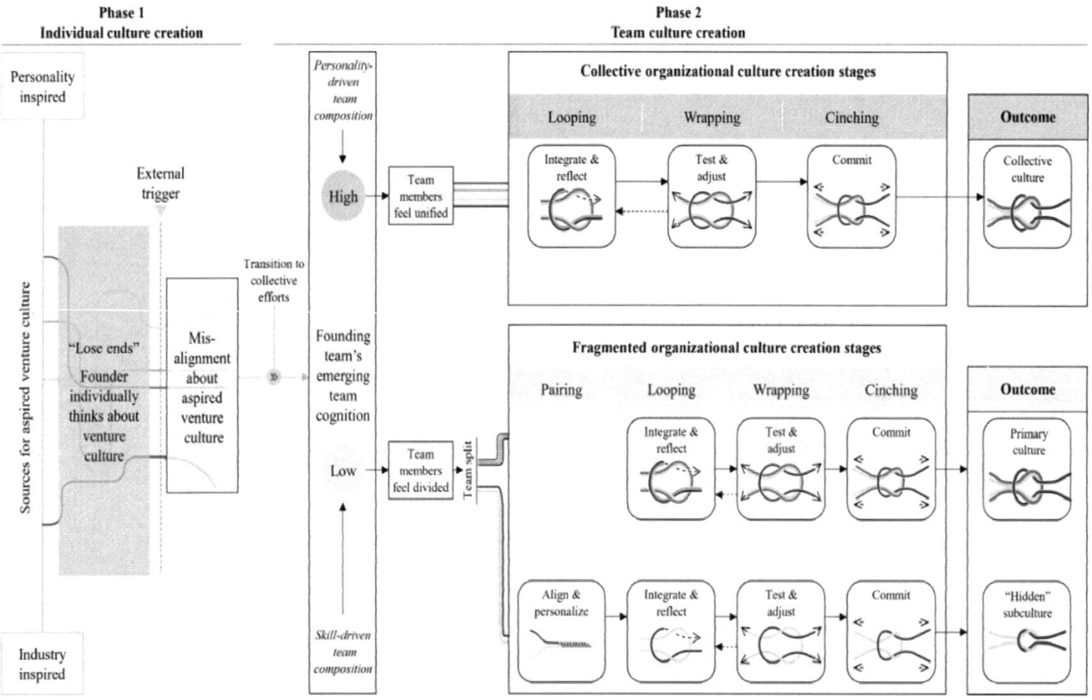

SUMMARY

LOSS OR GAIN SPIRAL? THE INFLUENCE OF ONLINE SOCIAL SUPPORT ON THE EXCHANGE OF OFFLINE SOCIAL SUPPORT BETWEEN CO-FOUNDERS AND THEIR WELL-BEING

Marc Dal-Canton, Rhein Main University of Applied Science, Germany
Gunnar Lang, RheinMain University of Applied Sciences, Germany
Mette Søgaard Nielsen, University of Southern Denmark, Denmark
Ute Stephan, King's College London, United Kingdom
Theresa Treffers, Seeburg Castle University, Austria

Principal Topic

The well-being of entrepreneurs affects the future of their ventures (Gielnik et al., 2012), yet research on their well-being remains 'under-socialized' (Stephan, 2018). Despite a major shift of entrepreneurial work to the digital space (Olanrejawu et al., 2020), the potential drawbacks of social media –where entrepreneurs can gain or lose social support– have been neglected in current research (Chen & Liu, 2023). Using the Conservation of Resources Theory (COR) (Hobfoll, 1989), we propose that entrepreneurs with low (high) online social support are more likely to reduce (increase) their investment in both online and offline social support, impacting their co-founders. Since provided social support depends on the support received (Halbesleben, 2015), the co-founders will reciprocate by reducing (increasing) their provided social support, creating a loss (gain) spiral. As social support predicts well-being, we suggest that this loss (gain) spiral will further decline (increase) the co-founder's well-being over time.

Method

We conducted an Experience Sampling Method (ESM) study with dyadic entrepreneurial teams (N = 24). They completed weekly questionnaires for 5 weeks regarding their well-being and online / offline social support. We measured well-being using the WHO well-being index (Brähler et al., 2007) and adapted versions of the Social Support Scale (Santiago et al., 2023) for online and offline social support. We used Latent Change Score Modeling (LCS) to examine the development of gain and loss spirals over time. In particular, we examined the relationship between the change in one variable (entrepreneurs online social support) and the development of another variable (co-founders offline social support and well-being).

Results and Implications

Our current results do not show a significant relationship between the entrepreneur's online social support and the development of the co-founders offline support and well-beings. In contradiction to COR, which postulates that losses (gains) in resources cause loss (gain) spirals. This could be explained by the fact that online and offline social support not only differ in the way that they are transported but that there are also perceived as separate resources. Another possible explanation is the currently limited sample size but we are actively continuing our data collection process. Since entrepreneurs often use social media daily, these findings are of high practical relevance as they show that the online social support of an entrepreneur does potentially not affect the relationship with the co-founders and their well-being.

CONTACT: Marc Dal-Canton; marc.canton@hs-rm.de; Rhein Main University of Applied Science; König-Adolf-Straße, Wiesbaden, Hessen 65191, Germany

SUMMARY

PREDICTING CO-FOUNDER EXITS: A FRAMEWORK FOR UNDERSTANDING SUBTLE INTERACTION CUES

Jennifer Ettner, Technical University Munich, Germany
Tessa West, New York University, United States

Principal Topic

Although teams possess qualities that enhance venture success (Ensley et al., 2006), solo-founded ventures often outperform team-founded ones (Greenberg & Mollick, 2018). Poor collaboration and early co-founder exits may explain this paradox, as early co-founder exits jeopardize the success of the company (Guenther et al., 2016). Although research highlights the role of conflict, particularly in a venture's early stages (Brattström, 2024; Ensley et al., 2002) studies on the subtle behavioral cues that shape these conflicts – such as interruptions or speech distribution – remain scarce in entrepreneurial literature. This study introduces a framework linking these subtle interactional cues to venture outcomes, offering new insights into the dynamics of entrepreneurial teams.

Method

To examine real-life interactional behaviors, data were collected from 111 dyadic new venture teams. Founders engaged in a video-recorded conflict discussion and completed questionnaires measuring team satisfaction (adapted from Foo et al., 2006) and exit intentions at the time of the lab session, and three and six months later. Behavioral cues, including talk time and interruptions, were coded for the first two minutes, as prior research indicates that 1.5-minute slices are sufficient to capture nonverbal patterns (Murphy et al., 2015).

Results and Implications

Three months after the discussion, 63.8% of participants had never considered leaving the company, while 5.5% of teams had dissolved. By six months, 12.2% had dissolved, and only 44.8% had never considered leaving. Using an Actor-Partner Interdependence Model for dyadic repeated measures, we examined satisfaction at the study's outset and exit intentions after three months as predictors of turn-taking duration. We found two partner effects: the longer a respondent's co-founder talked during that person's turns, the more satisfied the respondent was, wald χ^2 (1) = 11.435, p < .001 (B = .1811, SE = .536) and the less likely they wanted to leave the venture wald χ^2 = 4.182, p .= 041 (B = -177.54, SE = 86.82). No actor effects or effects related to relative talk length or talk overlap were found. These findings suggest that allowing a co-founder to hold the floor is linked to positive short- and long-term outcomes. Our findings have important implications for entrepreneurship scholars, founders, and investors. For scholars (e.g., Breugst et al., 2015; Ensley et al., 2002), this study highlights how subtle interactional cues like allowing a partner to finish impact venture progress. For founders, it underscores the benefits of letting co-founders speak uninterrupted. For investors, respectful turn-taking may signal promising venture teams.

CONTACT: Jennifer Ettner; jennifer.ettner@tum.de; Technical University Munich; Oettingenstr. 33A, Munich, Bayern 80538, Germany

SUMMARY

FRIENDS AND FOES IN NETWORKS OF NECESSITY ENTREPRENEURS

Johanna Fischer, Leibniz University Hanover, Germany
Julia Brennecke, University of Potsdam/University of Liverpool, Germany
Christiana Weber, Leibniz University Hanover, Germany

Principal Topic

Network-based research in entrepreneurship has extensively examined the beneficial role of positive social network ties for entrepreneurial activities and performance (Hoang & Antoncic, 2003; Elfring et al., 2021; van Burg et al., 2022). However, according to social ledger theory and insights from general management research, understanding work-related realities requires considering both positive and negative ties (Labianca & Brass, 2006). The limited focus on negative ties in entrepreneurship research is problematic, as insights from general management may not directly translate to entrepreneurial or specific cultural settings. This is particularly relevant in Sub-Saharan Africa, where especially necessity entrepreneurs rely heavily on social networks due to weak formal institutions (Sydow et al., 2022). Therefore, we investigate the following research question: How and why do positive and negative network ties lead to different kinds of financial and non-financial outcomes in the context of necessity entrepreneurship in non-Western settings?

Method

To address our research question, we conducted an exploratory qualitative study in six rural Tanzanian villages, including 31 focus group discussions with 176 necessity entrepreneurs, alongside expert and village representative interviews. The qualitative data were analyzed according to grounded theory. Additionally, we collected ego-network data from 789 necessity entrepreneurs, employing network mapping techniques to enhance data reliability.

Results and Implications

Our findings indicate a high prevalence of negative ties among necessity entrepreneurs. Ego-network data reveal that, on average, 32% of network ties are negative—significantly higher than in typical management contexts (Labianca & Brass, 2006). Through our qualitative analysis, we identified various types of (affective, behavioral, and cognitive) positive, negative, and multiplex ties, and investigated what they mean for the entrepreneurs in our context. Subsequently, we uncovered several context-specific mechanisms that explain how these different ties impact financial and non-financial outcomes, like life satisfaction. A key mechanism is the "crabs-in-a-bucket" mentality, where affective positive ties, such as friendships, lead to negative financial consequences. For example, within the Gogo culture, many individuals withhold support out of fear that others will outshine them. This illustrates how even positive ties can turn into liabilities. Our study challenges and extends social ledger theory by demonstrating that positive ties can impose liabilities, while negative ties may yield benefits. This nuanced perspective enhances scholarly knowledge on how exactly ties influence entrepreneurial outcomes. Additionally, we contribute to necessity entrepreneurship research by introducing social ledger theory to the debate on theorizing necessity entrepreneurship.

CONTACT: Johanna Fischer; johanna.fischer@ufo.uni-hannover.de; Leibniz University Hanover; Koenigsworther Platz 1, Hanover, Niedersachsen 30167, Germany

SUMMARY

ENTREPRENEURIAL FINANCING: HUMAN CAPITAL DIVERSITY AND GENDER COMPOSITION OF NEW VENTURE TEAMS

Lani Faith Gacula, University of Massachusetts Lowell, United States

Principal Topic

Despite extant literature on how solo founders affect funding performance, the effect of founding teams is less understood. Financing is critical for the success of new ventures (Zhang et al., 2010; Ko & McKelvie, 2018). Adequate financing enables investment in essential resources (Fraser et al., 2015), increases survivability (Robb & Robinson, 2014), improves the likelihood of profitability and signals credibility (Colombo, 2021). We draw upon information processing theory regarding the acquisition, exchange and application of information within teams (Deeter-Schmelz & Ramsey, 2003), considering gender composition, to examine how founding teams leverage human capital to obtain financing.

Method

This study employed logit regression analysis to investigate the relationship between independent variables and funding success, modeling average and diversity of both education and industry experience, and gender composition. Hypotheses were tested using data from the Entrepreneurship Database Program at Emory University (2013-2019), including 13,702 for-profit entities across 168 countries.

Results and Implications

Results indicate that new venture founding teams with higher average education levels are more likely to secure funding. This suggests that investors view higher educated teams as more likely to succeed. Founding teams with greater education-level diversity are only slightly more likely to secure funding, indicating that capital providers may consider a range of education levels as producing a more dynamic and functional team. In contrast, teams with higher average industry experience are less likely to secure funding. Though counterintuitive, founding teams with high average industry experience may lead potential investors to question the entrepreneurial abilities of the team. A diverse range of industry experience within a founding team shows no significant impact on funding success, underscoring that investors may place less weight on industry experience overall. Lastly, all-female founding teams are less likely than mixed-gender teams to achieve funding success, while both are less likely than all-male teams to achieve funding success. This suggests a persistence of gender bias in the funding decision-making process, with all-male teams being viewed most favorably. This study contributes to information processing theory, adding to our understanding of how founding teams may benefit from greater information processing capabilities achieved through higher and more varied perspectives. This study also contributes to the literature on human capital highlighting its crucial role within new venture teams. Human capital enriches entrepreneurial teams by bridging knowledge gaps that enhance decision-making and improve team dynamics, thereby increasing the likelihood of funding success (Çağa et al., 2024).

CONTACT: Lani Faith Gacula; lanifaith.gacula@student.uml.edu; University of Massachusetts Lowell; 922 North Street, Tewksbury, Massachusetts 01876, United States

SUMMARY

ARE WE REALLY IN LOVE? UNRAVELLING MUTUAL ATTRACTION IN CO-FOUNDER SEARCH AND SELECTION IN ENTREPRENEURIAL TEAMS

Dorothea Georgiadou, Technical University of Denmark, Denmark
Anna Brattström, Lund University, Sweden
Argyro (Iro) Nikiforou, Technical University of Denmark, Denmark

Principal Topic

Choosing the right co-founder is crucial for entrepreneurs with growth ambitions (Forbes et al., 2006; Lazar et al., 2020), as selecting the wrong one or failing to find a suitable match can severely hamper a venture's performance (Brattström, 2024). Despite its importance, we know surprisingly little about the actual matchmaking (dating) process between initial and potential co-founders and the critical stopping rules that ultimately determine whether they "fall in love" and decide to partner or part ways. Addressing this need for theory development, we ask: how does the mutual attraction of initial and potential co-founders shape the matchmaking process, and how do they select one another?

Method

We conducted an inductive, qualitative case study involving seven startups originating from a Scandinavian university. We collected data through 27 semi-structured interviews with initial and potential co-founders at three key points: (1) before their initial contact, (2) after each interaction, and (3) at the decision to team up or not. We followed this process to the extent possible for four cases, while we collected data retrospectively for three cases. To triangulate the data, we conducted 4 interviews with two university technology transfer facilitators.

Results and Implications

Given the importance of the co-founder selection (Forbes et al., 2006; Patzelt et al., 2021), one would expect initial founders to invest significant time and effort. However, our preliminary analysis reveals that initial founders often procrastinate the co-founder selection, only initiating the search under significant time pressure. Instead of systematically and deliberately searching for potential co-founders, initial founders often prioritize immediate availability over finding a co-founder with the 'ideal' founder characteristics and they go for the first-best alternative, following a satisficing search (Simon, 1955). While the intensity and duration of interactions between initial and potential co-founders vary, the dating process is largely driven by the potential co-founder, leaving initial founders in a position where they must "dress the bride" to impress and secure a partnership. Our study complements prior research on entrepreneurial team formation (Lazar et al., 2020) by emphasizing the critical role of time pressure in finding a co-founder and showing that ideal selection strategies often prove difficult to implement in practice due to constraints entrepreneurs face. Second, in parallel with the job search theory (Faggian, 2021; Mortensen, 1986), our analysis highlights a trade-off in the co-founder selection, as the longer entrepreneurs spend searching the higher the costs associated with identifying a suitable partner.

CONTACT: Dorothea Georgiadou; dorge@dtu.dk; Technical University of Denmark; Jernbane Allé 16, Denmark, Hovedstaden 2720, Denmark

SUMMARY

EVALUATING STARTUP SUCCESS: THE ROLE OF NEURODIVERSITY, GENDER, AND RACE IN ENTREPRENEURIAL TEAMS

Mi (Kelly) H. Tran, Syracuse University, United States
Arielle Newman, Syracuse University, United States
Johan Wiklund, Syracuse University, United States

Principal Topic

Diversity in academic entrepreneurial ecosystems remains limited, restricting the benefits of varied perspectives (Hamilton, 2020; Mickey & Doerr, 2022). This study examines diversity's impact within the NSF's I-Corps program, the world's largest academic research commercialization effort. I-Corps trains STEM researchers in entrepreneurship, emphasizing customer discovery (Blank & Eckhardt, 2023). Using Social Identity Theory (Tajfel & Turner, 1986), we explore how gender, race, and neurodiversity (ADHD) influence startup evaluations. While prior studies address these factors separately (Wang, 2019; Wheadon & Duval-Couetil, 2017; Wiklund, 2017), we examine their combined effect on startup success within I-Corps.

Method

We surveyed participants in NSF's Northeast Regional I-Corps courses and matched responses with instructor evaluations. Our sample includes 345 entrepreneurs from 195 startups across 29 cohorts. To complement quantitative analysis, we conducted qualitative interviews with participants and instructors. We aim to expand the dataset to 700 entrepreneurs (350 startups) and interview 50 participants and 10 instructors in six months. Startup success potential, our dependent variable, was measured by instructors rating eligibility for the National NSF program on a 0–2 scale. Our independent variable captures team-level ADHD intensity using ASRS-6 scores. We also measured female and Non-White founder proportions and interaction effects (ADHD x Gender, ADHD x Race). We used ordered logit regressions, controlling for cohort fixed effects, team size, team age, and the number of customer interviews conducted (a success proxy).

Results and Implications

Our results indicate that higher ADHD intensity in startups correlates with lower instructor ratings, particularly in teams with more female founders. Although the ADHD-race interaction was not statistically significant, the effect's direction suggests potential significance with a larger sample. This study contributes to the neurodiversity and entrepreneurship literature by highlighting biases against neurodiverse startups. While ADHD has been linked to entrepreneurial potential (Wiklund et al., 2017), our findings reveal barriers neurodiverse founders face, challenging optimistic views (Verheul et al., 2016; Yu et al., 2021). Examining ADHD at the team level is crucial since many startups are team-based. We also contribute to diversity research by demonstrating compounded challenges for neurodiverse female founders. Our study extends Social Identity Theory by incorporating neurodiversity as an invisible identity that interacts with gender and race, showing how overlapping biases shape external evaluations and influence entrepreneurial outcomes.

CONTACT: Mi (Kelly) H. Tran; mhtran@syr.edu; Syracuse University; 200 Comstock Avenue, Syracuse, New York 13210, United States

SUMMARY

WHAT THEY THINK MATTERS: UNPACKING THE BOARD MEMBERS' IMPLICIT THEORIES OF BOARD ROLES IN NEW VENTURES

Krystal Zhao-Jansson, Stockholm School of Economics, Sweden
Mattias Nordqvist, Stockholm School of Economics, Sweden
Roxana Turturea, Stockholm School of Economics, Sweden

Principal Topic

In new ventures, board members often hold overlapping positions as executives, owners and investors, creating governance complexities that require balancing competing interests. Prior governance research, largely grounded in agency and resource dependence theories, focuses on board functions like monitoring, resourcing and strategizing, assuming members uniformly fulfill board roles based on shareholder expectations. However, this overlooks cognitive processes shaping how board members perceive, interpret, and enact their roles. This study applies implicit theories as cognitive lenses to explore how board members develop and evolve their role mindsets in new ventures. By introducing two role-mindset frameworks, it challenges the static view of board functions.

Method

We conducted a multi-case study using semi-structured interviews to explore board members' role mindsets in six deep-tech life sciences ventures over 16 months. Across two interview waves, we conducted 39 interviews (48hours) and analyzed 250 pages of secondary data. Our abductive analysis iterated between emerging data-driven themes and theoretical insights. Using a two-cycle coding approach, the first cycle involved vivo coding to identify key constructs related to role mindset framing, while the second cycle explored dynamic tensions between role mindsets over time.

Results and Implications

We identify two distinct role mindsets among board members in new ventures—dynamic and anchored—shaped by four interrelated constructs: personal attributes, motivations, role perceptions, and effort beliefs. Board members with dynamic mindsets perceive their roles as flexible and adaptable, driven by entrepreneurial attributes, learning-oriented motivations, flexible role perceptions and proactive effort beliefs. Conversely, those with anchored mindsets view their roles as stable, predefined, and anchored to conventional governance norms. They exhibit managerial attributes, performance-oriented motivations, fixed role perceptions and reactive effort beliefs. These constructs form a continuum—from inherent attributes to malleable effort beliefs—highlighting heterogeneity in how members perceive and enact roles. It challenges the one-size-fits-all view, showing that diverse role mindsets create a balanced board that adapts to evolving venture needs. We also emphasize the fluidity of board roles, with members shifting between operational and governance functions throughout the venture lifecycle. Tensions arise when members struggle to align perceived roles with enacted behaviors, particularly when anchored-role members take on operational tasks. Early-stage boards favor dynamic role mindsets but shift to anchored approaches as ventures mature. During crises, boards often revert to dynamic mindsets for agility and rapid decision-making. This challenges traditional views of board roles, shifting focus from external expectations to members' self-framing, showing that roles are heterogenous, context-dependent and shaped by the evolving venture needs.

CONTACT: Krystal Zhao-Jansson; krystaldan.zhaojansson@hhs.se; Stockholm School of Economics; Norrtullsgatan 2, Stockholm, Stockholms Lan 113 29, Sweden

INTERACTIVE PAPER

IT IS NOT A MATCH: THE EFFECTS OF EXAGGERATED ENTREPRENEURIAL HUSTLE ON CO-FOUNDER WILLINGNESS TO JOIN

Gianfranco Ciurlia, Technical University of Dortmund, Germany
Bastian Kindermann, Technical University Braunschweig, Germany
Steffen Strese, Technical University Dortmund, Germany

Principal Topic

Finding the right co-founder is crucial for a startup's success (MacMillan et al., 1985), leading entrepreneurs to use assertive impression management (IM) techniques to create positive impressions (Rudic et al., 2021; Bolino et al., 2008). Kindermann et al.,(2023) show that Entrepreneurial Hustle (EH) as an IM technique can enhance organizational attractiveness. Prior research also suggests that potential co-founders assess organizational attractiveness based on interpersonal connections (Forster and Jansen, 2010). However, little is known about how potential co-founders react to entrepreneurs using EH as an IM technique, and even less is understood about their reactions to exaggerated EH. This could significantly impact a co-founder's role, given the criticism of exaggerated hustle culture and its negative effects on team members (Forbes, 2023). We hypothesize that exaggerated EH negatively affects co-founder willingness to join, with "interpersonal attraction" mediating this relationship and "co-founder narcissism" acting as a moderator.

Method

To introduce the concept of exaggerated EH, we first conducted a qualitative pre-study involving interviews with 10 venture capital investors to identify exaggerated EH traits. We then manipulated exaggerated EH by adapting a pitch based on these identified traits. Using a pre-registered, randomized between-subject experiment with experimental vignette methods (Auguinis & Bradley, 2014), we gathered 337 final observations from participants recruited via Prolific. Participants assumed the role of potential co-founders and evaluated an entrepreneur's pitch through a manipulated audio file. After listening to the pitch, they assessed their willingness to join and answered questions about their perceptions of the founder and themselves, including items from the Narcissistic Personality Inventory (Ames et al., 2006). We tested our hypotheses using OLS regressions and bootstrapping techniques (Preacher & Hayes, 2008).

Results and Implications

Bootstrapping analysis reveals that exaggerated EH negatively affects co-founder willingness to join through reduced interpersonal attraction. Additionally, we find that co-founder narcissism positively moderates the relationship between exaggerated EH and interpersonal attraction. Our research contributes to EH, entrepreneurship, and IM literature in three significant ways. First, it enhances IM theory by identifying the risks of overusing IM techniques with potential co-founders. Second, it uncovers the impact of exaggerated EH on co-founder willingness to join by reducing interpersonal attraction, offering deeper insights into EH dynamics. Third, it extends IM theory and interpersonal attraction theory by demonstrating that co-founder narcissism strengthens the link between exaggerated EH and interpersonal attraction. Our managerial implications indicate that entrepreneurs should avoid excessive self-promotion when aiming to attract co-founders and focus on effective self-presentation strategies.

CONTACT: Gianfranco Ciurlia; gianfranco.ciurlia@tu-dortmund.de; Technical University of Dortmund; Färbergraben 4, Munich, Bayern 80331, Germany

INTERACTIVE PAPER

PASSION IN PROGRESS: MAPPING THE EVOLUTION OF ENTREPRENEURIAL TEAM PASSION ACROSS VENTURE STAGES

Sonja Franzke, University of Bayreuth, Germany
Rebecca Preller, University of Bayreuth, Germany

Principal Topic

This paper explores how entrepreneurial passion evolves within teams across different stages of a venture's lifecycle. Our study seeks to enhance the understanding of how passion in entrepreneurial teams shifts in response to venture development. Entrepreneurial passion is widely recognized as a critical driver of venture success, fueling motivation, resilience, and sus-tained engagement (Cardon et al.,2009; Newman et al.,2021). Since many ventures are founded by teams, entrepreneurial passion must also be understood as a collective phenomenon, playing a pivotal role in team dynamics (Kier et al.,2022; Klotz et al.,2014; Knight et al.,2020; Lazar et al.,2022). As new ventures enter a new phase, changes, e.g., concerning the founding team or stakeholder involvement, affect team collaboration and decision-making (Boone et al.,2020; Preller et al.,2020). Such changes might necessitate adjustments to their team entrepreneurial passion to maintain a shared identity and team cohesion (Su et al.,2024). Although existing re-search underscores the importance of team entrepreneurial passion and its varied development within ventures, we know little about what a functional team entrepreneurial passion entails and how it is connected to the different stages of venture growth. This paper addresses this short-coming by investigating team entrepreneurial passion as new ventures progress through devel-opment stages, offering new insights into its evolving nature. Thus, we pose the research ques-tion: "How and why does team entrepreneurial passion change throughout venture stages?".

Method

We follow an inductive, qualitative research approach using life diagrams to better un-derstand how passion in entrepreneurial teams emerges and evolves (Söderström, 2020). We are in the process of collecting data from 16 new venture teams using semi-structured interviews with all founders in the new venture teams, aided by life diagrams to map important changes in the new venture.

Results and Implications

With this research, we aim to add to research on team entrepreneurial passion. We build on previous work (Cardon et al., 2017) to explore the diversity of team entrepreneurial passion, enhancing our understanding of potential passion foci for entrepreneurial teams (Boone et al., 2020). Further, recent research has raised the question of whether entrepreneurs' passion should be adapted to fit the venture's situational needs (Kakarika et al.,2022), assuming that it is the individual's passion that is crucial for venture success. However, less is known about what de-fines functional team entrepreneurial passion and how it relates to venture development.

CONTACT: Sonja Franzke; sonja.franzke@uni-bayreuth.de; University of Bayreuth; Nürnberger Str. 38, Bayreuth, Bayern 95448, Germany

INTERACTIVE PAPER

HELP, MY WORK KEEPS CHANGING!: THE IMPACT OF PIVOTING ON EMPLOYEE WELL-BEING AND TURNOVER

Evy Van Lancker, Ghent University, Belgium
Alessia Di Stasi, Ghent University, Belgium
Marie-Laurence Gyselinck, Ghent University, Belgium
Mirjam Knockaert, Ghent University, Belgium
Yasmine Van Heghe, Ghent University, Belgium

Principal Topic

Scholars have increasingly paid attention to wellbeing in the entrepreneurship context (e.g., Obschonka et al., 2023). The literature has however largely neglected joiners – early nonfounding startup employees (Roach & Sauermann, 2015) – whose wellbeing is likely also affected by the entrepreneurial context (Van Lancker et al., 2023). This is surprising as joiners face many challenges linked to early-stage entrepreneurial firms, such as high levels of uncertainty and continuous change (Burton et al., 2019). Only recently, scholars have started to generate insights into wellbeing of employees, but in SMEs (Stephan et al., 2024) and growing firms (Genedy et al., 2024).

Method

In this paper, we study the impact of pivoting (i.e., fundamentally changing business model aspects (Burnell et al., 2023)) on joiner turnover intentions. Building on the Job-Demands Resources (JD-R) model (Demerouti et al., 2001) , we develop a multi-level model, proposing a cross-level indirect effect of pivoting on joiner turnover intentions via joiner emotional exhaustion, contingent on founder relational energy (i.e., "a heightened level of psychological resourcefulness generated from interpersonal interactions that enhances one's capacity to do work" (Owens et al., 2016: 37), here in the interactions between a joiner and founder). To achieve our research objectives, we used a time-lagged multilevel research design and collaborated with local accelerator programs to invite Flemish entrepreneurial firms to participate. In specific, we surveyed both founders and joiners by means of two survey waves, with a three-month time-lag in between.

Results and Implications

With this study, we aim to contribute to the entrepreneurship literature. In specific, our objective is to provide insights into the influence of firm-level dynamics, such as pivoting, as well as founder-level factors, such as founder relational energy, on joiner behavior. As such, we hope to add to the ongoing conversation on wellbeing in entrepreneurship by bringing a multilevel perspective to it (Wach et al., 2021; Waldman et al., 2024). Similarly, we hope to provide insights into joiner retention, a highly strategic, yet understudied aspect of HRM in entrepreneurial firms (Van Lancker et al., 2023; Waldman et al., 2024).

CONTACT: Evy Van Lancker; Evy.VanLancker@UGent.be; Ghent University;
Tweekerkenstraat 2, Ghent, Oost-Vlaanderen 9000, Belgium

INTERACTIVE PAPER

REPLACING THE STICK: IDENTIFYING THE KEYS TO UNLOCKING THE POTENTIAL OF ADVISORY BOARDS IN THE ABSENCE OF FORMAL POWER

Stephan Weemaes, VIVES University of Applied Sciences, Belgium
Johan Bruneel, KU Leuven & IESEG Lille, Belgium
Ann Gaeremynck, KU Leuven, Belgium
Filip Vanderbeken, VIVES University of Applied Sciences, Belgium

Principal Topic

Small and medium-sized enterprises (SMEs) often face the challenge of business leaders -typically sole shareholders- carrying the burden of all strategic decisions, leading to a sense of isolation. To address this, many business leaders seek external advice to gain diverse perspectives and improve decision-making. External advice can be informal, such as consultations with family and friends, or mentors; or formal, like engaging advisors (e.g., accountants or lawyers) or forming boards. Two common board types are boards of directors (BoDs) and advisory boards (ABs). This paper explores the mechanisms that contribute to the effectiveness of ABs, despite their informal nature and lack of decision-making power.

Method

A qualitative approach was adopted, involving 39 semi-structured interviews with both business owner-managers and AB members. The sample included 16 businesses, half of which were startups, and half were more established firms. To compare and recognize repeating patterns and to understand in depth the determinants for AB effectiveness within and across cases, a multiple case study method was used. Interviews were inductively coded, and data was iteratively analyzed while contrasting the current literature. Data triangulation was achieved by comparing assessments from primary decision-makers and AB members. Disagreements among the authors about interpretations of data were debated until consensus was reached.

Results and Implications

Our analysis demonstrates the strong advisory potential of ABs, which stems from their lack of monitoring duties, decision-making power, and legal liability. This structure enables ABs to focus entirely on providing strategic guidance while fostering psychological safety for business leaders, thereby creating a trusted forum for open discussions. However, our research indicates that an AB's potential is not automatically realized. We identify seven essential elements for AB effectiveness, all of which must be addressed simultaneously, with the business leader playing a central role: (1) a clearly defined purpose for the AB, (2) strong belief in and commitment to the AB format by the business leader, (3) AB composition aligned with its purpose and external environment, (4) alignment between business leaders and advisors through well-defined roles and expectations, (5) voluntarily adopted effort norms to enhance effectiveness, (6) regular evaluation and open communication to maintain alignment, and (7) the creation of psychological safety as a catalyst for AB effectiveness. Compared to BoDs, ABs offer greater flexibility, presenting both opportunities and risks. Ultimately, the business leader's decisions and engagement determine whether an AB reaches its full potential or dissolves prematurely.

CONTACT: Stephan Weemaes; stephan.weemaes@vives.be; VIVES University of Applied Sciences; Doorniksesteenweg 145, Kortrijk, West-Vlaanderen BE-8500, Belgium

AUTHOR INDEX

Babson College
231 Forest Street
Babson Park, MA
02457

CURRAN ASSOCIATES INC.
proceedings.com

ISBN 979-8-3313-2884-9

9 798331 328849